INTRODUCTION TO NANOSCIENCE

Introduction to Nanoscience

S.M. LINDSAY

Arizona State University

OXFORD
UNIVERSITY PRESS

Great Clarendon Street, Oxford OX2 6DP

Oxford University Press is a department of the University of Oxford.
It furthers the University's objective of excellence in research, scholarship,
and education by publishing worldwide in

Oxford New York

Auckland Cape Town Dar es Salaam Hong Kong Karachi
Kuala Lumpur Madrid Melbourne Mexico City Nairobi
New Delhi Shanghai Taipei Toronto

With offices in

Argentina Austria Brazil Chile Czech Republic France Greece
Guatemala Hungary Italy Japan Poland Portugal Singapore
South Korea Switzerland Thailand Turkey Ukraine Vietnam

Oxford is a registered trade mark of Oxford University Press
in the UK and in certain other countries

Published in the United States
by Oxford University Press Inc., New York

British Library Cataloguing in Publication Data

Data available

Library of Congress Cataloging in Publication Data

Data available

Typeset by Newgen Imaging Systems (P) Ltd., Chennai, India
Printed in Great Britain
on acid-free paper by
CPI Antony Rowe, Chippenham, Wiltshire

ISBN: 978–019–954420–2 (Hbk)
ISBN: 978–019–954421–9 (Pbk)

10 9 8 7 6 5 4 3 2 1

Preface

Nanoscience is not physics, chemistry, engineering, or biology. It is all of them, and the first motivation for this text is to present an integrated description. A second motivation for this book lies with the *complexity* of nanostructures, an issue that is not widely addressed. Richard Feynman described the remarkable consequences of scale and quantum effects in his visionary essay "There's Plenty of Room at the Bottom" and I think enough of it to have purchased the rights to reproduce it here (as Appendix B). But nanoscale objects, where hundreds, thousands, or hundreds of thousands of atoms make up systems, are complex enough to show what is called "emergent behavior": Quite new phenomena arise from rare configurations of the system. We will encounter this in the Kramers theory of reactions in Chapter 3, the Marcus theory of electron transfer in Chapter 8, and several times again in the final chapter on Nanobiology.

I teach the class on which this book is based to upper division undergraduates and beginning graduate students, and do not impose prerequisites for enrollment. I do this precisely because of the integrative nature of the material. I would like biologists to be able to get the gist of quantum mechanics and statistical mechanics. To this end, much of the material is conceptual, and conceptual questions have been included at the end of Chapter 2 (quantum mechanics) and Chapter 3 (statistical mechanics). It is possible to teach to a broad audience: a social sciences undergraduate (from the Center for Nanotechnology in Society at Arizona State University) has scored better on the conceptual quantum test than some physics graduate students!

The class does not have to be taught this way and much of the introductory material may be skipped in the appropriate settings. But physics majors will want to review the material on reaction kinetics and chemical equilibria. Some of the material on Brownian motion may not have been covered in the undergraduate curriculum. The chemical aspects of molecular electronics (Chapter 8) will probably be completely new too. Chemistry majors may want to review aspects of quantum mechanics, and the basics of statistical mechanics and fluctuations. The use of a partition function to justify what is normally done on the basis of mass action will be new to most chemistry students. The properties of electrons in periodic solids (Chapter 7) are not part of the regular chemistry curriculum either.

The mathematics has been kept to a minimum. Introductory calculus and some knowledge of linear algebra is a sufficient background for the student who wants to follow all of the mathematical detail presented in the text and its appendices. Bibliographies and references to the primary research literature at the end of every chapter also make this a suitable starting point for research in nanoscience. One last goal was to make this the "handbook" containing the stuff that I want the students in my lab to know, and I hope I have succeeded.

Solutions to most of the problems are given in Appendix M. They include detailed derivations where needed for support of the main text. A CD contains Powerpoint presentations for the lectures, as well as color figures and movies.

A book produced to a deadline, its production taking second place to all the other demands of an academic job, is bound to contain mistakes. They are far fewer in number because of the critical input of many of my colleagues and students. Mark Ratner read most of the first draft, providing valuable comments. I received valuable feedback from Nongjian Tao, Timothy Newman, Otto Sankey, John Spence, Larry Nagahara, Peiming Zhang, and Ralph Chamberlin. Students in my class helped me refine the problems at the end of each chapter and caught many errors in this version. They include Ashley Kibel, Eric Alonas, Shreya Bhattacharya, Kevin Brown, Jared Burdick, Di Cao, Shuai Chang, Eric Dailey, Shaoyin Guo, Kaushik Gurunathan, Billie Harvey, Shuao Huang, Deepthi Jampala, Parminder Kaur, Steven Klein, Lisha Lin, Hao Liu, Dan and Alise Martin, Chelsea Mcintosh, Douglas Moorhead, Jeff Moran, Pei Pang, Peter Pelletier, Suman Ranjit, Kamil Salloum, Dan Shea, Nathaniel Sylvain, Matthijs Smith, Jill Stock, Tim Lomb and Nick Teodori. Hao Liu also created the cover art. Hosam Yousif checked the problems and produced clean solutions for Chapters 1 through 5. I am grateful to Maggie Black for tracking down permissions and taking care of copyright issues. Health issues did not get in the way of this project thanks to the diligent care of Dr. Alan Wachter M.D. Finally, the debt I owe to my remarkable wife, Christine, is enormous. She is constantly supportive, loving, and kind.

Tempe, AZ, 2008

Contents

Part II: Tools

Part III: Applications

What is Nanoscience?

1

1.1 About size scales

Nanoscience is about the phenomena that occur in systems with nanometer dimensions. Some of the unique aspects of nanosystems arise solely from the tiny size of the systems. Nano is about as small as it gets in the world of regular chemistry, materials science, and biology. The diameter of a hydrogen atom is about one-tenth of a nanometer, so the nanometer scale is the very smallest scale on which we might consider building machines on the basis of the principles we learn from everyday mechanics, using the 1000 or so hydrogen atoms we could pack into a cube of size 1 nm $\times$ 1 nm $\times$ 1 nm. If this is all that there was to nanoscience, it would still be remarkable because of the incredible difference in scale between the nano world and the regular macroscopic world around us. In 1959, Richard Feynman gave a talk to the American Physical Society in which he laid out some of the consequences of measuring and manipulating materials at the nanoscale. This talk, "There is plenty of room at the bottom," is reproduced in its entirety in Appendix B. It does a far better job than ever I could of laying out the consequences of a technology that allows us to carry out routine manipulations of materials at the nanoscale and if you have not already read it, you should interrupt this introduction to read it now.

The remarkable technological implications laid out in Feynman's talk form the basis of most people's impression of Nanoscience. But there is more to Nanoscience than technology. Nanoscience is where atomic physics converges with the physics and chemistry of complex systems. Quantum mechanics dominates the world of the atom, but typical nanosystems may contain from hundreds to tens of thousands of atoms. In nanostructures, we have, layered on top of quantum mechanics, the statistical behavior of a large collection of interacting atoms. From this mixture of quantum behavior and statistical complexity, many phenomena emerge. They span the gamut from nanoscale physics to chemical reactions to biological processes. The value of this rich behavior is enhanced when one realizes that the total number of atoms in the systems is still small enough that many problems in Nanoscience are amenable to modern computational techniques. Thus studies at the nanometer scale have much in common, whether they are carried out in physics, materials science, chemistry, or biology. Just as important as the technological implications, in my view, is the unifying core of scientific ideas at the heart of Nanoscience. This book seeks to build this common core and to demonstrate its application in several disciplines.

In this introductory chapter we will start with technology, that is, those applications that flow from the ability to manipulate materials on the nanometer scale. We will then go on to examine the scientific phenomena that dominate nanoscale systems.

In order to appreciate the technological implications of working at the nanoscale, one must appreciate the incredible scale difference between our regular microscopic world and the atomic world. There are wonderful Web sites that allow the user to zoom in from astronomical scales to subatomic scales by stepping through factors of 10 in size. This exercise is highly recommended, but here we will look at size scales from a chemical perspective.

One mole of any material (e.g., 28 g of silicon) contains Avogadro's number of atoms (i.e., 6.023×10^{23} atoms). This is a fantastic number of atoms. Exponential notation has hardened us to large numbers, so we will look a little further into what numbers as large as this mean.

I like the story of Caesar's last breath: Julius Caesar was murdered on March 15 in 44 B.C. Caesar's lungs, like our lungs, probably exhaled about 1 L of gas with each breath. Since one mole of an ideal gas at standard temperature and pressure occupies 22.4 L, Julius Caesar breathed out about 0.05 mole of N_2 gas as he fell to the ground uttering his famous last words "Et tu Brute?" In the intervening two millennia, these (mostly nitrogen) molecules have had ample time to disperse themselves evenly throughout the atmosphere of the earth. The mass of the Earth's atmosphere is 5×10^{18} kg, of which 80% is N_2 gas which has an atomic weight of 28 g/mole (2×14). There are therefore $0.8 \times 5 \times 10^{18}/0.028 \approx 1.4 \times 10^{20}$ moles of nitrogen gas in the earth's atmosphere, of which 0.05 mole was exhaled by Caesar. So each mole of our atmosphere contains 3.6×10^{-22} moles of Caesar's nitrogen, or about 200 molecules. Thus in each of our breaths, we inhale about 0.05×200 or 10 molecules of Caesar's last breath! The number is probably a little less because of sequestration of nitrogen by plants, but we can see that the enormous size of Avogadro's number draws us to the conclusion that we share each breath with the last lungful exhaled by Caesar. We should remember Caesar as we inhale on the Ides of March. This vast size of Avogadro's number underlies much of the technology discussed by Feynman in his 1959 talk.

1.2   History

Feynman's 1959 talk is often cited as a source of inspiration for Nanoscience, but it was virtually unknown outside of his small audience at the time and only published as a scientific paper in 1992.[1] Nanoscience really sprang into the public consciousness sometime after the invention of the scanning tunneling microscope (STM) in 1981.[2] Here was an amazing tool that could image and manipulate atoms. Atomic scale imaging had been possible in the past with multimillion-dollar transmission electron microscopes, but the STM was a benchtop tool that a graduate student could assemble for a few hundred dollars. The First International Conference on Scanning Tunneling Microscopy was held in Santiago De Compostela, Spain, July 14–18, 1986 (becoming a yearly fixture thereafter with an exponential growth in the number of papers presented). The impact of the STM on surface science was already so great at this time that the 1986 Nobel Prize in physics was awarded to Gerd Binning and Heinrich Rohrer for their invention of the STM. (It was shared with Ernst Ruska, one of the inventors of the electron microscope.) Fueled by the low cost and ease of building atomic resolution microscopes, interest in Nanoscience spread like wildfire, and by the time of the Second International Conference on Scanning Tunneling Microscopy, papers were presented that reported images of biological molecules.[3,4] The invention of the atomic force microscope (AFM) in 1986[5] greatly extended the reach of these tools. Now insulating materials could be imaged with atomic resolution and new types of direct measurement and manipulation on the atomic scale were made possible. The chemists could image directly some of the fantastic structures they had only dreamed of making, and a bright young biochemist in New York, Ned Seeman, was able to realize his dream[6] of building nanoscale machines with self-assembly of DNA molecules. Thus was born the field of DNA nanotechnology.[7] Major government funding of Nanoscience as a separate discipline began in the 1990s, with considerable subsequent impact on inventions in the development of technology.[8] Those of us who were swept up in this field found biologists, engineers, chemists, and materials scientists (and even geologists) knocking on the doors of our physics labs. The days of research within the narrow confines of one traditional discipline were over.

1.3 Feynman scorecard

If you have not already done so, you now have to read Feynman's talk in Appendix B in order to follow this discussion.

How well did Feynman do? A detailed analysis shows a remarkable success rate. I count 10 successes and 5 questionable predictions. A good grounding in physics makes for a better futurist, at least in this area. Here we take each specific prediction in turn and see how it has turned out.

Electron beam lithography: Talking of using electron beams, demagnified in an electron microscope, to write small features: "We can reverse the lens of an electron microscope in order to demagnify as well as magnify . . . This, when you demagnify it 25,000×, it is still 80 Å in diameter – 32 atoms across." Current e-beam technology allows features as small as about 10 nm to be written,[9] close to the prediction. Figure 1.1 shows a tiny Fresnel lens made by

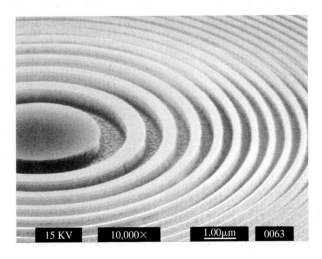

Fig. 1.1 Fresnel lens for focusing X-rays. Submicron features were patterned into a resist using an electron beam (Courtesy of C. David, PSI).

electron beam lithography. It was built to focus X-rays, so the distance between the rings is tens of nanometers. We will discuss this technique in Chapter 5.

Stamping out nanostructures: Feynman predicted that nanostructures could be "printed" directly on to surfaces: "We would just have to press the same metal plate again into the plastic and we would have another copy." George Whitesides has developed a stamping technology based on silicone rubber stamps that are cured on top of silicon nanostructures, leaving an imprint of the nano-features on the surface of the stamp. The stamp can then be used to print out multiple copies of the original (laboriously manufactured) nanostructure very rapidly.[10] Stamp technology is covered in Chapter 5.

Using ions to etch structures: "A source of ions, sent through the lens in reverse, could be focused to a very small spot." Focused ion beams (FIBs) are now used as "nanoscale milling machines," rapidly prototyping nanoscale structures by selectively bombarding away surface atoms with a beam of focused high-energy ions.[11] The FIB is another tool dealt with in Chapter 5.

Three-dimensional high-density storage: "Now, instead of writing everything, as I did before, on the *surface* of the head of a pin, I am going to use the interior of the material as well." Modern integrated circuits have complex three-dimensional structures, built up layer by layer,[12] but high-density information storage in three dimensions has yet to be realized on the nanoscale.

Solving the atomic structure of complex biological molecules by direct imaging: "The wavelength of an electron is only 1/20 of an Å. So it should be possible to see the individual atoms." Electron microscopes can image at the atomic level, but beam damage limits the resolution obtained with biological molecules. Structures have been solved to the nanometer scale by averaging images from large arrays of molecules embedded in ice.[13] Figure 1.2 shows the structure of a complex protein assembly reconstructed from low-dose electron microscopy of a layer of protein frozen in a thin film of ice. Nanoscale microscopy is covered in Chapter 4.

Chemical analysis by imaging atoms: "It would be very easy to make an analysis of any complicated chemical substance; all one would have to do is look at it and see where the atoms are." The very highest resolution electron microscope images have given insight into chemical bonding, but only in very special circumstances.[14] But atomic scale analysis is possible via electron beam induced X-ray emission in dry samples, and the AFM can identify single molecules via specific interactions with molecules tethered to the probe.

Machines the size of biological motors: "Consider the possibility that we too can make a thing very small, which does what we want—that we can manufacture an object that maneuvers at that level! . . . Consider any machine— for example, an automobile—and ask about the problems of making an infinitesimal machine like it." Some tiny nanomotors have been constructed[15,16] but, as we shall see, real biological motors operate on very different principles from the motors humans build. A motor that rotates on a carbon nanotube shaft is shown in Fig. 1.3. Biological motors are touched on in Chapter 10.

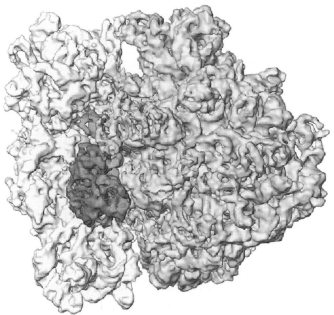

Fig. 1.2 Three-dimensional reconstruction of a protein machine that makes proteins from their RNA code. The complex is about 20 nm in diameter and contains thousands of atoms. This image was obtained by cryoelectron microscopy (LeBarron et al., 2008, made available by Joachim Frank).

Fig. 1.3 Tiny motor bearing made from a carbon nanotube shaft supported inside a larger nanotube that acts as a bearing (Courtesy of Professor A. Zettl).

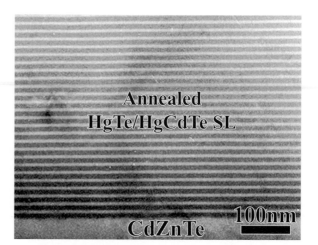

Fig. 1.4 Semiconductor superlattice made from alternating layers of two different compound semiconductors that are each just a few atomic layers thick (Reprinted from The Journal of Crystal Growth, Aoki et al.[18], copyright 2004 with permission from Elsevier).

Miniaturizing computer components to build super computers: "For instance, the wires could be 10 or 100 atoms in diameter . . .. If they had millions of times as many elements, they could make judgments . . .." These changes have already come about, but as a result of conventional lithography—the smallest features in modern chips are perhaps 1000 atoms across—and computer power has surpassed even this predicted increase. Any one who has struggled with Windows will agree that operating systems have a long way to go before *judgment* is an attribute that springs to mind.

Making atomic scale structures by evaporating layers of atoms: "So, you simply evaporate until you have a block of stuff which has the elements . . . What could we do with layered materials with just the right layers?" Molecular beam epitaxy[17] (MBE) is almost exactly what Feynman had in mind: layers of atoms are formed by projecting hot vapors onto a substrate. Different types of atoms can be projected to form layered structures with nanometer thickness. An example of structure made by layering materials this way is shown in Fig. 1.4. It is a semiconductor "superlattice" made by growing layers of mercury telluride that alternate with layers of mercury cadmium telluride. MBE is discussed in Chapter 5.

Lubricating tiny machines: "But actually we may not have to lubricate at all!" It turns out that many small mechanical structures fail because the closeness of their components leads to parts sticking together through long-range (van der Waals) interatomic forces, and the development of nonstick surfaces for small mechanical components is an important endeavor. But the whole science of lubrication received a boost from the application of AFM methods.[19] Now, at last, we have a probe to study friction where it arises—at the atomic scale at interfaces.

Nanoscale robots that operate inside humans: "It would be interesting if you could swallow the surgeon." Nothing like this has come about, though surgeons do manipulate some remarkable robots remotely. But nanoscale techniques, such as dyes, that report chemical motions in cells (molecular beacons)

Fig. 1.5 Chemical reaction induced by pushing molecules together with a scanning tunneling microscope (Courtesy of Professor Wilson Ho). See Color Plate 1.

and supramolecular assemblies that carry drugs (e.g., dendrimers) are some of the new technologies emerging from the NIH-sponsored Nanomedicine Initiative.

Exponential manufacturing: machines that make machines and so ad infinitum: "I let each one manufacture 10 copies, so that I would have a hundred hands at the 1/16 size." This idea, of making small machines, that make more even smaller machines that, in turn, make yet more even smaller machines is intriguing, but not realized. However, the idea of exponential growth through copying copies is what lies behind the amazing polymerase chain reaction, the biochemical process that yields macroscopic amounts (micrograms) of identical copies of just one DNA molecule. One molecule is replicated to two, two to four, and so on.

Doing synthesis of complex organic molecules by "pushing atoms together": "We can arrange atoms the way we want." STM has been used to construct some remarkable structures by pushing atoms together (Fig. 1.5),[20] but it is not a very practical way to make new materials because the quantities are so small.

Resonant antennas for light emission and absorption: "It is possible to emit light from a whole set of antennas, like we emit radio waves." This is the modern field known as "nanophotonics." For example, arrays of metal nanoparticles can be used to guide light.[21] Some applications of nanophotonic structures are discussed in Chapter 9.

Using quantum (atomic scale) phenomena in electronic devices: "We could use, not just circuits, but some system involving quantized energy levels, or the interaction of quantized spins." This was a remarkable observation: quantum mechanics offers us completely novel ways to do computations. Electron spin valves have become the dominant readout device in the disk drives.[22] Even more remarkable is the proposal to use the fundamental properties of quantum measurement as a way to do massively parallel computing. The field is called "quantum computing"[23] and it exploits

fundamental aspects of quantum measurement (some of which will be introduced when we survey quantum mechanics in the next chapter). Chapter 8
describes the use of molecules as electronic devices and Chapter 7 covers
nanoscale solid-state electronic devices.

Feynman started his talk by stating that we should look back from the vantage
of the year 2000 and wonder why it took anyone till 1960 to point these things
out. I suspect most of us are amazed at just how much Richard Feynman got
right in 1959.

1.4   Schrödinger's cat—quantum mechanics in small systems

Atoms are governed by the laws of quantum mechanics, and quantum mechanics
is essential for an understanding of atomic physics. The interactions between
atoms are governed by quantum mechanics, and so an understanding of quantum
mechanics is a prerequisite for understanding the science of chemistry. We
will start our study with a survey of quantum mechanics in Chapter 2. It is
a conceptually and mathematically challenging topic, to be sure. But modern
computational techniques allow for packaged programs that do many of the hard
calculations for us. So a conceptual appreciation may be all that is needed to
try out some quantum chemical calculations. To that end, I describe the modern
technique of *density functional theory* very briefly at the end of Chapter 2. It is
the basis of several quantum chemistry packages available over the Web or for
downloading on to a laptop computer.

So what has this to do with cats? When many atoms are put together to
make a macroscopic object like, for example, a cat, common sense tells us
that cats obey the laws of classical physics: when dropped from a tall building
they fall according to Newton's laws of motion. I chose the example of the
cat as a classical object because of the famous paradox in quantum mechanics
known as "Schrödinger's cat." One of the rules of quantum mechanics is that
all possible states of a system must be considered in predicting the final state of
the system, the final state only emerging on measurement. Quantum mechanics does not specify the result of a measurement, but only the probability a
particular final state will be measured. We will examine this rule in detail in
the next chapter. "Schrödinger's cat" takes this rule to the point of absurdity.
In Schrödinger's thought experiment, a cat is locked in a box together with
an instrument for detecting radioactive decay that, once a decay is detected,
breaks a vial of poison, killing the cat. In quantum mechanics, the state of the
atom is some combination of the decayed and nondecayed state before measurement. Therefore, it could be argued that the quantum description of the
entire system of cat, instrument, and radioactive atom is some combination of a
live cat and a dead cat. Both exist simultaneously until a measurement is made
by opening the box. Quantum mechanics is very subtle here: it is not actually the act of making a measurement that destroys quantum effects. Rather
it is an interaction of the quantum system that destroys the quantum effects.
And, of course, live cats interact with the outside world very strongly in many
ways. Atoms interact with the outside world in many ways as well. Thus, the
laws of quantum mechanics predict the way in which atoms form chemical
bonds, even though chemical reactions are usually carried out in solutions,

where atoms of the solvent are constantly bombarding the atoms as they are undergoing a reaction. The quantum mechanical description works because the energy of this bombardment is much less than the binding energy of electrons to atomic nuclei. Having said that, it turns out that the bombardment plays an important role in chemical reactions in solution. We will discuss the mechanism whereby electron transfer reactions occur in solution in some detail later in Chapter 8. This is, however, an excellent point to turn to the other aspect of physics at the microscopic level that will form a theme of this book: This is the role of fluctuations in complex systems, and particularly in small systems.

1.5 Fluctuations and "Darwinian Nanoscience"

Fluctuations play a large role in small systems simply because they are relatively larger in smaller systems. The total energy contained in the bonds that hold a collection of atoms together decreases with the number of atoms in the system. Importantly, smaller objects have relatively more atoms on their surfaces than in their interior, and these surface atoms are usually more weakly bound than the atoms in the interior. But if the system is in thermal contact with the environment (i.e., in a gas or liquid at room temperature), the average amount of the energy associated with random thermal motion of each atom remains constant (on average) as the system size is decreased at a constant temperature. If the energy of the bonds that hold the system together is comparable to the amount of thermal energy, the system will start to fall apart as a result of random thermal agitation when the total number of bonds becomes small enough. Tables do not fall apart spontaneously, and then reassemble. But proteins fold and unfold, bind with other proteins and randomly unbind. Even small molecules undergoing a chemical reaction that binds them together will spontaneously separate (infrequently if the bond is strong). These fluctuations are what drive the assembly of biological systems and establish the equilibrium between the relative amounts of reactants and products in a system of chemicals undergoing reactions. Such processes are most important in small systems.

But there is more to this story, and this is the part I have called "Darwinian Nanoscience." Darwin's theory of evolution of the species in biology (through selection of the fittest members of a randomized population) is an example of the central role of fluctuations in determining the macroscopic properties of biological systems. This might not seem like a problem of small numbers, but if you are the last dinosaur on the look out for a mate or the virus carrying the first mutant for virulent Spanish Flu, your numbers (and hence the size of the systems) are indeed small. At the end of this book (Chapter 10) we will take a look at how fluctuations drive nanoscale machines such as molecular motors, and how fluctuations in gene expression in small systems (e.g., cells) drive variability in biochemical signaling pathways. We will examine the role of fluctuations in reactions and of solvent fluctuations in mediating electron transfer reactions in solution (Chapter 8). These processes have certain things in common. Fluctuations *dominate* the outcome, and only some special (and rare) set of circumstances lead to the desired end result. The requirement for the processes to be robust is only that the fluctuations sample enough states to make the probability of the rare (but critically important) state very high over some

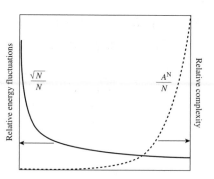

Fig. 1.6 Occurrence of "rare" phenomena as a function of system size. The relative size of energy fluctuations decreases as the square root of the number of particles, while the relative number of arrangements of the system increases approximately exponentially. Here they are plotted as a function of the system size.

period of time. Fluctuations are not just annoying departures from some average properties in these systems. They are the very essence of the process itself. This is illustrated schematically in Fig. 1.6. In Chapter 3, we will use statistical mechanics to derive the result that while the total energy of a system grows with the number of particles, N, in it, *deviations* from this average increase with the square root of the number of particles, $\sqrt{N}$. Thus the fluctuations relative to the mean energy scale as $\sqrt{N}/N$. It is a very general result for many quantities besides the energy of a system. To see this, recall that the standard deviation of the Poisson distribution is given by the square root of its mean,[24] so the fluctuation in repeated counting of N particles is just $\sqrt{N}$. Thus, repeated sampling of the opinions of 100 voters (for example) would return results that varied by 10% most times, purely owing to the random nature of the sampling. $\sqrt{N}/N$ is a function that decreases most rapidly for small values of N (solid curve in Fig. 1.6). For example, an electron transfer reaction requires an energy that is about 50 times the available thermal energy. Such a fluctuation in average energy would require about $(50)^2$ or 2500 particles. Given a fluctuation of the right magnitude of energy to drive some process, what about the probability that it is just the right type of fluctuation? That is to say that the components of the system line up exactly as needed to make some event happen. This will scale with the number of possible arrangements of the system, $N!$, if the particles are all distinct. In general, this complexity is a rapidly increasing function of N, and is shown as an exponential, normalized to the system size also, A^N/N (dashed line in Fig. 1.6).

Thus, the probability of the "right" fluctuation occurring increases very rapidly with the number of particles, so that even a modest increase in system size (over the minimum estimated on energy grounds) guarantees a high probability of success. Returning to the example of thermal fluctuations driving an electron transfer reaction, the requirement turns out to be that the molecules surrounding the site of electron transfer spontaneously line up their electric polarizations in the right way. Suppose that the units involved in the electron transfer reaction just discussed were water molecules each occupying about 10^{-2} nm^3, then the volume occupied by 2500 of them is about 25 nm^3 or a cube of sides about 3 nm. The critical size scale where fluctuations are big enough and the system is complex enough is indeed the *nanoscale*.

Biology is complex on a whole series of scales, starting with the nanoscale. A second level of complexity is reached when atoms and molecules are replaced with *genes*, another when atoms and molecules are replaced with *cells*, and yet another when cells are replaced with *whole organisms*.

1.6 Overview of quantum effects and fluctuations in nanostructures

The nanoscientist has no way of escaping the need for some knowledge of both quantum mechanics and statistical mechanics. These disciplines are essential for a conceptual understanding of what makes science so interesting on the nanoscale. On the other hand, there is no way of appreciating the reach of these aspects of physics without a detailed look at how they play out in subjects such as materials science, chemistry, and biology.

Table 1.1 lists many of the phenomena that dominate the nanoscale (and that will be covered in this book) and lists examples of the technologies and other consequences that follow from these phenomena.

Table 1.1 Some phenomena that dominate at the nanoscale and examples of their applications

Phenomenon	Examples
Below a certain length scale (that depends on interaction strengths) behavior must be described using quantum mechanics.	(a) Equilibrium chemistry, at the level of atomic bonding, is described by Schrödinger's equation. The periodic table reflects the allowed symmetries of quantum states in spherical potentials. (b) Quantum "dots" are small semiconductor particles with increased band gaps owing to "size" quantization. (c) Catalysts have novel electronic properties owing both to size and to the relative dominance of surfaces in small particles.
Even materials that interact strongly with the environment can (perhaps at low temperatures) show bulk quantum phenomena if they are small enough.	(a) Small wires and thin films transport electrons in a way that does not obey Ohm's law. This is an example of *mesoscopic* physics. (b) Electrons trapped in thin layers show bulk quantum-like behavior in magnetic fields.
Many processes depend on the number of available energy states per unit energy. This quantity varies with the dimensionality of the system.	Electronic and optical structures that have one (layered), two (wires), or three (dots) very small dimensions exhibit novel properties owing to this modification of the density of states.
Fluctuations are large in small systems. For example, the density of an ideal gas has a mean-square fluctuation given by the number fluctuation in a small volume: $\frac{\langle \Delta N^2 \rangle}{\langle N \rangle^2} = \frac{1}{\langle N \rangle}$.	(a) Fluctuations drive electron transfer reactions and control kinetics. (b) Fluctuations are exploited by enzymes and biological "motors" for function. (c) Fluctuations are important in cells where small numbers of molecules involved in gene expression can lead to large random changes. (d) Fluctuations operate at every level of biology. At the molecular level, random gene splicing generates antibodies and generates protein variations. In populations, they lead to the evolution of species.
Fluctuations in small systems destroy *quantum coherence*. Nano systems may exist at the quantum-classical boundary.	A very difficult and interesting problem: how much of a system is "quantum mechanical" (the chemical bonds in the cellulose that makes up wood) and at what scale is something classical (the table made of wood)? Herein may lie the answer to the Schrödinger's cat paradox.
The effective molarity of reactants that are *confined* in nanostructures may be very high.	(a) Chemical reactions that occur rarely in bulk may be driven by an increased concentration in nano-confinement. (b) Enzymes may *both* concentrate *and* provide the fluctuations that drive electron transfer reactions. (c) Critical spatial assembly of cofactors or *chaperones* may underlie highly specific processes in biology. (d) Diffusion is *geometrically constrained* in the nanostrucures inside a cell.
High information density opens the possibility of pattern analysis of complex many variable processes.	Nanoscale analysis for genomics and proteomics of complex systems like humans?

1.7   What to expect in the rest of this book

Our description of Nanoscience rests on the shoulders of quantum mechanics, statistical mechanics, and chemical kinetics. This essential background forms the materials for Chapters 2 and 3. Our survey is essentially conceptual with mathematics kept to an absolute minimum. I have tried wherever possible to use simple one-dimensional models. Conceptual problems are provided at the end of these chapters in addition to the usual numerical and mathematical problems.

The second part of this book deals with the tools that enable nanotechnology. Chapter 4 is about microscopy and single molecule manipulation. We deal first with scanning probe (atomic force and scanning tunneling) microscopy and then turn to electron microscopy. Finally, we deal with single molecule techniques such as optical tweezers and single molecule fluorescence. These topics give us an understanding of how we characterize nanostructures. The next two chapters deal with different approaches to making nanostructures. In chapter 5, we discuss the traditional "top down" approach to nanofabrication. These methods are extensions of the familiar procedures used in the semiconductor industry to make integrated circuits. Chapter 6 deals with an approach that is more intrinsically "nano," that is, the "bottom up" method based on self-assembly. It is, of course, the path that evolution has taken as replicating living systems have developed on this planet. In this context, the development of techniques that use synthetic DNA to make self-assembling nanostructures is particularly remarkable.

Part three of this book deals with applications. I hope it is more than just a shopping list of nano wizardry. Nanoscience reaches into many disciplines, and thus offers a wonderful basis for an interdisciplinary curriculum. Chapter 7 deals with electrons in nanostructures, and it starts out by presenting a survey of the electronic properties of condensed matter. Chapter 8 introduces the subject of molecular electronics. To do this, this chapter begins with a survey of the electronic properties of molecules from both a theoretical and experimental point of view. The theoretical approach is based on molecular orbital theory. The experimental approach is based on electrochemistry. This is regarded as a complicated topic even by chemistry students, but it is absolutely essential for understanding molecular electronics and Section 8.9 is a self-contained summary of the subject. This chapter is where we treat the Marcus theory of electron transfer, alluded to earlier in this introduction. Marcus theory describes the role of fluctuations in electron transfer, and is an example of what I have called "Darwinian Nanoscience." Chapter 9 is a small and selective survey of nanostructured materials. We have chosen a set of examples where novel material properties follow from restricting the density of states in one or more dimensions. We end with a look at several topics in nanobiology in Chapter 10. There is not room in a book like this for any meaningful survey of modern biology as introductory material in this chapter. But I have attempted to give a bird's eye view of some of what I consider to be some key issues. I hope that this last chapter both interests the reader and motivates a foray into the biological literature.

Fluctuations are a constant theme: We end our chapter on nanobiology with a brief discussion on the role of fluctuations in establishing the random networks that, when acted on by environmental pressures, become the fully formed brains of animals. Darwin's picture of how order emerges from chaos is seen to apply to processes that range from simple chemical reactions to, perhaps, the formation of conscious brains. If this seems a long way from physics, it is worth noting that Frank Wilczek, one of this generation's most talented theoretical physicists, has argued for a "Darwinian" mechanism for the development of the laws of physics that we currently observe.[25,26] Quantum fluctuations at the time of the "big bang" may allow for an infinite variety of universes based on, for example, different values of the electron mass. But only values of the electron mass close to what we observe today are compatible with the universe that supports life as we know it. Wilczek proposes that all possible universes can and do exist; we observe this one precisely because it supports life.

We have strayed a long way from the very small, that is, the subject of Nanoscience. But I hope the reader is convinced that concepts of statistical mechanics and quantum mechanics deserve a place at the center of Nanoscience.

1.8   Bibliography

These are some general texts dealing with nanoscience:

M. Wilson, K. Kannangara, G. Smith, M. Simmons and B. Raguse, Nanotechnology: Basic Science and Emerging Technologies. 2002, Boca Raton, FL: Chapman and Hall/CRC. A comprehensive survey at a fairly elementary level. Though written by a team, the book is well integrated. Chemical structures are emphasized.

M.A. Ratner, D. Ratner and M. Ratner, Nanotechnology: A Gentle Introduction to the Next Big Idea. 2002, Upper Saddle River, NJ: Prentice-Hall. A nontechnical introduction by a famous father and family team.

E.L. Wolf, Nanophysics and Nanotechnology—An Introduction to Modern Concepts in Nanoscience, 2nd ed. 2006, Weinheim: Wiley-VCH. As the title implies, this book is focused on physics and devices. It is aimed at physics undergraduates and does not make great mathematical demands.

G.A. Ozin and A.C. Arsenault, Nanochemistry—A Chemical Approach to Nanomaterials. 2005, Cambridge, UK: RSC Publishing. This book touches on just about every nanostructured material in the current literature. It is a comprehensive catalog.

M. Di Ventra, S. Evoy and J.R. Heflin (eds), Introduction to Nanoscale Science and Technology. 2004, Berlin: Springer. This is a collection of contributions by experts, valuable for plunging into selected topics in greater depth, though little has been done to integrate the material in the book.

1.9   Exercises

Appendix A contains a list of units, conversion factors, physical quantities, and some useful math.

1. *Encyclopedia Britannica on the head of a pin*: Assume that there are 26 volumes (A–Z) each of 1000 pages, each page being 6×10 inches. How many times does the entire surface area have to be demagnified to fit into the head of a pin (diameter 0.5 mm)? How does this compare with Feynman's estimate? Assuming 50 characters across a line of each

page, how many iron atoms would there be across each character (Fe atomic diameter = 0.25 nm)?

2. From the ideal gas law ($PV = nRT$) estimate the volume of one mole ($n = 1$) of an ideal gas at standard temperature and pressure (101.3 kPa and a temperature of 273.15 K). Check the assertion about the number of moles in a lungful of air at the start of this chapter.

3. Silicon has a density of 2330 kg^{-3} and an atomic weight of 28.05. Use these data to estimate the volume occupied by a silicon atom.

4. The Pentium IV chip has a die size of 217 mm × 217 mm. How many transistors could be fitted onto the surface of the chip if each could be made from 100 atoms of silicon in a monolayer? Use data from Question 3.

5. What is the size of the current fluctuations (shot noise) in a current of 100 pA over a counting period of one second? Use $e = 1.6 \times 10^{-19}$ C and the expression for number fluctuations in Table 1.1.

6. The electrostatic energy of an electron in the electric field of a nucleus is given by Coulomb's law:

$$E = \frac{1}{4\pi\varepsilon_0} \frac{e^2}{r}.$$

Take $r = 0.1$ nm and estimate this energy. Estimate the kinetic energy of the atom at room temperature (300 K) using $E = \frac{3}{2}k_B T$, where k_B is the Boltzmann's constant (1.38×10^{-23} J/K). Comment on the degree to which electronic structure might be changed by collisions of molecules at room temperature.

References

[1] Feynman, R.P., There's plenty of room at the bottom. J. MEMS, **1**: 60–66 (1992).

[2] Binnig, G., H. Rohrer, Ch. Gerber, and E. Weibel, Surface studies by scanning tunneling microscopy. Phys. Rev. Lett., **49**(1): 57–61 (1982).

[3] Lindsay, S.M. and B. Barris, Imaging deoxyribose nucleic acid molecules on a metal surface under water by scanning tunneling microscopy. J. Vac. Sci. Tech., **A6**: 544–547 (1988).

[4] Dahn, D.C., M.O. Watanabe, B.L. Blackford, M.H. Jericho, and T.J. Beveridge, Scanning tunneling microscopy imaging of biological structures. J. Vac. Sci. Technol., **A6**: 548–552 (1988).

[5] Binnig, G., C.F. Quate, and Ch. Gerber, Atomic force microscope. Phys. Rev. Lett., **56**(9): 930–933 (1986).

[6] Seeman, N.C., Nucleic acid junctions and lattices. J. Theor. Biol., **99**(2): 237–247 (1982).

[7] Seeman, N.C., DNA in a material world. Nature, **421**: 427–431 (2003).

[8] Huang, Z., H. Chen, L. Yan, and M.C. Roco, Longitudinal nanotechnology development (1991–2002): National Science Foundation funding and its impact on patents. J. Nanopart. Res., **7**: 343–376 (2005).

[9] Marrian, C.R.K. and D.M. Tennant, Nanofabrication. J. Vac. Sci. Technol., **A21**: S207–S215 (2003).

[10] Xia, Y. and G.M. Whitesides, Soft lithography. Angew. Chem. Int. Ed., **37**: 550–575 (1998).

[11] Giannuzzi, L.A. and F.A. Stevie, *Introduction to Focused Ion Beams*. 2005, New York: Springer.

[12] Baliga, J., *Chips go vertical*, in *IEEE Spectrum*. 2004. p. 42–47.

[13] Chiu, W., M.L. Baker, W. Jiang, M. Dougherty, and M.F. Schmid, Electron cryomicroscopy of biological machines at subnanometer resolution. Structure, **13**: 363–372 (2005).

[14] Zuo, J.M., M. Kim, M. O'Keeffe, and J.C.H. Spence, Direct observation of d-orbital holes and Cu–Cu bonding in Cu_2O. Nature, **401**: 49–52 (1999).

[15] Fennimore, A.M., T.D. Yuzvinsky, W.Q. Han, M.S. Fuhrer, J. Cumings, and A. Zettl, Rotational actuators based on carbon nanotubes. Nature, **424**: 408–410 (2003).

[16] Regan, B.C., S. Aloni, K. Jensen, R.O. Ritchie, and A. Zettl, Nanocrystal-powered nanomotor. Nano Lett., **5**(9): 1730–1733 (2005).

[17] Yarn, K.F., C.Y. Chang, Y.H. Wang, and R.L. Wang, Molecular beam epitaxy grown GaAs bipolar-unipolar transition negative differential resistance power transistor. Jpn. J. Appl. Phys., **29**(12): L2411–L2413 (1990).

[18] Aoki, T., M. Takeguchi, P. Boieriu, R. Singh, C. Grein, Y. Chang, S. Sivananthan, and D.J. Smith, Microstructural characterization of HgTe/HgCdTe superlattices. J. Crystal Growth, **271**: 29–36 (2004).

[19] Meyer, E., H. Heinzelmann, P. Grütter, T. Jung, H.R. Hidber, H. Rudin, and H.J. Güntherodt, Atomic force microscopy for the study of tribology and adhesion. Thin Solid Films, **181**: 527–544 (1989).

[20] Ho, W., N. Nilius, and T.M. Wallis, Development of one-dimensional band structure in artificial gold chains. Science, **297**: 1853–1856 (2002).

[21] Maier, S.A., M.L. Brongersma, P.G. Kik, S. Meltzer, A.A.G. Requicha, and H.A. Atwater, Plasmonics—a route to nanoscale optical devices. Adv. Mater., **13**: 1501–1505 (2001).

[22] Tanaka, M., Spintronics: recent progress and tomorrow's challenges. J. Crystal Growth, **278**: 25–37 (2005).

[23] Steane, A., Quantum computing. Rep. Prog. Phys., **61**: 117–173 (1998).

[24] Bevington, P.R. and D.K. Robinson, *Data reduction and error analysis for the physical sciences*. 1992, Boston: McGraw-Hill.

[25] Wilczek, F., *On absolute units, III: Absolutely not?*, in *Physics Today*. 2006. p. 10–11.

[26] Wilczek, F. Enlightenment, knowledge, ignorance, temptation. Summary talk at *Expectations of a Final Theory*, September 2005, Trinity College, Cambridge.

The Basics

Quantum mechanics

Quantum mechanics is conceptually challenging and requires mathematics, but it is an essential basis for understanding nanoscience. Though simplified, the material in this chapter covers what might be learned in an entire semester, so if you have not been exposed to the subject before, panic would be a reasonable reaction! The chapter is long because it (almost) serves as a stand-alone guide to the subject (taken in conjunction with Appendix C). This outline lists the essential "take-home" points needed to follow the rest of the book. If you grasp these, you can always come back to this chapter later for more background. Here is the list of things you should know:

- Quantum mechanics replaces firm predictions about where a particle goes with a probabilistic description of where it might be found. The prediction is made using a complex number called the probability amplitude. The probability amplitudes for all positions and times are given by a "wavefunction."
- The square of the wavefunction at any point yields the probability of finding a particle.
- The "wavefunctions" have a wavelike character to them, and for free particles, they repeat in time and space like a real wave does. The repeat distance (i.e., wavelength) is related to the momentum of a particle by a simple formula called the *deBroglie relation*. The probability of finding a free particle (given by the square of the wavefunction) is everywhere constant, as we will show.
- This same wavelike property of the probability amplitudes means that the position of a particle may not be absolutely defined. In quantum mechanics, only the product of the uncertainty in position with the uncertainty in momentum is well defined (and equal to Planck's constant). This is the *uncertainty relation*.
- No two electrons (electrons belong to a quantum mechanical class of particles called fermions) can have the same wavefunction. This is called the *Pauli exclusion principle.*
- The probabilistic description of the position of a particle means that it might be found in a region where it would not be allowed to be classically (because it would not have the energy to get there). This process is called tunneling, and you should understand the simple formula for electron tunneling and where it comes from.
- Confining a particle to a region of space imposes conditions on the associated wavefunction that lead to *energy quantization*: the particle can only exist in certain states with certain energies. You should know the

states and energies for a particle trapped in a deep one-dimensional (1-D) box ("particle in a box").

- The wavelike behavior of the probability amplitude for free particles means that even though their energies are not constrained similar to a particle in a box, the number of available states per unit energy is not constant. The quantum mechanical predictions for the number of states per unit momentum are very important results.
- The positively charged atomic nucleus forms a very important kind of box for electrons. The mathematics is much more challenging than that for a rectangular box, but the end results form our basis for understanding atoms and the periodic table.
- The approximate transition rates from one quantum state to another are given by a very useful result called *Fermi's golden rule.*
- Approximate solutions to quantum mechanical problems can be obtained using *perturbation theory.* In one application, we show how a covalent bond forms as atoms are brought together.

The tool for carrying out most of these calculations is the Schrödinger equation. This is a differential equation that relates time and space derivatives of wavefunctions. It is difficult to solve for all but the simplest problems, and if this chapter gives you a permanent aversion to the associated mathematics, you can take consolation from the knowledge that some very powerful computer programs solve the Schrödinger equation approximately at the push of a button. We end the chapter with an outline of one such approach called Density functional theory (DFT).

2.1   Why physics is different for small systems—the story of the Hitachi experiment

Smallness is not a prerequisite for quantum mechanical behavior. But we usually think of small systems—atoms, molecules, nanostructures—as obeying the laws of quantum mechanics, and big things—baseballs, cars, planets—as obeying classical mechanics. The full story is a little more complicated than that, as we shall see. Many problems in nanoscience straddle the boundary between the classical and quantum worlds, so we get to confront the big question of "when is a system quantum mechanical and when is it classical?"

Quantum mechanics was invented in response to the problems posed by very small systems through the attempts of early twentieth century physicists to understand the behavior of atoms. Much of this history is given in the introductory chapters to books on quantum mechanics.[1,2] "The Making of the Atomic Bomb"[3] is an enthralling and utterly readable popular account of the birth of the atomic age. Quantum mechanics describes the physics of atoms and it is an essential language for chemistry. But we are going to have a look at the subject by describing an experiment on a big scale—one involving beams of electrons in an electron microscope. This remarkable experiment was carried out at Hitachi labs in Japan.[4] The experimental arrangement is illustrated in Fig. 2.1. A beam of electrons was fired at a detector that could identify the point

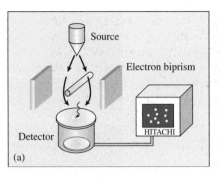
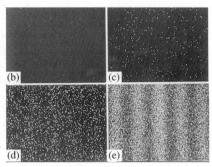

Fig. 2.1 (a) The Hitachi electron diffraction experiment. Monoenergetic electrons from a field emission gun can pass either side of an electron biprism (a wire that is negatively charged to repel the electrons). The biprism serves to divide the beam into two parts. The divided beam continues on to hit a detector plate that is imaged by a sensitive CCD camera. Each electron detected is plotted as a bright spot on a screen. (b, c , d, and e) Shows how a pattern builds up in the distribution of detected electrons over time. The Hitachi experiment appears to show that a beam of only one electron can be divided (Courtesy of Dr. Akira Tonomura).

at which each electron landed on the detector (as a bright spot on the computer screen). The electron beam passed through an electron biprism. This device is a negatively charged wire that deflects electrons to the right or the left of it. A key feature of the experiment was the use of a very low electron beam current. The current was so low that the probability of even one electron being in the apparatus at any time was tiny. The probability that two were in the device at the same time was therefore vanishingly small. Electrons were observed landing at random on the screen, one at a time, as shown in Fig. 2.1(b). As time passed, the number of detected electrons increased, resulting in a greater density of spots on the screen. As only one electron was in the column at any one time, the passage to the left or the right of the biprism was surely a random series of events? If the electrons behave as independent particles, then they must hit the screen randomly with some overall distribution determined by the electron source. But this is not how the electrons behaved! As time progressed, some regions were seen to acquire more spots—the pattern is discernable in Fig. 2.1(d), becoming much clearer on longer exposure as shown in Fig. 2.1(e).

2.1.1   Particles and waves?

If you have already had exposure to quantum mechanics, then you will recognize this pattern as the interference pattern that results from the wavelike behavior of the probability amplitude for finding electrons. deBroglie proposed, and many experiments have verified, that electrons with a momentum of magnitude p_e (when specifying both magnitude and direction we will use the vector symbol $\boldsymbol{p}_e$) have a "wavelength" λ associated with the probability of finding the particle given by

$$\lambda = \frac{h}{p_e}, \tag{2.1}$$

where h is the Planck's constant, 6.63×10^{-34} Js, and p_e is the momentum, mv, of an electron of mass m_e ($m_e = 9.1 \times 10^{-31}$ kg) undergoing linear motion with velocity, v(m/s). The classical definition of momentum (as mass × velocity) is valid at speeds much less than that of light. The spacing of the fringes observed

Fig. 2.2 Geometry of the electron biprism experiment. The middle positions of the two sources formed by the biprism (i.e., the parts of the beam that pass to the left or the right of the center of the biprism) are a distance, d, apart. Beams that travel straight to the screen travel the same distance. However, at an angle, θ, the lower beam travels an extra distance "x" to get to the same region as the top beam (in the far-field approximation that the detector is so distant that all beams are parallel at the detector).

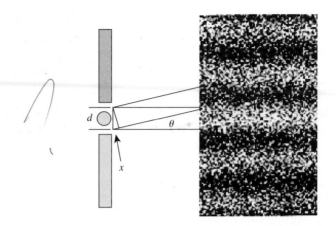

in Fig. 2.1(e) turns out to be exactly that expected from the interference of waves with a wavelength given by Equation 2.1, emerging from the two "sources" in the biprism. By analogy with optics, we will call each of the sources a "slit," acting as a source of "light" in a "double slit" experiment.

In the center of the screen, waves from both slits travel the same distance to the screen, so they have the same phase on reaching the screen. This is illustrated by the two beams that pass from the center of each side of the biprism on to the screen in Fig. 2.2. However, at a point off to one side of the center of the screen (as measured by an angular displacement of the beam, θ) the lower beam travels an extra distance "x" given by $d \sin \theta$ (see the little triangle in Fig. 2.2—d is the spacing between the middle of the two sources). Whenever x is an integral number of wavelengths, the beams arrive exactly in phase and a maximum (bright fringe) is seen. The angular spacing of the fringes is thus given by

$$\theta_n = \sin^{-1}\left(\frac{n\lambda}{d}\right) = \sin^{-1}\left(\frac{nh}{dp_e}\right), \tag{2.2}$$

where λ is obtained from Equation 2.1. Therefore, the statistical distribution of these point-like particles follows an interference pattern like waves. But we know that the electrons are not themselves "waves" and cannot be split like waves because there is only one electron in the apparatus at a time.

Exercise 2.1: If the electrons in the Hitachi experiment are accelerated by a 50 kV potential, then what is (a) their speed, calculated classically; (b) the wavelength of their probability amplitude; and (c) the angular spacing between fringes if the spacing between emerging beams, d, was 100 μm? (The actual experimental geometry was little more complicated.[4])

Answers:

(a) The charge on an electron (Appendix A) is 1.6×10^{-19} C, so the kinetic energy of the electrons as accelerated by 50 kV is $50{,}000 \times 1.6 \times 10^{-19}$ Joules $= 8 \times 10^{-15}$ J. The classical kinetic energy is $E = \frac{1}{2}mv^2$, so the speed of the electrons is given by $\sqrt{\frac{2E}{m}}$ or 1.3×10^8 m/s, where we have used 9.1×10^{-31} kg for m_e (this speed is a significant fraction of the speed of light, so we should worry about relativity).

(b) Their momentum is 9.1×10^{-31} kg $\times 1.3 \times 10^8$ m/s $= 1.2 \times 10^{-22}$ m-kg/s. Using Equation 2.1 with $h = 6.63 \times 10^{-34}$ Js gives $\lambda = 5.52 \times 10^{-12}$ m.

(c) The first minimum occurs at $\theta \sim \lambda/d = 5.52 \times 10^{-12}\, m/100\,\mu m = 55$ nanoradians.

A more efficient way to calculate λ would be to solve the deBroglie relation (Equation 2.1) using the classical expressions for energy and momentum to obtain $\lambda = h/\sqrt{2}$ (meV).

2.1.2 But there is only one electron in the microscope at a time!

Recall that the intensity in the Hitachi experiment was so low that there was only one electron passing through the apparatus at a time. How can an electron interfere with another electron that passed through the apparatus at a different time? This apparent difficulty illustrates a key feature of quantum mechanics: the electron is *not* a wave. It is a particle, but only probabilistic predictions can be made about its position. These probabilistic predictions can have wavelike features when many measurements are averaged.

The problem here arises because of our classical understanding of particles and probability. *Our classical intuition is simply wrong.* Before we examine what is going on in more detail, let us consider a key experiment. We might hypothesize that the interference arises from some kind of hidden correlation in the motion of the electrons. So let us dream up a way of measuring where each electron goes to see if we can understand this better. To do this, we need to illuminate the biprism with a powerful laser and put detectors near each side of the biprism, as shown in Fig. 2.3. The upper beam emerging from the biprism is designated as source 1 and the lower beam as source 2. If an electron passes

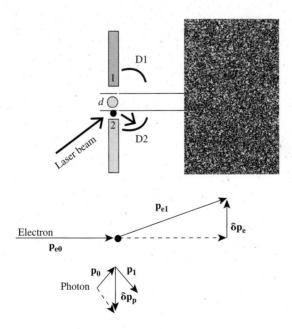

Fig. 2.3 Apparatus to determine where each electron came from. A powerful laser illuminates the biprism. Light scattered off the electron gives a bigger signal in D1 if the electron passed through region 1 and a bigger signal in D2 if the electron passed through region 2. An electron is shown in slit 2 scattering light into D2. However, scattering of the photon changes the initial momentum of the electron, $\mathbf{p_{e0}}$, to $\mathbf{p_{e1}}$. The vector difference between these two momenta, $\boldsymbol{\delta p_e}$, must equal the change in momentum for the photon, $\boldsymbol{\delta p_p}$ (the vector difference between $\mathbf{p_0}$ and $\mathbf{p_1}$), to conserve momentum in the collision. The resulting disturbance owing to the attempt to observe the electron's path destroys the interference fringes.

through 1, it will scatter light into detector D1 and a counter will register the source of the corresponding spot on the screen as source 1. Similarly, detector D2 will tell us when an electron passes through source 2.

There is a catch, however. We need to use the light of wavelength much less than the distance, d, between the two sources if our "laser microscope" is to determine where the electron came from. Light of a wavelength longer than d would give similar signals in each detector, regardless of the source of the electron. We need to use a photon of wavelength $\lambda_p \ll d$ if our optical microscope is going to resolve from which source the electron came. But photons carry momentum that increases as the wavelength decreases (i.e., as the energy increases). The energy, E, of a photon is related to the photon frequency ν (or wavelength λ_p) by Planck's constant also:

$$E = h\nu = \hbar\omega = \frac{hc}{\lambda_p}, \tag{2.3}$$

where c is the speed of light, ω is the angular frequency, and $\hbar = h/2\pi$. Relativity gives the energy of a photon in terms of the magnitude of its momentum, p, by $E = pc$, so Equation 2.3 can be rearranged to give the wavelength of a photon in terms of the magnitude of its momentum as

$$\lambda_p = \frac{h}{p}. \tag{2.4}$$

This is exactly the same as the deBroglie expression for the wavelength of a material particle (Equation 2.1). (deBroglie's suggestion was inspired by this result for photons.) Therefore, to make our laser microscope work, we need a photon with a momentum of magnitude

$$p \gg \frac{h}{d}. \tag{2.5}$$

In order to give a distinctive signal, the photon must be scattered through a significant angle to reach the detector, D1 or D2. But if the photon is deflected, then so must the electron also be deflected, so that the total momentum of the system is conserved. Thus the electron must have its own momentum changed by an amount on the order of h/d. Referring to the scattering diagrams on the bottom of Fig. 2.3, if the original momentum of the photon is $\mathbf{p_0}$, changing to $\mathbf{p_1} = \mathbf{p_0} + \delta\mathbf{p_p}$ after it is scattered into the detector, then the electron must change its momentum by an equal and opposite amount $\delta\mathbf{p_e} = \mathbf{p_{e1}} - \mathbf{p_{e0}}$. If the magnitude of $\delta\mathbf{p_e}$ is a significant fraction of p (given by Equation 2.5) then $|\delta\mathbf{p_e}| \sim h/d$. So the electron changes its direction by an angle which is approximately h/dp_e radians or

$$\delta\theta \approx \sin^{-1}\left[\frac{h}{dp_e}\right]. \tag{2.6}$$

But this is the same as the angle for the first-order ($n = 1$) fringe as given by Equation 2.2. Thus, our attempts to observe the path of the electron through the instrument disturb the electron by just enough to destroy the interference fringes that signaled quantum behavior (Fig. 2.3).

Note that the loss of quantum interference was not tied to the act of observation per se (as some popular accounts of quantum mechanics would have it) but rather to the physical disturbance associated with our attempts to observe the path of the electron. Such physical disturbances occur all the time (poor vacuum in the microscope, broad range of thermal energies for the electrons, stray electric fields, etc.) and usually have nothing to do with an attempt at observation.

2.2   The uncertainty principle

In the example above, measurement of the position of the electron imparted just enough momentum to alter its direction by about the amount of its angle of diffraction before the measurement. Heisenberg, arguing in a similar but more mathematical manner, derived his famous "uncertainty principle" relating the product of the uncertainty in position, Δx, and the uncertainty in momentum, Δp, in a measurement made on a quantum system as follows:

$$\Delta x \Delta p \geq \frac{h}{4\pi}. \tag{2.7}$$

(As written here, this is a relationship between the magnitudes of these vector quantities.) Note that the units of Planck's constant—Joule-seconds—are equivalent to position times momentum ($\mathrm{kg\,(m/s)^2 \times s = kg\,(m/s) \times m}$).

Exercise 2.2: Equation 2.7 tells us that a particle confined to a small space must have a large momentum, even at zero temperature, as described by the uncertainty in momentum. What is the speed of an electron confined to a hydrogen atom (of diameter approximately 1 Å)?

Answer: Equation 2.7 gives the uncertainty in momentum as $\Delta p \sim h/4\pi\,\Delta x$ or $6.63 \times 10^{-34}/4\pi \times 10^{-10} = 5.27 \times 10^{-25}$ kg m/s. With $m = 9.1 \times 10^{-31}$ kg we get $\Delta v = \Delta p/m = 5.8 \times 10^5$ m/s.

The discussion about observing the electron's path above can be recast in terms of Equation 2.7. The uncertainty in position in the Hitachi experiment (with a light detector) would be the spacing between fringes on the screen or

$$\Delta x = z\delta\theta \sim \frac{hz}{dp_e},$$

where z is the distance between the biprism and the screen. The uncertainty in momentum is on the order of the angular difference between the direction of the beams emerging from source 1 or source 2 times the magnitude of the incident momentum or

$$\Delta p \approx \frac{dp_e}{z}$$

so that $\Delta p \Delta x \approx h$ in accordance with the uncertainty principle (Equation 2.7).

The dimensions of Planck's constant—Joule-seconds—suggests an equivalent statement of Heisenberg's uncertainty principle in terms of energy and time:

$$\Delta E \Delta t \sim h. \tag{2.8}$$

This form of the uncertainty principle has profound implications for measurements of energy in a quantum system. Energy does not have to be conserved (to within ΔE) over time intervals Δt when $\Delta E \Delta t \sim h$. For very short time intervals, this can amount to a large amount of energy that is "not conserved." In this context, it is useful to state a value for Planck's constant in units of electron-volts-seconds: $h = 4.14 \times 10^{-15}$ eV-s.

Exercise 2.3: An atom can jump into its excited state for a short time, even at the absolute zero temperature. How long is the lifetime of this spontaneous excitation if the energy between ground and first excited state is 4 eV?

Answer: According to Equation 2.8, $\Delta t \sim h/\Delta E$ and using $h = 4.14 \times 10^{-15}$ eV-s and $E = 4$ eV, we get $\Delta t \sim 10^{-15}$ s. This sets the timescale associated with quantum mechanical electron transfer processes.

2.3   The Hitachi microscope as a quantum system

The key to the observation of quantum behavior in the Hitachi experiment lies in the fact that the electrons do not interact with any other particles as they transit the possible paths through the instrument. The only interaction occurs *after* the biprism when the electrons hit the detection screen. Even though the electron appears to be a particle at that point, inside the apparatus they are something else—something that could have gone through source 1 or through source 2, somehow "aware" of the other source so that they can carry the information about *both paths* to the screen (where the electrons revert to particle-like behavior). This "ghostly" behavior is the true description of the path of the electron. Any attempt to probe it in more detail ruins quantum effects as discussed above. We will elaborate on the rules for predicting the average behavior of quantum systems in this chapter, but suffice it to say that the key requirement is not that a system should be small but rather that it should not interact with the "outside world" in a significant way (until, of course, the system interacts with the device being used to make a measurement)—where the meaning of "significant" is defined by the uncertainty principle and the scale is set by the magnitude of Planck's constant.

Quantum behavior is most evident in small systems precisely because it is very hard to avoid interactions with other atoms, photons, and molecules over any great distance in big systems. In fact, in the Hitachi experiment,[4] the distance over which the electrons could be thought of being in a "pure" quantum state (their coherence length) was only about 100 μm. The thermal energy spread in the energy of emitted electrons was what set this particular limit. But it is possible to observe quantum effects even over distances of kilometers if the experiment is set up so as to avoid interactions that disturb the path of "particles" through the instrument.[5]

To reiterate: an atomic "particle" is not really a "particle" at all, but rather a quantum system obeying what, to our classical intuition, is a set of very strange rules. It can be easy to disturb these systems so that they end up appearing to be composed of classical particles, but this is an artifact of disturbing the quantum system. However, the notion of "classical particles" is so intuitive that quantum behavior can appear to be illogical (Bernard d'Espagnat has written a very accessible article about this in *Scientific American*[6]). The apparent breaches of logic stem from our desire to treat everything as composed of classical particles. This view is just not true.

The scale of important interactions is given by the uncertainty principle. Particles bound together in a very small space, such as the neutrons and protons that make up an atomic nucleus, have enormous momentum (hence energy) because of their confinement. Consequently, it takes a lot of energy to perturb the quantum behavior of a nucleus (the size of a nucleus is on the order of an fm—corresponding to MeV of energy, by analogy with Exercise 2.2, taking account of the 2000-fold greater mass of a neutron or proton compared with an electron). Thus, a nucleus is a "quantum particle" for most purposes. Only if probed with high-energy beams (tens of millions of electron volts) does it appear to be composed of individual neutrons and protons.

An atom is much bigger (cf. Exercise 2.2) so the scale of energy associated with quantum behavior is much smaller—electron volts. But this is still a lot of energy (compared to thermal energy at room temperature which is 1/40 of an electron volt) so an atom also appears to be a good quantum object, with behavior predicted by the rules of quantum mechanics.

As we get larger in scale, the energy scale of the interaction between particles, relative to the energy scale of their other interactions with the "outside world," is what determines whether the particles can be considered as a quantum system or not. This give rise to the notion of a "quasiparticle" in solid-state physics: if two particles interact more strongly with each other than that with the "rest of the world," they can be (and are) considered to be an entirely new type of particle put together from their "components" according to the laws of quantum mechanics, something we elaborate on at the end of this chapter.

2.4 Probability amplitudes and the rules of quantum mechanics

The mechanics of classical particles is straightforward and described by Newton's second law of motion. The acceleration of a particle is given by the ratio of the net force acting on it to its mass. Once its initial position and velocity are given, its subsequent motion is completely determined. However, quantum mechanics tells us that the very idea of a classical particle is an artifact of a complicated world where interactions have smeared out the underlying quantum behavior.

The uncertainty principle tells us that we can never know exactly and simultaneously both the position and the velocity of a "particle." Because of this, the underlying laws of motion for a quantum particle have to be framed in such a way that lets us make predictions only for quantities that are the *average* of many individual measurements. We can predict what position and momentum would be found by averaging many repeated measurements in nominally identical, but repeated experiments. But the outcome of the measurements on any one "particle" is stochastic, with a spread of data given by the uncertainty principle. Therefore, instead of predicting the position and momentum of any one particle, quantum mechanics tells us how to predict the *probability* that a "particle" has a certain position (or a certain momentum). Considering the example of the Hitachi experiment again, quantum mechanics tells us nothing about where and when each individual flash will occur on the detector screen. Each flash has no meaning in terms of where the corresponding electron came from, so long as the system is undisturbed enough to remain truly quantum mechanical. But quantum mechanics *does* predict the points of high and low probability that make up the "interference fringes" (Fig. 2.2).

To codify this process, a new variable is introduced, called the "probability amplitude," ψ. The following list provides some rules for making predictions based on ψ:

a. *Probability amplitudes and wavefunctions:* The probability that a particle is at a certain point in space and time is given by the square of a *complex number* called the probability amplitude, ψ. Being a complex number means that it has two components in general: one a real number and the other an imaginary number. These components behave similar to the orthogonal components of a 2-D vector. A function that encodes all the (real and imaginary) values of ψ at various points in space and time is called the *wavefunction*, $\psi(\mathbf{r}, t)$. It is important to realize that the probability amplitude is not a probability (it cannot be, because it is, in general, a complex number). Rather, it is a tool used to calculate the probability that a measurement will yield a particular value.

Engineers often use complex numbers as a convenient way to represent two quantities at once, for example, a wave possessing both an amplitude and a phase, and the two components of the probability amplitude can be interpreted this way. However, we will see that the equation that governs the time evolution of a quantum system contains the symbol "i" ($i = \sqrt{-1}$), so the complex nature of the probability amplitude is more than just a mathematical convenience.

b. *Calculating probability amplitudes for one "path":* Even though ψ is not a probability per se, we can calculate the overall probability amplitude (i.e., a particle goes from A to C via B) as the product of the probability amplitude for going from A to B with the probability amplitude for going from B to C. This is analogous to the multiplication rule for calculating the probability of classical independent events, but the quantum mechanical rules apply only to a quantum system where the "particle" is not disturbed significantly as it passes from A to C (via B). Consider a concrete example. Suppose that we knew (or could calculate) the probability amplitude for a "particle" to "go" from a to b (i.e., from the electron source to the biprism in the Hitachi experiment) ψ_{ab}. Similarly, suppose we also know the probability amplitude, ψ_{bc}, to "go" from b to c (e.g., from the biprism to the screen in the Hitachi experiment). The overall probability amplitude is given by the product of the two probability amplitudes:

$$\psi_{ac} = \psi_{ab}\psi_{bc} \tag{2.9}$$

The double quotes on the word "go" in the above discussion remind us that we are really calculating the probability that a measurement will find the particle at some position. In general, probability amplitudes are complex, $\psi_{ab} = x + iy$, $\psi_{bc} = u + iv$, so the product, ψ_{ac} is also complex: $xu - yv + i(xv + yu)$.

c. *Calculating probabilities from probability amplitudes:* The probability that a particular outcome is measured is given by the square of the corresponding probability amplitude. In the example just given, the probability to go from a to c is $|\psi_{ac}|^2$. Since ψ is a complex number, the square is formed by taking the product of ψ with its complex conjugate, ψ^*. That is, if $\psi_{ac} = a + ib$, then $|\psi_{ac}|^2 = (a+ib)(a-ib) = a^2 + b^2$ (a real number, as a probability should be).

d. *More than one possible "path" through the apparatus:* Now we turn to the rule that will let us understand the interference fringes observed in the Hitachi experiment. To work out the probability amplitude for some outcome when there is more than one possible path, *we must add the probability amplitudes for each possible pathway.* This is like classical probability (where we would add the probabilities) but applied to the probability *amplitude*. Once again, this rule only applies if the system remains undisturbed up to the point of measurement of the outcome. The probability of an outcome is the square of this sum of probability amplitudes. It is this addition of two (or more) complex numbers, followed by squaring, that gives the "quantum interference" pattern. Illustrating the process with the Hitachi experiment, suppose that one possible way through the apparatus in the Hitachi experiment corresponds to a probability amplitude:

$$\psi_{ac1} = x + iy \tag{2.10a}$$

(say from the electron source, via side 1 of the biprism—Fig. 2.3—to the screen). And let us write the probability amplitude for the second possible way as follows:

$$\psi_{ac2} = u + iv. \tag{2.10b}$$

Then according to the rules just stated, the probability to arrive at some point of the screen, $\psi(\mathbf{r})$, is given by

$$\begin{aligned}
|\psi(\mathbf{r})|^2 &= (\psi_{ac1} + \psi_{ac2})^*(\psi_{ac1} + \psi_{ac2}) \\
&= [(x + u) - i(y + v)][(x + u) + i(y + v)] \\
&= (x + u)^2 + (y + v)^2 = x^2 + u^2 + 2xu + y^2 + v^2 + 2yv. \quad (2.11)
\end{aligned}$$

We ignored time as a variable here, assuming that the apparatus stays the same as data are collected.

In contrast, in classical probability theory, we would add the probabilities that the electron went via **1** (which would be $x^2 + y^2$) or **2** (which would be $u^2 + v^2$):

$$P_{\text{classical}}(1 \text{ and } 2) = |\psi_{ac1}|^2 + |\psi_{ac2}|^2 = x^2 + y^2 + u^2 + v^2, \tag{2.12}$$

which is quite different from the quantum mechanical result in Equation 2.11. x^2, y^2, u^2, and v^2 must, as squares, all be positive numbers in Equation 2.12. But the extra product terms, $2xu$, $2yv$ in the quantum result (Equation 2.11) could be either positive or negative, adding or subtracting depending on their sign, to give the "interference" pattern shown in Fig. 2.2. We will work out this pattern using wavefunctions for free electrons later.

e. *Probability amplitudes for more than one particle:* What we did was correct for one electron in the apparatus (which is how the Hitachi experiment was conducted). But what would happen if two electrons were in the apparatus at the same time? How do the probability amplitudes for events involving two or more particles behave? We know that one electron cannot be in precisely the same place as a second electron at a given time. In addition to considering the two possible paths for each of the two electrons, we must then modify the resultant probability amplitude for the pair of particles to avoid combinations that have both electrons in the same place at the same time. This is not a universal requirement, for if we were discussing photons (in the case of an experiment with interfering light beams) we know that any number of photons can occupy the same state at the same time. Thus, probability amplitudes (for more than one particle) must be combined in different ways for different types of particles. This is not as bad as it sounds, for it turns out that there are just two kinds of particles: fermions (like electrons) cannot be in the same state at the same time. Bosons (like photons) are not constrained in this way. We shall see that these two possibilities correspond to the two signs that the probability amplitude can have ($\pm$) before it is squared.

It is not surprising that quantum states that describe waves (light waves for photons and sound waves for "phonons") allow an arbitrary number of particles to be in the same state, for this number is just the intensity of the wave. What is less obvious is that material particles can also be bosons. Even more surprisingly, nuclei of one element can have one isotope that is a fermion and another that is a boson. It turns out that the controlling factor is a quantum mechanical quantity (which arises naturally when relativity and quantum mechanics are combined) called "spin." It is called "spin" because the quantum mechanical rules for manipulating it are reminiscent of the rules for adding the angular momenta of spinning tops. Electrons have "spins" of $\pm\frac{1}{2}\hbar$ (where $\hbar = h/2\pi$). In units of $\hbar$, they are said to have "half-integer spin." Photons have "spins" of $\pm\hbar$ and are said to have "integer spin" (in terms of classical optics, the two states correspond to left and right circularly polarized light). Particles with integer spin values are bosons, while those with half-integer spin values are fermions.

The rules for the probability amplitudes of two or more particles are as follows: (1) For bosons, the probability amplitudes for all combinations of the particles that can contribute to an outcome are added. (2) For fermions, each combination in which a pair of particles change place must change its sign (from + to − or − to +). We will illustrate this below and will see later how this restricts the occupation of any one state to just one fermion.

2.5   A word about "composite" particles

Our discussion of bosons and fermions is a convenient place to remind ourselves that what we call a "particle" depends on how we measure its properties (i.e., perturb the system). The observed "particle" may be a mixture of other particles, but from the point of view of quantum mechanics, it is one particle so long as it is not disturbed significantly in an attempt to find out where its components are. This is what is meant by a composite particle in the quantum mechanical sense. We discussed the example of atomic nuclei above. Protons and neutrons (collectively called nucleons) have half-integer spins of value $\pm\frac{1}{2}\hbar$. The spin of the nucleus is the sum of the magnitude of the spin for each nucleon. So the spin of a ^{3}He nucleus is $\pm\frac{3}{2}\hbar$, a half-integer spin. Thus the ^{3}He nucleus is a fermion. But the spin of a ^{4}He nucleus is $2\hbar$, an integer spin which makes it a boson. From the viewpoint of classical physics, it seems very strange that adding one neutron (a fermion) to ^{3}He (also a fermion) turns it into another kind of object, a boson with the new property that a ^{4}He nucleus can occupy the same state at the same time as another ^{4}He nucleus. It is not quite as strange if we realize that the two isotopes are particles made up of "unobservable" quantum components. This distinction has real experimental consequences. ^{4}He liquid can form a superfluid, flowing without viscosity, in part because its nuclei are bosons.

Electrons can also form composite particles, though it is difficult for them to do because they repel each other strongly via their electrical charge and they lack the strong attractive interaction that binds nucleons together. However, at low enough temperatures, in certain materials, very weak attractive interactions can lead to particles composed of two electrons with oppositely paired spins so that the net spin is zero. These "Cooper pairs" (named after their discoverer) are bosons and form the basis of superconducting materials (materials in which current flows without electrical resistance).

2.6 Wavefunctions

We have kept the discussion simple thus far by referring to a "probability amplitude" for one specific outcome (i.e., to calculate the probability that the electron goes from a to c, we calculate $|\psi_{ac}|^2$). But what about the probability that the electron goes from the source at a to any point on the screen (denoted by a vector **r**)? The resulting (infinite) set of complex numbers is the wavefunction, $\psi(\mathbf{r})$. We will work with quite simple wavefunctions (leaving real calculations to computers that can store $\psi(\mathbf{r})$ as a very large list of complex numbers). We will often simplify the problem even further by concentrating on properties that do not change with time (corresponding to the so-called stationary states of the system).

Wavefunctions are usually *normalized* based on the fact that the probability of finding the particle *somewhere* is unity. In a system with time-independent probabilities, this is equivalent to requiring that the integral of the probability summed for all points in space be unity:

$$\int_{\mathbf{r}} \psi^*(\mathbf{r})\psi(\mathbf{r})\mathrm{d}^3\mathbf{r} = 1, \tag{2.13}$$

where the integral over **r** means the sum of the integrals over x, y, and z for each spatial coordinate running from $-\infty$ to $+\infty$. We take the complex conjugate of the wavefunction in the product (this is what is meant by the "*" in $\psi^*(\mathbf{r})$) because it is a function of complex numbers.

Exercise 2.4: For the simple 1-D, real wavefunction $\psi(x) = A\exp(-k|x|)$, calculate A in terms of k using Equation 2.13.

Answer: In just one dimension, Equation 2.13 becomes

$$\int_{-\infty}^{\infty} A^2 \exp(-2k|x|)\mathrm{d}x = 2\int_{0}^{\infty} A^2 \exp(-2kx)\mathrm{d}x = 2A^2 \left.\frac{-1}{2k}\exp(-2kx)\right|_{0}^{\infty} = 1$$

giving $A = \sqrt{k}$. If this had been a 3-D problem with x replaced with r, we would have had to include the factor $\mathrm{d}^3\mathbf{r} = r^2 \sin\theta\mathrm{d}\theta\mathrm{d}\varphi\mathrm{d}r = 4\pi r^2\mathrm{d}r$ when ψ is not a function of θ and ϕ.

2.7   Dirac notation

Integrals like Equation 2.13 occur so often in quantum mechanics that a shorthand notation invented by Dirac is often used. By definition, Dirac notation is

$$\int_{-\infty}^{\infty} \psi^*(x)\varphi(x)\mathrm{d}x \equiv \langle\psi|\varphi\rangle. \tag{2.14}$$

So Equation 2.13 in Dirac notation is written as

$$\langle\psi|\psi\rangle = 1. \tag{2.15}$$

Often, we will be concerned with how much one wavefunction "overlaps" another, an important concept in quantum mechanics. This is performed by calculating an integral like Equation 2.14 taken between two *different* wavefunctions (ψ_n and ψ_m for example). Different wavefunctions that satisfy the same Schrödinger equation (to be introduced in Section 2.10) have the property of being *orthogonal*. If they are also normalized, they are said to be *orthonormal*. In Dirac notation, orthonormality is defined by

$$\langle\psi_n|\psi_m\rangle = \delta_{nm}. \tag{2.16}$$

The "Kronecker delta function," δ_{nm}, has the property of being 1 when $m = n$ and 0 otherwise. The fact that different wavefunctions that solve the same Schrödinger equation have this property can be verified by carrying out the integration in Equation 2.14 on some of the solutions of the Schrödinger equation given later in this chapter.

We are also often interested in the numerical value of the integrals like $\int \psi_1^* \hat{H}\psi_2 \mathrm{d}x$, where $\hat{H}$ is some mathematical operation (such as a derivative) carried out on ψ_2.

In Dirac notation, these integrals are written as

$$\int \psi_1^* \hat{H}\psi_2 \mathrm{d}x \equiv \langle\psi_1|\hat{H}|\psi_2\rangle. \tag{2.17}$$

When $\psi_1 = \psi_2$, the number produced by such an integral is called the "expectation value" of $\hat{H}$. Often, wavefunctions are expressed as $\langle\psi|$. No operation is implied until the wavefunction is multiplied by another to form a closed "bra-ket" combination (as in 2.16).

2.8 Many particle wavefunctions and identical particles

Here we will take a look at what happens when two "particles" interact. How do we write down a wavefunction for the system of two (or more) particles incorporating the rules for fermions or bosons?

To do this, we will consider a scattering experiment. In this type of experiment, a beam of particles is shot at a target, and the probability of finding scattered particles is measured as a function of angle about the line connecting the incident particle and the target particle (we will assume that the density of particles in the target and the number of particles in the beam are low enough so that we only have to think about one beam particle interacting with one target particle). We need a *two particle wavefunction* to predict this outcome of this experiment.

For simplicity, we will work in the center of mass frame of reference. This is a frame of reference in which the total momentum of the system is zero, as viewed by an observer moving with just the right speed to see the initial total momentum as zero. This transformation is illustrated for two particles of equal mass in Fig. 2.4. Because the particles have equal mass in this case, the center of mass frame of reference is one in which each particle has equal and opposite velocities to begin with. Thus, if the observer is moving toward particle 1 with velocity $-v/2$ directed along the line joining the particles (in the lab frame of reference) the scattering experiment appears as illustrated in the right panel of Fig. 2.4 in the center of mass frame of reference. The particles must also move in opposite directions with equal speeds after the scattering event in order to keep the total momentum zero. Now consider a scattering experiment as shown in Fig. 2.5. A particle, P1, could scatter into one of the two detectors, D1 or D2, by scattering off a second particle P2 (which, by conservation of momentum,

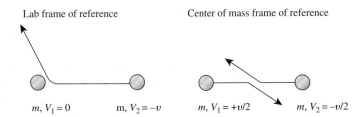

Lab frame of reference   Center of mass frame of reference

$m, V_1 = 0$ $m, V_2 = -v$ $m, V_1 = +v/2$ $m, V_2 = -v/2$

Fig. 2.4 (Left panel) The Lab frame of reference: a scattering experiment viewed in the normal laboratory frame of reference where the incoming particle strikes a stationary target. Particle 2 is approaching particle 1 with a velocity, $-v$. (Right panel) The center of mass frame of reference: viewed by an observer moving with a velocity that makes the sum of the momenta zero, particle 1 is initially moving with velocity $+v/2$ and particle 2 is moving with a velocity of $-v/2$ (in this special case, where both particles have the same mass).

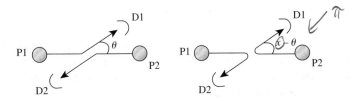

Fig. 2.5 Particle scattering experiment in the center of mass frame of reference showing two detectors arranged on opposite sides of the scattering volume, and two types of event that lead to a count in detector D1.

would scatter into the opposite detector, D2 or D1, respectively). One possible outcome is shown in Fig. 2.5.

We are now going to analyze this experiment for different combinations of particles. Let us start with the case where both particles are identical bosons, for example, ^{4}He nuclei. Given the probability amplitude as a function of angle, $\psi(\theta)$, what is the probability that some particle (P1 or P2) scatters into the detector D1? Since the particles P1 and P2 are identical, and we are not going to interfere with the experiment to try to determine what direction the particle came from before scattering, we know that we must add the probability amplitudes for all the possible ways of getting a particle into D1, and then square this quantity to get the probability, P(D1). One way is that P1 scatters into D1, corresponding to the probability amplitude $\psi(\theta)$ (the event shown on the left panel of Fig. 2.5). A second way is that P2 scatters into D1, as shown on the right panel of Fig. 2.5. The corresponding probability amplitude is $\psi(\pi - \theta)$ so

$$P(D1) = |\psi(\theta) + \psi(\pi - \theta)|^2. \tag{2.18}$$

Evaluating this for the simplest case where $\theta = \pi/2$

$$P(D1) = 4|\psi(\pi/2)|^2 \quad \text{(for identical bosons)}. \tag{2.19}$$

Now let us consider what happens when the two particles are identical fermions (e.g., electrons). Again there are two ways that the (identical) particles can reach the detectors. In the first event, particle 1 scatters into D1 and particle 2 scatters into D2. In the second, particle 1 scatters into D2, and particle 2 scatters into D1. The second event is the same as the first, but with the particles having exchanged states. We said in Section 2.4(e) that exchanging fermions between states requires the sign of the probability amplitude change. So we will call the first case ($+$) and the second case ($-$) so that

$$P(D1) = |\psi(\theta) - \psi(\pi - \theta)|^2. \tag{2.20}$$

Now evaluating this for $\theta = \pi/2$ we find

$$P(D1) = 0 \quad \text{(for identical fermions)}. \tag{2.21}$$

This is a completely different result from that obtained for bosons (2.19) and it tells us that there is *no* scattering at 90° in the center of mass frame for identical fermions. This prediction seems strange indeed, but it is exactly what is observed in scattering experiments. It is a result that can only be understood in terms of the rules of quantum mechanics.

So far, we have considered *identical* particles in the experiments above. What if P1 and P2 are different particles altogether? For the sake of argument, let us say that one is a proton and the other is a neutron. These particles have essentially the same mass, so our scattering diagram does not have to change, but we could add a charge sensor to our detectors, so we could know, in principle, which detector caught the neutron, and which detector caught the proton. We do not have to actually make the measurement of charge: it is sufficient that the interaction of the neutron and the proton with the matter in the detectors is different. The two possibilities cannot now be tied together as a "quantum system" because each pathway is distinguishable. *This means that now the final probability must obey classical rules: the probability that a particle enters D1 must be the sum of the probability (not the probability amplitude) that the proton enters D1 and the probability that the neutron enters D1.* This is given by

$$P(D1) = |\psi(\theta)|^2 + |\psi(\pi - \theta)|^2. \qquad (2.22a)$$

Evaluating this for $\theta = \pi/2$ we find

$$P(D1) = 2|\psi(\pi/2)|^2 \quad \text{(for nonidentical particles).} \qquad (2.22b)$$

This is exactly half the result for identical bosons (Equation 2.19).

The discussion above was framed in terms of very simple wavefunctions that gave the probability amplitude in terms of just one variable, the scattering angle in the plane of scattering. But the results apply to all wavefunctions:

a. For *identical* bosons, the wavefunction for an event is obtained by adding the wavefunctions for all the combinations that could lead to the event. The probability of the event is obtained by squaring the result.
b. For *identical* fermions, the wavefunction for an event is obtained by adding all the combinations of wavefunctions that could lead to the event, changing the sign (relative to one of the states taken as a reference) whenever any pair of particles are exchanged in the wavefunction. The probability of the event is obtained by squaring the result.
c. For nonidentical particles (or nominally identical particles whose path is labeled in some way), the probability of an event is just the sum of the probabilities of each of the ways the final outcome can be achieved.

2.9 The Pauli exclusion principle

This famous principle for identical fermions follows directly from the rules above. Consider two identical fermions which we will call particles 1 and 2. Let particle 1 be in a state ψ_1 and let particle 2 be in a state ψ_2, leading to a total probability amplitude that is the product of these two: $\psi_1(1)\psi_2(2)$. Since the particles are identical, we have to add in the probability amplitude that particle 1 was in ψ_2 and particle 2 was in ψ_1, that is, $\psi_1(2)\psi_2(1)$. But wait! These are fermions, and exchanging particles between states requires that we change the sign of the probability amplitudes (or wavefunctions). So the total wavefunction is

$$\Psi_{\text{total}} = A[\psi_1(1)\psi_2(2) - \psi_1(2)\psi_2(1)], \qquad (2.23)$$

where A is a normalization factor calculated using Equation 2.13. Now if the two wavefunctions are the same wavefunction, $\psi_1 = \psi_2$, then Equation 2.23 tells us that $\Psi_{\text{total}} = 0$. No two fermions can be in the same state (at the same time). This is the famous Pauli exclusion principle. Electrons can have two spin states (corresponding to spin values $\pm\frac{1}{2}\hbar$—see Section 2.5), and spin should really be written in as part of the wavefunction (though we will never do so explicitly). So any given state (by which we will always mean only describing space and time variables) *can hold up to two electrons* (spin up and spin down). This simple principle has some quite unexpected consequences: for example, electrons in metals have, on average, an enormous energy, even at zero temperature. This is because the Pauli principle requires that the large number of electrons in a metal fill the states from the lowest energy up, no more than two occupying each state.

what is a state?

2.10 The Schrödinger equation: a tool for calculating probability amplitudes

The key to predicting the quantum behavior of a system lies in working out the probability amplitudes (and using them in a way that is consistent with the rules outlined above). Erwin Schrödinger discovered a second-order differential equation that is to probability amplitudes what Newton's second law is to classical particles. In his paper,[7] the equation was introduced as a way of adapting equations from classical mechanics so as to give results in accordance with what was known about quantum mechanics at the time. There is no a priori derivation of the equation. It is a basic "law" of quantum mechanics. The Schrödinger equation works well, giving excellent predictions for slow-moving particles where relativistic effects can be ignored. It states a relationship between the second derivative of the wavefunction, Ψ, with respect to position, the potential field in which a particle moves, $U(x, y, z)$ (i.e., a position-dependent energy which results in a force on the particle if it is not at some minimum of the potential), and the first derivative of the wavefunction with respect to time. The Schrödinger equation for a particle of mass (m) moving in one dimension (x) is

$$-\frac{\hbar^2}{2m}\frac{\partial^2\Psi(x,t)}{\partial x^2} + U(x)\Psi(x,t) = i\hbar\frac{\partial\Psi(x,t)}{dt}. \qquad (2.24)$$

In this equation, $\hbar = h/2\pi$, and the ∂ symbol specifies a partial derivative (i.e., the derivative only for terms in which the variable appears explicitly). m is the mass of the particle (we will almost always be dealing with an electron in this book). We will generally deal with situations where U is independent of time, so the Schrödinger equation can be solved with a product of wavefunctions that depend only on space or time:

$$\Psi(x,t) = \psi(x)\phi(t). \qquad (2.25)$$

The functions $\psi(x)$ are the stationary states. Substituting the product solution 2.25 into the Schrödinger equation 2.24 yields

$$\frac{1}{\psi(x)}\left[-\frac{\hbar^2}{2m}\frac{\partial^2\psi(x)}{\partial x^2} + U(x)\psi(x)\right] = \frac{1}{\phi(t)}i\hbar\frac{\partial\phi(t)}{dt}. \qquad (2.26)$$

Each side is now a function of only one variable (x or t) and the statement that one side of 2.26 equals the other side can only be true in general if each side equals the same constant, E. So

$$i\hbar\frac{\partial\phi(t)}{dt} = E\phi(t). \tag{2.27}$$

This differential equation is solved by the exponential function

$$\phi(t) = \exp\left(\frac{E}{i\hbar}t\right). \tag{2.28}$$

So "E" is not just a constant but many energy

This describes a sinusoidally oscillating amplitude with an angular frequency, $\omega = E/\hbar$. But recall that for a photon, $E = \hbar\omega$ (Equation 2.3), where E is the energy of the photon. So we see that in this case the constant, E, is the energy of the "particle."

This is a general result for systems in which the potential is independent of time. The wavefunction can be separated into a spatial component, $\psi(x)$, and a time-dependent component given by Equation 2.28 with E being the (constant) total energy of the particle:

$$\Psi(x,t) = \psi(x)\exp\left(-\frac{iE}{\hbar}t\right) \tag{2.29}$$

Systems with time-independent potentials are "stationary systems" and the corresponding wavefunctions, $\psi(x)$, are the "stationary states." Note that the probability function is not time dependent:

$$|\Psi(x,t)|^2 = |\psi(x)|^2\exp\left(\frac{iE}{\hbar}t\right)\exp\left(-\frac{iE}{\hbar}t\right) = |\psi(x)|^2. \tag{2.30}$$

Armed with Equation 2.29, we can write down the *time-independent Schrödinger equation*:

$$-\frac{\hbar^2}{2m}\frac{\partial^2\psi(x)}{\partial x^2} + U(x)\psi(x) = E\psi(x) \tag{2.31a}$$

or

$$H\psi(x) = E\psi(x) \tag{2.31b}$$

where H is the Hamiltonian operator:

$$H \equiv -\frac{\hbar^2}{2m}\frac{\partial^2}{\partial x^2} + U(x). \tag{2.31c}$$

The word "operator" means that it is a mathematical set of operations to be carried out on a function placed to its right in this case (direction matters here). Specifically, the wavefunction is multiplied by $U(x)$, and this term is added to the second derivative of the wavefunction multiplied by $-\frac{\hbar^2}{2m}$. The Hamiltonian contains the physics of the problem, and, once specified, the

Schrödinger equation can, in principle, be solved for the wavefunctions and allowed energies of the system. The wavefunctions are all those functions that, when operated on by H, return a number, E, times the original wavefunction. These values of E constitute the allowed energies.

There is, in general, an infinite series of functions that will solve the Schrödinger equation, each one corresponding to a different energy. If we have just one electron, then only the state corresponding to the lowest energy is of interest to us (if we do not put extra energy into the system). This state is called the *ground state*. The series of functions that solve Equation 2.31 are called *eigenfunctions* from the German word for "characteristic," because they make the left-hand side equal to the right-hand side when inserted into the Schrödinger equation. For each eigenfunction, the corresponding energy is called its *eigenenergy*. We will look at a series of eigenfunctions when we consider a particle trapped in a box below.

So we are now ready to solve problems involving one electron in some potential, $U(x, y, z)$. To do this, we solve the three sets of equations (for x, y, and z) given by 2.31 to get $\psi(x, y, z)$ (at the same time obtaining the energy of the eigenstate) and then square the result to find the time-averaged probability of finding an electron at any position in space. This probability as a function of position is equivalent to a normalized charge distribution for the electron in the potential when the probability is multiplied by e, the charge on the electron. It is normalized because, for one electron,

$$\int_{\mathbf{r}} \psi^*(\mathbf{r})\psi(\mathbf{r})\mathrm{d}^3\mathbf{r} \equiv \langle \psi | \psi \rangle = 1. \tag{2.32a}$$

Finally, we shall see that there are many states that solve the Schrödinger equation for a given potential, and these are labeled with quantum numbers, n, m, etc. Solutions corresponding to two *different* quantum numbers have the property of being *orthogonal*:

$$\int_{\mathbf{r}} \psi_m^*(\mathbf{r})\psi_n(\mathbf{r})\mathrm{d}^3\mathbf{r} \equiv \langle \psi_m | \psi_n \rangle = 0 \quad \text{for } m \neq n, \tag{2.32b}$$

or, expressed in terms of the Kronecker delta function, ($\delta_{nm} = 1$ for $n = m$, 0 for $n \neq m$)

$$\int_{\mathbf{r}} \psi_m^*(\mathbf{r})\psi_n(\mathbf{r})\mathrm{d}^3\mathbf{r} \equiv \langle \psi_m | \psi_n \rangle = \delta_{mn}. \tag{2.32c}$$

We will practice using the Schrödinger equation to solve some simple problems in this chapter.

2.11   Problems involving more than one electron

The Schrödinger equation becomes a quagmire when we treat even a few electrons. The exception is for the special case of *independent* (noninteracting) electrons (how it might be that negatively charged particles do not interact with one another is a good question addressed in Chapter 7). If the potential contains no terms in the sum or difference of coordinates for different electrons (e.g., terms like $r_1 \pm r_2$) we can write down a trial wavefunction for all the electrons as a product of the wavefunction for each electron, separating the Schrödinger equation just as we did for time-independent potentials above. That is

$$\Psi^{\text{trial}}(x_1, x_2, \ldots) = \psi_1(x_1)\psi_2(x_2)\ldots, \qquad (2.33a)$$

where x_1, x_2 are the coordinates of the first and second electrons, etc. and the ψ_n are trial single particle wavefunctions. The Schrödinger equation then separates into separate equations for each electron. ψ^{trial} is obviously not correct as written above because electrons are fermions and the Pauli principle (Section 2.9) requires that the wavefunction changes sign when any pair of particles are exchanged. This can be corrected by taking sums and differences of the products of the $\psi(x_n)$ using a *Slater determinant* as described in quantum texts.[1] This way of combining wavefunctions is straightforward to illustrate for just two particles, let us say an electron at $\mathbf{r}_1$ and another at $\mathbf{r}_2$. The product wavefunction (2.33a) is just $\psi_1(r_1)\psi_2(r_2)$. The correct linear combination is

$$\Psi(r_1, r_2) = \frac{1}{\sqrt{2}}[\psi_1(r_1)\psi_2(r_2) - \psi_2(r_1)\psi_1(r_2)] \qquad (2.33b)$$

This meets the requirement of the Pauli exclusion principle that the wavefunction goes to zero for $\psi_1 = \psi_2$ (both fermions cannot simultaneously be in the same state—here we include spin as a variable in describing the state implicitly in 2.33b—i.e., when we say that the wavefunctions are equal we mean that both the spatial parts and the spin parts are equal). This minus sign in front of the second term is an illustration of the requirement that the product wavefunction changes sign on exchange of particles between wavefunctions.

The equivalent procedure for bosons is to add the product functions:

$$\Psi(r_1, r_2) = \frac{1}{\sqrt{2}}[\psi_1(r_1)\psi_2(r_2) + \psi_2(r_1)\psi_1(r_2)], \qquad (2.33c)$$

which *increases* the probability that two particles will occupy the same state. This, in essence, is the quantum mechanics that underlies *Bose condensation*, manifest in lasers, superconductivity, and superfluid behavior, where particles fall into the same quantum state.

These simple independent particle wavefunctions would appear not to be very useful at first glance, because most problems involve *interacting* particles. But it is a good starting point for many situations involving interacting particles as we will discuss at the end of this chapter. It turns out to be a useful way to describe metals and semiconductors (Chapter 7). Tools are available for handling systems of interacting particles explicitly, and these include Hartree–Fock theory, quantum Monte Carlo algorithms, and DFT (discussed at the end of this chapter).

2.12 Solution of the one-electron time-independent Schrödinger equation for a constant potential

Suppose that the potential does not depend on position, but is everywhere a constant, i.e., $U(x) = V$.

The time-independent Schrödinger equation becomes

$$-\frac{\hbar^2}{2m}\frac{\partial^2 \psi(x)}{\partial x^2} + V\psi(x) = E\psi(x),$$

i.e.,

$$-\frac{\partial^2 \psi(x)}{\partial x^2} = \frac{2m(E-V)}{\hbar^2}\psi(x). \qquad (2.34)$$

This is solved by

$$\psi(x) = A\exp ikx, \qquad (2.35)$$

where

$$k^2 = \frac{2m(E-V)}{\hbar^2}. \qquad (2.36a)$$

For a "free particle," $V = 0$ and

$$E = \frac{\hbar^2 k^2}{2m}. \qquad (2.36b)$$

Including the time dependence (Equation 2.29) the probability amplitude for a single free particle in a region of constant potential is given by

$$\Psi(x,t) = A\exp(i[kx - \omega t]), \qquad (2.37)$$

where k is given by 2.36a and $\omega = E/\hbar$. The probability amplitude has the form of plane wave of wavelength $2\pi/k$. This is consistent with the deBroglie relation—see the exercise below.

Exercise 2.5: Use the classical expressions for energy and momentum to show how the wavelength of the probability amplitude given in Equation 2.37 is the deBroglie wavelength.

Answer: From Equation 2.36, $k^2 = \frac{2mE}{\hbar^2} = \frac{p^2}{\hbar^2}$ where we used $E = \frac{p^2}{2m}$. Since $k = \frac{2\pi}{\lambda}$ for a plane wave, we get $\lambda = \frac{h}{p}$, the deBroglie result.

2.13   Electron tunneling through a potential barrier

We are now in a position to consider what happens when a "free" particle hits an "impenetrable barrier." This situation is illustrated in Fig. 2.6. The spatial part of the wavefunction on the left, where $E > V$, is the sinusoidal wave described by Equation 2.35. (The figure shows a snapshot at one point in time of the time-dependent wavefunction 2.37.) At $x = 0$, the potential jumps up to a value $V > E$, so that, according to classical mechanics, the electron does not have the energy to cross. This is a classically forbidden region for an electron of energy less than V.

However, the Schrödinger equation does not allow the wavefunction to go to zero abruptly. A discontinuity in the wavefunction would lead to an infinite derivative (and, of course, an infinite second derivative), thereby leading to an infinite energy. Thus, the conditions that the wavefunctions need to satisfy at every point are

$$\psi \text{ (just to the left of a boundary)} = \psi \text{ (just to the right of a boundary)}$$
$$(2.38a)$$

and

$$\frac{\partial \psi}{\partial x} \text{ (just to the left of a boundary)} = \frac{\partial \psi}{\partial x} \text{ (just to the right of a boundary).}$$
$$(2.38b)$$

Thus if the wavefunction has some nonzero time average value just to the left of $x = 0$, it must have the *same* time average value just to the right. To the right

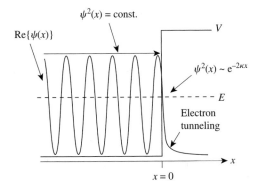

Fig. 2.6 The real part of the wavefunction as a free particle (left of $x = 0$) impinging on a region where $V > E$ and the electron is not classically allowed to enter is shown. The wavefunction has two components, a real part, oscillating like $\cos(kx)$ and an imaginary part oscillating like $\sin(kx)$ on the left of the barrier, and both swing between positive and negative values. The probability distribution is the square of these components, and it is constant on the left of the barrier (recall $\sin^2 + \cos^2 = 1$). The wave-matching requirements of quantum mechanics imply some probability of finding the electron on the classically forbidden side of the barrier—a phenomenon called tunneling. In the classically forbidden region on the right-hand side of the barrier, both the wavefunction and its square decay exponentially.

of the barrier, $V > E$ so

$$k = \sqrt{\frac{2m(E-V)}{\hbar^2}} = i\sqrt{\frac{2m(V-E)}{\hbar^2}}, \qquad (2.39)$$

where we multiplied $(E-V)$ by -1 and took the $\sqrt{-1}$ (by definition $\sqrt{-1} = i$) outside the $\sqrt{}$. So the previously complex wavefunction (Equation 2.35 which has ikx in the exponent) becomes a real exponential decay:

$$\psi(x) = A\exp(-\kappa x), \qquad (2.40)$$

where

$$\kappa = \sqrt{\frac{2m(V-E)}{\hbar^2}}. \qquad (2.41)$$

Once again, we have ignored the time dependence. The probability of finding an electron in the forbidden region decays like $\psi(x)^2 = A^2\exp{-2\kappa x}$, so the distance over which the probability falls to $1/e$ of its value at the boundary is $1/2\kappa$. For the sorts of barrier we will encounter in many chemical problems, $(V-E)$ is on the order of few electron volts (eV), and so the corresponding distance over which significant tunneling occurs is a few Å.

Exercise 2.6: Calculate $1/2\kappa$ for $V-E = 5$ eV (characteristic of the barrier that holds an electron inside gold). The number will give you some idea of how far electrons "leak" out of a metal surface owing to tunneling. The calculation is easier if we use some strange but useful units. The mass of an electron, in units appropriate to Einstein's formula, $m = E/c^2$ is 511 keV/$c^2 \cdot \hbar$ in units of eV-s is 6.6×10^{-16} eV-s and c, the speed of light in Å/s is 3×10^{18} Å/s.

Answer:

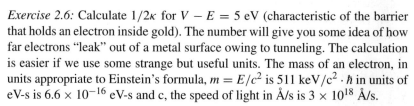

$$\kappa = \sqrt{\frac{2m\Delta E}{\hbar^2}} = \sqrt{\frac{2(mc^2)\Delta E}{c^2\hbar^2}} = \sqrt{\frac{2 \times 511 \times 10^3 \times 5}{(3 \times 10^{18})^2(6.6 \times 10^{-16})^2}} = 0.51\sqrt{5}.$$

$2\kappa = 1.02\sqrt{5}$, so the decay length is about 0.44 Å.

2.14 The Hitachi experiment with wavefunctions

We will now use the constant potential result (Equation 2.35) to calculate a wavefunction for the Hitachi experiment. There are a number of approximations involved in doing this. The biprism is not a region of constant potential,[4] but we will pretend that it can be modeled by two sources, **1** and **2**, a distance d apart as shown in Fig. 2.7. Point sources do not produce plane waves (like Equation 2.40) either, but we will assume that the screen is so far away ($L \gg d$) that Equation 2.40 is a good description (because circular wavefronts look plane far from the source). In this limit, the fact that the two paths converge on one

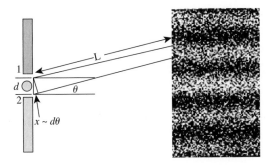

Fig. 2.7 Double slit experiment of Fig. 2.2 showing the "far-field" geometry. The screen is far from the slits relative to the slit spacing so that the two "interfering beams" can be treated as parallel.

another at the screen can also be ignored and we will describe the two paths, slit **1** to the screen and slit **2** to the screen, as parallel lines.

There was only one electron in the apparatus at a time, so we will write down a one-electron wavefunction for the experiment (we do not have to worry about the fact that the electrons are fermions). The measured data come from repeating the one-electron experiment many times, but this will mirror the probability distribution for each single electron. Since we do not know which of the two slits the electron "passed through," we are obliged to add together the wavefunctions for the two possible pathways to get the complete wavefunction, which we then square (i.e., multiply by the complex conjugate) to obtain the probability distribution.

Given that the distance from slit **1** to the screen is L at some angle θ, then the distance from slit **2** to the screen is $L + d \sin \theta$ (see Fig. 2.7). In the far-field limit, we can approximate $\sin \theta$ by θ, so the wavefunction at the screen is

$$\psi(\theta) = A \exp(ikL)(1 + \exp[ikd\theta]). \tag{2.42}$$

The probability for an electron to be at this point (at an angle θ) is given by

$$\psi(\theta)\psi^*(\theta) = A^2(1 + \exp[ikd\theta])(1 + \exp[-ikd\theta]) = 2A^2(1 + \cos(kd\theta)). \tag{2.43}$$

At $\theta = 0$, this has a (maximum) value of $4A^2$. Every time $kd\theta$ reaches a multiple of π, the probability for finding a particle falls to zero. It becomes a maximum again whenever $kd\theta$ reaches a multiple of 2π. This explains the form of the "interference fringes" shown in Fig. 2.2.

2.15   Some important results obtained with simple 1-D models

2.15.1   Particle in a box

An electron trapped inside a box is a first approximation to an atom, so it is a useful model to study. In a real atom, a positively charged nucleus causes the electron to be confined to a small space in its vicinity (giving rise to a large ground state energy as we estimated by using the uncertainty principle in an earlier exercise).

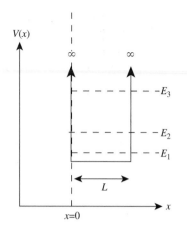

Fig. 2.8 Particle in a box of width L. The potential is taken to be infinitely large outside the box so the wavefunction goes to zero at its edges. We will set $V = 0$ in the box. The first three eigenenergies are shown schematically as horizontal dashed lines.

Consider a particle confined to a box of width L with infinitely steep walls (an infinite potential well) as shown in Fig. 2.8. Because the potential barrier outside the box is infinitely high, the tunneling distance $(1/2\kappa)$ is zero, and thus the wavefunction just outside the box must be zero. The continuity conditions for the wavefunction require that it must be zero right at the edge of the box also. The free-particle wavefunction (Equation 2.40)

$$\psi = A \exp(ikx) = A \cos kx + iA \sin kx$$

would not work for this problem because $\cos(kx)$ is not zero at $x = 0$. However, the sine term has the required property. Thus, we find that the time-independent Schrödinger equation is satisfied if we use $\psi = B \sin kx$, and we obtain for the energy of the particle

$$E = \frac{\hbar^2 k^2}{2m}. \tag{2.44}$$

Requiring that the wavefunction goes to zero at the edges of the box requires that $\sin kL = 0$ so

$$kL = n\pi, \quad \text{i.e.,} \, k = \frac{n\pi}{L}, \quad n = 1, 2, 3, \dots. \tag{2.45}$$

The constant B is found by integrating ψ^2 to get 1 so that the probability of finding the electron anywhere is unity. The integral is from zero to L, but changing the variable to $y = n\pi x/L$ causes the n dependence to cancel and we get $B = \sqrt{2/L}$ (problem 24) so

$$\psi_n = \sqrt{\frac{2}{L}} \sin\left(\frac{n\pi x}{L}\right). \tag{2.46}$$

These are the *eigenstates* (set of wavefunctions that satisfy the time-independent Schrödinger equation) for the particle in a box. For each n, there is a corresponding eigenenergy given by (2.44) and (2.45):

$$E_n = \frac{n^2 \hbar^2 \pi^2}{2mL^2}. \tag{2.47}$$

The allowed energy states start from the ground state energy

$$E_1 = \frac{\hbar^2 \pi^2}{2mL^2} \tag{2.48}$$

and increase as n^2. The first three eigenenergies are shown schematically in Fig. 2.8.

The energy difference between the nth and the $(n + 1)$th eigenstates is

$$\Delta E_{n,n+1} = \frac{(2n + 1)\hbar^2 \pi^2}{2mL^2}. \tag{2.49}$$

If we use this system to model a molecule or small crystal, with states up to the nth filled up with electrons (assuming that the one electron picture still

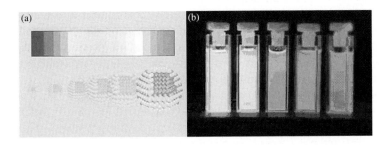

Fig. 2.9 Quantum size effects: semiconductor nanoparticles have gaps between ground and excited states that are modified by confinement ($\Delta E \propto 1/L^2$). This is illustrated schematically in (a). This leads to a change in fluorescence as the particles get bigger. Fluorescence from solutions of dots gets progressively redder as the dots get larger (shown here in grayscale) from left to right are shown in (b). Quantum dots are used as tags for bioimaging. ((a) Courtesy of Dylan M. Spencer. (b) Source: Philips). See Color Plate 2.

holds), the energy gap between the highest filled state and the first unfilled state above it is size dependent. (In molecules this gap is called the HOMO-LUMO gap and it will be discussed in Chapter 8.) The label, n, is called a *quantum number*. For a given n, the energy gap between adjacent states falls with $1/L^2$. If the system has some intrinsic gap ΔE_g (cf. Chapter 7) in the bulk ($L \to \infty$), then quantum size effects cease to be important when $\Delta E_g > \Delta E_{n,n+1}$ given by Equation 2.49. Semiconductor crystals that are just a few nm in diameter (called "quantum dots") are so small that their gap is controlled primarily by their size. They fluoresce at progressively shorter wavelength as the particle size is decreased. This is because the energy of the light emitted by electrons that drop back to the ground state depends directly on the size of the gap. Smaller particles have larger gaps (Equation 2.49) and thus they emit light that lies toward the blue end of the spectrum. These quantum dots make handy colored "tags" for chemical labeling and are used in biological light microscopy to identify the tagged molecules. Their structure and use are illustrated in Fig. 2.9.

2.15.2 Density of states for free particles

The discussion above implies that a free particle ($L \to \infty$ in Equation 2.49) has eigenstates so close together that quantum effects are irrelevant. In fact, quantum mechanics has great deal to say about the density of these states as a function of the particle's momentum (or, equivalently, wave vector, $k = p/\hbar$). The density of states means the number of quantum states available per unit energy or per unit wave vector. Quantum effects are observed even for free particles, so the density of states is an important quantity to calculate. These results will prove useful in our discussions of statistical mechanics (Chapter 3), nanoelectronics (Chapter 7), and nanomaterials (Chapter 9).

The kinetic energy of a free particle of mass m and velocity v in classical mechanics is

$$\frac{1}{2}mv^2 = \frac{p^2}{2m},$$

where p is the classical momentum (mv). The momentum of a plane wave is given by the wave vector, k, and the quantum mechanical expression for the kinetic energy of a (plane wave) particle of mass, m, follows from the expression from classical mechanics with the substitution $p = \hbar k$ (Equation 2.36b):

$$E = \frac{\hbar^2 \mathbf{k}^2}{2m} = \frac{\hbar^2}{2m}(k_x^2 + k_y^2, k_z^2). \tag{2.50}$$

Here we have used the 1-D plane wave result for a free particle for each of the x, y, and z directions. We saw above how confinement in one direction into a box of length, L, led to the constraint on k that, for the x component, for example, $k_x = \frac{n\pi}{L}$ (for a particle confined to a box with a wavefunction $\sin(kx)$). The result is a little different for a *free* particle where the wavefunction, e^{ikx} (Equation 2.35) contains both sine and cosine terms. To treat this case, we have to introduce *periodic boundary conditions*. In a big enough box, it must be possible to find a point where the wavefunction repeats its value. In the case of the particle in a real box with an infinitely large confining potential, this point is at $x = 0$ and $x = L$ were $L \rightarrow \infty$ for a completely free particle. Another way to think about this is that we could curve the material along some direction ever so slightly so that it eventually forms a closed circle of circumference L. Then the wavefunction at $x = 0$ and $x = L$ would have to be the same. So we can conceive of ways to make the wavefunction repeat itself after some distance L. This leads to the following expression if $L_x = L_y = L_z$:

$$\exp(-ik_x L) = \exp(-ik_y L) = \exp(-ik_z L) = \exp(ik_{x,y,z}0) = 1, \quad (2.51)$$

(where $L \rightarrow \infty$ for a truly free particle). The choice of L is, of course, arbitrary at this point, but do not worry—it will not matter in the end! The conditions (2.51) are met by requiring

$$k_x = \frac{2n_x \pi}{L}, \quad k_y = \frac{2n_y \pi}{L}, \quad k_z = \frac{2n_z \pi}{L}. \quad (2.52)$$

For this 3-D case, we now have three sets of quantum numbers, n_x, n_y, and n_z, compared with the single n we obtained for the problem of a particle in a 1-D box. Note that allowed values of k differ from the values allowed for the particle in a box (Equation 2.45) by a factor of 2. This is because the wavefunctions are different, reflecting a real physical difference between the two cases. The particle in a box is stationary. Changing k to $-k$ makes no difference except to change the sign of the wavefunction (a difference that vanishes when probability is calculated). The "free particle" changes its direction of propagation when the sign of k is changed, a very important difference.

It is convenient to think about the allowed states as points in a space with coordinates k_x, k_y, and k_z, which we call *k-space*. According to 2.52, k-space is filled with a uniform grid of points each separated in units of $2\pi/L$ along any axis. It will turn out that L disappears from the quantities we will carry through calculations with this idea, so the difficult questions of "how big is the box?" and "does quantum mechanics still apply at some boundary where the wavefunction repeats?" do not really need to be addressed in the end. The grid of points corresponding to the allowed states in k-space is shown in Fig. 2.10.

The volume of k-space occupied by each point is

$$\left(\frac{2\pi}{L}\right)^3. \quad (2.53a)$$

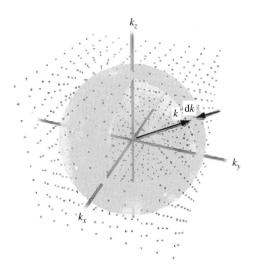

Fig. 2.10 Allowed states in k-space. Points are evenly spaced by $2\pi/L$, where L is the distance over which periodic boundary conditions apply to the free particles. In practice, the quantity L disappears when measurable quantities are calculated.

As an aside, we note that in one dimension, the distance between allowed k points is

$$\left(\frac{2\pi}{L}\right)^1 \tag{2.53b}$$

and in two dimensions, the area around each allowed k point is

$$\left(\frac{2\pi}{L}\right)^2. \tag{2.53c}$$

Returning to the 3-D case and referring to Fig. 2.10, if we assume that the allowed points are closely spaced enough so that we can treat them as "continuous" (on the scale of, for example, thermal energy) then we can write a differential expression for the density of allowed states at some values of k. The number of allowed states between k and $k + dk$ in 3-D is

$$dn = \frac{4\pi V k^2 dk}{8\pi^3} = \frac{V k^2 dk}{2\pi^2}. \tag{2.54}$$

L^3 has become the volume of the box, and it turns out that we can get rid of it by incorporating it into quantities like the number of electrons per unit volume in calculations. But even at this stage, note the nontrivial consequence of quantum mechanics: the number of states per unit wave vector increases as the square of the wave vector.

2.15.3 A tunnel junction

We will now consider a more complicated version of tunneling, and this is tunneling from one classically allowed region to another classically allowed region separated by a classically forbidden region. This is a model for the scanning tunneling microscope (STM) where electrons tunnel between a metal probe (an allowed region) and a metal surface (also an allowed region) via a

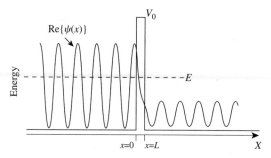

Fig. 2.11 Electron penetration of a thin barrier as a model for the scanning tunneling microscope. As in Fig. 2.6, just one component of the wavefunction is shown. The vertical axis is energy (and also probability amplitude–the wavefunction oscillates between maximum + and − amplitudes). An electron is incident from the left on a classically forbidden region ($V > E$) but a second allowed region ($V < E$) is placed close enough such that there is a finite probability that electrons will propagate through the barrier by tunneling. The form of the wavefunction is shown for one particular time.

vacuum gap (a forbidden region). This discussion is also a starting point for discussing *quantum well* structures (Chapter 9).

Consider a free electron of energy, E, incident on a potential barrier of height, V_0, at $x = 0$ as shown in Fig. 2.11. As we will see later, this corresponds to an electron in a metal where the barrier to escape, $V_0 - E$, is called the workfunction of the metal (and typically has a value of a few eV). A second allowed region exists at $x > L$ on the right where the electron would also propagate as a free particle. This diagram corresponds to an STM probe a distance L from the first metal surface. In the region $0 < x < L$, $E < V_0$ so the electron is classically forbidden.

As we shall see, the electron will generally be reflected at a potential barrier, so we will write the solutions of the Schrödinger equation in the form

$$\psi = const. \exp(\pm ikx) \tag{2.55}$$

to allow for both forward (+) and reflected (−) waves. Taking the incident wave to have an amplitude equal to 1, then the wavefunction on the left of the barrier is

$$\psi(x)_L = \exp(ikx) + r \exp(-ikx), \tag{2.56}$$

where r is a reflection coefficient.

In the middle of the barrier, $0 < x < L$, substitution of (2.55) into the Schrödinger equation produces

$$\psi(x)_M = A \exp(\kappa x) + B \exp(-\kappa x), \tag{2.57}$$

where A and B are to be determined and

$$\kappa = \sqrt{\frac{2m(V_0 - E)}{\hbar^2}}. \tag{2.58}$$

We have repeated the trick we used when describing the simple tunnel barrier. When we solve the Schrödinger equation for this region ($V_0 > E$) we find that

$ik = \sqrt{2m(E - V_0)/\hbar^2}$. However, the argument of the square root is negative, giving the root an imaginary value. So by changing to the real variable, κ, we get rid of the imaginary value. As a result, the oscillating wavefunction, $\exp(ikx)$, in the allowed region becomes a decaying, $\exp(-\kappa x)$, or growing, $\exp(+\kappa x)$, exponential in the forbidden region. The $\exp(+\kappa x)$ solution is unphysical because the amplitude cannot grow in the forbidden region. Thus, the electron penetrates the "forbidden" region with an exponentially decaying wavefunction. If the gap L is small enough, there is some reasonable probability that the electron will be found on the right-hand side of the classically forbidden barrier. This is the phenomenon of quantum mechanical tunneling (the electron is said to have tunneled through the barrier). This phenomenon appears widely in chemistry. For example, in some limits we can describe molecules as atoms linked together by electrons that tunnel between them.

From the decaying wavefunction, and noting that the probability of charge crossing the barrier decays as ψ^2, we expect that the current through this tunnel gap should decay as

$$\exp(-2\kappa x) = \exp\left(-2\sqrt{\frac{2m(V_0 - E)}{\hbar^2}}x\right). \tag{2.59}$$

This guess has the correct form (see below). The constants were evaluated in Exercise 2.6. Setting the current through the junction, $i(L)$, as directly proportional to the transmission coefficient (with a proportionality constant $i(0) = i_0$—a quantity that we calculate in Chapter 9) we get for the current, i

$$i(L) = i_0 \exp(-1.02\sqrt{\phi}L), \tag{2.60}$$

where $\phi = V_0 - E$ is the workfunction of the metal (in eV) and L is the gap distance in Å. For gold, $\phi = 5$ eV, so the tunnel current decays by $\exp(-2.28)$ (about a factor of 10) for each Angstrom the gap is increased (see Exercise 2.6).

We guessed the form of the transmission coefficient, but we can calculate it as follows: On the right of the gap, the wavefunction must correspond to a rightward traveling wave only:

$$\psi(x)_R = t\exp(ikx), \tag{2.61}$$

where t is the transmission amplitude. So, to solve 2.56, 2.57, and 2.61, we need four unknowns, t, r, A, and B. Now we will keep the term $B\exp(-\kappa x)$ in 2.57, because this describes the decay of the wave reflected from the far side of the barrier, important if the barrier is thin. We have four equations to solve for these four unknowns. These come from the conditions on the wavefunction and its derivative (2.38a and 2.38b) evaluated at $x = 0$ and $x = L$. The transmission of charge through the barrier is given by the square of the amplitude transmission coefficient, $T = t^2$. After a considerable amount of algebra, the result (Problem 25) is that:

$$T = \frac{1}{1 + (V_0^2/4E(V_0 - E))\sinh^2(\kappa L)}. \tag{2.62}$$

For small tunneling rates ($2\kappa L \gg 1$), this reduces to an expression (Problem 25) similar to the result we stated without proof as Equation 2.60. The more complicated form (2.62) reflects interference effects that become important in thin barriers.

This is a good point to divert and take a look at the STM.[8] To do this, we need to steal one result from Chapter 7: that is, the resistance of a very small contact is ~12.6 kΩ. "Very small" means so small that electron scattering at the contact can be neglected. Such contacts are called quantum point contacts. In this limit, with 1 V applied across a point contact between a sharp probe and a metal surface, a current of ~0.1 mA will flow. Equation 2.60 and the discussion below showed that, for typical metals, the current decays about an order of magnitude for each Angstrom that the gap is increased. Thus, for a gap of ~5 Å, the current, with 1 V applied, is ~1 nA, an easily measurable quantity. The STM works by holding a conducting probe a fixed distance above a metal surface and scanning it over the surface in a raster pattern (Fig. 2.12). A servo system records the current flowing through the probe and adjusts the height of the probe so as to keep the tunnel current constant. Because of the strong dependence of current on distance (Equation 2.60), even a very simple servo control system can hold the tip height constant to within a tiny fraction of an Angstrom (if the workfunction remains constant over the surface). The amount of the height correction signal generated by the servo is plotted as the spot intensity (or color or height) on a computer screen on which the spot is swept across the screen synchronously with the raster scan signals that move the probe across the sample. A consequence of the strong distance dependence of the tunnel current signal is that the very *end atom* of the probe can dominate the current signal. This makes it possible to map details of the surface *with atomic resolution*. Figure 2.13 shows a beautiful image of iron atoms arranged in a circle on a nickel surface.[9]

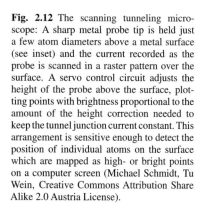

Fig. 2.12 The scanning tunneling microscope: A sharp metal probe tip is held just a few atom diameters above a metal surface (see inset) and the current recorded as the probe is scanned in a raster pattern over the surface. A servo control circuit adjusts the height of the probe above the surface, plotting points with brightness proportional to the amount of the height correction needed to keep the tunnel junction current constant. This arrangement is sensitive enough to detect the position of individual atoms on the surface which are mapped as high- or bright points on a computer screen (Michael Schmidt, Tu Wein, Creative Commons Attribution Share Alike 2.0 Austria License).

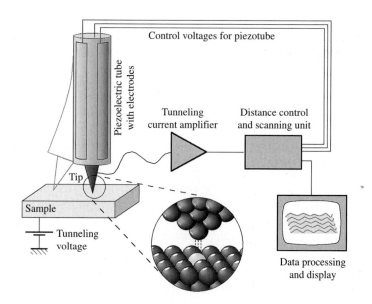

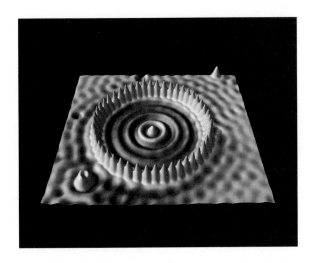

Fig. 2.13 The "Quantum Corral." An IBM team deposited a sub-monolayer of iron atoms on a very clean, cold Ni surface, and then used the STM tip to "push" the atoms into the form of a circle. The iron atoms appear in a pseudo color STM image as the blue peaks (for color see Color Plate 3). The "ripples" on the surface are caused by confinement of the Ni electrons by the circle of Fe atoms. The consequences of quantum mechanics are being imaged directly. (Photo courtesy of IBM.)

2.16   The hydrogen atom

The hydrogen atom consists of one electron bound to one (positively charged) proton by electrostatic attraction. This problem is a very important example of an electron trapped inside a 3-D potential well, and we will outline the solution of the problem here. The one-electron atom serves as a guide to the properties of all the elements, as we shall see.

The two-body problem (of proton and electron) can be reduced to a one-body problem (essentially the electron's motion about a fixed positive charge) by using the reduced mass

$$\mu = \frac{M_p m_e}{M_p + m_e}$$

and describing the motion in terms of motion about the center of mass (coordinates are illustrated in Fig. 2.14). The potential that binds the electron to the proton is the Coulomb interaction

$$V(r) = \frac{\kappa e^2}{r} \quad \text{where } \kappa = \frac{1}{4\pi\varepsilon_0}. \tag{2.63}$$

(ε_0 is the permittivity of free space, 8.85×10^{-12} F/m). Here, r is the distance between the electron and proton in the center of mass frame of reference, but this is essentially the same as the distance in the laboratory frame of reference because the mass of the electron is so small (see Fig. 2.14).

The Schrödinger equation is best solved using the spherical coordinates also illustrated in Fig. 2.14. In these coordinates, the Coulomb potential, Equation 2.63, appears only as a function of one variable, r. However, the expression for the second spatial derivative of the wavefunctions must be transformed to these new coordinates and they will yield differential equations in the variables θ and ϕ, as well as r. The equations for θ and ϕ contain no reference to the potential, of course, but they do impose constraints on the wavefunction. To

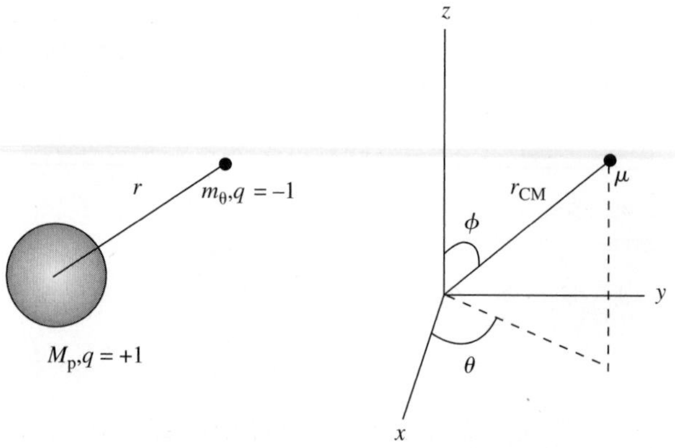

Fig. 2.14 Coordinates for the hydrogen atom. The two-body problem of the proton (mass M_p) and electron (mass m_e) is replaced by the one-body problem of the motion of the center of mass of a single particle of reduced mass ($\mu = M_p m_e / [M_p + m_e]$) orbiting about a fixed point. Since $M_p \gg m_e$, r_{CM} is essentially equal to r and μ is essentially equal to m_e.

be specific, they require certain symmetry properties for the wavefunction for rotations about the axes that define these angular coordinates. The angular parts of the wavefunctions must satisfy these constraints. This results in new quantum numbers for the angular eigenfunctions. The process is quite technical and it is set up and outlined in Appendix C. The reader who wants to "look under the hood" should consult this. This solution is discussed qualitatively here, quoting results from the appendix as needed.

In this 3-D problem, we expect that we will need three quantum numbers to describe the energy eigenstates, just as we required n_x, n_y, and n_z for the case of a free particle moving in a constant potential in three dimensions. What are the appropriate eigenstates in the central (i.e., spherically symmetric) potential of Equation 2.63?

In order to answer this question, it is useful to digress and take another look at the particle in a constant potential. The wavefunctions are

$$\psi(x) = A \exp(ik_x x) \tag{2.64a}$$

$$\psi(y) = A \exp(ik_y y) \tag{2.64b}$$

$$\psi(z) = A \exp(ik_z z). \tag{2.64c}$$

And the kinetic energies are

$$E = \frac{\hbar^2 \mathbf{k}^2}{2m} = \frac{\hbar^2}{2m}(k_x^2 + k_y^2 + k_z^2),$$

where the k's are quantized according to

$$k_x = \frac{2n_x \pi}{L}, \quad k_y = \frac{2n_y \pi}{L}, \quad k_z = \frac{2n_z \pi}{L}.$$

Energy is conserved as the particle is translated along the x-, y-, or z-axis, because the potential is not a function of position. This invariance of properties of the particle can be stated another way. Defining a momentum operator, $\hat{p}$,

$$\hat{p} = -i\hbar \left(\hat{x}\frac{\partial}{\partial x} + \hat{y}\frac{\partial}{\partial y} + \hat{z}\frac{\partial}{\partial z} \right) \equiv -i\hbar\nabla, \qquad (2.65)$$

where $\hat{x}, \hat{y}, \hat{z}$ are unit vectors, and ∇ is the gradient operator. The x component of momentum, for example, is given by

$$p_x = -i\hbar\frac{\partial}{\partial x},$$

which yields the result that the free-particle wavefunctions 2.64 are eigenfunctions of the momentum operator, e.g.,

$$-i\hbar\frac{\partial}{\partial x}A \exp ik_x x = \hbar k_x A \exp ik_x x \qquad (2.66)$$

with eigenvalues given by $\hbar k_x$, etc. Linear momentum, **p**, is a conserved quantity because of the translational invariance of the potential. Furthermore, it has the same eigenfunction as the Hamiltonian operator for the free particle. Operators that have the same eigenfunctions *commute*, meaning that the order in which they operate on a function does not matter. This follows directly from the fact that operating on an eigenfunction returns a number times the original eigenfunction. In the case of the Hamiltonian for a free particle and the momentum operator

$$H\hat{p}f(x) - \hat{p}Hf(x)$$

$$\equiv [H,\hat{p}]f(x) = -\frac{\hbar^2}{2m}\frac{\partial^2}{\partial x^2} \bullet - \left(i\hbar\frac{\partial f(x)}{\partial x} \right) + i\hbar\frac{\partial}{\partial x} \bullet - \left(\frac{\hbar^2}{2m}\frac{\partial^2 f(x)}{\partial x^2} \right) = 0,$$

because the same overall result is returned no matter the order of the operations (the square brackets above are shorthand for the commutation operation). This is a general result: *operators that commute have the same eigenfunctions.*[1] On the other hand, position and momentum operators do not commute:

$$[x,\hat{p}]f(x) = -i\hbar x\frac{\partial f(x)}{\partial x} + i\hbar \left(f(x) + x\frac{\partial f(x)}{\partial x} \right) = i\hbar f(x). \qquad (2.67a)$$

Or, in shorthand

$$[x,\hat{p}] = i\hbar. \qquad (2.67b)$$

This means that the same particle cannot simultaneously have eigenfunctions of position and momentum. This is a mathematical restatement of the famous uncertainty principle.

Armed with these ideas, we can now turn to the spherically symmetric central potential. We expect that confinement of the particle within a sphere would, by analogy to the particle in a box, give rise to one quantum number, n. This is called the *principal quantum number*. The potential is only a function of r, and thus we expect that the eigenenergies of the particle do not depend on rotations about the x-, y-, and z-axes. We further expect to find conserved quantities analogous to the linear momentum for each of these three rotations. These conserved quantities are called the *angular momenta* and they are given by a vector product (cross product) of the position with the linear momentum operator, as described in Appendix C. The three components of angular momentum, L_x, L_y, and L_z, must surely be conserved, so we might ask if there are three additional quantum numbers associated with these three components of angular momentum? If this was indeed true, then added to the "particle in a sphere" quantum number, n, this would give a total of four quantum numbers, one more than we expect. In fact, as shown in the appendix, the components of angular momentum do not commute. Specifically with

$$\hat{L}_x = -i\hbar \left(y\frac{\partial}{\partial z} - z\frac{\partial}{\partial y} \right) \tag{2.68a}$$

$$\hat{L}_y = -i\hbar \left(z\frac{\partial}{\partial x} - x\frac{\partial}{\partial z} \right) \tag{2.68b}$$

$$\hat{L}_z = -i\hbar \left(x\frac{\partial}{\partial y} - y\frac{\partial}{\partial x} \right) \tag{2.68c}$$

(where $\hat{L}_x$ and $\hat{L}_y$ are the operators for the x and y components of the angular momentum),

$$\lfloor \hat{L}_x, \hat{L}y \rfloor = i\hbar \hat{L}_z. \tag{2.69}$$

Thus, an angular momentum eigenfunction for the x component of momentum cannot be an eigenfunction for the y component. Thus it is only possible to measure *one* of the three components of angular momentum, a quantity with which we associate a second quantum number, m. This is called the *magnetic quantum number*.

It turns out (see Appendix C) that the operator for the total squared magnitude of the momentum, $\hat{L}^2$, does commute with any one of the three components of angular momentum (it is conventional to choose $\hat{L}_z$). Thus the angular momentum eigenfunctions are also eigenfunctions of $\hat{L}^2$. A third quantum number, ℓ, labels the eigenfunctions of $\hat{L}^2$. ℓ is called the azimuthal quantum number.

Each quantum number is constrained to a range of values by the parent differential equation that contained it as a parameter (as elaborated in Appendix C). This is straightforward to understand in the case of the eigenfunctions of $\hat{L}_z$ which are exponentials in the angle ϕ (Fig. 2.14):

$$\psi(\phi) = \exp\left(\frac{i\mu\phi}{\hbar} \right), \tag{2.70}$$

where μ represents the (to be determined) eigenvalues of $\hat{L}_z$. In order to describe a physical situation, we require that this function be invariant on rotation of the argument by 2π or

$$\psi(\phi + 2\pi) = \psi(\phi), \tag{2.71}$$

which requires

$$\psi(\phi) = \exp(im\varphi), \quad m = \pm 1, \pm 2, \ldots \tag{2.72}$$

giving the eigenvalues of $\hat{L}_z$ as

$$\mu = m\hbar. \tag{2.73}$$

The eigenvalues of $\hat{L}^2$ are λ, and it can be shown that

$$\lambda = \ell(\ell + 1)\hbar^2 \quad \text{with } \ell = 0, 1, 2, \ldots \tag{2.74}$$

$$-\ell \leq m \leq \ell. \tag{2.75}$$

Thus, the azimuthal quantum number ℓ that describes the eigenstates of L^2 (according to 2.74) is limited to integer values greater than or equal to m. This second condition occurs as a consequence of the fact that the magnitude of a component of a vector cannot exceed the magnitude of the vector itself.

The azimuthal quantum numbers are further constrained by the values of principle quantum number, according to which

$$0 \leq \ell \leq n - 1. \tag{2.76}$$

Finally, confinement to within some radius (which depends on the particular eigenstate) leads to a series of energy eigenvalues that depend on n according to

$$E_n = -\frac{me^4}{2(4\pi\varepsilon_0)^2\hbar^2 n^2} \approx -\frac{13.6 \text{ eV}}{n^2}. \tag{2.77}$$

The eigenstates are fixed by n, l, and m (Equation 2.78). They are not, in general, spherically symmetric, which seems strange because the confining potential (Equation 2.63) is only a function of r and not θ or ϕ. However, the total energy depends only on n, and measurements of energy would have to be averaged over all of the states of equal energy (*degenerate states*). This average will be over all possible orientations in θ or ϕ and so the complete set of states corresponding to a given energy are indeed spherically symmetric.

The eigenstates take the form

$$\psi_{n,\ell,m}(r, \theta, \phi) = A_{n\ell} Y_{\ell,m}(\theta, \phi) R_{n,\ell}(r). \tag{2.78}$$

The $A_{n\ell}$ are normalization constants that depend on n and ℓ. The $Y_{\ell,m}(\theta, \phi)$ are special functions called spherical harmonics. The $R_{n,\ell}(r)$ are the product of exponential functions of r with a series of polynomials in r called "Laguerre polynomials."

Table 2.1 Some hydrogen atom wavefunctions

n	ℓ	m	$\psi_{n\ell m}$
1	0	0	$\left(\sqrt{\pi a_0^3}\right)^{-1} \exp(-r/a_0)$
2	0	0	$\left(\sqrt{8\pi a_0^3}\right)^{-1} (1 - r/2a_0) \exp(-r/2a_0)$
2	1	0	$\left(\sqrt{8\pi a_0^3}\right)^{-1} (r/2a_0) \cos\theta \exp(-r/2a_0)$
2	1	1	$\left(\sqrt{\pi a_0^3}\right)^{-1} (r/8a_0) \sin\theta \exp(i\phi) \exp(-r/2a_0)$
2	1	−1	$\left(\sqrt{\pi a_0^3}\right)^{-1} (r/8a_0) \sin\theta \exp(-i\phi) \exp(-r/2a_0)$

Some of the wavefunctions (2.78) for low values of the quantum numbers are given explicitly in Table 2.1. They are expressed in units of the Bohr radius:

$$a_0 = \frac{4\pi \varepsilon_0 \hbar^2}{me^2} \approx 0.529 \text{ Å}. \tag{2.79}$$

We do not actually measure the wavefunction itself. Rather, we are interested in the probability of finding an electron at some given position with respect to the proton. Referring to Table 2.1, the wavefunction for the ground state, ψ_{100}, is maximum at the origin (where $\exp(-r/a_0) = 1$) but the probability of finding an electron at exactly that point is miniscule (because the radius of the corresponding orbit is zero, so the volume is zero). To correct for this geometrical effect, we need to form the product of the wavefunction squared (which gives the probability per unit volume) with the available volume element, $4\pi r^2 \mathrm{d}r$,

$$P(r)\mathrm{d}r = 4\pi r^2 |\psi_{n,\ell,m}(r,\theta,\varphi)|^2 \mathrm{d}r. \tag{2.80}$$

For the ground state ($\psi \propto \exp -r/a_0$), Equation 2.80 has a maximum value in a shell located at the Bohr radius. This radius of orbit and the energies given by Equation 2.77 are also found from a much simpler theory of the hydrogen atom (due to Bohr) that preceded the full development of quantum mechanics. The Bohr theory is described in detail in the introductory sections of quantum texts.[1,2]

The probability distribution given in 2.80 is analogous to a "charge density" for the state, and it is often referred to as a "charge distribution." This terminology is fine, so long as one remembers what is meant by this expression in terms of measurements of this distribution.

Chemists describe the atomic eigenstates by orbitals: pictures of the charge probability distributions. For historical reasons, orbitals corresponding to $\ell = m = 0$ are called s orbitals (so the first entry in Table 2.1 is the 1s orbital, the second is the 2s, and so on). For $n = 2$, there are three orbitals corresponding to $\ell = 1 (m = -1, 0, +1)$. These are called the p orbitals. We will enumerate

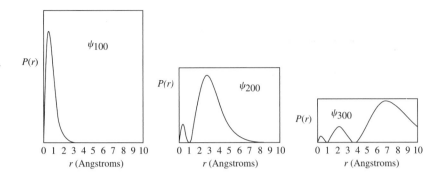

Fig. 2.15 Probability distributions for the first 3s orbitals of the hydrogen atom. These are radially symmetric and correspond to a series of "shells" of high probability for finding an electron. The number of "zeroes" (nodes) in the radial distribution function is equal to $n - 1$.

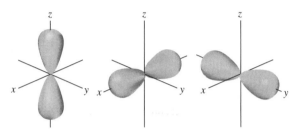

Fig. 2.16 The three p orbitals (probability densities as defined in Equation 2.80) corresponding to $\ell = 1, m = 0, \pm 1$. The $m = 0$ orbital has no angular momentum about the z-axis and is called the p_z orbital. The $m = \pm 1$ orbitals are combined to produce standing waves around the x- or y-axes. Such combinations are required because, even in the presence of electron–electron interactions, the $m = \pm 1$ orbitals are degenerate.

this series more fully below, but some orbital shapes (according to Equation 2.80) are shown in Figs. 2.15 and 2.16.

2.17 Multielectron atoms

The remainder of the periodic table is composed of elements that are much more complicated than hydrogen. But some qualitative features of the hydrogen atom are shared by all atoms. When atoms exist in free space (but not in compounds), the potential that keeps the outermost (valence) electrons tethered to the nucleus is primarily the Coulomb interaction:

$$V(r) = -\frac{\kappa Z_{\text{eff}} e^2}{r}. \tag{2.81}$$

As written here, it is identical to Equation 2.63, save for the introduction of a quantity, Z_{eff}. This is the effective charge that an outermost electron sees owing to the positive nucleus (charge $+Z$ for an atom of atomic number Z) as screened by the remaining $Z - 1$ electrons (in the case where there is only 1 valence electron). Z_{eff} is generally bigger than 1 because there are a number of electrons occupying the valence states furthest from the nucleus and because different states "penetrate" into the electron core to different degrees. This model is the simplest example of a "pseudopotential" used to reduce a multielectron problem to an equivalent one-electron problem.

To the extent that Equation 2.81 describes the potential seen by the valence electrons, the states of multielectron atoms will look just like those of the hydrogen atom (but with more nodes as they must be orthogonal to the states of the core electrons). The features that carry over are these:

a. The atoms have an allowed series of energy states corresponding to the allowed set of values of the *principal* (n), *azimuthal* (ℓ), and *magnetic* (m) quantum numbers. These are integers and subject to the same constraint found above for the hydrogen atom:

   - n is a positive integer
   - $0 \leq \ell \leq n - 1$
   - $-\ell \leq m \leq +\ell$.

   Unlike the hydrogen atom, states with the same principle quantum number, n, are no longer degenerate, but have energies that depend on ℓ and m.

b. Each energy state corresponds to a particular wavefunction or orbital (or, equivalently, charge density distribution, proportional to the probability distribution of the state). The states for $\ell = 0$ are called s-states and they are spherically symmetric. $\ell = 1$ yields three "p" states (p_x, p_y, and p_z, corresponding to $m = \pm 1, 0$); $\ell = 2$ yields five "d" states ($m = \pm 2, \pm 1, 0$); and $\ell = 3$ yields seven "f" states ($m = \pm 3, \pm 2, \pm 1, 0$). The charge distributions for s- and p-states resemble the hydrogenic distributions shown in Figs. 2.15 and 2.16. The more complex shapes of the charge (probability) distributions for d and f states can be found in chemistry texts.[10]

c. According to the Pauli principle, each orbital can hold up to two electrons, spin up and spin down. In the hydrogen atom, the energy of an orbital depends only on the principal quantum number n, but this is not true for multielectron atoms. States are filled up from the lowest energy on upward as electrons are added (at the same time as protons and neutrons are added to the nucleus to "build up" the element). The states are labeled according to $n\ell^k$, where $n = 1, 2, 3, \ldots$ (denoting the principle quantum number); $\ell = s, p, d, f$ (denoting the azimuthal quantum number); and k is the number of electrons occupying the particular orbital designated by n and ℓ. Thus, for example, the electronic configuration of silicon is

$$1s^2 2s^2 2p^6 3s^2 3p^2$$

showing how the 14 electrons of silicon are distributed among the 7 distinct orbitals (1s, 2s, $2p_x$, $2p_y$, $2p_z$, 3s, and $3p_{x,y, \text{or } z}$).

   Where an orbital can hold more than two electrons with the same energy (e.g., in orbitals corresponding to the different ℓ and m values for a given n), the states are filled up one electron at a time first to minimize electron–electron interactions, with spin up–spin down pairs formed only after all states corresponding to the different ℓ and m values for a given n have been singly occupied. This generally corresponds to the relative order of energies when electron–electron interactions are included. This result is called Hund's rule.

d. In addition to Hund's rule, electron–electron interactions can lead to other complications: for example, atoms with occupied 3d states generally have higher electron energy in these states than the energies of the

4s states because of the strong electron–electron interactions in the 3d states. Thus, the 3d states are filled *after* the 4s states.

2.18   The periodic table of the elements

The sequence of orbitals generated by these rules is summarized in Table 2.2. It is the basis of the periodic table (Fig. 2.17), when taken together with the effects of electron–electron interactions. Remarkably, the periodic table was first published in something like its current form by Dmitri Mendeleev in 1869, long before the discovery of quantum mechanics and even well before

Table 2.2 Quantum numbers and orbitals. Note that 3d states usually fill up *after* the 4s states as a consequence of increased electron–electron interaction energies

n	ℓ	Name	m	Number of orbitals in subshell	Maximum electrons in subshell	Total orbitals (n^2)	Maximum electrons $(2n^2)$
1	0	s	0	1	2	1	2
2	0	s	0	1	2		
	1	p	$-1, 0, +1$	3	6	4	8
3	0	s	0	1	2		
	1	p	$-1, 0, +1$	3	6		
	2	d	$-2, -1, 0, +1, +2$	5	10	9	18
4	0	s	0	1	2		
	1	p	$-1, 0, +1$	3	6		
	2	d	$-2, -1, 0, +1, +2$	5	10		
	3	f	$-3, -2, -1, 0, +1, +2, +3$	7	14	16	32

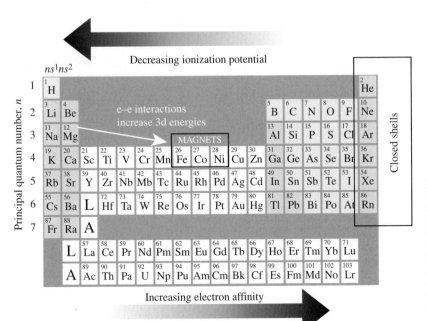

Fig. 2.17 The periodic table of the elements. Originally composed based on similarities between the chemical properties of elements in each column, it can be understood in terms of the series of eigenstates filled as an electron (and proton) are added to each element to make the next one in the table. Elements just one or two electrons short of a closed shell most easily accept additional electrons (they have a greater electron affinity) while elements with just one or two electrons outside a full shell most easily donate an electron (they have the smallest ionization potential).

J.J. Thompson's discovery of the electron. Mendeleev noticed that when elements were listed in order of their atomic weight, their properties tended to repeat periodically. He arranged his table so that elements in each vertical column have similar chemical properties. We now know that the controlling factor is the number of electrons (or protons) in an element (the atomic number) so the modern periodic table is slightly (but not much) changed from Mendeleev's original. The reason for the similarity of elements in a given column is the identical number of valence electrons (electrons outside a closed, i.e., full shell) and the similarity in the nature of the states occupied by those electrons. Unlike the case of hydrogen, the various angular eigenstates (corresponding to ℓ and m quantum numbers) now have different energies because of electron–electron interactions.

Turning to take a closer look at Fig. 2.17, the two 1s states correspond to the first row, hydrogen with one electron ($1s^1$) and helium with 2 ($1s^2$). As $n = 1$, ℓ and $m = 0$ so there are no other states in the first row. The two 2s plus three 2p states (taking 6 electrons) give the second row. For example, lithium is $1s^2,2s^1$; beryllium is $1s^2,2s^2$; and boron is $1s^2,2s^2,2p^1$. The d states add the 10 columns (five states, two electrons each) in the row below, while the 14 electrons in f states fill out the Lanthanide and Actinide series. When each shell (corresponding to a principal quantum number) is filled up, the atom is particularly stable (the noble gasses He, Ne, Ar, etc., labeled "Closed shells" on the right). Adding just one more electron and proton produces the alkali metal in the next row down on the far left. The alkali metals donate electrons easily in chemical reactions (ending up as ions with a closed shell configuration). Atoms with one electron short of a full shell (i.e., the halogens F, Cl, Br, etc.) accept an electron in chemical reactions (to gain a closed shell configuration). These effects are evident when the first ionization potential (potential for removal of one electron) is plotted as a function of atomic number, as shown in Fig. 2.18. In summary, elements with a smaller ionization potential tend to donate electrons to elements that readily accept an extra electron. This ability to accept an electron is characterized by the *electron affinity*, which is the energy gained on adding an electron to the neutral atom. The halogens have a high electron affinity because so much energy is gained by forming a closed shell configuration. Bonds between elements that are close in the periodic table (e.g., GaAs)

Fig. 2.18 Ionization potentials across part of the periodic table. Atoms with closed shell configurations (all states corresponding to a principal quantum number just filled) are the most stable (rare gasses listed at top). Atoms with just one electron outside a full shell are easiest to ionize (alkali metals listed at the bottom).

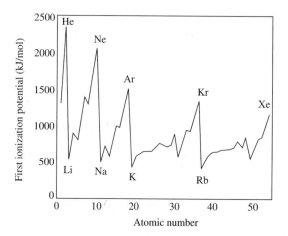

tend to involve little transfer of charge, whereas bonds between well-separated donors and acceptors (e.g., NaCl) involve a lot of charge transfer (so Na^+Cl^- is a good description of common salt). Note that the d states corresponding to $n = 3$ actually lie in row 4 because electron–electron interactions increase the energy of these states so that they lie higher in energy than the 4s states and so are occupied after the 4s states, even though their principal quantum number is smaller.

We can even gain some insight into exotic effects like magnetism. Hund's rule requires that electrons fill up highly degenerate states (like the d states) with spins all in one direction. Thus the maximum spin polarization of a transition metal will occur when the d-shell is half full. Inspection of Fig. 2.17 shows that this is exactly where magnetic materials are found. (There is more to the story than atomic polarization, and some aspects of magnetic materials will be discussed in Chapter 9.)

Exercise 2.7: Write out the electronic configuration of zinc. List the values of quantum numbers for each occupied state.

Answer: Zn is $1s^2$, $2s^2$, $2p^6$, $3s^2$, $3p^6$, $4s^2$, $3d^{10}$ for a total of 30 electrons, corresponding to $n = 1$, $\ell = 0$ ($1s^2$); $n = 2$, $\ell = 0$ ($2s^2$); $n = 2$, $\ell = 1$, $m = -1, 0, +1$ ($2p^6$); $n = 3$, $\ell = 0$ ($3s^2$); $n = 3$, $\ell = 1$, $m = -1, 0, +1$ ($3p^6$); $n = 4$, $\ell = 0$ ($4s^2$); $n = 3$, $\ell = 2$, $m = -2, -1, 0, +1, +2$ ($3d^{10}$).

2.19 Approximate methods for solving the Schrödinger equation

The degree to which the many electron atom problem maps on to the hydrogen atom problem illustrates the fact that it is often possible to find approximate solutions for a new Hamiltonian, $\hat{H}$ in terms of the Hamiltonian for some other problem, $\hat{H}_0$ if the two are related by a small additional term, $\hat{H}'$. The method for doing this is referred to as *perturbation theory*. The Hamiltonian for the modified Schrödinger equation is

$$\hat{H} = \hat{H}_0 + \hat{H}', \tag{2.82}$$

where $\hat{H}'$ represents the perturbation of the original system. To construct perturbation theory, we introduce a small arbitrary parameter, λ ($\lambda \ll 1$),

$$\hat{H} = \hat{H}_0 + \lambda\hat{H}' \tag{2.83}$$

on the grounds that as $\lambda \to 0$, the new eigenenergies, E_n, and eigenfunctions, ψ_n, can be expanded as a power series using the unperturbed energies, $E_n^{(0)}$, and eigenfunctions, $\psi_n^{(0)}$, to which are added corrections in successively higher order terms in λ:

$$E_n = E_n^{(0)} + \lambda E_n^{(1)} + \lambda^2 E_n^{(2)} + \cdots \tag{2.84}$$

$$\psi_n = \psi_n^{(0)} + \lambda \psi_n^{(1)} + \lambda^2 \psi_n^{(2)} + \cdots \tag{2.85}$$

The $E_n^{(1)}$, $E_n^{(2)}$ are corrections to the energy and $\psi_n^{(1)}$, $\psi_n^{(2)}$ are corrections to the wavefunction (both to be calculated). Inserting 2.84 and 2.85 into the Schrödinger equation for the perturbed system ($\hat{H}\psi_n = E_n\psi_n$) and equating terms on the left and right in the same order of λ (which must separately be equal because λ is an arbitrary parameter) returns the unperturbed Schrödinger equation from terms in λ^0, and, from terms in λ^1,

$$E_n^{(1)}\psi_n^{(0)} = \hat{H}'\psi_n^{(0)} + (\hat{H}_0 - E_n^{(0)})\psi_n^{(1)}. \tag{2.86}$$

The solutions of the Schrödinger equation form a complete orthonormal set, a mathematical property that allows any other function to be expressed in terms of some linear combination of these solutions (much as any periodic function can be constructed from the sum of sine and cosine terms):

$$\psi_n^{(1)} = \sum_m a_{nm}\psi_n^{(0)}. \tag{2.87}$$

Substituting 2.87 into 2.86 gives

$$E_n^{(1)}\psi_n^{(0)} = \hat{H}'\psi_n^{(0)} + \sum_m a_{nm}(E_m^0 - E_n^{(0)})\psi_m^{(0)} \tag{2.88}$$

(where we have used the Schrödinger equation to substitute for $\hat{H}_0\psi_m^{(0)}$). Multiplying from the left by $\psi_n^{(0)}*$ and integrating,

$$E_n^{(1)}\langle\psi_n^{(0)}|\psi_n^{(0)}\rangle = \langle\psi_n^{(0)}|\hat{H}'|\psi_n^{(0)}\rangle + \sum_m a_{nm}(E_m^0 - E_n^{(0)})\langle\psi_n^{(0)}|\psi_m^{(0)}\rangle. \tag{2.89}$$

Using the orthogonality properties (2.16) and the fact that the term in $(E_m^0 - E_n^{(0)})$ vanishes for $m = n$, we have the result of first-order perturbation theory:

$$E_n^{(1)} = \langle\psi_n^{(0)}|\hat{H}'|\psi_n^{(0)}\rangle. \tag{2.90}$$

The correction to the unperturbed energy for state n is just the expectation value (defined by the integral in 2.90) of the perturbing Hamiltonian taken with the *unperturbed* wavefunctions for state n.

A similar line of reasoning leads to the following result for the first-order correction to the wavefunction:

$$\psi_n = \psi_n^{(0)} + \sum_{m \neq n} \frac{H'_{mn}\psi_m^{(0)}}{E_n^{(0)} - E_m^{(0)}} \quad \text{where } H'_{mn} = \langle\psi_m^{(0)}|\hat{H}'|\psi_n^{(0)}\rangle. \tag{2.91}$$

The matrix elements H'_{mn} have nonzero values to the extent that the unperturbed wavefunctions are not eigenfunctions of $\hat{H}'$ (if they were eigenfunctions of $\hat{H}'$ the matrix element would reduce to the overlap integral $\langle\psi_m^{(0)}|\psi_n^{(0)}\rangle$ which would be zero).

Equation 2.91 breaks down when there is an accidental degeneracy ($E_m^0 = E_n^{(0)}$) for $m \neq n$ (as would occur for states of different magnetic quantum number, m but the same n in the hydrogen atom) because the corresponding term in the sum (2.91) blows up. This problem is solved by forming a single new wavefunction from a linear combination of all the degenerate wavefunctions by finding a combination that solves the perturbing Hamiltonian (an example is given in the next section).

This problem of perturbation theory with degenerate states illustrates a very powerful general feature of quantum mechanics: If some states are very close in energy, a perturbation generally results in a new state that is a linear combination of the originally degenerate unperturbed states. To take a concrete example, the electronic configuration of silicon is $1s^2 2s^2 2p^6 3s^2 3p^2$, which might lead one to believe that silicon should act as though it has two valence electrons. However, the 3s and 3p states are very close in energy, and the perturbation generated when two Si atoms approach to form a bond causes the 3s and 3p states to combine into a new linear combination of four linear combinations of s and p orbitals called sp^3 hybridized orbitals. For this reason, chemical bonds in silicon (and also in carbon) correspond to a valency of four, and the bonds in many such compounds are arranged in a tetrahedral pattern around the atom.

An even simpler approach to perturbation theory is to evaluate the expectation value given by Equation 2.90 for some arbitrary (guessed) combination of wavefunctions, varying the mixing parameter (a_{mn} in 2.87) so as to find a value that minimizes the total energy of the system. This is easy to implement on a computer and is described as *variational* approach to perturbation theory.

First-order perturbation theory can also be extended to the time-dependent Schrödinger equation in order to address the following question: If a perturbation $\hat{H}'$ is turned on abruptly at time $t = 0$, what is the probability, $P(k, t)$, that the system will make a transition from an initial state ψ_m^0 to a final state ψ_k^0 in some time interval t? The result is that[1]

$$P(k, t) = \left[\frac{1}{i\hbar} \int_0^t \langle \psi_m^0 | \hat{H}' | \psi_k^0 \rangle \exp i \left(\frac{E_m^0 - E_k^0}{\hbar} t' \right) dt' \right]^2 . \qquad (2.92)$$

Note that this formula does not (and cannot) predict when a system undergoes a transition, but it will give the probability that a transition occurs in a given interval in time, a quantity that has to be determined experimentally by repetition of many identical measurements. Evaluation of Equation 2.92 for a cosinusoidal perturbation ($\hat{H}'(r, t) = \hat{H}(r) \cos \omega t$) shows that the probability peaks sharply for values of ω such that

$$\omega = \frac{|E_m^0 - E_k^0|}{\hbar} . \qquad (2.93)$$

This is a statement of conservation of energy: a photon will be absorbed by an atom if it has the same energy as a transition between eigenstates of the atom. (But there is some probability of a transition at the wrong energy, consistent with the uncertainty principle as stated in Equation 2.8.)

These results can be turned into a powerful tool called Fermi's golden rule. The integral over time can be replaced by a probability per unit time (i.e., a rate of transitions from state m to state k, $\frac{dP}{dt}(m, k)$) and the fact that the integral peaks only when the resonance condition (2.93) is met leads to

$$\frac{dP}{dt}(m, k) = \frac{2\pi}{\hbar}|\langle \psi_m^0|\hat{H}'|\psi_k^0\rangle|^2 \delta(E_m^0 - E_k^0 - \hbar\omega), \qquad (2.94a)$$

where $\hbar\omega$ is the energy supplied by the perturbation that drives the transition from $k \rightarrow m$. It could be the energy of a photon driving an electron from a ground state k to an excited state, m. The delta function is infinite for $E_m^0 - E_k^0 + \hbar\omega$ and zero otherwise (this "Dirac" delta function has the property that its *integral* is unity). If there are many states very close to E_k in energy, the rate of transitions into all of them is the sum of the rates into any one of them. If the number of states in an small energy interval is represented by the density of states function, $\rho(E_k)$, then another version of the golden rule (arising from integrating 2.94 over energy) is

$$\frac{dP}{dt}(m, k) = \frac{2\pi}{\hbar}|\langle \psi_m^0|\hat{H}'|\psi_k^0\rangle|^2 \rho(E_k) \qquad (2.94b)$$

The energy difference between adjacent states in this case is very small, so the perturbation energy, $\hbar\omega$, no longer appears. We will use Equation 2.94b in later chapters.

2.20   Chemical bonds

Two (or more) atoms can form chemical bonds when the energy of the combination is lower than the energy of the two atoms in isolation. This generally entails an enhanced probability of finding an electron between the atoms, an effect involving charge transfer from what was (in the unperturbed system) a spherical electron distribution. Chemical bonds are generally classified according to the amount of redistribution of the probability density that occurs on bonding:

- In *ionic* bonds (e.g., NaCl, KCl, etc.), an electron is transferred almost completely from the donor (e.g., an alkali metal) to the acceptor (e.g., a halogen) so the resulting salt is represented by Na^+Cl^- where the donor atom is sodium and the acceptor atom is chlorine. The energy to form these ions in isolation is very large (cf. Fig. 2.18), but it is offset by the fact that each Na^+ is surrounded by six Cl^- ions in a sodium chloride crystal, for example (and vice versa for the Cl^-). Importantly, Na^+ becomes an ion by losing its 3s electron to form the noble gas valence configuration of Ne ($2s^22p^6$). Cl^- becomes an ion by gaining an electron to form the Ar valence structure $3s^23p^6$. The formation of closed shell electronic configurations in the ions results in a lowering of electronic energy. Bonds between elements on the left of the periodic table and the non-noble gasses on the right tend to be ionic in character.
- In covalent bonds, electrons are mostly shared and transferred only partially. Divalent molecules such as O_2 and N_2 have a symmetrical charge distribution about the central plane of the molecule, with an increased probability for finding electrons in the middle. Bonds formed between elements on the right of the periodic table tend to be covalent.

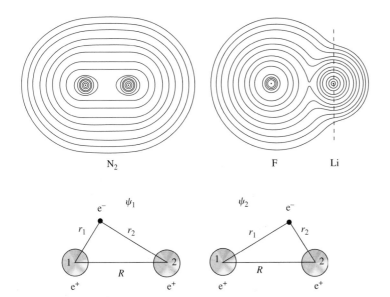

Fig. 2.19 Calculated charge density contours for a covalent (N_2) and an ionic (LiF) bond. Most charge distributions in bonds lie between these extremes (Courtesy Richard Bader).

Fig. 2.20 The dihydrogen molecular ion showing two equivalent arrangements. One, ψ_1, corresponds to an initial hydrogen 1s state on proton 1 perturbed by an interaction with proton 2. The second, ψ_2, corresponds to an initial hydrogen 1s state on proton 2 perturbed by an interaction with proton 1.

- Elements on the left of the periodic table tend to form metallic bonds where the charge is spread out over a distance larger than the size of the individual atoms in the solid (more of this in Chapter 7).
- Even neutral rare gas elements experience a weak attractive force based on the *fluctuations* in the otherwise spherical charge distributions associated with closed shell configurations. Fluctuations from a spherical shape lead to a dipole moment, and two dipoles can attract one another. These weak bonds are called van der Waals interactions.

Examples of charge distributions for archetypical covalent and ionic bonds are shown in Fig. 2.19.

We will illustrate the quantum mechanical nature of the chemical bond with the simple (one-electron) example of the hydrogen molecular ion, H_2^+, illustrated in Fig. 2.20. It consists of a single electron "shared" by two protons. The electronic Hamiltonian for the problem is

$$\hat{H}_0(r_1, \theta, \phi) - \frac{e^2}{4\pi \varepsilon_0 r_2}, \tag{2.95}$$

where $\hat{H}_0(r_1, \theta, \phi)$ is the hydrogen atom Hamiltonian, and the second term is the additional interaction of the electron with the second proton (the proton–proton repulsion is ignored here). Figure 2.20 shows how there are two equivalent descriptions of the problem, which we have labeled ψ_1 and ψ_2. By this point, we understand that the answer must be a linear superposition of these two states. But let us look at this from the standpoint of the perturbation theory we have just described, taking the perturbation potential to be

$$\hat{H}' = \frac{-e^2}{4\pi \varepsilon_0 r_2}. \tag{2.96}$$

The unperturbed Hamiltonian, $\hat{H}_0$, is solved (for the ground state) by the 1s wavefunction given in Table 2.1 with the ground state energy, E_0, given by

Equation 2.77 with $n = 1$. There are two ground states,

$$\psi_1 \equiv \psi^{1s}(1) \tag{2.97a}$$

$$\psi_2 \equiv \psi^{1s}(2) \tag{2.97b}$$

corresponding to the electron on the left proton or the right proton. These are the unperturbed wavefunctions, and they both have the same energy, E_1. We see straightaway that we cannot use the simple perturbation theory developed above, because the states are degenerate and Equation 2.91 would blow up. We have to form new linear combinations (two of them) from the unperturbed wavefunctions such that the corresponding terms in the expansion for the coefficients of the wavefunction disappear. This is equivalent to solving Schrödinger's equation for the subset of degenerate states (in general), that is, the two degenerate states in 2.97.

We will start by writing down one linear combination

$$a\psi_1 + b\psi_2, \tag{2.98}$$

where a and b are to be determined. The Schrödinger equation is

$$[\hat{H}_0 + H'](a\psi_1 + b\psi_2) = E(a\psi_1 + b\psi_2), \tag{2.99}$$

where $E(E \neq E_0)$ is the eigenenergy of the perturbed system. Collecting together terms and using the fact that $\hat{H}_0\psi_{1,2} = E_0\psi_{1,2}$,

$$(a\psi_1 + b\psi_2)(E_0 - E) + a\hat{H}'\psi_1 + b\hat{H}'\psi_2 = 0. \tag{2.100}$$

At this point, we are going to cheat a little to illustrate the way that degenerate perturbation theory is usually carried out. Usually, we use different solutions of the unperturbed Schrödinger equation as a basis, and these have the property that $\langle \psi_n^0 | \psi_m^0 \rangle = \delta_{nm}$. This is not the case here because ψ_1 and ψ_2 are the same functions, albeit centered on different points. For now, we will take the amount of overlap between them to be small and suppose that $\langle \psi_1 | \psi_2 \rangle \approx \delta_{12}$. (It is straightforward to include overlap between the functions and this is done in Section 8.4.) Multiplying from the left by ψ_1^*, integrating and using our "fake" orthogonality condition:

$$a(E_0 - E) + a\langle \psi_1 | \hat{H}' | \psi_1 \rangle + b\langle \psi_1 | \hat{H}' | \psi_2 \rangle = 0. \tag{2.101a}$$

A similar type of equation is produced by multiplying from the left by ψ_2^*:

$$b(E_0 - E) + a\langle \psi_2 | \hat{H}' | \psi_1 \rangle + b\langle \psi_2 | \hat{H}' | \psi_2 \rangle = 0. \tag{2.101b}$$

The remaining integrals over the perturbation Hamiltonian produce numbers, and, from the symmetry of the problem,

$$\langle \psi_2 | \hat{H}' | \psi_1 \rangle = \langle \psi_1 | \hat{H}' | \psi_2 \rangle \equiv \Delta \tag{2.102a}$$

$$\langle \psi_1 | \hat{H}' | \psi_1 \rangle = \langle \psi_2 | \hat{H}' | \psi_2 \rangle \equiv \delta \tag{2.102b}$$

leading to

$$(E_0 + \delta - E)a + \Delta b = 0 \qquad (2.103a)$$

$$\Delta a + (E_0 + \delta - E)b = 0. \qquad (2.103b)$$

The condition for these to have a nontrivial solution (i.e., $a, b \neq 0$) is that the determinant of the coefficients of a and b vanishes or

$$(E_0 + \delta - E)^2 - \Delta^2 = 0$$

giving

$$E = E_0 + \delta \pm \Delta. \qquad (2.104)$$

This is a very important result. The perturbation has lifted the degeneracy, giving rise to two new eigenstates that differ in energy by 2Δ (as well as being shifted from the original by an amount δ). In an attractive interaction similar to that shown in Fig. 2.20 the overall energy is lowered, so Δ must be a negative number. Thus, the lower energy eigenstate corresponds to the $+$ sign in 2.104. Substituting $E = E_0 + \delta + \Delta$ in 2.103a or b yields $a = b$, so the appropriate (normalized) eigenstate is

$$\psi_B = \frac{1}{\sqrt{2}}(\psi_1 + \psi_2) \quad \text{for } E = E_0 + \delta + \Delta. \qquad (2.105)$$

Substituting $E = E_0 + \delta - \Delta$ yields for the second eigenstate

$$\psi_A = \frac{1}{\sqrt{2}}(\psi_1 - \psi_2) \quad \text{for } E = E_0 + \delta - \Delta. \qquad (2.106)$$

If δ is small compared to Δ, the ground state energy of H_2^+ is lowered compared with the free atom for ψ_B, for the *bonding state* described by 2.105. The higher energy state (2.106) is unoccupied, and is called the *antibonding state*. This lifting of the initial degeneracy is illustrated schematically in Fig. 2.21, where the shape of the corresponding wavefunctions is also shown.

   This attractive interaction between the two atoms only tells us why they come together in the first place. It does not explain why there is a distinct bond length (R in Fig. 2.20). In the case of the dihydrogen molecular ion, we missed out an important Coulomb *repulsion* between the two protons. It is screened (to some extent) by the buildup of electronic charge between the two centers, but it gets more important as the two protons are pushed closer together. (In the case of molecules with two or more electrons, electron–electron repulsions become important, as pushing electrons together raises their potential energy rapidly.) These effects give rise to a strong repulsion between atoms at short distances. Thus bond lengths correspond to the equilibrium distance (lowest potential) in a plot of energy against the distance between the atom centers. An example is shown in Fig. 2.22. This interaction reflects the total electronic energy as well as the Coulomb interaction between nuclei and it comes from solving the Schrödinger equation. However, it is often very useful to forget about its quantum mechanical origin and just call it an "interaction potential"

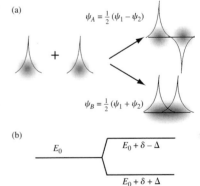

Fig. 2.21 (a) Wavefunctions (lines) and charge distributions (shadings) for the unperturbed and perturbed dihydrogen molecular ion. (b) The corresponding eigenenergies. The perturbation lifts the degeneracy between the two isolated atoms by an amount 2Δ, where Δ is the value of the interaction matrix element. This lowering of the energy of the state occupied by the single electron is what drives the chemical bonding.

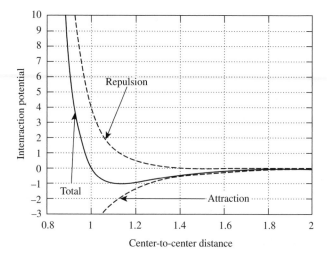

Fig. 2.22 Components of the interaction potential between the protons in the dihydrogen molecular ion including both attractive and repulsive interactions. Potential is plotted in atomic units and distances in units of the "hard-sphere radius," σ.

between two atoms. One useful parametric form of this type of potential vs. distance curve is the *Morse potential*:

$$\phi(r) = D[1 - \exp(-a[r - r_{\mathrm{e}}])]^2, \qquad (2.107\mathrm{a})$$

where D is the depth of the potential well binding the electron and r_{e} is the equilibrium center-to-center distance. Equation 2.107a can be difficult to manipulate, and much simpler version, based on replacing the exponential repulsion with a twelve-power term is the *Lennard-Jones 6-12* potential:

$$\phi(r) = 4\varepsilon \left[\left(\frac{\sigma}{r_{ij}} \right)^{12} - \left(\frac{\sigma}{r_{ij}} \right)^6 \right]. \qquad (2.107\mathrm{b})$$

This has a minimum at $r_{ij} = 1.12\sigma$ where the binding energy is ε. Quantities like the cohesive energy and bulk modulus of a material may be estimated from σ and ε using this model, once the parameters are evaluated from, for example, crystallographic and melting point data. These effective pairwise interaction potentials between atoms are a great simplification (compared with the real quantum mechanical interaction) and can be very useful for simulations (like molecular dynamics—see Chapter 3).

2.21 Eigenstates for interacting systems and quasiparticles

We are now in a position to revisit some of the concepts introduced at the beginning of this chapter. We will use the language of degenerate perturbation theory presented in the last section to illustrate some of the terms of quantum mechanics. The simultaneous equations (2.103) to be solved for the coefficients of the new mixed wavefunction can be rewritten in matrix form as follows:

$$\begin{pmatrix} \hat{H}_0 + \hat{H}'_{11} - E & \hat{H}'_{12} \\ \hat{H}'_{21} & \hat{H}_0 + \hat{H}'_{22} - E \end{pmatrix} \begin{pmatrix} a \\ b \end{pmatrix} = 0, \qquad (2.108)$$

where δ and Δ have been replaced by their original definition. Stated this way, the Schrödinger equation is solved as a problem in matrix algebra (once the elements are calculated from integrals like those in 2.102). This is called the *matrix representation* of quantum mechanics and it represents an alternative to the wavefunction representation. We found the eigenvalues from the secular determinant of our algebraic equations, and this corresponds to the determinant of the matrix in 2.108. Inserting these energies for E and using the correct linear combinations of wavefunction coefficients in 2.108 puts it into a form where the off-diagonal components are zero. In this approach, solving the Schrödinger equation is equivalent to *diagonalizing* the matrix of the elements obtained from all of the expectation values of the Hamiltonian. This can be seen for the two-state system just discussed by direct substitution. Thus, the problem becomes one of finding the correct linear combinations of the *basis states* (ψ_1^0 and ψ_2^0 for the two-state system) that result in a diagonal Hamiltonian matrix, and there are efficient mathematical tools available for doing this. With no off-diagonal terms, the product of the Hamiltonian matrix with the (column vector) containing the coefficients of the wavefunctions (i.e., the a_{mn} in Equation 2.87) produces just the sum of the eigenenergies for each eigenstate with the coefficients for that eigenstate, the matrix equivalent of the Schrödinger equation $\hat{H}\psi_n = E_n\psi_n$.

Taking a system of noninteracting particles (with product wavefunctions like 2.33) as a basis, it is often possible to include interactions as a perturbation. To the extent that the perturbation theory is correct, the new linear combinations diagonalize the Hamiltonian that includes the perturbations. These linear combinations are the new eigenstates of the system, and they thus correspond to new types of "effective particles" that are composites of the wavefunctions for the independent particles in the unperturbed Hamiltonian.

The generation of new types of particles turns out to be possible even in strongly interacting systems where perturbation theory cannot be applied. For example, an elastic solid is composed of atoms held together by strong bonds. However, the vibrations of the solid are well described by plane waves. Since a plane wave corresponds exactly to the wavefunction for a free particle, the quantum description of vibrations in solids is of *free particles* called *phonons*.[11] Excitations in strongly interacting systems that behave like a gas of free particles are called *quasiparticles*. It turns out that other strongly interacting systems (like electrons in metals) can have excitations that behave like free particles, a topic we will return to in Chapter 7.

2.22 Getting away from wavefunctions: density functional theory

Working with wavefunctions is clearly difficult to do, even for the simple problems we have looked at so far. We would like to go much further. For example, we would like to be able to

- Solve problems involving many interacting electrons, going beyond perturbation theory.
- Find the wavefunctions that satisfy the Pauli principle, i.e., changing sign when we exchange pairs of electrons between states.
- Work with arbitrarily complicated potentials.

These are very difficult goals, but the theory of many electron systems has undergone a revolution in the past half century. In 1964, Hohenberg and Kohn[12] proved a remarkable theorem. They showed that the ground state energy of a system of interacting particles could be expressed as a unique functional (function of a function) of the electron density $n(\mathbf{r})$, that is, as $E[n(\mathbf{r})]$. The electron density, $n(\mathbf{r})$, is the sum of the squares of the wavefunctions at $\mathbf{r}$. The correct electron density is that which minimizes the ground state energy of the system, that is, it is found from the condition that

$$\frac{\delta E[n(\mathbf{r})]}{\delta n(\mathbf{r})} = 0, \tag{2.109}$$

where the symbol $\delta n(\mathbf{r})$ means "varying the function $n(\mathbf{r})$." The program is to search for that distribution $n(\mathbf{r})$ of electron density that minimizes the ground state energy of the system. The output is the electron density distribution, $n(\mathbf{r})$, and the ground state energy of the system. This is a remarkable result, for it sidesteps wavefunctions altogether and goes directly to the desired quantities (electron density and energy). The theorem proved the existence of such a functional, but it did not yield the form of the functional. At present, DFT is carried out with functionals that are based on trail and error. The "Holy Grail" functional has yet to be found.

 We will give an example of a functional here. I have deliberately chosen a rather old-fashioned functional, because the terms in it are easier to understand. Modern theories use much better functionals. We start with an expression for the kinetic energy of a uniform electron gas (density a constant, n, everywhere). This can be expressed in terms of the electron density, $n(= N/V)$, because the number of states per unit volume, V, can be calculated as shown in Section 2.15b; therefore, the approach is to fill the states up, two electrons at a time, until the desired electron density is reached, and then calculate the average electron energy from this distribution. This leads[11] to a kinetic energy proportional to $n^{5/3}$. Next, we consider the terms for the potential: The Coulomb energy between electrons at r_1 and r_2 is proportional to $n(r_1)n(r_2)/r_{12}^2$ and the potential energy of the electrons owing to nuclear charges is $n(r)V(r)$, where $V(r)$ is the user input to the problem (i.e., the effective charges of nuclei and their positions). Effective potentials for an electron interacting with an atom and all its core electrons have been developed over the years. They are called pseudopotentials and they are built into programs for calculating charge densities using DFT. Terms like these lead to an energy functional:

$$E[n(\mathbf{r})] = C \int n(\mathbf{r})^{5/3}d\mathbf{r} + \frac{1}{2}\iint \frac{n(\mathbf{r}_1)n(\mathbf{r}_2)}{r_{12}}d\mathbf{r}_1 d\mathbf{r}_2 + \int n(\mathbf{r})V(\mathbf{r})d\mathbf{r} + XC, \tag{2.110}$$

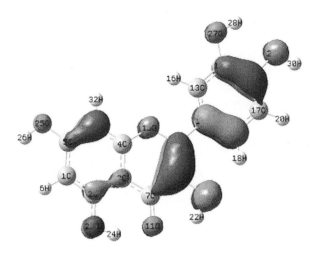

Fig. 2.23 DFT calculation of the electron density associated with the lowest energy unoccupied state of a complex antioxidant molecule (quercetin). (Reprinted from the Journal of Molecular Structure: THEO-CHEM, Mendoza-Wilson and Glossman-Mitnik,[14] copyright 2005 with permission from Elsevier.)

where *XC* is an "exchange-correlation" term which we will discuss below. Once again, this particular functional is given only as an illustration—much better ones are now available. Using a sum of independent particle kinetic energies, Kohn and Sham[13] found a solvable system of equations to implement (2.109). These yielded $n(\mathbf{r})$ as well as the ground state energy of a system. This program is widely implemented today in software (as a web search on the words "density functional software" will reveal) and programs exist for use on laptop computers. These programs can solve for the band structure of a complicated crystal in a matter of minutes! They rely on solving an equation somewhat like the Schrödinger equation for the "Kohn–Sham" states of the problem. These are not the eigenstates of the system, but they are often used in place of eigenstates in calculations.

The most common approach is to work in the local density approximation (LDA). This says that in some region of space the nonlocal effects of "exchange" (the need to keep the overall wavefunction antisymmetric under exchange of particle labels) and correlation (an artifact of treating interacting electrons as single particles interacting with an effective environment) can be represented by some function of the local electron density. These effects are incorporated in the term *XC* in Equation 2.110, and the LDA uses a simple (local) function of $n(\mathbf{r})$. Dirac showed that for a uniform electron gas, the requirement of anti-symmetric wavefunctions could be incorporated by adding a term to the energy of independent electrons proportional to $n^{4/3}$. In the LDA, this is replaced by a similar function of the local density, $n(\mathbf{r})^{4/3}$. Because of the existence of a number of "turnkey" programs, DFT is widely used in electronic structure calculations in nanoscience and chemistry. Figure 2.23 shows a plot of electron density on complex molecule (an antioxidant in plants). Importantly, DFT can be used to calculate structures for materials. The calculation starts with a "best guess" for a structure for which DFT returns a ground state energy. The atoms are moved (DFT programs provide a way of doing this in a way that is better than a random guess) and the ground state energy recalculated. This process is reiterated until the structure with the lowest ground state energy is found.

2.23   Bibliography

P.C.W. Davies and D.S. Betts, Quantum Mechanics, 2nd ed. 1994, London: Chapman and Hall. A great introduction to quantum mechanics without too much mathematics.

R.L. Liboff, Introductory Quantum Mechanics. 1980, Oakland, CA: Holden-Day Inc. A standard upper division physics text with lots of good problems.

R.P. Feynman, R.B. Leighton and M.I. Sands, Feynman Lectures in Physics III: Quantum Mechanics. 1965, Readwood City, CA: Addison-Wesley. Misleadingly labeled a freshman text, this is a beautifully elegant description of quantum mechanics with much discussion on fundamental concepts.

Here are some more modern texts recommended to me by younger physicists:

D.J. Griffiths, Introduction to Quantum Mechanics, 2nd ed. 2004, NJ: Prentice-Hall.

A. Peres, Quantum Theory: Concepts and Methods. 1995, Berlin: Springer.

J.J. Sakurai, Modern Quantum Mechanics. 1994, NY: Addison-Wesley.

2.24   Exercises

a. Conceptual questions:

1. The probability that a quantum particle makes it from a source to a slit is 10% and the probability that it makes it from the slit to the detector is also 10%. What is the probability that the particle makes it from source to detector if there is no other path through the system and the arrival of the particle at the slit is not measured?

2. If there had been a second slit in the experiment above and the probabilities were the same for the second path, would this be enough information to determine the probability of arrival at the detector?

3. Two electrons occupying the same state must have opposing spins, and thus no net spin (e.g., the ground state of He, $1s^2$). What about two electrons occupying *different states*, as in the first excited state of a helium atom, $1s^1 2s^1$? How many ways can the electron spins be arranged? (See McQuarrie and Simon,[10] Chapter 8).

4. Suppose that the electron spins were polarized (using a magnet) in the scattering experiment shown in Fig. 2.5 and the detectors are capable of detecting spins, which one of the three results, 2.19, 2.20, and 2.21, would apply for the case of (a) both spin up and (b) one spin up, the other spin down?

5. Discuss the relative kinetic energies of s-states and p-states in terms of the curvature of the wavefunction. What is the probability of finding an electron at $r = 0$ for both types of wavefunction?

6. What is the wavelength of an electron in a 1-D potential for E exactly equals V?

7.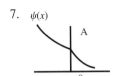

The vertical line at x = 0 indicates a point on the x-axis where the potential changes abruptly by some finite amount. Which of the wavefunctions (A, B, C) are unphysical in terms of continuity of the wavefunction and its derivative?

8. If the size of a quantum dot is doubled, what happens to the frequency of the fluorescent light emitted by it?

9. Which of the two energies of an electron propagating in free space will most likely have a quantum state nearby: $E = 0.1$ eV or $E = 10$ eV (these are free particles, so this is a question about the density of states in terms of the wave vector)?

10. Write out the electron configuration of Neon in the form $1s^2$..., etc. Which is easier to ionize: Ne or Na?

11. States of an unperturbed system will be mixed more by a perturbation if the unperturbed eigenenergies are (a) near in energy or (b) far apart in energy. Which is true, (a) or (b)?

12. If the interaction energy between two atoms is ΔeV, what will be the split between the bonding and antibonding states on bringing them together?

13. Can the energy split mentioned in Question 12 be the sole explanation for atoms forming well-defined chemical bonds?

14. The "exchange interaction" between electrons is a consequence of the requirement that the multiparticle wavefunction for electrons change sign on exchange of any two particles, *wherever* they may be in the system. What approximation in DFT makes this problem much easier to handle?

15. The theory of relativity states that information about the state of a system cannot be transmitted more quickly than the speed of light. Suppose an atom is excited to a state that decays by emitting pairs of photons in opposite directions with opposite circular polarization (left circularly polarized and right circularly polarized). The polarization of any one photon is entirely random, but once one has been measured, the polarization state of the other is known. Now suppose that an apparatus is set up to measure the polarization of both photons very quickly, more quickly than the distance between the detectors divided by the speed of light (the atom is halfway between). Two possible outcomes are that (a) once the polarization of one is measured, the polarization of the other is immediately determined or (b) no measurement can be made until after the time it takes light to propagate from one detector to the other. State what you think will happen and why?

16. Find a free DFT program on the Web (or elsewhere if you can) and run it to calculate the charge distribution in the molecules H_2O and CCl_4; otherwise, just find a result for the charge density for these two molecules. Using these distributions offer an explanation for why water dissolves ions well and carbon tetrachloride does not?

b. Simple numerical questions

(Use data from Appendix A where needed)

17. What is the energy in Joules and in electron volts of a photon light of wavelength 488 nm?
18. What is the wavelength of an electron accelerated by a potential of 10 kV in an electron microscope? How is this related to the maximum possible resolution of the microscope? Why do electron microscopes usually not achieve such resolution?
19. What is the wavelength of a proton accelerated by 10 kV?
20. Estimate the deBroglie wavelength of a car traveling at 100 km/h.
21. A covalent bond energy can be on the order of 2 eV per pair of atoms, X, forming molecules X_2. Calculate the heat released when 1 mole of X_2 is formed (in calories).
22. Look up the workfunction for Cesium and calculate the decay of tunnel current according to (2.60) for every 1 Å increase in distance between a Cs surface and a Cs tip in an STM.
23. Calculate the energy required to charge proton to make a hydrogen atom by adding an electronic charge to a sphere the size of a hydrogen atom in free space ($\varepsilon = 1$). The Bohr radius is 0.529 Å. The energy to charge a capacitor (C, Farads) with an electronic charge is $e^2/2C$, and the capacitance of a sphere of radius, a, is given as $4\pi\varepsilon\varepsilon_0 a$.

c. Mathematical problems

24. Carry out the integration of the square of the wavefunction to get the factor $\sqrt{\frac{2}{L}}$ in Equation 2.46.
25. Using Equations 2.55, 2.56, and 2.57, apply the boundary conditions to get four equations that relate $A, B, t,$ and r in terms of $k, \kappa,$ and L. Eliminate $Ae^{-\kappa L}$ and show that

$$t\exp ikL = \frac{2B\kappa\exp\kappa L}{\kappa + ik}.$$

26. Show how 2.62 becomes a simple exponential expression in the limit that $2\kappa L \gg 1$ (where $V_0^2/4E(V_0 - E)$ is approximately unity).
27. Show, by direct integration, that the hydrogen 1s ($\sqrt{\pi a_0^3}\exp(-r/a_0)$) and 2s ($\sqrt{8\pi a_0^3}(1-r/2a_0)\exp(-r/2a_0)$) wavefunctions are orthogonal.

References

[1] Liboff, R.L., *Introductory Quantum Mechanics.* 1980, Oakland, CA: Holden-Day, Inc.
[2] Davies, P.C.W. and D.S. Betts, *Quantum Mechanics, 2nd ed.* 1994, London: Chapman and Hall.
[3] Rhodes, R., *The Making of the Atomic Bomb.* 1986, New York: Touchstone Press.

[4]   Tonomura, A., J. Endo, T. Matsuda, T. Kawasaki, and H. Ezawa, Demonstration of single-electron buildup of an interference pattern. Am. J. Phys., **57**: 117–120 (1989).

[5]   P.R. Tapster, J.G. Rarity, and P.C.M. Owens, Violation of Bell's inequality over 4 km of optical fiber. Phys. Rev. Lett., **73**: 1923 (1994).

[6]   d'Espagnat, B., *The quantum theory and reality, in Scientific American.* 1979. p. 158.

[7]   Schrodinger, E., Quantisierung als Eigenwertproblem. Ann. Physik, **79**: 361–376 (1926).

[8]   Binnig, G., H. Rohrer, C. Gerber, and E. Weibel, Surface studies by scanning tunneling microscopy. Phys. Rev. Lett., **49**(1): 57–61 (1982).

[9]   Crommie, M.F., C.P. Lutz, and D.M. Eigler, Confinement of electrons to quantum corrals on a metal surface. Science, **262**: 218–220 (1993).

[10] McQuarrie, D.A. and J.D. Simon, *Physical Chemistry: A Molecular Approach.* 1997, Saucilito, CA: University Science Books.

[11] Ashcroft, N.W. and N.D. Mermin, *Solid State Physics.* 1976, New York: Holt, Rinehart and Winston.

[12] Hohenberg, P. and W. Kohn, Inhomogeneous electron gas. Phys. Rev., **136**: B864–B871 (1964).

[13] Kohn, W. and L.J. Sham, Self-consistent equations including exchange and correlation effects. Phys. Rev., **140**: A1133–A1138 (1965).

[14] Mendoza-Wilson, A.M. and D. Glossman-Mitnik, CHIH-DFT study of the electronic properties and chemical reactivity of quercetin. J. Mol. Struct. THEOCHEM, **716**: 67–72 (2005).

3 Statistical mechanics and chemical kinetics

Nanosystems are interesting precisely because they generally contain too many atoms to be thought of as simple mechanical systems, but too few to be described by the bulk properties we assign to materials that contain vast numbers of atoms. This chapter starts with a review of the properties of bulk systems, as described in terms of the possible arrangements of the atoms of which they are composed. We then review fluctuations in these bulk properties and show how their relative importance depends on the size of the system. Fluctuations drive chemical reactions, so we follow this discussion with a review of the thermodynamics of chemical reactions, and a discussion of the factors that control the rates (i.e., kinetics) of reactions. At the end of this chapter, we touch on questions such as the relationship between thermodynamic equilibrium and kinetic phenomena and the tricky subject of how thermal fluctuations affect quantum mechanical phenomena.

As in Chapter 2, this chapter covers much more ground than can be reasonably covered in the time available in an introductory course. So here is a list of the essential points needed in order to appreciate what follows in this book:

- The concept of entropy, and how entropy determines the equilibrium configuration of an isolated system.
- The concept of a system in contact with a "heat bath" and how this alters the distribution of states away from that found in an isolated system. The resulting Boltzmann distribution is the most important result of the chapter.
- The concept of equipartition of energy.
- The concept of a partition function as a way of counting states available to a system, and the use of a partition function for deriving thermodynamic quantities.
- The free energy as a quantity that expresses the balance between interaction energies and the contribution of entropy. The Helmholtz and Gibbs free energies are compared side by side in Appendix E.
- The Fermi–Dirac result for the thermal occupation of states by fermions.
- Derivation of fluctuations in an ideal gas, Brownian motion, and diffusion.
- The role of concentrations in altering the free energy that drives a chemical reaction, and the definition and meaning of the equilibrium constant and the dissociation constant.
- The results of the Kramers model for reaction rates.
- How molecular dynamics calculations are set up and carried out.
- The qualitative picture of how fluctuations affect quantum phenomena.

3.1 Macroscopic description of systems of many particles

Macroscopic amounts of liquids, solids, and gasses consist of $\sim10^{23}$ atoms* or molecules, each one of which exist in many possible quantum states. Yet they have well-defined properties that appear to be intrinsic to the bulk material, in spite of ferocious complexity at the atomic level. Statistical mechanics and thermodynamics replace very complex microscopic phenomena with a few simple macroscopic parameters. In this section, we will look at how a few  well-defined macroscopic properties (temperature, pressure, work, changes of phase, chemical energy, and viscosity) are explained in microscopic terms.

As an example of this replacement of complex microscopic descriptions with a macroscopic parameter, we know that putting more energy into a material (by compressing it, for example) makes it "hotter," so we can think of temperature as a measure of the energy content of a material. At a microscopic level, a hotter material is one in which the particles have a higher average kinetic energy than they would in a colder material (Fig. 3.1(a)). For a very large sample (but not for a nanoscale system), fluctuations about this average energy are completely negligible, and we can specify its internal energy *precisely* in terms of temperature. This concept is taken even further when we consider a "system" of particles that is at a constant temperature. We analyze the system as though it was isolated, but imagine that heat can pass freely between the system and the outside world to maintain the temperature of the system.

Fluctuations in the rate at which the individual particles collide with the walls of a container are also negligible in a large sample, so the resulting force on a unit area of the container is precisely defined by a pressure (Fig. 3.1(b)). If

*Avogadro's number, the number of atoms or molecules in a gram–mole, is 6.02×10^{23}.

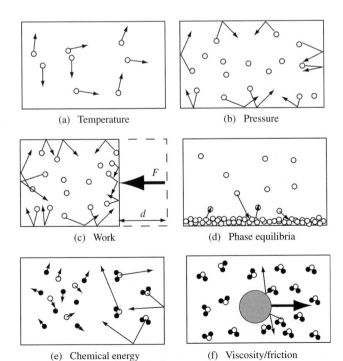

(a) Temperature

(b) Pressure

(c) Work

(d) Phase equilibria

(e) Chemical energy

(f) Viscosity/friction

Fig. 3.1 Illustrating the microscopic view of some macroscopic phenomena. (a) Temperature as the kinetic energy content of a collection of particles. (b) Pressure as the force generated on the walls of a container by particles colliding with it. (c) Work (force, F, applied over a distance, d) as raising the speed and density of particles. (d) Equilibrium between liquid and vapor phase as a constant rate of exchange of particles between the phases. (e) Release of chemical energy as atoms combine to become a compound of more energetic molecules heated by the reaction. (f) Viscosity as a transfer of kinetic energy from a large sphere to translational and rotational degrees of freedom of small molecules surrounding it.

work is done against this pressure to compress the gas, the energy produced goes into raising the energy content of the gas, either by raising temperature or pressure, or both (Fig. 3.1(c)). This statement of conservation of energy for thermodynamic quantities is the basis of the first law of thermodynamics.

If two different phases (e.g., solid–liquid, liquid–gas) of a material are in equilibrium with one another, at some temperature and pressure, then atoms pass from one phase to the other in both directions (Fig. 3.1(d)) maintaining an equilibrium between the amount of each phase at a given temperature. Thus, the equilibrium water content of air (the saturation vapor pressure) in contact with liquid water is a well-defined function of temperature and pressure. For example, at 24°C and one atmosphere of pressure (1013.25 millibars), the partial pressure owing to water vapor is 29.6 millibars. Despite the enormous complexity of water in terms of molecular structure, it always boils at 100°C at one atmosphere of pressure. The very complex molecular phenomena of evaporation and boiling can be quantitatively described by a few thermodynamic variables.

The first law of thermodynamics is a statement of the conservation of energy that incorporates heat flow: the change in energy of a system, ΔE, the sum of the work done on it, ΔW, plus the heat transferred into it from a hotter body, ΔQ,

$$\Delta E = \Delta Q + \Delta W. \tag{3.1}$$

By definition, $\Delta E > 0$ corresponds to an increase in the internal energy of a system (work done on the system and/or heat put into it). Conservation of energy requires that the energy content of a system is always the same in a given set of conditions, regardless of the manner in which the system acquired the energy. Thus, the amount of energy in 1 kg of water at 100°C (relative to the amount at 0°C, say) is always the same, whether the water was heated by an electric heater or by mechanical compression. Variables, like energy content, that characterize the state of a system are called *state variables*. Heat and work are *not* state variables because many different combinations of heat and work could yield the same change in energy content.

It is very useful to extend the concept of energy content to include chemical potential energy. In these terms, 1 kg of dynamite has a lot more internal energy than 1 kg of sand. Figure 3.1(e) illustrates the burning of hydrogen in oxygen to produce very energetic water molecules. The energy released (or taken in) to produce 1 mole of a compound from its constituent elements at standard temperature and pressure (25°C and 1 atmosphere) is called the *standard enthalpy of formation*, and the data for many compounds are available in chemistry texts. The standard enthalpy is a state function, and because its value is independent of the route by which a particular compound is made, data for a few compounds can be used to predict values for many others, by inventing conceptual routes to synthesis of the unknown compound in terms of the enthalpies of formation for known intermediates.

Viscosity is not an equilibrium quantity, because viscous forces are generated only by movement that transfers energy from one part of the system to another. But viscosity is another very useful example of isolating one part of a system to describe it simply. Using the concept of viscosity, we can consider energy transfer from a slow moving large particle to a large number of small molecules as simply a loss or dissipation of energy from the large particle. A system consisting of a sphere moving through a liquid or gas (Fig. 3.1(f)) conserves energy of course, but if we consider just the sphere alone, we would say that it is losing energy as it moves through the liquid. The force on the sphere is directly proportional to its speed and a well-defined quantity called the viscosity. Specifically for a sphere of radius, a, moving at a speed, v, in a medium of viscosity, η

$$F = 6\pi a \eta v. \tag{3.2}$$

This result, valid for nonturbulent flow far from a surface, is called Stokes' law, after George Stokes who first derived it. The concept of viscosity allows us to hide all the messy molecular details. This simplification requires that the small molecules rearrange themselves on very short timescales compared with the timescale of the motion of the sphere. We will see that we can often hide a lot of complexity if the timescales we are interested in are much longer than the timescales of molecular motion. An overview of the response time of a simple mechanical system in a viscous medium is given in Appendix D.

Exercise 3.1: A system absorbs 140 J of heat from the surroundings and does 85 J of work on the surroundings. What is the net energy change of the system?

Answer: $\Delta Q = +140$ J and $\Delta W = -85$ J, so $\Delta E = +55$ J.

Exercise 3.2: With what speed will a sphere of radius 0.05 m fall in water (viscosity 10^{-3} Pa-s)? The sphere is twice the density of water. Remember to take account of buoyancy.

Answer: In equilibrium, the gravitational force will balance the viscous force so $\Delta mg = 6\pi a \eta v$, leading to $v = \frac{\Delta mg}{6\pi a \eta}$. $\Delta m = \frac{4}{3}\pi a^3 (\rho_{\text{object}} - \rho_{\text{water}})$ giving $v = \frac{2a^2 g (\rho_{\text{object}} - \rho_{\text{water}})}{9\eta}$. The density difference, $(\rho_{\text{object}} - \rho_{\text{water}})$ is 1000 kg/m^3 (see Appendix A for the density of water) and $g = 9.81$ m/s^2, so $v = 5.4$ km/s.

3.2 How systems get from here to there: entropy and kinetics

The first law of thermodynamics tells us nothing about the direction of a process. For this, we need the second law of thermodynamics. The elementary statement of this law is well known in the form "heat always passes from the hotter body to the colder body." Figure 3.2 illustrates this microscopically. Here, we show particles moving on a *potential surface*. This is a very important concept that can be thought about in simple mechanical terms. Imagine that the energy minimum at "a" in Fig. 3.2(a) corresponds to the bottom of a fruit bowl. Now imagine that the second energy minimum at "b" corresponds to a deep groove cut into the bowl at some distance from the center. Then the curve in Fig. 3.2(a) would represent the gravitational potential energy for a ball bearing rolled out from the center of the bowl.

If we imagine putting particles (like ball bearings) into a potential well (i.e., the minimum at "a" in Fig. 3.2(a)), they will not stay at the bottom of this potential because of the thermal agitation from the other (not shown in the figure) particles that are constantly bombarding them. In the case of the ball bearings,

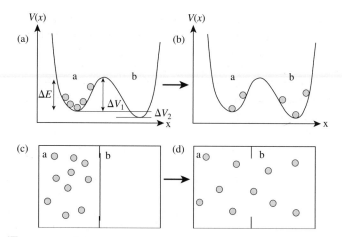

Fig. 3.2 Statistical mechanics in pictures. The top two panels (a and b) show particles moving on a potential surface under the influence of random thermal forces. Initially, all the particles are placed in energy minimum, a, but they move up and down the potential well because they have an average thermal energy on the order of ΔE. Fluctuations in energy that are greater than ΔV_1 will take the particles into the second potential well, b. Eventually, we expect to see slightly more particles (on average) in "b" because its energy minimum is lower than that of "a" (by an amount ΔV_2). Part of this redistribution is driven purely by probability. In the two boxes shown below (c and d), the potential of particles in "a" is the same as in "b," but all the particles are placed in "a" initially. After a partition is removed, we expect to find that the particles are equally distributed between "a" and "b," simply because this is the most probable arrangement. This is the origin of the concept of entropy—processes tend to move to the most probable distribution, the one that maximizes entropy. At very high temperature—so high that the small difference between the minima at "a" and "b" in (a) and (b) is insignificant, there will be an equal number of particles on each side and entropy dominates (so that the potentials in (a) and (b) in effect resemble the situation shown in (c) and (d) where the most probable state dominates). The effects owing to entropy are *dependent on temperature*.

we can imagine shaking the fruit bowl to simulate a finite temperature. So the particles will move up and down the potential surface over a range ΔE (which is directly proportional to the temperature of the system). If the potential well "a" is connected to a second well "b" by a potential barrier ΔV_1, thermal fluctuations will cause some particles to fall into this second well, and also eventually cause some of the particles in "b" to flip back to "a." Eventually, we expect to see the particles almost evenly divided between the two wells, with a few more (on average) in "b" than "a," because the bottom of "b" is at a lower potential than the bottom of "a," by an amount ΔV_2. A significant part of this redistribution occurs because, in the presence of thermal shuffling, the initial state (all the particles in "a") is a very unlikely arrangement. This tendency for a system to move to the most probable state is illustrated in the two lower panels of Fig. 3.2. Gas molecules, initially all placed in compartment "a," will rapidly distribute between compartments "a" and "b" after a dividing partition is removed. There is no potential energy difference between the two compartments, but the system will move toward the most probable distribution of particles. This probability-driven effect is given the name "*entropy*." Two systems that have come into equilibrium with each other are, by definition, at the same temperature.

Entropy can completely dominate the intrinsic energy difference between states (ΔV_2 in Fig. 3.2(a)). If the temperature is so high that the effects of this small energy difference between "a" and "b" in Fig. 3.2(a) can be ignored, then the equilibrium state will be the same as in Fig. 3.2(d)—equal numbers of particles on side "a" and side "b." *Thus entropy adds a temperature-dependent component to the energy differences that determine the final equilibrium state of a system.* For an isolated system (i.e., a system from which and into which no heat flows), the equilibrium state is always the state of maximum entropy. This is a statement of the second law of thermodynamics.

The description of the equilibrium states of the system (as controlled by ΔV_2 in the case illustrated in Fig. 3.2(a)) is given by the subject of *thermodynamics*. The microscopic description of thermodynamics is given by *statistical mechanics*. The way in which a system moves toward equilibrium is described by its *kinetics*. In the system shown in Fig. 3.2(a) and (b), the kinetics (rate of movement toward the equilibrium state) is controlled by the intermediate barrier, ΔV_1.

Thermodynamics describes the equilibrium state of a system in terms of potentials that take account of the most probable distribution by adding an extra state variable over and above the energy content of the system. The thermodynamic term for the degree of likeliness of a given state is *entropy* (entropy is proportional to the logarithm of the likeliness of a particular state). Thermodynamic potentials, modified by probabilities, are called *free energies*. They are so named because they can be used to calculate the maximum amount of work that might be extracted from a system. For example, the high-energy state (shown in Fig. 3.2(a)) cannot give up an amount of energy ΔV_2 per particle because it is extremely improbable that all particles end up in well "b."

There are alternative ways of stating the free energy of a system, each form useful for a particular type of problem (i.e., constant pressure, constant volume, etc.—see Appendix E). Thermodynamic calculations usually involve obtaining an expression for the free energy of a system, using the appropriate potential expressed in terms of macroscopic variables such as pressure and temperature. The equilibrium state of the system is found by minimizing the free energy with respect to the various macroscopic parameters. Statistical mechanics follows the same program, but microscopic variables are used.

The different types of free energy, and the types of statistical averaging they correspond to, can be confusing. The confusion arises because we use the more complicated free energies to hide some complexities. The thermodynamics of a completely isolated system is very simple: its energy stays constant and it moves toward the state of maximum entropy (the most probable state). It is only when we couple the system to a heat bath to allow heat to pass to maintain a constant temperature that we have to use a more complicated "free energy," called the Helmholtz free energy. An even more complicated free energy (the Gibbs free energy) is needed if particles can pass to and from the system. Once again, these "free energies" allow us to make calculations for the system we are interested in while ignoring the "outside world" with which it interacts.

Many systems, particularly chemical systems, are not in thermodynamic equilibrium, but are rather in a steady state. For example, very large concentrations of reactants, A and B, might be producing a product, AB:

$$A + B \rightarrow AB. \tag{3.3a}$$

The dissociation of the compound, AB,

$$AB \rightarrow A + B. \tag{3.3b}$$

is highly unlikely (because it is a very stable compound) so essentially all of A and all of B will be used up as the reaction proceeds to completion. However, if the change in the concentration of A and B is very slow, then the reaction will proceed at an essentially constant rate. Calculation of these rates is the subject of chemical kinetics.

In this chapter, we will begin by outlining a statistical mechanics view of thermodynamics, focusing on the simple problem of an *ideal gas* (a gas of noninteracting identical particles).

Special issues for nanoscale systems are (a) the significance of fluctuations because the number of particles in a system is much less than Avogadro's number (perhaps only tens of atoms in some cases) and (b) the importance of surfaces, so that thermodynamic quantities no longer scale with the number of atoms in a system because the energy associated with the surface may be a significant fraction of the total.

3.3   The classical probability distribution for noninteracting particles

The *Boltzmann distribution* is the most important function in classical statistical mechanics, but before discussing it in detail, let us remind ourselves of what we mean by a "probability distribution." Suppose that four pennies are tossed repeatedly. We know that getting four heads or four tails are the most unlikely outcomes. Getting two heads and two tails (in any order) is the most likely outcome, because there are more ways of doing this (HHTT, HTHT, TTHH, and so on). So, if we make a histogram of the number of occurrences of each combination on the vertical axis against the number of heads on the horizontal axis, it will be peaked in the middle (at two heads) and smallest at each end (no heads or four heads). Now suppose we repeated the process for a very large number of coins. The histogram will now look more like a continuous curve. This curve is the probability distribution, and the highest point on it will be the state of maximum entropy (50% heads in this case).

The Boltzmann distribution gives the probability of finding a system of noninteracting particles in a state of energy, E, at a temperature, T. It is the distribution of energies for a system in contact with a heat bath (see Appendix E and the description of the *canonical ensemble*). This is the distribution we are usually interested in. For example, we might want to know the thermal energies of nanoparticles in contact with air at 300 K, or the distribution of energies for biomolecules in a solution at 310 K (body temperature is 37°C).

It is fairly straightforward to derive the Boltzmann distribution. Suppose that some part of a system of noninteracting particles is in state *1* with a total energy $E(1)$, and that another part is in state *2* with energy $E(2)$. The probability that states *1* and *2* occur simultaneously is the product of the probabilities of each of the states occurring (because they do not interact and so are independent). Thus

$$P(1,2) = P(1)P(2). \tag{3.4}$$

The total energy of the system is the sum of the individual energies of the parts (because the particles do not interact so there is no energy change on bringing the systems together):

$$E(1,2) = E(1) + E(2). \tag{3.5}$$

The two requirements, 3.4 and 3.5, force us to conclude that the probability that the system has an energy $E, P(E)$, must be an exponential function of energy:

$$P(1) = Const. \exp(-\beta E(1)), \quad P(2) = Const. \exp(-\beta E(2));$$
$$P(1,2) = Const. \exp\left[-\beta(E(1) + E(2))\right], \tag{3.6}$$

where β is a constant to be determined. The two conditions (3.4 and 3.5) could also be satisfied with a "+" sign in front of β but that would lead us to the physically unrealistic conclusion that infinite energy is most probable. Since the energy of a system is also proportional to the absolute temperature, we must have

$$\beta \propto \frac{1}{T}$$

or

$$\beta = \frac{1}{k_B T}, \tag{3.7}$$

where k_B is the Boltzmann constant and its value has been determined by experiment to be

$$k_B = 1.381 \times 10^{-23} \text{ Joules/}^\circ\text{K} = 8.62 \times 10^{-5} \text{ eV/}^\circ\text{K}. \tag{3.8}$$

At room temperature ($\sim$300 K), $k_B T = 4.14 \times 10^{-21}$ J $= 25.86$ meV. The thermal energy of a mole of molecules, $N_A k_B T$ (where N_A, Avogadro's number, is 6.025×10^{23}), at 300 K is 593 calories (i.e., thermal energy at room temperature is 0.6 kcal per mole of particles).

We can now write the Boltzmann distribution as

$$P(r) = Const. \exp\left(-\frac{E_r}{k_B T}\right), \tag{3.9}$$

where $P(r)$ is the probability of finding the system in state r with energy, E_r, at absolute temperature, T.

The Boltzmann distribution (Equation 3.9) is plotted in units of $E/k_B T$ in Fig. 3.3. The effect of changing temperature is simulated by adjusting the value of $\beta (\equiv \frac{1}{k_B T})$ and the distribution has been normalized to have unit area by setting the constant in front of the exponential in 3.9 equal to β. For high temperatures (i.e., energies well below $k_B T$) almost all energy states are sampled. At low temperatures, the energy distribution is sharply peaked at low energy.

Exercise 3.3: An oxygen molecule has a mass of 32 amu (1 amu $= 1.661 \times 10^{-27}$ kg). Assume that the edge of the atmosphere is defined by the point where the density of the atmosphere falls to $1/e$ of its value at the earth's surface. The potential energy at a height h owing to gravity is *mgh*. For simplicity, take the temperature and g to have the same value throughout the atmosphere as at the surface of the earth and estimate the height of the atmosphere.

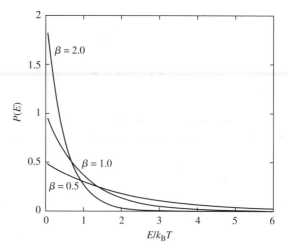

Fig. 3.3 Plots of the Boltzmann distribution (normalized to have unit area under each curve) as a function of the energy in units of k_BT. At low temperature ($\beta = 2$), the distribution peaks sharply at low energy. At high temperature ($\beta = 0.5$), all energies become almost equally likely.

Answer: In thermal equilibrium, the probability of finding an air molecule at a given height will be given by the Boltzmann distribution. Since the density of the atmosphere is proportional to this probability, and the density at the surface of the earth ($h = 0$) is ρ_0, the density at some height h is given by $\rho = \rho_0 \exp\left(-\frac{mgh}{k_BT}\right)$. The density will fall to ρ_0/e at a height h_0 given by $h_0 = k_BT/mg$ or $4.14 \times 10^{-21}\text{J}/(32 \times 1.661 \times 10^{-27} \times 9.81 \text{ N})$, i.e., 7.9 km (close to the actual effective height of the atmosphere).

3.4   Entropy and the Boltzmann distribution

It is instructive to approach the counting of states in a quantum mechanically correct way by assuming that they are discrete eigenstates with eigenenergies E_r. Doing so brings out the connection between the Boltzmann distribution and the entropy of a system explicitly. The constant of proportionality for a system having eigenstates, labeled r, each with an energy, E_r, is found from evaluating the single particle *partition function*, Z. Thermodynamic quantities can be derived from microscopic models using the partition function. It is the reciprocal of the normalization constant in Equation 3.9, and it is

$$Z = \sum_r \exp -\frac{E_r}{k_BT}. \tag{3.10}$$

A sum (instead of an integral) is written here to imply that we are counting distinct eigenstates of the system. The normalized ($\sum_r P(r) = 1$) version of the Boltzmann distribution is

$$P(r) = \frac{1}{Z} \exp\left(-\frac{E_r}{k_BT}\right) \tag{3.11}$$

and it is independent of the total number of particles. It is useful to reframe this in terms of the probability of finding the system with a particular energy (rather than in some state). It may be that several different states have the same energy (such states are said to be *degenerate*). The number of discrete states, n_r, with a given energy, E_r, is called the *degeneracy*, $g(r)$, of the state r. Thus, for a given $E_r, p(E_r) = g(r)p(r)$ or

$$P(E_r) = \frac{g(r)}{Z} \exp\left(-\frac{E_r}{k_B T}\right). \qquad (3.12)$$

This is another form of the Boltzmann distribution which shows explicitly how a large number of states in some energy range enhances the probability in that range.

The degeneracy of a state leads naturally to a definition of entropy. A set of highly degenerate energy states are more likely to be occupied, at a given temperature, than a nondegenerate energy level (recall the example of tossing four coins where a state of 50% heads was most likely).

Formally, the entropy, s, of a system in state r is proportional to the number of ways that state r can occur. This is called the statistical weight of state r, $\Omega(r)$ (or, in many textbooks, the number of microstates in the macrostate r).

Boltzmann defined the entropy, s, of a system in state r as

$$S(r) = k_B \ln \Omega(r). \qquad (3.13a)$$

The logarithm is introduced so that when two systems are combined, the statistical weight of the whole system is the product of the statistical weight of each individual system while the entropy is the sum of the two entropies. For example, if there are 10 ways of arranging system 1 and 10 ways of arranging system 2, then there must be 100 ways of arranging the two systems (i.e., the first arrangement of system 1 taken with any of the 10 arrangements of system 2 and so on). The entropy is simply the sum of the entropies of the individual systems because of the logarithm, exactly as probability and energy are related by the Boltzmann distribution. Boltzmann did this to produce a thermodynamic potential (entropy) that is additive (like energy is). Defined this way, entropy behaves just like other thermodynamic variables. Second, it is consistent with the idea that entropy is maximized in equilibrium, the state with the most microstates per macrostate. The introduction of the Boltzmann constant in front of the definition gives the product of entropy and temperature units of energy. Entropic contributions to the energy of a system always appear as the product of entropy and temperature (see Exercise 3.4).

It can be shown (Appendix F) that the statistical weight and probability of the rth state are related in a way that yields another definition of entropy in terms of the probability of the rth state, p_r, as

$$S = -k_B \sum_r p(r) \ln [p(r)]. \qquad (3.13b)$$

This definition is valid in terms of any distribution, $p(r)$, as shown in Appendix F. For an ensemble of systems at constant temperature $p(r)$ follows the Boltzmann distribution.

Exercise 3.4: (A simple model of rubber): Consider a very long chain of n links each one of which has a length a and can randomly lie only in the $+x$ or $-x$ direction. Find the tension on the chain as a function of its extended length,

L, if the chain is isolated (constant energy) given the expression for tension, $\tau = -T \left(\frac{\partial S}{\partial L}\right)_E$. (This expression is derived from force = energy change per unit length change.) Assume that the energy of the links is the same whether they point forward or backward. Consult Appendix F for help with the math. As an aid to visualization, the illustration below shows the two microstates for a chain of 4 links with a macrostate 2 links long.

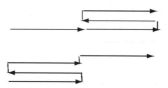

Answer: If the chain takes n_+ steps forward and n_- steps backward, $L = (n_+ - n_-)a$ where $n = n_+ + n_-$ so $n_\pm = (na \pm L)\frac{1}{2a}$. The statistical weight of the macrostate with n_+ steps forward and n_- steps backward is $\Omega = \frac{n!}{n_+! \times n_-!}$. As in Appendix F, we use $S = k_B \ln \Omega$ and Stirling's approximation $\ln n! = n \ln n - n$

$$S = k_B n \left[\ln 2 - \frac{1}{2}\left(1 + \frac{L}{na}\right) \ln\left(1 + \frac{L}{na}\right) - \frac{1}{2}\left(1 - \frac{L}{na}\right) \ln\left(1 - \frac{L}{na}\right) \right].$$

From which

$$\tau = -T \left(\frac{\partial S}{\partial L}\right)_E = \frac{k_B T}{2a} \left[\ln\left(1 + \frac{L}{na}\right) - \ln\left(1 - \frac{L}{na}\right) \right],$$

where we used the fact that $\frac{d \ln x}{dx} = \frac{1}{x}$, the chain rule to differentiate the products, and changed variables to L/na to simplify the derivatives. Note that in the limit that $L \ll na$ this result reduces to Hooke's law: $\tau \approx \frac{k_B T}{na^2} L$. This is an example of an *entropic* force. The mean length of the chain under zero tension is zero, and the work has to be done to stretch it away from this most probable configuration. The higher the temperature, the greater the energy cost of putting the system into an improbable configuration, even though there is no intrinsic energy difference between the $+$ and $-$ configurations. This illustrates the quantitative temperature dependence of the entropic contribution to the "energy" of a system, as discussed qualitatively in Section 3.2.

3.5 An example of the Boltzmann distribution: ions in a solution near an electrode

A charged surface in contact with a salt solution accumulates ions that neutralize its charge. In the absence of thermal fluctuations, ions (of the opposite sign to the surface charge) would simply adsorb onto the surface at whatever density was required to produce a neutral system. Thermal fluctuations spread this "accumulation layer" out. This process is illustrated in Fig. 3.4. Far from the electrode surface, the concentration of positive and negative ions (e.g., Na$^+$, Cl$^-$) is equal to the bulk concentration, C_0. At the electrode surface, a net charge accumulates that is equal and opposite to the electrode charge, and it

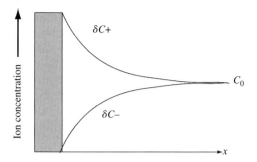

Fig. 3.4 Distribution of ions in electrolyte near a charged surface. In the bulk, negative and positive ions have equal concentrations, C_0, but near the surface an excess of one type of ion accumulates to screen the surface charge. This figure is based on a negative surface charge screened by an accumulation of positive ions.

consists of an increase in the concentration of one type of ion (positive ions if the surface charge is negative as in Fig. 3.4) and a decrease in the concentration of the other (negative ions in this case). This accumulation of net charge, $\delta Q(x)$, gives rise to an electric field, with a corresponding electric potential, $\phi(x)$, given by Poisson's equation (Gauss's Law re-expressed in terms of potential, and written here in just the x direction):

$$\frac{d^2\varphi}{dx^2} = -\frac{\delta Q(x)}{\varepsilon \varepsilon_0}. \tag{3.14}$$

The electrostatic energy of an ion in this potential is the product of its charge, Ze (for a Z valent ion), with the potential, so the excess charge is distributed according to the Boltzmann distribution (Equation 3.9):

$$\delta Q(x) = 2ZeC_0 \left(\exp\left(-\frac{Ze\varphi(x)}{k_B T} \right) - 1 \right), \tag{3.15}$$

where the factor -1 ensures that the excess charge falls to zero in the bulk (where $\phi(x)$ is zero) and the factor 2 takes account of the contribution to the net charge from the fall in concentration of one type of ion and the rise of the other (Fig. 3.4). Equations 3.14 and 3.15 combine to give a nonlinear differential equation for $\phi(x)$ called the Poisson–Boltzmann equation. This cannot be solved analytically. However, in the limit of very small electrostatic energies, i.e., small surface charge on the electrode, $Ze\varphi \ll k_B T$, the exponential can expanded in a Taylor series, which, taken to first order in φ gives, as an approximation to 3.15:

$$\frac{d^2\varphi(x)}{dx^2} = -\frac{2C_0 Z^2 e^2}{\varepsilon \varepsilon_0 k_B T} \varphi(x). \tag{3.16}$$

This is solved with an exponential, $\varphi(x) = A\exp -\frac{x}{\ell_D}$, where ℓ_D (the Debye length) is found by substituting this exponential into 3.16 to get

$$\ell_D = \sqrt{\frac{\varepsilon \varepsilon_0 k_B T}{2C_0 Z^2 e^2}}. \tag{3.17}$$

Thus, the electric field owing to the surface charge is significant over a distance that varies as $1/\sqrt{C_0}$. In 1 M monovalent salt (e.g., NaCl) at 300 K, this distance is ∼0.3 nm. It follows from this that interactions caused by surface charges (on, for example, proteins or DNA) require close encounters (a few Å) for the surface electric field to be felt given the strong screening owing to ions (salt concentrations are typically tenth molar in biological environments).

Exercise 3.5: Take the dielectric constant of water to be 80, use values for the other constants from Appendix A and calculate the Debye screening lengths in 1 M NaCl, 0.01 M NaCl, and 0.01 M MgCl$_2$. (A 1 molar solution contains Avogadro's number of each ion and there are 1000 liters in a cubic meter.)

Answer: $\ell_D = \sqrt{\frac{\varepsilon\varepsilon_0 k_B T}{2e^2}} \times \sqrt{\frac{1}{C_0 Z^2}} = \frac{0.3}{Z}\sqrt{\frac{1}{C_0}}$ nm. This gives $\ell_D = 0.3$ nm for 1 M NaCl, 3 nm for 0.01 M NaCl, and 1.5 nm for 0.01 M MgCl$_2$.

3.6   The equipartition theorem

We are now going to turn to a very important theorem for classical particles (that can have any value of energy as opposed to the discrete levels of their quantum counterparts). We will illustrate it with a simple oscillator consisting of a mass m on a spring of spring constant κ (like Fig. D.1 in Appendix D, but without the damping). The energy of the spring consists of the kinetic energy of the moving mass plus the energy stored in the spring:

$$E = \frac{1}{2}m\dot{x}^2 + \frac{1}{2}\kappa x^2, \tag{3.18}$$

where x is the displacement away from the equilibrium position and $\dot{x}$ is the velocity, $\frac{dx}{dt}$. Since a classical oscillator can have any value of energy, the sum in Equation 3.10 can be replaced with an integral and the Boltzmann probability distribution (Equation 3.11) becomes

$$P(E) = \frac{\exp\left(-E/k_B T\right)}{\int_0^\infty \exp\left(-E/k_B T\right) dE}. \tag{3.19}$$

Taking just the potential energy term and carrying out integrals over all values of x, we get the following expression for the thermal average of the potential energy (we will take the limits to be $\pm\infty$, though the integrals will only be nonzero over some much smaller range in practice).

$$\left\langle \frac{1}{2}\kappa x^2 \right\rangle = \int_{-\infty}^{\infty} \frac{1}{2}\kappa x^2 P(E(x)) dx = \frac{\int_{-\infty}^{\infty} \frac{1}{2}\kappa x^2 \exp\left(-\frac{\kappa x^2}{2k_B T}\right) dx}{\int_{-\infty}^{\infty} \exp\left(-\frac{\kappa x^2}{2k_B T}\right) dx}. \tag{3.20}$$

The notation $\langle x \rangle$ means "take the thermal average of x." The top integral becomes $k_B T \sqrt{\frac{2k_B T}{\kappa}} \int_{-\infty}^{\infty} t^2 \exp -t^2 dt$ with the substitution $t = \sqrt{\frac{\kappa}{2k_B T}}x$. With the same substitution, the bottom integral becomes $\sqrt{\frac{2k_B T}{\kappa}} \int_{-\infty}^{\infty} \exp -t^2 dt$. The two integrals are standard (see Appendix A), one yielding $\frac{\sqrt{\pi}}{2}$ and the other $\sqrt{\pi}$

so we obtain finally

$$\left\langle \frac{1}{2}\kappa x^2 \right\rangle = \frac{1}{2}k_\mathrm{B}T. \qquad (3.21a)$$

We could have carried out exactly the same analysis for the kinetic energy term in Equation 3.18 using $\dot{x}$ as the variable instead of x and the result would be

$$\left\langle \frac{1}{2}m\dot{x}^2 \right\rangle = \frac{1}{2}k_\mathrm{B}T. \qquad (3.21b)$$

Many expressions for the energy of a system take the form $\frac{1}{2}Ay^2$, where y is some variable. For example, the rotational energy of a body of moment of inertia I is $\frac{1}{2}I\dot{\theta}^2$, where $\dot{\theta}$ is the angular velocity of the rotating body. The equipartition theorem says that the average thermal energy associated with each degree of freedom (each variable contributing energy in the form $\frac{1}{2}Ay^2$) is $\frac{1}{2}k_\mathrm{B}T$. This is a very powerful general result for all types of motion that are low enough in energy that we do not have to worry about quantum effects (which we will discuss later in this chapter).

The theorem assumes that all degrees of freedom are in equilibrium with the heat bath, and, for a complex system like a protein bathed in water, this might be quite valid. But a rather interesting paradox arises when a strictly linear mechanical system is considered. Degrees of freedom take the form $\frac{1}{2}Ay^2$ precisely because they are independent of one another. This is a property of linear systems (like systems of springs obeying Hooke's law where $F = -\kappa x$ leads to $E = \frac{1}{2}\kappa x^2$). However, it takes coupling between the degrees of freedom to "spread" the thermal energy out evenly and this requires something other than a linear response. The question of how a system "thermalizes" (i.e., comes into equilibrium when the temperature of the heat bath jumps) is therefore a very challenging problem, the investigation of which has lead to some remarkably interesting physics.[1]

Exercise 3.6: (The Dulong and Petit law): An ideal gas consists of point particles with no internal degrees of freedom. What is its molar specific heat at temperatures for which quantum effects can be ignored?

Answer: There are no interactions between the particles, so there is no potential energy. The kinetic energy is $E = \frac{1}{2}m(\dot{x}^2 + \dot{y}^2 + \dot{z}^2)$ so each particle has $\frac{3}{2}k_\mathrm{B}T$ of thermal energy, giving an energy per mole of $\frac{3}{2}N_A k_\mathrm{B}T$ (where N_A is Avogadro's number). The specific heat (at constant volume) is given by $C_V = \frac{\partial E}{\partial T} = \frac{3}{2}N_A k_\mathrm{B}$.

3.7   The partition function

The partition function (Equation 3.10) is a quantity with uses that go well beyond normalizing the Boltzmann distribution. It contains information about the number of states per unit energy, weighted by the Boltzmann factor. *Once a partition function is calculated, all other equilibrium thermodynamic properties follow from minimizing the free energy of the system expressed in terms of the partition function.*

For a system at constant temperature (i.e., in contact with a heat bath—Appendix E), the appropriate free energy is the *Helmholtz free energy*, F,

$$F(T, V, N) = -k_B T \ln [Z(T, V, N)], \tag{3.22}$$

where $Z(T, V, N)$ is the partition function defined in Equation 3.10. If we substitute the Boltzmann distribution (3.11) into the definition of entropy (3.13), we obtain

$$S(T, V, N) = k_B \ln Z + \frac{\bar{E}}{T}. \tag{3.23}$$

To get this result, we used the fact that the sum of the Boltzmann distribution over all r is a constant and that the sum of $E_r P(r)$ is $\bar{E}$, the average energy of the system. Equations 3.22 and 3.23 taken together yield

$$F = \bar{E} - TS, \tag{3.24}$$

where F is the Helmholtz free energy, stated here in terms of the macroscopic variables, $\bar{E}$ the average energy of the system, T its temperature, and S its entropy. The equilibrium state of a system in contact with a heat bath is found by minimizing F.

In contrast, for a system in complete isolation (e.g., Fig. 3.2(c) and (d)) with no work or heat input, the equilibrium state is found by maximizing the entropy (note the minus sign in front of TS in Equation 3.24). In this case, the entropy is the appropriate thermodynamic potential, as assumed implicitly when we discussed what happens when the dividing partition is removed from the container shown in Fig. 3.2(c) and (d).

The Helmholtz free energy, as expressed in Equation 3.24, now tells us explicitly how the trade-off between internal energy and entropy works (cf. the discussion in Section 3.2). At low temperatures ($k_B T < \bar{E}$), the internal energy dominates. At high temperatures, the term in TS dominates, and the system goes to its most probable state, independent of the internal energy cost (which can be ignored if $TS >> \bar{E}$).

We need to use yet another potential when considering chemical processes, which usually take place at constant pressure. In the most general case, we want to allow for particles being exchanged between the system and the rest of the world. The appropriate thermodynamic potential to minimize (Appendix E) is the *Gibbs free energy*,

$$G = \bar{E} + PV - TS + \mu N, \tag{3.25}$$

where the quantity

$$\bar{E} + PV = H \tag{3.26}$$

is called the *enthalpy* of the system. The enthalpy includes the work done in changing the volume of the system at constant pressure (which is how chemical reactions are normally carried out). Since particle number is no longer held constant, a new quantity μ (the chemical potential) is introduced to describe the change in free energy as particles are added to or taken from the system (see

Appendix G). As shown below, μ contains contributions both from the change in energy of the system on adding a particle as well as the change in entropy caused by the change in the number of particles per unit volume. μ replaces N in the partition function (which is then called the *grand partition function*, $\mathbf{Z}(T, V, \mu)$). A derivation of the grand partition function is given in Appendix G. The Gibbs potential is derived from the grand partition function, $\mathbf{Z}$, just as the Helmholtz free energy was derived from the partition function, Z (Equation 3.22) according to

$$G = -k_B T \ln [\mathbf{Z}(T, V, \mu)].$$

(3.27)

This is the link between microscopic models and thermodynamic quantities for systems at constant pressure where particle exchange can occur. It is a very useful expression and we will make good use of it.

3.8 The partition function for an ideal gas

Here, we will illustrate the use of a partition function with a discussion of an ideal gas of noninteracting point particles. Simple though this model is, the calculation of the appropriate partition function is quite technical, and some of the details have been squirreled away in Appendices H and I. There are two parts of the calculation that are nonobvious. The first has to do with the requirement that we work out a partition function for a single point particle. The second has to do with how we count up the possible arrangements of the particles. We will treat the case of the low-density classical limit of a quantum system for simplicity. Here our classical particles are identical (this affects how we count states)—only for a true quantum mechanical system do they become fundamentally indistinguishable. So the quantum limit is used here only for convenience.

Turning to the first point—the partition function for a point particle—we might ask why we need one at all? One might think that for one particle, the sum in Equation 3.10 is trivial. The answer is *no* and the origin of the complication lies in quantum mechanics.

Recall that in Section 2.15.2 we worked out the density of states for a free particle. Even if the states are very close in energy, the number of states per unit energy (or the number of states per unit wave vector) are *not* constants, but depend on E (or k since $E = \frac{\hbar^2 k^2}{2m}$). If quantum levels are closely spaced compared to $k_B T$, then the equipartition theorem tells us that the average thermal energy (which is only kinetic energy for a noninteracting particle) is $\frac{1}{2}k_B T$ for each of the motions along x, y, and z, with corresponding quantum mechanical wave vectors k_x, k_y, and k_z. Setting

$$\frac{1}{2}k_B T = \frac{\hbar^2 k_x^2}{2m}$$

(3.28)

for the x direction, for example, yields the following result for the de Broglie wavelength of the probability amplitude:

$$\lambda = \frac{2\pi}{k_{x,y,z}} = 2\pi \sqrt{\frac{\hbar^2}{mkT}}.$$

(3.29)

Thus, one particle occupies a "quantum volume" of about λ^3 or $(2\pi)^3 \left(\frac{\hbar^2}{mk_B T}\right)^{3/2}$. This estimate is out by a factor $\sqrt{2\pi}$. Taking proper account of the distribution of energy (Appendix H) yields

$$v_Q = \left(\frac{2\pi \hbar^2}{mk_B T}\right)^{3/2}. \tag{3.30}$$

Exercise 3.7: Calculate the quantum volume for an electron at 300 K. What concentration of electrons would fill up all available quantum states? How does this compare with the concentration of electrons in sodium metal, assuming that each sodium atom contributes one "free" electron to the metal and the interatomic spacing is $\sim$4 Å?

Answer: $k_B T = 4.14 \times 10^{-21}$ J at 300 K, with $m_e = 9.110 \times 10^{-31}$ kg and $\hbar = 1.055 \times 10^{-34}$ Js, so we get 7.9×10^{-26} m^3. Two electrons can occupy each state (spin up and spin down), so a concentration of electrons of about $(1.3 \times 10^{25})/2\,\mathrm{m}^{-3} = 6.3 \times 10^{24}\,\mathrm{m}^{-3}$ would fill up all available quantum states. The density of electrons in sodium is about $(4 \times 10^{-10})^{-3}$ or $1.6 \times 10^{28}\,\mathrm{m}^{-3}$ (a more accurate value that takes account of crystal structure is $2.6 \times 10^{28}\,\mathrm{m}^{-3}$). This is far above the critical quantum density. Electrons in sodium are far from being a gas of noninteracting particles.

If the system consists of fermions (like electrons in the example just given), once all the available quantum states at a given energy are filled *no further electrons can be added at that energy*. This has the remarkable consequence that for systems more dense than the quantum concentration (i.e., $1/v_Q$), the additional electrons are forced to go into states that are higher in energy than $k_B T$. This is what happens in electron-rich materials like metals (cf. Exercise 3.7). Even at room temperature, most of the electrons are in a strange quantum state called a *degenerate Fermi gas*.

For systems less dense than the quantum concentration, we can now see that many states may be available to each particle. Specifically, if the density corresponds to one particle in a volume V, the number of available states is V/v_Q. We assumed that the energy spacing between allowed levels was much less than $k_B T$, so each of the exponentials in the partition function is unity and the partition function for a single particle is

$$Z(T, V, 1) = \frac{V}{v_Q} = \left(\frac{mk_B T}{2\pi \hbar^2}\right)^{3/2} V, \tag{3.31}$$

where V is the volume per particle (i.e., ρ^{-1} where ρ is the number of particles per m^3). Exotic though the quantum arguments may appear to be, they lead to a nontrivial dependence of the partition function for a single particle on density and temperature, a dependence that is observed experimentally.

Now that we have the partition function for a single particle, we might try to wirte down the partition function for N noninteracting particles as

$$Z(T, V, N) = \left[\sum_r \exp(-\beta E_r)\right]^N = [Z(T, V, 1)]^N, \tag{3.32}$$

but we would be wrong.

The argument for 3.32 is that, for each of the states counted in $Z(T, V, 1)$, the first particle could be in any one of Z (thermally weighted) states, the second particle could also be in any one of Z (thermally weighted) states, and so on. So the total number of available states would appear to be the product of the single particle partition functions. But in fact this is wrong. This is a good example of the traps hidden in forgetting that sums written out using the $\sum$ notation are shorthand for something more complicated. The line of reasoning above has just become the victim of such a trap. The way forward is write out a few terms and see what happens. So, doing this for just two particles with states r and s, respectively, yields

$$\left[\sum_r \exp(-\beta E_r)\right]\left[\sum_s \exp(-\beta E_s)\right]$$

$$= \sum_r \exp(-2\beta E_r) + \sum_r \sum_{\substack{s \\ (r \neq s)}} \exp\left[-\beta(E_r + E_s)\right]. \qquad (3.33)$$

The first term on the right-hand side of 3.25 is for the two particles in the same state. The second term is for the particles in different states, *but each pair of different states is counted twice if the particles are identical* (e.g., $r = 1, s = 2$ is the same state as $r = 2, s = 1$).

For a low-density system that is far from being a degenerate quantum gas (cf. Exercise 3.7), we show in Appendix I that the correct result is

$$Z(T, V, N) = \frac{1}{N!}\left[\sum_r \exp(-\beta E_r)\right]^N = \frac{1}{N!}\left[Z(T, V, 1)\right]^N. \qquad (3.34)$$

The factor $\frac{1}{N!}$ deals with the overcounting problem.

3.9 Free energy, pressure, and entropy of an ideal gas from the partition function

This section illustrates how thermodynamic results follow from statistical mechanics, once the partition function is calculated. We use the simplest example of the ideal gas for which the partition function is given by Equation 3.34. Some of the results, like the ideal gas law, will be utterly familiar, while others, like the entropy of an ideal gas, will probably prove surprising at first glance.

For a gas at constant temperature, the Helmholtz free energy (Equation 3.22) is the appropriate potential, so putting the partition function of an ideal gas (Equation 3.34) into this expression gives

$$F = -k_B T \ln Z = -k_B T \ln\left[Z(T, V, 1)^N\right] + k_B T \ln N! \qquad (3.35)$$

The Stirling approximation for $\ln N!$ is $N \ln N - N$ (for large N), and using Equation 3.31 for the single particle partition function

$$F = -Nk_B T \ln\left(\left[\frac{mk_B T}{2\pi\hbar^2}\right]^{3/2} V\right) + k_B T(N \ln N - N). \qquad (3.36)$$

The thermodynamics of the ideal gas follow from 3.36. For example, the pressure is the change in free energy with volume at constant temperature

$$p = -\left(\frac{\partial F}{\partial V}\right)_T. \tag{3.37}$$

Applied to Equation 3.36 this yields the ideal gas law:

$$p = \frac{Nk_BT}{V}. \tag{3.38}$$

In arriving at 3.38, we used the fact that $\ln(ab) = \ln(a) + \ln(b)$ and $\frac{d}{dx}\ln(x) = \frac{1}{x}$.

The entropy is obtained from the free energy by differentiating $F = \bar{E} - TS$ (Equation 3.24):

$$S = -\left(\frac{\partial F}{\partial T}\right)_V. \tag{3.39}$$

Carrying out this operation on Equation 3.36 yields

$$\frac{\partial F}{\partial T} = -Nk_BT\frac{\partial}{\partial T}\ln\left(\left[\frac{mk_BT}{2\pi\hbar^2}\right]^{3/2}V\right)$$

$$- Nk_B\ln\left(\left[\frac{mk_BT}{2\pi\hbar^2}\right]^{3/2}V\right) + k_B(N\ln N - N). \tag{3.40}$$

Which simplifies to

$$S = Nk_B\left[\ln\left(\frac{n_Q}{n}\right) + \frac{5}{2}\right]. \tag{3.41}$$

(For the details of the calculation, do Problem 24). In 3.41, N/V has been replaced by the density (or concentration) of the gas, $n = \frac{N}{V}$, and $n_Q = \frac{1}{v_Q}$ is the quantum concentration, the reciprocal of the quantum volume given by Equation 3.30,

$$n_Q = \left(\frac{mk_BT}{2\pi\hbar^2}\right)^{3/2}. \tag{3.42}$$

Equation 3.41 is a very important result. It says that the entropy of an ideal gas scales with N as it should (entropy is an extensive thermodynamic property). But it also depends on the *concentration* or *density*, i.e., the number of particles per unit volume. Because this is an ideal gas, a nonzero chemical potential *cannot* arise from interactions between the particles. Rather it reflects the change in *entropy* of the system. Adding more particles to a fixed volume decreases the ways in which particles can be arranged for a given macrostate. This decreases the entropy and increases the free energy of the state.

This result has important implications for chemistry. One might think that the driving force for a chemical reaction would depend completely on the difference in enthalpy between reactants and products (as determined by the change in heat energies as the reactants combine to form products). But this is not the case. At the finite temperatures required to carry out chemical reactions, the entropies of the reactants and products matter. Equation 3.41 can be rewritten in terms of concentration, C, as

$$S \propto - \ln C, \tag{3.43}$$

so that the Gibbs free energy of the products of a reaction ($G \propto H - TS$) goes up as their concentration increases. If it increases so far that the free energy of the products equals the free energy of the reactants, the reaction will stop, no matter how much electronic energy might be gained by continuing the reaction. Concentrations play a key role in dictating the extent and direction of chemical reactions. Specifically, for a chemical reaction to occur, the change in Gibbs free energy for the reaction must be negative

$$\Delta G = \Delta H - T\Delta S < 0. \tag{3.44}$$

The quantities ΔH and ΔS can be calculated from the differences between the *standard molar enthalpies*, H^0, and the *standard molar entropies*, S^0, for the reactants and products at molar concentrations and standard conditions of temperature and pressure (and, for ideal gasses, Equation 3.41 gives good values for S^0). For simplicity, let us assume that one mole of reactants, of free energy, G_R^0, combine to form one mole of products, of free energy, G_P^0, then the standard molar free energy for the reaction is $\Delta G^0 = G_R^0 - G_P^0$, which is negative if the products have a lower free energy than the reactants. The superscript means "as determined at a standard concentration" (e.g., one mole/liter). But reactions are almost never carried out at molar concentrations (or the appropriate multiples, depending on the stoichiometry of the reaction). If the actual concentration of the reactants is [R] times the standard concentration and the actual concentration of product is [P] times the standard concentration, then a free energy $RT \ln[R]$ (R is the gas constant, $R = N_A k_B$) is added to the reactant side of the equation and a free energy $RT \ln[P]$ is added to the product side. Defining the ratio $Q = [P]/[R]$ (Q is called the reaction quotient; see Section 3.17), the ideal gas correction to the entropy gives the modified free energy difference between reactants and products as

$$\Delta G = \Delta G^0 + RT \ln Q. \tag{3.45}$$

This correction is familiar to all who have studied thermochemistry. It is remarkable to realize that we derived it from the quantum mechanics of an ideal gas! The relative size of the concentration effect depends on ΔG_0. For one molar concentration, the entropy term is on the order of 2.5 kJ/mole. For a strong covalent reaction ($\Delta G_0 \approx 100$ kJ), the concentration makes little difference to the reaction. But for a weak driving free energy, a sufficiently high concentration of product could stop the reaction altogether.

Exercise 3.8: Calculate the Gibbs free energy difference for the reaction

$$N_2 + 3H_2 \rightleftharpoons 2NH_3$$

in standard conditions (25°C, 1 atmosphere) for a reaction mixture that consists of 1 atmosphere N_2, 3 atmospheres H_2, and 0.5 atmosphere NH_3. The standard molar enthalpy for the formation of NH_3 is -46.19 kJ/mole (standard molar enthalpies for the formation of elements are defined to be zero). The standard molar entropy of NH_3 is 192.5 J/mole-K, of N_2 is 191.5 J/mole-K, and of H_2 is 130.6 J/mole-K. Note that "atmospheres" here is a dimensionless unit of relative pressure because pressures are expressed relative to the pressure in standard conditions ($P = 1$ atmosphere).

Answer: The standard free energy difference is the difference in free energy between the reactants and products. Thus, for the enthalpy we have

$$\Delta H^0 = 2 \text{ moles} \times (-46.19 \text{ kJ/mole}) - [(1 \text{ mole} \times (0) + 3 \text{ moles} \times (0)]$$
$$= -92.38 \text{ kJ}.$$

For entropy, we have

$$\Delta S^0 = 2 \text{ moles} \times (192.5 \text{ J/mole} - K) - [(1 \text{ mole} \times (191.5 \text{ J/mole} - K)$$
$$+ 3 \text{ moles} \times (130.6 \text{ J/mole} - K)] = -198.3 \text{ J/K}.$$

(So $-T\Delta S^0$ at 298 K is $+59.1$ kJ—this reflects the entropy cost of converting N_2 gas and H_2 gas into its less-free compound NH_3.)
Thus, $\Delta G^0 = \Delta H^0 - T\Delta S^0$ yields $\Delta G^0 = -33.28$ kJ. To calculate the correction for the concentrations given (concentration being directly proportional to pressure), note that we add a term in $RT \ln C$ for the concentration of the products for each mole participating in the reaction, and subtract a term in $RT \ln C$ for the concentration of the reactants for each mole participating in the reaction, i.e.,

$$2RT \ln C(NH_3) - RT \ln C(N_2) - 3RT \ln C(H_2)$$

so

$$\Delta G = \Delta G^0 + RT \ln \frac{C(NH_3)^2}{C(N_2)C(H_2)^3} = \Delta G^0 + RT \ln \frac{(0.5)^2}{(1.0)(3.0)^3}$$
$$= \Delta G^0 + RT \ln(9.3 \times 10^{-3}).$$

The last term is -11.6 kJ, so $\Delta G = -44.9$ kJ. The free energy for the production of ammonia is decreased because the product is present in a much smaller amount than would be the case in equilibrium conditions.

3.10   Quantum gasses

We saw in Exercise 3.7 that systems such as metals are degenerate quantum gasses; therefore, the distribution function for this state is needed to understand electrons in metals. As we will see in Chapter 7, metals can be modeled remarkably well as an ideal gas of noninteracting electrons.

The state of an ideal quantum gas is completely specified by the number of particles, n_r, in each of the single particle states r with energy, ε_r. For the classical gas, we assumed that we could neglect the cases where more than one particle was in a given state. In the case of the (high-density) quantum gas, we cannot do this. We have to consider the case where more than one particle is in the same state. For bosons (see Section 2.4), any occupation number is allowed and

$$n_r = 0, 1, 2, 3, \ldots \tag{3.46}$$

While, for fermions

$$n_r = 0, 1, \tag{3.47}$$

reflecting the fact that no two fermions can occupy the same single particle state (n can go up to 2 when spins are taken into account). For both bosons and fermions, the normalization condition for a gas of N particles

$$\sum_r n_r = N. \tag{3.48}$$

The partition function is

$$Z(T, V, N) = \sum_{n_1, n_2 \ldots} \exp\left\{-\beta \sum_r n_r \varepsilon_r\right\} = \sum_r \exp\left\{-\beta\left((n_1 \varepsilon_1 + n_2 \varepsilon_2 + \cdots)\right)\right\}, \tag{3.49}$$

where $\sum_{n_1, n_2 \ldots}$ means the sum over all sets of occupation numbers satisfying 3.48 and either 3.46 or 3.47.

Evaluating the Helmholtz free energy from this set of equations is not straightforward. It is easier to work with the more general *Gibbs distribution*, the generalization of the Boltzmann distribution that includes variable particle number (Appendix G). The Gibbs distribution is

$$P_{Nr} = \frac{\exp\left[\beta\left(\mu N - E_{Nr}\right)\right]}{\mathbf{Z}}, \tag{3.50}$$

where the grand partition function $\mathbf{Z}$ is

$$\mathbf{Z}(T, V, \mu) = \sum_{Nr} \exp\left[\beta\left(\mu N - E_{Nr}\right)\right], \tag{3.51}$$

where μ is the chemical potential (Section 3.7), the energy change of the system on addition of a particle. The state of an ideal gas (point particles with no internal excitations) is fully specified by a set of occupation numbers, $n_1, n_2, \ldots, n_i$, of single particle states, each with an eigenenergy, $\varepsilon_1, \varepsilon_2, \ldots, \varepsilon_i$. The sum over N in 3.51 is replaced with a sum over all allowed values of occupation numbers,

$n_1, n_2, \ldots, n_i$:

$$\mathbf{Z}(T, V, \mu) = \sum_{n_1, n_2, \ldots} \exp[\beta(\mu(n_1 + n_2 + \cdots + n_i + \cdots)$$

$$- (n_1\varepsilon_1 + n_2\varepsilon_2 + \cdots + n_i\varepsilon_i + \cdots))], \qquad (3.52)$$

so that the probability of a given state is

$$P_{Nr} = P(n_1, n_2, \ldots)$$

$$= \frac{\exp[\beta(\mu(n_1 + n_2 + \cdots + n_i + \cdots) - (n_1\varepsilon_1 + n_2\varepsilon_2 + \cdots + n_i\varepsilon_i + \cdots))]}{\mathbf{Z}}.$$

$$(3.53)$$

The numerator can be rearranged as a product of terms:

$$\exp \beta n_1(\mu - \varepsilon_1) \times \exp \beta n_2(\mu - \varepsilon_2) \times \cdots$$

as can the denominator:

$$\sum_{n_1} \exp[\beta n_1(\mu - \varepsilon_1)] \times \sum_{n_2} \exp[\beta n_2(\mu - \varepsilon_2)] \times \cdots$$

so that the probability distribution 3.53 factors out into the product of probability distributions for each single particle state:

$$p(n_1, n_2, \ldots) = \prod_{i=1}^{\infty} \frac{\exp[\beta(\mu - \varepsilon_i)n_i]}{Z_i}, \quad Z_i = \sum_n \exp[\beta n(\mu - \varepsilon_i)].$$

$$(3.54)$$

At this point, we can limit our discussion to the probability of occupation of a particular one-particle state (e.g., ith) using 3.54 to calculate the partition functions for a collection of particles.

We have for Fermi–Dirac (FD) statistics $n_i = 1$ or 0, so

$$Z_i = 1 + \exp[\beta(\mu - \varepsilon_i)]. \qquad (3.55)$$

Thus, the thermally averaged occupation of the ith state is

$$\bar{n}_i = \sum_{n_i} n_i p_i(n_i) = \frac{0 \times \exp[\beta(\mu - \varepsilon_i)] + 1 \times \exp[\beta(\mu - \varepsilon_i)]}{1 + \exp[\beta(\mu - \varepsilon_i)]} \qquad (3.56)$$

leading to

$$\bar{n}_i = \frac{1}{\exp[\beta(\varepsilon_i - \mu)] + 1}. \qquad (3.57)$$

This is the Fermi–Dirac expression for the thermal average occupation of the ith single particle state. As ε approaches μ in the Fermi–Dirac distribution (Equation 3.57), the occupation number approaches unity. A given state can

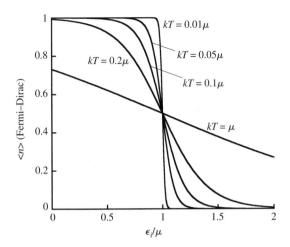

Fig. 3.5 Average occupation number of a state of energy ε_i (in units of the chemical potential, μ) for fermions for various temperatures (expressed in term of μ) from μ/k_B down to 0.01 μ/k_B. The chemical potential of electrons in a metal is very high, so states are almost fully occupied up to $\varepsilon_i = \mu$ at all reasonable temperatures.

only be occupied with one particle (or two if they have opposite spins). At $T = 0$, the function switches abruptly from 1 to 0 as ε exceeds μ. Thus, in a system where $\mu \gg k_B T$, particles fill up states one at a time. In the case of electrons in a metal, only the electrons close in energy to unfilled states can move. This important energy (the chemical potential at $T = 0$) is called the *Fermi energy*. Electrons much below the Fermi energy do not contribute to conduction. The result is that the electronic properties of most conductors are dominated by quantum statistics. The Fermi–Dirac thermal average occupation number is plotted as a function of the state energy for various temperatures in Fig. 3.5.

For the case of bosons (such as photons), we follow the reasoning above but sum the occupation numbers from 0 to ∞ (with the condition that $\mu < \varepsilon i$ so that the sums converge). Doing this yields the Bose–Einstein (BE) result for the thermal average occupation of a single particle state for bosons:

$$\bar{n}_i = \frac{1}{\exp\left[\beta\left(\varepsilon_i - \mu\right)\right] - 1}. \tag{3.58}$$

In the classical limit (much less than one particle per available state), $n_i \rightarrow 0$ so that the exponential in 3.57 and 3.58 is $\gg 1$ and both distributions become identical to the Gibbs distribution (or Boltzmann distribution at constant particle number). But for low temperatures or densities approaching the quantum concentration, the two distributions are quite different. As ε approaches μ in the Bose–Einstein distribution (3.58), the occupation number approaches infinity. This leads to the phenomenon of *Bose condensation*. Systems of bosons condense into one quantum state at very low temperatures.

The occupation number for bosons is plotted as a function of temperature in Fig. 3.6. Since the Bose–Einstein function (3.58) blows up when $E = \mu$, the chemical potential of bosons must be zero or less (negative μ means that free energy is gained by adding a boson to a system). The plot (Fig. 3.6) has been carried out for $\mu = 0$ and the peaked nature of the distribution at low temperatures is clear (a log scale has been used because of the large range of values of the occupation numbers). Thermal vibrations are bosons with no

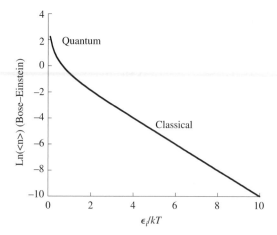

Fig. 3.6 Natural log of the average occupation of a state of energy ε_i (in units of $k_B T$) for bosons (for $\mu = 0$). The straight line at high energy corresponds to the Boltzmann distribution, and the occupation of states is fractional, in line with our assumptions about ideal gasses. The distribution departs from the classical case at low energies where states become multiply occupied.

chemical potential, and Equation 3.58 states that the occupation number for phonons goes to zero as temperature goes to zero. This is the origin of the third law of thermodynamics which states that the entropy of a perfectly ordered crystalline system goes to zero at the absolute zero of temperature.

Exercise 3.9: The Fermi energy (chemical potential at $T = 0$) of sodium metal is 3.24 eV. For what energy of electron is the thermally averaged occupation level equal to 0.5 (1 if we include the possibility of spin up and spin down)? The distribution approaches the classical Boltzmann result when $k_B T = \mu$. How hot must sodium be to become "classical"?

Answer: Equation 3.57 becomes equal to 0.5 for $\varepsilon_i = \mu$, so the occupation number is 0.5 for electrons of 3.24 eV. $k_B T = 0.025$ eV at 300 K, so 3.24 eV is equivalent to 38,880 K!

3.11   Fluctuations

In a system of 10^{23} particles, fluctuations from thermodynamic average quantities are completely negligible. The situation is quite different for nanoscale systems. These can be dominated by fluctuations. For example, the motors that drive biological motion—molecular motors—are essentially driven by random fluctuations. The chemical energy consumed by them goes into "locking in" the desired fluctuations, a mechanism called a "thermal ratchet." This allows directed work to come out of random fluctuations.

In this section, we analyze fluctuations using a statistical mechanics approach. We will focus on the specific example of the relative magnitude of energy fluctuations in a system as a function of its size, but the fluctuations in other thermodynamic quantities can be calculated in a similar way. First, we show that the average energy of a system can be expressed in terms of the first derivative of the partition function with respect to β. Then, we show that the variance in energy can be obtained from the second derivative of the partition function with respect to β.

We begin with the definition of the average energy of a system

$$\bar{E} = \frac{\sum_r E_r \exp -\beta E_r}{Z}. \qquad (3.59)$$

This expression can be restated in terms of a derivative of the logarithm of the partition function

$$\bar{E} = -\frac{\partial \ln Z}{\partial \beta} \qquad (3.60)$$

because $-\frac{\partial \ln Z}{\partial \beta} = -\frac{\partial \ln Z}{\partial Z}\frac{\partial Z}{\partial \beta}$. The first term in this product, $\frac{\partial \ln Z}{\partial Z} = \frac{1}{Z}$. The second term in the product pulls down a term $-E_r$ in front of each exponential in the sum.

Taking the second derivative of 3.60, we obtain

$$\frac{\partial^2 \ln Z}{\partial \beta^2} = \frac{\partial}{\partial \beta}\left(\frac{1}{Z}\frac{\partial Z}{\partial \beta}\right) = \frac{\partial}{\partial Z}\frac{\partial Z}{\partial \beta}\left(\frac{1}{Z}\right) \cdot \frac{\partial Z}{\partial \beta} + \frac{1}{Z}\frac{\partial^2 Z}{\partial \beta^2}$$

$$= -\frac{1}{Z^2}\left(\frac{\partial Z}{\partial \beta}\right)^2 + \frac{1}{Z}\frac{\partial^2 Z}{\partial \beta^2}$$

$$= -\left(\frac{\sum_r E_r \exp -\beta E_r}{Z}\right)^2 + \frac{\sum_r E_r^2 \exp -\beta E_r}{Z} \equiv \left\langle E^2 \right\rangle - \langle E \rangle^2 \qquad (3.61)$$

This says that the second derivative of Z with respect to β yields the difference between the thermal average of the square of the energy and the square of the thermal average (the bracket symbols, $\langle\ \rangle$, mean "take the thermal average"— do not get this confused with Dirac notation here). The right-hand side of 3.61 is equivalent to an expression for thermal average of the square of the fluctuations in energy. This can be seen by expanding the expression for the mean-square fluctuations in energy

$$\left\langle \Delta E^2 \right\rangle \equiv \left\langle (E - \langle E \rangle)^2 \right\rangle = \left\langle E^2 \right\rangle + \langle E \rangle^2 - 2\langle E \langle E \rangle\rangle = \left\langle E^2 \right\rangle - \langle E \rangle^2, \quad (3.62)$$

where $\langle E \langle E \rangle\rangle = \langle E \rangle^2$ because taking the thermal average of a quantity that is already a thermal average does not change its value.

From (3.60), (3.61), and (3.62) it follows that

$$\left\langle \Delta E^2 \right\rangle = -\frac{\partial \langle E \rangle}{\partial \beta} = -\frac{\partial T}{\partial \beta}\frac{\partial \langle E \rangle}{\partial T} = -\frac{1}{k_B}\frac{\partial}{\partial \beta}\left(\frac{1}{\beta}\right)C_V = k_B T^2 C_V, \quad (3.63)$$

where $C_V \equiv \frac{\partial \langle E \rangle}{\partial T}$ is the specific heat of the gas and we have used $\beta \equiv \frac{1}{k_B T}$. For an ideal gas, $\langle E \rangle = \frac{3}{2}Nk_B T, C_V = \frac{3}{2}Nk$ so that the relative size of

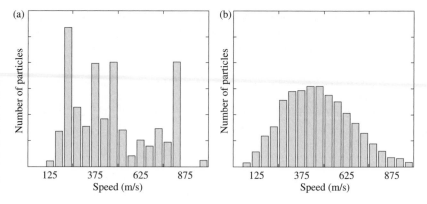

Fig. 3.7 Simulated distribution of speeds in a population of 23 atoms of an ideal gas in a box of volume 1.25×10^{-24} m^3 at 300 K (corresponding to a pressure of 8.6×10^5 Pa). (a) Snapshot from one frame of the distribution of speeds. (b) Shows an average over half a million frames. The distribution in (b) resembles the result of averaging the kinetic energy with a Boltzmann factor. The distribution in (a) is dominated by fluctuations.

energy fluctuations is

$$\frac{\Delta E}{\langle E \rangle} = \frac{\sqrt{kT^2 C_V}}{\langle E \rangle} \approx \frac{1}{\sqrt{N}}. \qquad (3.64)$$

Thus, the relative size of energy fluctuations scales as $\sqrt{1/N}$ for an ideal gas of N particles. In 1 mole of gas, fluctuations in energy are only 1 part in $\sqrt{1/(6 \times 10^{23})}$, and are therefore completely negligible. In a collection of 10 gas molecules, fluctuations in energy are $\pm 33\%$. Figure 3.7 shows the distribution of the speeds of molecules in a gas at 300 K for a tiny sample of $N = 23$. The distribution at any one time is far from uniform. After averaging half a million distributions, a much more uniform distribution emerges with a clear peak speed.

Exercise 3.10: Double-stranded DNA is quite stiff, but it can be modeled as a chain of freely jointed links that are each quite rigid. The length of these links (the "persistence" length) is 50 nm, equivalent to 147 of the "base pairs" that make up DNA. The end-to-end length of a random coil of n segments each of length a is given by $L = a\sqrt{n}$. If we assume that the stored energy in this "spring" is proportional to L^2, how long (in base pairs) must a DNA chain be for the fluctuations in its length to be 10% of its average length?

Answer: $\frac{\Delta E}{E} = \frac{1}{\sqrt{n}} = \left(\frac{\Delta L}{L}\right)^2$. For $\frac{\Delta L}{L} = 0.1$, $n = 10^4$ persistence lengths or 1.47×10^6 base pairs. This is a long piece of DNA—10^{-3} of an entire human genome.

3.12   Brownian motion

Brownian motion is the best known example of fluctuations in a small system. Looking through a high-powered microscope at smoke particles in air or micron-sized plastic beads suspended in water, the random motion of the small particles is immediately obvious. Robert Brown, the English botanist who first observed this motion in pollen grains suspended in water, ruled out the possibility that the pollen was "alive" by repeating the experiment with dust particles.

Einstein proposed the first mathematical theory of Brownian motion. Objects in water (or floating in air) are bombarded on all sides by water (or air) molecules. If the particles are micron-sized, the number of molecules hitting the particle on any one side per unit time is small enough that the fluctuations in momentum transferred to the particle are obvious under an optical microscope.

A key point in analyzing Brownian motion is to realize that many small collisions occur before a significant alteration occurs in the position and velocity of the large particle, because the intrinsic momentum transfer between particles of very different masses is small. Therefore, thermal averaging techniques can be applied to model the forces that lead to small changes in position and momentum (i.e., changes that are small enough to justify the use of calculus).

One way to describe this motion is to use Newton's second law with a random driving force to mimic the fluctuations in the "environment." The resulting equation of motion is called the Langevin equation,[2]

$$m\frac{d\mathbf{v}}{dt} = -\alpha \mathbf{v} + F(t),$$

(3.65)

where α is a friction coefficient (given by $\alpha = 6\pi \eta a$ for a sphere of radius a in a medium of viscosity η—Equation 3.2). $F(t)$ is a random force that obeys the following constraints:

$$\langle F(t) \rangle = 0$$

$$\langle F(t)F(t') \rangle = F\delta(t - t')$$

(3.66)

These equations state that the time average of the force is zero (as it must be for a randomly directed force) and that the force is finite only over the duration of a single "effective" collision (corresponding to many atomic collisions in reality). Multiplying both sides of (3.65) by x and using the notation $\mathbf{v} = \dot{x}$

$$m x\ddot{x} = m \left[\frac{d}{dt}(x\dot{x}) - \dot{x}^2 \right] = -\alpha x\dot{x} + xF(t).$$

Rearranging and taking thermal averages gives

$$m\frac{d}{dt}\langle x\dot{x} \rangle = m\langle \dot{x}^2 \rangle - \alpha \langle x\dot{x} \rangle + \langle xF(t) \rangle.$$

(3.67)

The last term in (3.67) is zero and the equipartition theorem gives $\frac{1}{2}m\langle \dot{x}^2 \rangle = \frac{1}{2}k_B T$. This yields the following differential equation for $\langle x\dot{x} \rangle$

$$\frac{d}{dt}\langle x\dot{x} \rangle = \frac{k_B T}{m} - \frac{\alpha}{m}\langle x\dot{x} \rangle.$$

(3.68)

This is solved with

$$\langle x\dot{x} \rangle = A \exp(Bt) + C.$$

(3.69)

Substituting (3.69) into (3.68) yields $C = \frac{k_B T}{\alpha}$ and $B = -\frac{\alpha}{m}$. The remaining boundary condition is a little trickier. Note that

$$\langle x\dot{x} \rangle = \frac{1}{2} \frac{d}{dt} \langle x^2 \rangle \tag{3.70}$$

and taking x to be 0 at $t = 0$ we have, from (3.69), $A = -\frac{k_B T}{\alpha}$ and so

$$\frac{1}{2} \frac{d}{dt} \langle x^2 \rangle = \frac{k_B T}{\alpha} \left[1 - \exp -\frac{\alpha t}{m} \right]. \tag{3.71}$$

This is integrated by putting $-\frac{\alpha t}{m} = \theta$, $dt = -\frac{m}{\alpha} d\theta$ and solving for the constant of integration by setting $\langle x^2 \rangle = 0$ at $t = 0$. The solution is

$$\langle x^2 \rangle = \frac{2k_B T}{\alpha} \left[t - \frac{m}{\alpha} \left(1 - \exp -\frac{\alpha t}{m} \right) \right]. \tag{3.72}$$

In the long time limit ($t >> \frac{m}{\alpha}$), the solution becomes

$$\langle x^2 \rangle = \frac{2k_B T t}{\alpha} = \frac{k_B T t}{3\pi \eta a}. \tag{3.73}$$

The mean-square displacement grows linearly with time in this limit. The constant of proportionality is

$$D' = \frac{k_B T}{3\pi \eta a}. \tag{3.74}$$

This behavior is illustrated in Fig. 3.8. This is a characteristic of a "random walks" in general.[†]

Exercise 3.11: Calculate the number of air molecules hitting a dust particle of 1 micron in diameter at 300 K along the $+x$ direction in 1 ms. Use the equipartition theorem to calculate the average x component of the velocity and

[†]This is quite a general result: suppose a system is taking N random steps (magnitude a) per unit time in directions $\mathbf{r}_i$, then the mean-square displacement is

$$\langle r^2 \rangle = \left\langle \sum_{i,j} \mathbf{r}_i . \mathbf{r}_j \right\rangle = Na^2 + \left\langle \sum_{i \neq j} \mathbf{r}_i . \mathbf{r}_j \right\rangle,$$

where the Na term comes from the sum of all terms for which $i = j$ and, for a Markov (i.e., random) process the second sum is zero. This is the result quoted in Exercise 3.10.

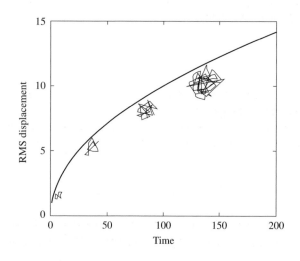

Fig. 3.8 Projection of three-dimensional random walk trajectories onto a plane showing the increase in root-mean-square end-to-end distance with time (the average displacement is always zero). The solid line is the $\sqrt{\langle x^2 \rangle} = \sqrt{D't}$ result for an average over a large number of random walks.

the fact that 1 mole of air occupies 22.4 L in standard conditions. Assume air is pure N_2 gas of mass 28 amu. How much bigger are the pressure fluctuations on the dust particle than on a 1 m^2 surface, assuming that the fluctuations in the concentration of air scale with fluctuations in energy (which scale as $1/\sqrt{N}$)?

Answer: The x-speed of a molecule follows from the equipartition theorem, $v_{xRMS} = \sqrt{(k_B T)/m}$. With $m = 28 \times 1.66 \times 10^{-27}$ kg, $v_{xRMS} = 298$ m/s. 1 m^3 of air (i.e., 10^3 L) contains 44.6 moles or 2.69×10^{25} molecules, of which half will have a component along the $+x$ direction. The cross-sectional area of the dust particle (diameter 1 micron) is $\frac{\pi d^2}{4} = 7.85 \times 10^{-13}$ m^2. Thus, the number of molecules hitting the dust particle in 1 ms will be 1.35×10^{25} (molecules) $\times$ 7.85×10^{-13} (area) $\times$ 298(velocity) $\times 10^{-3}$(time) $= 3.14 \times 10^{12}$. The root-mean-square fluctuations will be $\sqrt{(3.14 \times 10^{12})}$ or 1.7×10^6 for a fractional fluctuation in momentum transferred from air to the particle of 5.6×10^{-7}. In contrast, the number hitting 1 m^3 in 1 ms is 4×10^{24} for a fractional variation of 5×10^{-13}, six orders of magnitude smaller than for the dust particle.

3.13 Diffusion

The Brownian motion just discussed is intimately related to the motion of many things under the influence of random forces: small particles or solute molecules in solution, or even gradients in the random motion itself as in the diffusion of heat. Here, we look at the problem from another angle: the macroscopic flux of particles or heat through an elementary volume of the material.

Consider the motion of particles passing though an imaginary box with faces of area A separated by a distance Δx (Fig. 3.9). If $J(x)\Delta t$ particles enter the box in a time Δt and $J(x + \Delta x)\Delta t$ particles leave per unit time, the change in the number of particles in the volume $A\Delta x$ of the box (ΔC) must be

$$\Delta C = \left[\frac{AJ(x + \Delta x)\Delta t - AJ(x)\Delta t}{A\Delta x} \right]. \tag{3.75}$$

In deriving this expression, we assumed a flux only in the x direction (see below). In the limit $\Delta x, \Delta t \to 0$

$$\frac{\partial C(x,t)}{\partial t} = -\frac{\partial J(x,t)}{\partial x}. \tag{3.76}$$

The minus sign follows from the fact that the concentration will increase if the flux decreases as the flow crosses the box.

The flux through a surface must be equal to the change in concentration across the surface multiplied by the speed of the particles toward the surface, i.e.,

$$J = \Delta C \frac{\Delta x}{\Delta t} = -\frac{\partial C}{\partial x} \frac{(\Delta x)^2}{\Delta t}. \tag{3.77}$$

The minus sign indicates that the concentration falls more across the box as the flux goes up. From 3.73 and 3.74, we can replace $\frac{\langle \Delta x^2 \rangle}{\Delta t}$ with D, the diffusion

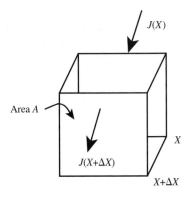

Fig. 3.9 Flux of particles through a small box. The number accumulating per unit time must equal the difference in flux across two faces (shown here for a flux only in the x direction).

constant, to obtain:

$$J = -D\frac{\partial C}{\partial x}.$$

(3.78)

This result, that the flux across a surface is proportional to the concentration gradient, is called Fick's first law. Equation 3.76 is called Fick's second law. Taken together, they yield the *diffusion equation*:

$$\frac{\partial C(x,t)}{\partial t} = D\frac{\partial^2 C(x,t)}{\partial x^2}.$$

(3.79)

Diffusion usually occurs in three dimensions, for which the diffusion equation is

$$\frac{\partial C(\mathbf{r},t)}{\partial t} = D\left[\frac{\partial^2 C(\mathbf{r},t)}{\partial x^2} + \frac{\partial^2 C(\mathbf{r},t)}{\partial y^2} + \frac{\partial^2 C(\mathbf{r},t)}{\partial z^2}\right] \equiv D\nabla^2 C(\mathbf{r},t).$$

(3.80)

This equation takes account of the fluxes across all six faces of the box in Fig. 3.9. For an initial point source (i.e., a delta function in concentration at $t = 0$), the solution of the diffusion equation in one dimension is

$$C(x,t) = \frac{A}{\sqrt{Dt}}\exp\left(-\frac{x^2}{4Dt}\right).$$

(3.81)

The constant A is determined by requiring that the integral of 3.81 be equal to the initial amount of material added at $t = 0$. The development of the distribution of a solute initially added as a point source as a function of time is shown in Fig. 3.10, as calculated for (3.81) with $A = 1$. As $t \to \infty$, the distribution becomes uniform, the point of "half-maximum concentration," $x_{1/2}$ material advancing with time according to

$$x_{1/2} = \sqrt{2Dt}$$

(3.82)

(we have ignored the small time dependence of the factor in front of the exponential).

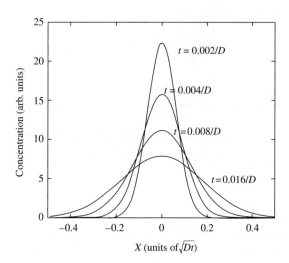

Fig. 3.10 Evolution of the distribution of a solute initially added to solvent as a point source as predicted by the diffusion equation.

3.14   Einstein–Smoluchowski relation

A very interesting result follows from comparing the time progress of our Brownian "walker" (Equations 3.73 and 3.74) with the spread of the concentration fluctuation described by Equation 3.82 using the macroscopic diffusion equation:

$$x_{1/2}^2 = 2Dt. \tag{3.83}$$

Written in terms of the diffusion constant for Brownian motions (D', Equation 3.74), Equation 3.83 is equivalent to

$$\left\langle x^2 \right\rangle = D't. \tag{3.84}$$

Thus, $D = \frac{D'}{2}$, so, for our Brownian motion, the D that appears in the diffusion equation (Equation 3.79) is

$$D = \frac{k_B T}{6\pi \eta a} \ (\text{m}^2 \ \text{s}^{-1}). \tag{3.85}$$

The diffusion constant is given by the ratio of $k_B T$ to the friction coefficient for the particle. As a reminder, Stokes' law gives the velocity of a particle acted on by a force F in a viscous liquid as $V = \frac{F}{6\pi a \eta} = \mu F$, where the mobility, $\mu = (6\pi a \eta)^{-1}$. In these terms, (3.85) relates a diffusion constant (describing thermal fluctuations) to a mobility (describing driven motion). This unexpected relationship (between driven motion and random Brownian motion) is called the Einstein–Smoluchowski relation, given in general as

$$D = \mu k_B T. \tag{3.86}$$

Thus, if the mobility of a driven system is known, the diffusion constant for random motion can be calculated.

Exercise 3.12: Use the Einstein–Smoluchowski relation and Stokes' law to show that the diffusion constant for a small ion (radius 0.1 nm, $T = 300$ K) in water (viscosity $= 1 \times 10^{-3}$Pa · s) is $\sim 2000 \ \mu\text{m}^2/\text{s}$.

Answer: $\mu = (6\pi a \eta)^{-1} = (6\pi \times 10^{-10} \times 10^{-3})^{-1} = 5.3 \times 10^{11}$ m/N − s. With $k_B T = 4.14 \times 10^{-21}$ J, $D = 2.19 \times 10^{-9}$ m^2/s or $2190(\mu\text{m})^2/\text{s}$.

Exercise 3.13: If a copper sulfate crystal is dropped into a test tube filled to a depth of 5 cm with water, how long will it take for the top of the solution to appear blue if the sample is not stirred? Use the diffusion constant derived above (2000 $\mu\text{m}^2/\text{s}$).

Answer: Using $\sqrt{\langle r^2 \rangle} = \sqrt{Dt}$ or $t = \frac{\langle r^2 \rangle}{D}$, we obtain $t = (5 \times 10^4 \ \mu\text{m})^2/$ 2000 $\mu\text{m}^2/\text{s}$ or 1.25×10^6 s. This is about a month, which is why we stir to mix solutions more quickly.

3.15   Fluctuations, chemical reactions, and the transition state

Chemical reactions generally proceed at rates that increase exponentially with temperature. We saw already that, for a reaction to occur spontaneously, the free energy of the products must be lower than the free energy of the reactants. So why do reactions not always occur spontaneously? One reason is diffusion. The reactants must diffuse about randomly until they collide. The diffusion rate will increase with temperature because $D = \mu k_B T$ and also because μ generally increases strongly with temperature. But this will not account for the observed exponential dependence of rate on temperature. The simplest model for this exponential dependence on temperature is *transition state theory*.

Reactants and products are clearly in a state of local equilibrium. N_2 gas and H_2 gas exist quite stably at room temperature. Likewise, NH_3 is quite stable at room temperature. But we saw in Exercise 3.8 that a free energy of 33.28 kJ is to be gained from the reaction of N_2 and H_2 to make 2 moles of ammonia. If we pull (or push) on the bonds in the N_2 molecules and the H_2 molecules, their energy will increase as they are distorted away from their equilibrium position. This is shown schematically on the left in Fig. 3.11. Likewise, if we distort the bonds in ammonia, its energy will increase. Now imagine distorting both the reactants and the products so that the resulting complex looks like an intermediate between them (as shown in the middle of Fig. 3.11). This intermediate clearly has a higher energy than the reactants. If one searched through all the possible distortions that could connect the reactants to the products, the one with the smallest increase in energy is the one that will dominate the rate of the reaction. This state is called the transition state, and it is shown in the figure as having a free energy ΔG^+ above the energy of the reactants. In reality, this reaction is much more complex than represented in the figure. The initial slow step is diffusion of the reactants into one place. Once positioned correctly, the

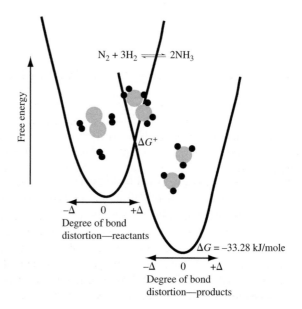

Fig. 3.11 Energy "landscape" connecting reactants to products in the Haber process for the production of ammonia. The products have a lower free energy than the reactants, but, once diffusion has brought the reactants together, the rate of reaction is controlled by the lowest energy intermediate state—the transition state (energy ΔG^+ above the reactant free energy).

reaction proceeds via a series of intermediate steps, not just the one transition state implied by the figure.

Now imagine a very complex, multidimensional coordinate (a function of all the bond lengths, and, for the ammonia molecules, the bond angles) that describes the distortions needed to morph from the reactants, to the transition state, and then back on to the products. Such a coordinate is called a *reaction coordinate*. A plot of free energy vs. the reaction coordinate will look something like the energy landscape shown in Fig. 3.2(a). The meaning of the plot is, however, much more complicated than the "marbles rolling in a grooved bowl" analogy given at the start of this chapter.

The Boltzmann factor for the occupation of the transition state, C^+, is

$$C^+ = C_R \exp\left(-\frac{\Delta G^+}{k_B T}\right), \tag{3.87}$$

where C_R is the occupancy of the reactant state. In what follows, we will confine our discussion to a simple first-order unimolecular reaction ($A \rightarrow B$). In the simplest approach to transition state theory, the rate of attempts at crossing the barrier posed by the transition state is taken to be a bond stretching rate, v, obtained by equating $E = hv$ with $E = k_B T$, so the rate of decomposition of the reactants is given by

$$-\frac{dC_R}{dt} = vC^+ = C_R \frac{k_B T}{h} \exp\left(-\frac{\Delta G^+}{k_B T}\right) \tag{3.88}$$

leading to a first-order rate constant for the reaction given by

$$k_1 = \frac{k_B T}{h} \exp\left(-\frac{\Delta G^+}{k_B T}\right). \tag{3.89}$$

On the basis of the experimental data on reaction rates, Arrhenius proposed a relation similar to Equation 3.39 in 1889, and it is known as the Arrhenius relation.

3.16 The Kramers theory of reaction rates

Transition state theory, as just presented, is *ad hoc* in the choice of parameters for the "attempt frequency" that appears in front of the Arrhenius law. A detailed theory for stochastic motion in a potential well was worked out by Kramers[3] in 1940 and his result is presented here. We will use a version of it in Chapter 4.

In order to show how the problem is set up, Fig. 3.12 illustrates the result of a simulation of motion in a potential with two minima. This problem is difficult to analyze numerically, because the thermal fluctuations and viscous collisions that allow the particle in the well to equilibrate with its surroundings occur on rapid (atomic vibration) timescales, but the motion of the particle over the barrier is much slower (if the barrier is high). Hänggi et al.[4] solved this computational problem by building an analog computer to analyze the motion of the system, and the results are presented in Fig. 3.12(b). The particle generally resides very close to the two minima at $x = \pm 1$, making only infrequent transitions

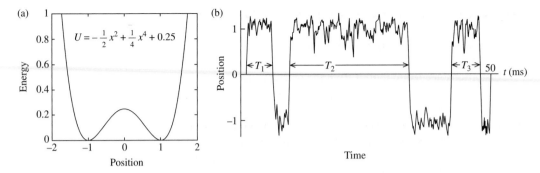

Fig. 3.12 (a) A one-dimensional potential model for noise-driven transitions between metastable states at $x = -1$ and $x = -1$ (generated by the quadratic potential shown inserted in (a)). Motion of a particle in this potential was simulated using an analog computer which yielded the trajectory shown in (b). The particle spends almost all of its time at $x = \pm 1$, making infrequent transitions between the two states (relative to the high frequency noise that drives the system). (Reprinted with permission from the American Physical Society, Reviews of Modern Physics, P. Hänggi[5]. Courtesy Professor Peter Hänggi, University of Augsburg.)

Fig. 3.13 Parameters in the Kramers model for a unimolecular reaction involving escape from A to C via a transition state at B with rate constant k^+. The reverse process occurs at a rate k^-. The frequencies, ω_a, ω_b, and ω_c, refer to the second derivative of the energy at the minima (x_a and x_c) or the saddle point (x_b). This gives a spring constant that can be expressed in terms of a frequency, given the particle mass. (Reprinted with permission from the American Physical Society, Reviews of Modern Physics, P. Hänggi[5]. Courtesy Professor Peter Hänggi, University of Augsburg.)

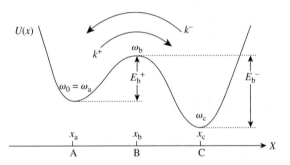

between the two. Transitions, when they occur, are rapid, as the time spent in the high-energy intermediate states is costly in terms of the associated Boltzmann factor. Notice, in particular, how the motion between the wells is slow (in terms of an overall transition rate) compared to the noise that drives, and thermally equilibrates, the particle (the small wiggles in position between transitions). This separation of timescales is what makes it appropriate to replace the very complex molecular motion with a reaction coordinate described by the potential in Fig. 3.12(a).

A double-well potential is shown in Fig. 3.13. Since the particles spend most of their time near the minima, an important simplification results from using just the curvature of the potential at the minima to describe the potential near the metastable positions. The saddle point at B is an unstable point, but its curvature controls the motion back down the slope, either for an escape from A to C, or a return to A. These curvatures can be expressed as frequencies according to

$$\omega_{a,b,c}^2 = \frac{1}{m} \left. \frac{\partial^2 U}{\partial x^2} \right|_{x=a,b,c}, \tag{3.90}$$

where we have used the result for the angular frequency of a mass m oscillating on a spring with a spring constant given by the second derivative of the energy with respect to position (evaluated at the minimum, or saddle point). Motion at the saddle point is not, of course, oscillatory, but the curvature parameter is still useful. In the limit of heavy viscous damping, the expression for the rate of escape from the potential minimum at A is[5]

$$k^+ = \frac{\omega_a \omega_b}{2\pi\gamma} \exp -\frac{E_{b+}}{k_B T}.$$ (3.91)

In this expression, γ is the viscous damping rate, given by the reciprocal of the viscous relaxation rate (Appendix D) and, for the expression to be valid, $\gamma \gg \omega_b$. A similar result holds for the reverse process (rate k^-) with E_{b-} replacing E_{b+} and ω_c replacing ω_a.

The Kramers result, Equation 3.91, has been put to experimental test using "laser tweezers" (see Chapter 4) to generate two adjacent potential minima, so that the motion of a micron-sized bead could be followed as it flipped between the two potential minima under the driving force of thermal fluctuations at room temperature in solution. The potential was mapped out precisely by measuring the force required to move the bead, so a three-dimensional version of Equation 3.91 could be used to predict the rate of transition between the two minima. The experimentally measured transition rate[6] was in precise agreement with Kramers theory, validating the Kramers calculation of the prefactor in 3.91.

3.17 Chemical kinetics

Thus far, we have avoided a detailed discussion of how models such as a particle in a double-well potential map on to real chemical reactions. The relationship is certainly not as simple as implied in Fig. 3.11, where a very complex, multiparticle reaction was mapped on to a rather simple potential. The "particle in a well" picture only really applies to a unimolecular reaction, a word used above without definition. Unimolecular reactions are the simplest of reactions in which a single particle (or molecule) changes state:

$$E_1 \underset{k_-}{\overset{k_+}{\rightleftharpoons}} E_2$$ (3.92)

This chemical equation means that E_1 converts into E_2 at a rate k_+ per second and E_2 converts back to E_1 at a rate k_- per second. Unimolecular reactions are rare beasts. The spontaneous rearrangement of methylisonitrile (CH_3CN) in solution is an example of one (the dots represent unpaired electrons – see Chapter 8):

The unimolecular reaction is straightforward to analyze. Using the notation that $[E_1]$ means the concentration of E_1 (in moles/liter) in the reaction 3.92, the rate of accumulation of E_2 is

$$\frac{d[E_2]}{dt} = k_+ [E_1] - k_- [E_2].$$ (3.93)

The solution of (3.93) is

$$\frac{[E_2](t)}{[E_2](t) + [E_1](t)}$$

$$= [E_1](0) + \left(\frac{k_+}{k_+ + k_-} - [E_1](0)\right)(1 - \exp[-(k_+ + k_-)t]). \quad (3.94)$$

The concentration approaches its steady-state value with a time constant $(k_+ + k_-)^{-1}$. In steady state $(t \to \infty, \frac{d[E_1]}{dt} \to 0)$ both Equations 3.93 and 3.94 yield

$$\frac{[E_2]}{[E_1]} = \frac{k_+}{k_-} \equiv K_{eq} = \exp\left[-\frac{\Delta G}{RT}\right], \quad (3.95)$$

where K_{eq} is called the equilibrium constant for the reaction. ΔG is the free energy difference between reactants and products (not the transition state energy) expressed here in calories per mole. Equation 3.95 is a statement of the Boltzmann factor for the ratio of $[E_2]$ to $[E_1]$. The Boltzmann constant has been replaced by the gas constant, R, because of the units used for the free energy (calories or Joules per mole). Equilibrium constants span an enormous range of values. For a strong covalent bond, $\frac{\Delta G}{RT}$ may be on the order of 40 (the ratio of 1 eV to 0.025 eV) so the ratio of product to reactant would have to be $\sim 10^{17}$ before back reactions become significant. Entropy plays little part in strong reactions. It cannot be ignored though. Some reactions are entirely driven by the increase in entropy of the products—so much so that the enthalpy for the reaction can be positive. In this case, the reaction can occur spontaneously while taking in heat purely because $T\Delta S > \Delta H$. These reactions are rare, but they do occur and they are known as *endothermic* reactions.

More common, by far, are reactions that involve multiple species. A simple example is

$$NO + O_3 \longrightarrow NO_2 + O_2$$

This called a bimolecular reaction, because the two reactants, NO and O_3, must first come together in order for a reaction to occur. This is a quite different physical process from the thermally activated hopping over a transition state (which is a second step once the reactants have come together). In fact many reactions occur via a series of intermediate steps, so must be broken down into a series of bimolecular (or, if three reactants are involved, termolecular) elementary reactions to be analyzed. Multicomponent reactions occur at rates that are proportional to the *product* of the reactant concentrations. This is because the collision rate for

$$A + B \to C$$

is proportional to both $[A]$ and $[B]$ so the overall rate of the forward reaction must be proportional to the product $[A][B]$ if the rate is determined by collisions. Although this result appears obvious when viewed in terms of probabilities for molecular collisions, it is worthwhile reviewing the reaction from a thermodynamic perspective. Recall (Equation 3.43) that the entropy of an ideal solution is proportional to the natural logarithm of the concentration of a reactant, $S \propto -\ln C$. Recall also that the free energy of the reactant contains a term in $-TS$, so the free energy of the reactants is

$$G = H - TS = H + S^0 + RT \ln[A] + RT \ln[B]. \tag{3.96}$$

The free energy of the reactants is raised by an amount $RT \ln[A] + RT \ln[B]$. This decreases the energy barrier at the transition state, leading to a more rapid rate constant for the forward reaction (and an equilibrium state with an increased amount of product relative to reactant). Since the rate is proportional to the exponential of the free energy, it must therefore be proportional to the product of the concentrations, a result argued for heuristically above.

The bimolecular reaction can still be described by forward and reverse rate constants:

$$A + B \underset{k_-}{\overset{k_+}{\rightleftharpoons}} C \tag{3.97}$$

only now the differential equation for $[C(t)]$ (cf. Equation 3.93 for the unimolecular reaction) is

$$\frac{d[C]}{dt} = k_+[A][B] - k_-[C]. \tag{3.98}$$

Note that the rate constant for this bimolecular reaction has different units from the rate constant for the unimolecular reaction. In this case, the units are liter-mole^{-1}s^{-1}. The system comes into equilibrium when

$$\frac{[C]}{[A][B]} = \frac{k_+}{k_-} \equiv K_{eq} = \exp{-\frac{\Delta G^0}{RT}}. \tag{3.99}$$

When ΔG^0 is negative, the concentration of C increases spontaneously if it is present in a smaller concentration that predicted by the equilibrium constant. But we seem to have a problem here: the result on the right-hand side is clearly dimensionless, but the left-hand side appears to have dimensions of moles^{-1}. In order to understand this equation, we need to detour to remind ourselves that we are using *corrections to the standard molar enthalpies and entropies* as defined in Equation 3.45, $\Delta G = \Delta G^0 + RT \ln Q$. The reaction quotient for the reaction 3.97 is

$$Q = \frac{[C]}{[A][B]} \tag{3.100}$$

so that the concentration-dependent term that adds to ΔG^0 provides a multiplier in front of the Boltzmann expression for the rate when the free energy is just ΔG^0, which is the free energy for molar concentrations. Thus, when used *for equilibrium calculations* (as opposed to kinetic rate calculations) "concentrations" *are dimensionless*, because they are the ratio of the actual concentration to one molar concentration. Thus, the equilibrium constant, K, appearing in Equation 3.99 is dimensionless (it does not have units of moles^{-1}). To make this distinction clear, chemists refer to concentrations used in equilibrium calculations as *activities* and activities, like equilibrium constants, are always dimensionless. To clarify this point, consider Equation 3.45 for a system where the reactants and products are in equilibrium so their free energy difference is zero:

$$0 = \Delta G^0 + RT \ln Q = \Delta G^0 + RT \ln \frac{[C]}{[A][B]} \equiv \Delta G^0 + RT \ln K_{eq}. \quad (3.101)$$

From which

$$K_{eq} = \exp -\frac{\Delta G^0}{RT}, \quad (3.102)$$

which is the right-hand side of Equation 3.99. Finally, when p multiples of a reactant R are consumed in a reaction, the activity appears as $[R]^p$. For example, the reaction

$$NO_2 + CO \longrightarrow NO + CO_2$$

occurs via a first slow (rate-limiting) step,

$$NO_2 + NO_2 \longrightarrow NO_3 + NO$$

The second step is

$$NO_3 + CO \longrightarrow NO_2 + CO_2$$

so the rate constant for the first step is $k_+[NO_2]^2$. Being a second-order rate constant (and not an equilibrium constant) k_+ here has units of liter $-$ mole^{-1}s^{-1}.

Generally, the rate-limiting step for a bimolecular reaction is not the thermally activated reaction itself (but it could be) but rather the time it takes for reactants to diffuse together. This is a relatively straightforward quantity to estimate, a problem discussed in what follows.

Exercise 3.14: Exercise 3.14: Dinitrogen tetroxide, N_2O_4, decomposes in the gas phase into a mixture of N_2O_4 and nitrogen dioxide, N_2O:

$$N_2O_4 \rightleftharpoons 2NO_2,$$

where the arrows imply an equilibrium interconversion between the two. Define an equilibrium constant for this reaction and calculate it if the equilibrium concentrations measured in the gas phase are 0.0172 M NO_2 and 0.0014 M N_2O_4.

Answer: The equilibrium constant for this process is $K = \frac{[NO_2]^2}{[N_2O_4]}$ giving $K = 0.211$. Note K has no units.

3.18   Acid–base reactions as an example of chemical equilibrium

An acid is a substance that readily donates protons in a chemical reaction, and a base is a substance that readily accepts protons. Hydrogen chloride (an acid) readily donates protons to water (water acting as a base in this reaction) to form a solution of hydrochloric acid:

$$HCl + H_2O \rightarrow H_3O^+ + Cl^-,$$

where H_3O^+ is the hydroxonium ion (water with a proton attached to the negatively charged oxygen) and the single arrow implies that the reaction is strongly biased toward the product. HCl is said to form a "strong acid" because it dissociates completely in water.

Water itself will act either as an acid (donating protons to strong bases) or as a base (accepting protons from strong acids). Protons in water are exchanged rapidly between water molecules, and a small fraction of the molecules exist as H_3O^+ or OH^- (hydroxide) ions at any one time:

$$2H_2O \rightleftharpoons H_3O^+ + OH^-.$$

For simplicity, we will consider just dissociation into protons and hydroxonium ions:

$$H_2O \rightleftharpoons H^+ + OH^-.$$

An acid and a base that differ only in the presence or absence of a proton are called a conjugate base pair. Thus, OH- is the conjugate base of H_2O and H_3O^+ is the conjugate acid of H_2O.

The equilibrium constant (see the previous section) for the dissociation of water is given by $\frac{[\mathbf{H^+}][\mathbf{OH^-}]}{[\mathbf{H_2O}]}$. Because the activity is defined as the ratio of the concentration to the standard state, water, as a pure (undiluted) material must always have an activity of unity. Thus, the equilibrium constant for the dissociation of water is

$$K_W = [H^+][OH^-]. \tag{3.103}$$

K_W has an experimentally determined value of 10^{-14} at 25°C. Charge neutrality implies that $[H^+] = [OH^-]$ so it follows that $[H^+] = [OH^-] = 10^{-7}$ M at 25°C. The equilibrium constant is little affected by the addition of solutes to water, so adding acid to water drives the concentration of H^+ up, and results in a corresponding drop in the concentration of OH^- according to (3.103).

Since the free energy associated with any one species changes as the logarithm of its activity (cf. 3.43), it is customary to measure the concentration of protons (or hydroxonium ions) on the pH scale:

$$pH = -\log[H^+]. \tag{3.104}$$

Thus, pure water has a pH of 7. Adding acid lowers the pH (more protons or hydroxonium ions) while adding base increases the pH (fewer protons or hydroxonium ions and more hydroxonium ions).

Many residues attached to the peptide backbone of proteins are weak acids or bases, weak, in this context, meaning that only a fraction of the dissolved species dissociate into ions. Their behavior is described by a quantity called the pK_a. This is equal to the pH at which half the species is dissociated (i.e., charged) and half is neutral. Lysine, which is basic, has a pK_a of 2.18. In neutral solution (pH = 7), it is fully protonated and carries a positive charge, so heavy concentrations of lysine residues are often found in the parts of proteins that are exposed to water because charged groups are generally more soluble.

Substances (like lysine) that only partially dissociate are called weak bases (or weak acids when the dissociation yields a proton). In general, the dissociation reaction is

$$HX \rightleftharpoons H^+ + X^-$$

so that the equilibrium constant (referred to as the "acid dissociation constant", K_a in this case) is

$$K_a = \frac{[H^+][X^-]}{[HX]}. \tag{3.105}$$

Solving for $[H^+]$ and taking logarithms yields the Henderson–Hasselbach equation:

$$pH = pK_a + \log \frac{[base]}{[acid]}, \tag{3.106}$$

where $pK_a \equiv -\log K_a$ analogous to the definition of pH, and we have replaced $\log \frac{[X^-]}{[HX]}$ with $\log \frac{[base]}{[acid]}$ because this holds for a conjugate base pair (see above). It follows from (3.106) that the pK_a is the pH at which HX is exactly 50% dissociated.

When a weak acid or a weak base share a common ionizable species (X) are mixed together with a salt containing that same ion, the result is a *buffer* solution. For example, we can make a buffer from the weak acid HX and the salt NaX where X is the acetate ion.

Adding base to a buffer solution

$$OH^- + HX \longrightarrow H_2O + X^-$$

consumes the acid component to produce water and the base component of the buffer, while adding acid to a buffer solution

$$H^+ + X^- \longrightarrow HX$$

consumes the base component. Either way, the pH does not change significantly from the pK_a until the amount of acid (or base) added exceeds the buffer concentration. This is illustrated in Fig. 3.14, which shows a titration curve for adding NaOH to acetic acid (CH_3COOH). The anion CH_3COO^- acts as the conjugate base to the acid ($H^+CH_3COO^-$). Adding base to acetic acid changes the molar ratio of base to acid according to

$$\frac{[base]}{[acid]} = \frac{\alpha}{1-\alpha} \tag{3.107}$$

where α is the amount of base added in relation to the original concentration of the acid (i.e., equivalents of base added). Using 3.107 in 3.106 with the known

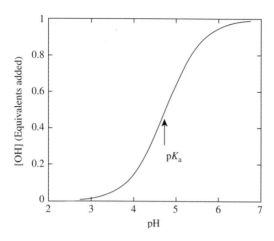

Fig. 3.14 Titration curve for adding NaOH to acetic acid calculated using the Henderson–Hasselbach equation. The pK_a of acetic acid is 4.75.

pK_a of acetic acid (which is 4.75) yields the calculated titration curve shown in Fig. 3.14. The buffering action is clear, because the pH does not change rapidly with added base in the vicinity of the pK_a. Buffers are important for maintaining the charge state of the ionizable groups in proteins (like lysine residues) and blood is an excellent buffer, maintaining a pH of 7.3.

Exercise 3.15: How many moles of NH_4Cl must be added to 2 L of 0.1 molar NH_3 to form a buffer of pH 9? Use the fact that in any mixture of an acid with its conjugate base, $K_a \times K_B = K_W$ where K_B is the dissociation constant for the conjugate base (1.8×10^{-5}) and K_W is the dissociation constant for water (10^{-14} at 25°C).

Answer: The dissociation constant of the salt NH_4Cl follows $pK_a + pK_B = 14$ with $pK_B = -\log(1.8 \times 10^{-5}) = 4.75$ so $pK_a = 9.25$. With $\log \frac{[\text{base}]}{[\text{acid}]} = \text{pH} - pK_a = -0.25$, $\frac{[\text{base}]}{[\text{acid}]} = 0.562$, so the acid concentration must be 0.178 M, meaning 0.356 moles must be added to 2 L of the ammonia solution.

3.19   The Michaelis–Menten relation and on-off rates in nano–bio interactions

Bimolecular reactions (Equations 3.97–3.99) are important in biology, particularly as reactions between an enzyme and its substrate. (An enzyme is a protein that catalyzes a chemical reaction by lowering the transition state energy for turning the reactant—the "substrate"—into one or more products.) Calling the enzyme, E, the substrate, S, and the product, P, the reaction can be represented by

$$E + S \underset{k_{\text{off}}}{\overset{k_{\text{on}}}{\rightleftharpoons}} ES \overset{k_2}{\longrightarrow} E + P. \qquad (3.108)$$

Generally, the first step, of enzyme binding substrate, is reversible, but the second step, usually the breaking of a covalent bond, is not. The solution of

Equation 3.108 is obtained from

$$\frac{d[ES]}{dt} = k_{on}[E][S] - (k_{off} + k_2)[ES]. \tag{3.109}$$

The solution of this equation can be found in Appendix 5.4 of the book by Howard.[7] In the long time limit and for the normal situation of high substrate concentration ($[S] >> [E]$), the concentration of bound substrate approaches an equilibrium value given by

$$[ES] = E_t \frac{[S]}{K_M + [S]}, \tag{3.110}$$

where $E_t = [E] + [ES]$ (a constant equal to the total enzyme concentration) and K_M, the Michaelis–Menten constant, is given by

$$K_M = \frac{k_{off} + k_2}{k_{on}}. \tag{3.111}$$

In this limit (of constant $[ES]$), the rate of formation of the product, given by $\frac{d[P]}{dt}$, is just

$$\frac{d[P]}{dt} = k_2[ES] = k_2 E_t \frac{[S]}{K_M + [S]} \tag{3.112}$$

Equation 3.112 is known as the Michaelis–Menten equation for the rate of an enzymatic reaction.

Two limits are interesting to consider. If the enzyme catalyzes the reaction very slowly ($k_{off} >> k_2$), then the concentration of the intermediate $[ES]$ is independent of k_2 and

$$K_M \approx \frac{k_{off}}{k_{on}} \equiv K_D. \tag{3.113}$$

K_D is called the dissociation constant, and it describes the equilibrium ratio of reactants to products for a simple bimolecular reaction in equilibrium. Specifically, in this limit

$$K_D = \frac{[E][S]}{[ES]}. \tag{3.114}$$

The dissociation constant is the reciprocal of the equilibrium constant (Equation 3.99). However, unlike the equilibrium constant, it is conventional to give the dissociation constant units (which can lead to confusion when expressions like Equation 3.99 are used). For example, if reactants and products are in equilibrium when the concentration of each is 1 mM, then $K_D = (1 \text{ mM})^2/(1 \text{ mM}) = 1$ mM (we have assumed that $[E] = [S]$ – this will be discussed further below). Many biological molecules bind their targets with $K_D \sim 1\ \mu\text{M}$. This means that if a solution is diluted below this concentration, the probability of finding bound molecules falls. This has many implications for analytical measurements and for making sensors. Special molecules designed to recognize specific targets might have $K_D \sim 1$ nM or even a few pM. Thus, they could recognize their targets in highly diluted samples (though they may recognize other molecules too—high affinity, i.e., small K_D, is not necessarily the same as high specificity).

One gets away with giving K_D units because it is always used in the context of bimolecular reactions (A binds B to form the bound complex AB). In these terms

$$K_D = 1 \times \exp\left(+\frac{\Delta G_0}{RT}\right) \text{ moles.} \qquad (3.115)$$

Thus, anything that makes the free energy of the bound state more negative decreases K_D (remember that a smaller K_D means stronger binding).

The other limit in the enzyme–substrate reaction is when enzyme catalysis is very rapid, i.e., $k_{off} \ll k_2$. The bound complex is converted to product essentially immediately, so K_M in this limit is the substrate concentration at which the enzyme spends half the time waiting to bind substrate.

The dissociation constant, K_D, is an important quantity in characterizing biomolecular reactions and reagents designed for sensing, so it is worth looking at the kinetic factors that control it. According to Equation 3.113, it is the ratio of k_{off} to k_{on}. k_{off} is relatively easy to understand (though calculating it exactly might be difficult). It is either the k_+ or the k_- of Kramers theory, calculated according to Equation 3.91. Which one (k_+ or k_-) is used depends on whether the bound state has a higher or lower free energy than the reactants (as determined from the equilibrium constant).

k_{on} is a trickier quantity to estimate because it depends on two factors. First, the reactants must collide with one another in solution. Then they must also pass through some transition state to get into the bound state, a second step also described by the Kramers equation (if it is really just one step).

The first step, the probability of a collision in solution, depends on the concentration of reactants and their rate of diffusion. Specifically, if the target is a sphere of radius, R, the number of molecules crossing the surface of that sphere per unit time (as given by a solution of the spherically symmetric diffusion equation is (see Howard,[7] Appendix 5.3)

$$I = 4\pi RDC_\infty \text{ particles/second.} \qquad (3.116)$$

Here C_∞ is the bulk concentration of molecules. But $D = \frac{k_B T}{6\pi \eta R}$ according to the Einstein–Smoluchowski relation (Equation 3.86), so the on-rate is independent of particle size. k_{on} is defined by $I = k_{on} C_\infty$, so that, in this simple model, k_{on} is given by

$$k_{on} = \frac{2}{3}\frac{k_B T}{\eta} \text{ (particles/m}^3)^{-1}\text{s}^{-1} \qquad (3.117)$$

which, when converted to units of $M^{-1}s^{-1}$ yields $k_{on} \sim 10^9$ $M^{-1}s^{-1}$ (using the viscosity of water and assuming $T = 300$ K). This calculation assumes that molecules bind as soon as they touch, but this is not the case, because there will also be an activation energy barrier for the formation of the transition state. In reality, k_{on} is on the order of 10^6 to 10^8 $M^{-1}s^{-1}$ for small-molecule protein binding and 10^6 to 10^7 $M^{-1}s^{-1}$ for protein–protein interactions.[8]

Finally, we return to the question of the ratio of reactants in a bimolecular reaction. The discussion of rates might seem to imply that collisions become more probable if the concentration of one species is raised. However, the concentration dependence of the free energy is dictated by reaction quotient, which must be 1:1 for a simple two-particle reaction. Thus, the value of K_D is determined by the smallest amount of reactant. A reactant present in excess certainly has a larger free energy, but, since the excess does not get turned into product, it cannot drive a change in free energy. Viewed from the kinetic perspective just discussed, the excess reagent undergoes unproductive collisions with itself. An interesting situation arises when we discuss interactions of a single, surface tethered molecule, but the discussion of that problem can be left until Chapter 4.

3.20   Rate equations in small systems

All of the discussion above has assumed that the concentrations (or activities) of reactants or products are smooth, continuous variables and that the change in any of the quantities is given by the appropriate rate constant in a differential equation. On the molecular scale, this is not true. The concentrations need to be replaced with a particle number (in a given volume) and the "rate constants" replaced with a *probability* that some reaction will occur. The overall numbers of reactants and products can then be calculated by simulating the progress of a reaction probabilistically. One might think that this would yield exactly the same result as the differential equations governing the bulk system. For very large systems, this holds true. But for small systems, consisting of, for example, a few hundred or a few thousand reactants, completely new phenomena can emerge. One example is time-dependent oscillations in the concentrations of reactants and products in small systems.[9] Rate equations for the bulk concentrations for the same systems predict that concentration fluctuations would be damped out.

3.21   Nanothermodynamics

We will examine self-assembly of nanosystems in Chapter 6, but suffice it to say that interactions between immiscible materials that spontaneously separate often lead to structures of nm to μm dimensions at room temperature. Examples are living cells, colloidal suspensions (like detergents in water), and protein–mineral layers (like nacre in oyster shells). The reason is a competition between bulk and surface energies of the phases that self-assemble. If an interfacial interaction is somewhat favorable between two otherwise incompatible materials, then stable nanostructures can form. Thus, oil and water separate into two macroscopically different phases. But oil that is linked to a detergent tends to form small colloidal particles, filled with oil in the center, but interacting with the surrounding water via the detergent on the surface of the droplet. If the droplet is small enough, it remains soluble (which is how a washing machine, loaded with detergent, can remove oil stains).

This observation poses a problem for conventional thermodynamics. Free energies are extensive quantities—doubling the volume of a material doubles its free energy. There are, therefore, no tools in conventional thermodynamics for dealing with equilibrium nanostructures, where interfacial interactions play an important role. In fact, the problem is dealt with in an *ad hoc* manner by simply adding terms that are functions of the surface area of a particle to the free energy, and we will do this in Chapter 6. However, Terrell Hill has worked out a self-consistent formulation of small-system thermodynamics, now called nanothermodynamics, where the free energy is no longer extensive, but depends on system size.[10] One result is that the canonical ensemble of Gibbs (see Appendix E) can be replaced with the generalized ensemble,[10] where each part of a system can change its size to form a thermodynamic equilibrium distribution of sizes.

Another result is that there is an additional term in the first law of thermodynamics. According to Gibbs, the first law of thermodynamics is

$$dE = T dS - V dP + \mu dN, \qquad (3.118)$$

which Hill generalizes to

$$dE = T dS - V dP + \mu dN + \left(\frac{\partial E}{\partial \eta} \right) d\eta, \qquad (3.119)$$

where $\partial E / \partial \eta$ is called the "subdivision potential" (by analogy with the chemical potential) and η is the number of subdivisions (i.e., the number of independent parts that comprise the system). Thus, all contributions to the energy that do not depend linearly on the number of particles—such as surface effects, length-scale terms, constant factors, and fluctuations that change the net energy—can be treated systematically by the subdivision potential.

The result is a thermodynamically consistent treatment of the equilibrium properties of nanosystems. However, Hill's thermodynamics is not currently widely taught (perhaps because the existing framework is hard enough) and we will stick to using *ad hoc* additions to represent interfacial energy when we discuss self-assembly. The generalized approach of Hill may, however, have consequences beyond introducing self-consistency into the thermodynamic treatment of colloids. Chamberlin has suggested that this formulation of thermodynamics can be used to understand the formation of dynamical inhomogeneities in otherwise homogeneous materials, offering an explanation of some remarkable general properties of disordered materials.[11]

3.22   Modeling nanosystems explicitly: molecular dynamics

Molecular dynamics modeling uses a computer simulation of the mechanics of the individual atoms that make up a system. Essentially, all current approaches are based on Newton's laws of motion applied to the individual atoms, and molecular dynamics is applicable to nanosystems simply because the number of atoms in a nanosystem can be small enough to make the approach practical. There is really no simple way to extend these calculations to real chemical reactions by somehow including Schrödinger's equation, because of the fact that there are many electrons, and even more quantum states for each atom, so the calculations are limited to modeling interactions that can be described by classical forces between pairs of atoms.

Molecular dynamics calculations begin by constructing a model of the forces between all the atoms. The complex force fields are replaced with "springs" that generate the same force for a small displacement as the real force field. Special models are needed for very nonlinear interactions (like hydrogen bonding). The approximation that the force between two atoms is proportional to the displacement away from the equilibrium bond distance (i.e., "spring"-like) is only valid for small displacements. Similar force models are made for rotations around bonds. The simulation begins with a structure that is believed to be a correct starting point and then calculates the motion of each atom using classical mechanics, after moving each atom a small random distance from equilibrium to simulate a finite temperature. Since the calculation yields the potential energy stored in the "springs" that connect atoms and the kinetic energy in the vibrations of the atoms, minimizing the total energy of the system should result in a correct structure. Once the energy-minimized structure is found, something interesting can be done to the system (like docking another molecule up against it, stretching it, etc.) and the new structure calculated.

Classical pairwise potential models ($U(r_{ij})$) are set up (often on the basis of empirical data) so that that the forces between any pair of atoms can be calculated from

$$F_{ij} = -\frac{\partial^2 U(r)}{\partial r_{ij}^2} \delta r_{ij}, \tag{3.120}$$

where δr_{ij} is the relative stretching of the bond between atom i and atom j. The system is set to a desired temperature using equipartition to distribute random particle velocities so that on average

$$\left\langle \frac{1}{2} m_i V_{i,x,y,z}^2 \right\rangle = \frac{1}{2} k_{\mathrm{B}} T. \tag{3.121}$$

One limitation of molecular dynamics lies in the fact that interatomic vibrations are rapid ($\sim 10^{13}$ Hz) so that the time step in the calculations must be much smaller ($\sim 10^{-15}$ s). When such calculations must involve all interacting pairs in tens of thousands of atoms, it becomes impractical to calculate out to times that are much more than a few ns from the starting time. Suppose that we are simulating a system of 1000 atoms (a tiny nanosystem, only 10 atoms on each side) using a very fast computer that processes one calculation per ns. The number of calculations needed scales as $\propto N^2$ (depending on the type of problem—if all possible pairs of particles need to be considered, the number of interactions would scale like 2^N). Therefore, we would have to carry out at least a million calculations, consuming a least a millisecond to model a femtosecond! Going out to a nanosecond would take 20 min of computer processor time, and to a microsecond about a month. The calculated energy of the system only approaches the true free energy when the calculation has run for long enough to sample all relevant fluctuations, so that "real" thermal averages can be calculated. For this reason, molecular dynamics calculations have limited applicability as tools to calculate thermodynamic quantities for all but the smallest nanosystems.

One current approach to solving this problem exploits the principle of least action. This alternative statement of Newton's law says that the path followed by a system is that which minimizes the action[12] defined by

$$\int_{t_1}^{t_2} (KE - PE)\, dt, \tag{3.122}$$

where KE means the kinetic energy of the system as a function of time and PE means the potential energy. Thus, one may make a guess about the trajectory of a system from t_1 to t_2 (corresponding to from position 1 to position 2), calculate the action integral according to Equation 3.122, vary the path a little, and recalculate the action until a path is found that minimizes the action, and therefore might indeed represent a reaction coordinate for the process. Once such a path is found, the energy and kinetics of the system can be calculated by carrying out a series of smaller simulations that compile statistics about fluctuations in the system for a series of carefully chosen points along the reaction pathway—an approach known as "milestoning."[13,14]

There is little doubt that methods for carrying out calculations like these will develop rapidly with improvements in techniques, in computer code, and in hardware. One useful new tool avoids the calculation of the atomistic dynamics altogether, using geometry to find the modes of a complex system that are intrinsically "floppy" and therefore most likely to correspond to the large fluctuations of the system.[15] This approach is compared with a molecular dynamics simulation in Fig. 3.15. The geometric approach generates

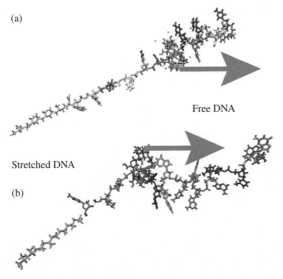

Fig. 3.15 (a) Shows a molecular dynamics simulations of an experiment in which a molecular ring is pulled over a single-stranded DNA molecule by an atomic force microscope. The left-hand side of the DNA is tethered to a fixed point and the red arrow shows the direction of the force applied to the ring that slides over the DNA. (A bath of 5000 water molecules and 13 Na ions is not shown). Note that the "free" DNA on the right-hand side of the pull is still lying close to is original path, even though single-stranded DNA is very flexible and should be wriggling around like a snake. This is because the calculation only ran out to ~0.5 ns, so that only local vibrational modes of the DNA were sampled. (b) Shows a simulation carried out with geometrical approach that randomly samples all possible fluctuations (albeit on an uncalibrated timescale). The geometry starts out like that in (a), but now the molecular ring becomes trapped against a loop of DNA that has formed spontaneously. Degrees of freedom like this were not caught in the molecular dynamics simulation.

qualitatively different dynamics from molecular dynamics, but it does not readily yield quantitative data.

3.23   Systems far from equilibrium: Jarzynski's equality

Processes in nanomachines like molecular motors are often far from thermodynamic equilibrium, so it would appear that the equilibrium thermodynamics approaches are of no use. However, in 1997, Jarzynski proved a remarkable theorem that connects the nonequilibrium dynamics of a system with its equilibrium properties. The implications of the theorem are still being worked out at the time of writing. Jarzynski's theorem states that the free energy difference between the initial and final states of a system can be obtained from a Boltzmann-weighted average of the *irreversible* work done along a reaction coordinate that connects the initial and final states of the system. Specifically, the free energy difference separating states of a system at positions 0 and z, $\Delta G(z)$, is related to the work done to irreversibly switch the system between two states by the following relationship:

$$\exp[-\beta \Delta G] = \lim_{N \to \infty} \langle \exp\left[-\beta w_i(z, r)\right] \rangle_N, \tag{3.123}$$

where $\langle\ \rangle_N$ denotes averaging over N work trajectories, $w_i(z, r)$ represents the work of the ith of N trajectories, and r is the switching rate. The mechanical work $w_i(z, r)$ required to switch the system between positions 0 and z under the action of a force F is

$$w_i(z, r) = \int_0^z F_i(z', r)\, dz', \tag{3.124}$$

where $F_i(z', r)$ is the external force applied to the system at position z' with switching rate r. Thus a system, switching between an initial and final state irreversibly at a rate r under the action of a force, can be described in terms of the equilibrium free energy difference between the initial and final states.

This equality was recently put to a direct test by Liphardt et al.[16] They studied the folding and unfolding of a small RNA molecule that was pulled apart by "optical tweezers" (see Chapter 4). When the force that was applied across the ends of the RNA molecule was ramped up slowly and then ramped down again slowly, the measured force vs. extension curve showed a little kink at the position where the RNA unfolded on the way up and refolded on the way down (Fig. 3.16(a), left curve). When the force was ramped more quickly, folding and refolding were not coincident, resulting in a hysteresis loop (right curve). The area under the loop represents the work put into the system that is lost on refolding away from equilibrium. Since the energy loss mechanism has nothing to do with the intrinsic refolding, one might think that the faster measurements would yield only an upper limit on the value of ΔG for the unfolding event. But when a series of nonequilibrium curves (Fig. 3.16(b)) was analyzed using Equations 3.123 and 3.124, the correct ΔG between folded and unfolded states was recovered.

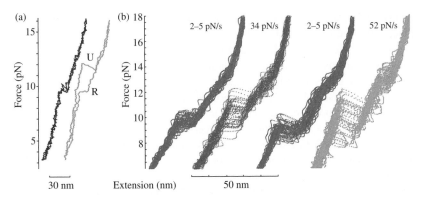

Fig. 3.16 A equilibrium force vs. extension plot for a small RNA molecule (left curve in (a)). The jump in extension at a force of 10 pN corresponds to the extra length yielded as hairpins and loops in the RNA unfold. The two overlapping curves are for increasing and decreasing force. When pulled more rapidly, the folding is not reversible (curves on right). A collection of data for irreversible folding (b) were used together with Jarzynski's theory to extract the equilibrium free energy difference between folded and unfolded states from this nonequilibrium data (from Liphardt et al.[16] Reprinted with permission from AAAS).

This experiment shows how Jarzynski's remarkable result could be used to extract information that might otherwise require impossibly long timescales to calculate or measure and suggests new ways that use short-timescale simulations to deduce the equilibrium behavior of a system.

3.24 Fluctuations and quantum mechanics

Other than a passing reference to time-dependent perturbation theory, we have left the subject of time-dependent quantum mechanics well alone, because it is a difficult subject. But quantum phenomena take time to develop (e.g., see the discussion of the time development of resonant tunneling in Chapter 7). Some estimate of this time comes from the uncertainty principle as stated in the form of Equation 2.8:

$$\Delta E \Delta t \sim h.$$

Planck's constant is 4.14×10^{-15} eV-s, so a strong covalent bond (a few eV) forms on timescales of $\sim 10^{-15}$ s. On the other hand, a very weak bond (on the order of $k_B T$ or 0.025 eV) develops on timescales of $\sim 10^{-13}$ s. This is roughly the frequency of typical molecular vibrations. Harmonic vibrational modes have no spin or chemical potential associated with them, so they are bosons, occupied according to Equation 3.58 (with $\mu = 0$)

$$\bar{n}_i = \frac{1}{\exp\left[\beta \varepsilon_i\right] - 1} = 0.58 \quad \text{for } \beta \varepsilon_i = 1.$$

Thus vibrational modes are $\sim 50\%$ occupied (when vibrational energy equals thermal energy), *so a system at room temperature is changing on the timescale of the buildup of quantum interference effects in weak bonds*. This is why we can talk of covalent bonds in terms of quantum mechanics, but of a whole protein, for example, as a classical object.

There is a formulation of quantum mechanics designed to permit thermal averaging of quantities calculated using quantum mechanics. It is based on the *density operator* (not to be confused with density functional theory). In this formulation, wavefunctions are expanded in terms of basis sets: For example, our simple hydrogenic 1s state, $\exp -r/a_0$ could be expressed as an infinite sum over sines and cosines. This makes the formulation much more complicated, but it allows for the possibility that the wavefunction evolves into something new over time as the system interacts with its environment.

The approach is as follows: We first express the wavefunctions of a "pure" quantum mechanical system (one not interacting with a heat bath) in terms of the density operator. We will then see how the density operator can be reformulated for a mixed state (where certain parts of the basis are changing owing to interactions), writing down an equation of motion for the density operator from which the time-evolution of the system can be obtained. The key point is that, under the application of significant randomizing influences, quantum interference effects can be destroyed, as discussed qualitatively above.

The expectation value of an operator $\hat{A}$ (e.g., angular momentum, kinetic energy, etc.) is defined in terms of density operator, $\hat{\rho}$, by the matrix operation

$$\langle A \rangle = Tr\hat{\rho}\hat{A}, \tag{3.125}$$

where Tr ("take the trace of") implies that that diagonal elements of the matrix are summed. Equation 3.125 defines the density operator. The requirement that the probabilities be normalized is equivalent to

$$Tr\hat{\rho} = 1 \tag{3.126}$$

For the case of a system whose wavefunctions are known, the expectation value of $\hat{A}$ is given by

$$\langle A \rangle = \langle \psi| A |\psi \rangle . \tag{3.127}$$

The wavefunction can always be expanded in terms of an appropriate set of basis states $\{|n\rangle\}$;

$$|\psi\rangle = \sum_n |n\rangle \langle n | \psi \rangle . \tag{3.128}$$

This expression uses the Dirac notation where a closed bra and ket imply an integration over all space, so the term $\langle n | \psi \rangle$ is the coefficient weighting how much of state $|n\rangle$ appears in ψ. Inserting this expansion into the definition of the expectation value (3.127) and using 3.125,

$$\langle A \rangle = \sum_q \sum_n \langle \psi | q \rangle \langle q| \hat{A} |n\rangle \langle n | \psi \rangle \equiv \sum_q \sum_n \rho_{nq} A_{qn} = Tr\hat{\rho}\hat{A}. \tag{3.129}$$

Here, we have defined the elements of the density matrix according to

$$\rho_{nq} = \langle \psi | n \rangle \langle q | \psi \rangle \equiv a_n^* a_q. \tag{3.130}$$

The sum of the diagonal elements, $a_n^* a_n$, does indeed correspond to the probability of finding the system in one of the states, justifying the statement that the trace of the density operator is unity.

What have we achieved here? We have put the wavefunction into a rather unwieldy form that expresses it in terms of all its projections onto a complete basis. This is useful as we now turn to consider mixed states. To do this, we will consider an ensemble of density operators, each corresponding to a different interaction with the environment and we will replace the matrix elements with values that are averaged over this ensemble:

$$\rho_{nq} = \left\langle a_q^* a_n \right\rangle_{\text{ensemble}}. \tag{3.131}$$

(here the brackets mean "the average of" as in statistical physics—they are not Dirac bras and kets). Operating on each of the states $|n\rangle$ with the Schrödinger equation leads to[17]

$$i\hbar \frac{\partial \rho_{nl}}{\partial t} = \sum_k (H_{nk}\rho_{kl} - \rho_{nk}H_{kl}), \tag{3.132}$$

which, written in the commutator notation introduced in Appendix C, is

$$i\hbar \frac{\partial \hat{\rho}}{\partial t} = \left[\hat{H}, \hat{\rho}\right]. \tag{3.133}$$

This is the equation of motion for the density operator. Thus, we have a set of tools for making quantum mechanical predictions for systems, using not wavefunctions, but rather sets of time averaged numbers that represent the projection of the system wavefunction onto a basis set.

Using this approach, we can consider what happens when the phase of the individual components is fluctuating rapidly owing to interactions with a heat bath. Writing a_n as

$$a_n = c_n \exp i\phi_n \tag{3.134}$$

we see that the elements of the density matrix are given by

$$\rho_{nm} = \left\langle c_m^* c_n \exp i(\phi_n - \phi_m) \right\rangle_{\text{ensemble}}$$
$$= \left\langle c_m^* c_n \right\rangle \left\langle \cos(\phi_n - \phi_m) + i \sin(\phi_n - \phi_m) \right\rangle \tag{3.135}$$

The sine and cosine terms are rapidly varying with time and will average to zero, except for the diagonal terms for $n = m$ where the exponentials are unity. Thus, in the presence of strong, rapid fluctuations, the system appears to lose all interference terms: it looks like a classical particle. Putting a hydrogen atom into a pot of boiling oil will not make it classical, because the fluctuation states are not occupied at energies as high as 13.6 eV, but quantum effects that depend on much weaker interactions that are occupied at the temperature of boiling oil will be destroyed. Some small parts of the system which are interacting strongly will remain "quantum mechanically pure" over a sufficiently long time that we can describe electron transfer (for example) as occurring according to Fermi's golden rule. However, the energies of the states will be fluctuating on a scale of a few times $k_B T$. Indeed, these fluctuations play a central role in chemical reactions as was reviewed earlier in this chapter. How do we cope with this mixed "quantum-classical" behavior? The solution to this problem lies in realizing that the mass of electrons is so much smaller than the mass of nuclei, that the problem can be worked by solving the quantum mechanical problem of the electronic structure *with all the nuclei fixed in place*, allowing the nuclei to move to a new position and then solving the electronic structure again with the nuclei fixed. This is called the Born–Oppenheimer approximation.

3.25   Bibliography

C. Kittel and H. Kroemer, Thermal Physics, 2nd ed. 1980, New York: W.H. Freeman. The standard undergraduate physics text for statistical mechanics. Very clear, but do not be confused at first glance by the choice of normalized temperature ($\tau = T/k_B$) used in the book.

F. Mandl, Statistical Physics. 1988, New York: Wiley. I used this book as an undergraduate and liked it.

D.A. McQuarrie and J.D. Simon, Physical Chemistry: A Molecular Approach. 1997, Sausilito, CA: University Science Books. The "bible" of graduate physical chemistry. Comprehensive and very clear, it has lots of examples of statistical mechanics calculations related to chemistry.

J. Howard, Mechanics of Motor Proteins and the Cytoskeleton. 2001, Sunderland, MA: Sinauer Associates. This lovely little book is referred to often in this chapter, because it puts statistical mechanics and chemical kinetics in a biophysical context and is a much easier read than some of the standard texts.

Here are some newer texts recommended to me by colleagues:

J.P. Sethna, Statistical Mechanics: Entropy, Order Parameters and Complexity. 2006, Oxford: Oxford University Press.

S.J. Blundell and K.M. Blundell, Concepts in Thermal Physics, 2nd ed. 2010, Oxford: Oxford University Press.

M. Glazer and J. Wark, Statistical Mechanics: A Survival Guide. 2001, Oxford: Oxford University Press.

3.26   Exercises

a. Conceptual questions:

1. A heater is placed inside a sealed cylinder containing a gas and a fixed amount of heat put into the gas by operating the heater at a fixed current and voltage for a fixed time. In one experiment, the container has a fixed size. In the second experiment, one wall is a free moving piston that maintains the pressure inside the cylinder equal to atmospheric pressure. In which experiment does the gas get hotter? Explain your answer using the first law of thermodynamics.

2. Assuming that the acceleration owing to gravity is constant up to 30,000 ft, explain how the velocity of an object dropped from a high flying plane changes with time (refer to the Stokes' formula for viscosity).

3. How does the boiling point of water depend on the volume of gas available above the water?

4. Based on the number of reactants and products, predict the sign of the entropy change:

$$2NO + O_2 \longrightarrow 2NO_2$$

5. Sketch the entropy of 1 mole of water as a function of temperature schematically from the absolute zero of temperature to above the boiling point of water. In particular, comment on the behavior of the entropy at the melting and boiling points.

6. The following is an endothermic reaction. Identify and discuss the molecular change that makes this reaction spontaneous despite the fact that the bonds that are reformed require that heat is taken *into* the reactants.

$$2NH_4SCN(l) + Ba(OH)_2.8H_2O(s)$$
$$\longrightarrow Ba(SCN)_2(aq) + 2NH_3(aq) + 10H_2O$$

In this reaction, "l" means liquid, "aq" means aqueous solution, and "s" means solid. The ".8H$_2$O" means that eight water molecules are crystallized along with each Ba(OH)$_2$.

7. How different is the specific heat of a gas of diatomic molecules from a monatomic gas? Can you give a quantitative answer and discuss what was assumed about the types of motion that are thermally excited?

8. What is the effect (increase or decrease) on ΔG for the following reactions of increasing the partial pressure of H_2?

$$N_2 + 3H_2 \longrightarrow 2NH_3$$
$$2HBr \longrightarrow H_2 + Br_2$$
$$2H_2 + C_2H_2 \longrightarrow C_2H_6$$

9. The *work function* of a metal is the work done in removing an electron from the bulk of the metal to far from the metal. How would you express the work function in terms of the free energy of the electrons in the metal, and, in particular, what free energy would you use?

10. Can a metal ever become hot enough to behave classically? (*Hint*: The boiling point of titanium is ~3650 K).

11. Cargo in cells diffuses along molecular tracks called microtubules. How do you expect diffusion along these tracks to compare to free diffusion of the cargo in the cytoplasm of the cell? To tackle this problem, sketch the projection of a number of 2-D random walks onto a fixed 1-D axis and compare with the effect of taking the same number of steps in 1-D. Then extrapolate to 3-D.

12. In semiconductor materials, charge is carried by two types of carrier: electrons and positively charged quasiparticles called holes. Typically, the mobility (the speed per unit electric field) of holes is much lower than the mobility of electrons. If the electric field is turned off, will electrons diffuse randomly (a) at the same rate as holes, (b) faster than holes, or (c) more slowly than holes?

13. Will the endothermic reaction in Question 6 occur at the absolute zero of temperature? Explain your answer.

14. Using sketches of the energy landscape for a chemical reaction that turns reactants into products, show what a catalyst must do in terms of the landscape.

15. Rank the following in terms of the $\Delta G/RT$ for the accompanying reaction:

$$K_{eq} = 10^{15}, \quad K_{eq} = 10^{-15}, \quad K_{eq} = 10, \quad K_d = 1 \text{ nM}.$$

b. Simple numerical questions:
(Use data from Appendix A where needed)

16. Consider nine identical particles allocated into energy states $\varepsilon_1 < \varepsilon_2 < \varepsilon_3$ according to

$$n(\varepsilon_1) = 4$$
$$n(\varepsilon_2) = 3$$
$$n(\varepsilon_3) = 2.$$

Calculate the statistical weight, Ω of the system, recalling that the number of distinct ways of putting N objects into groups of n_1, n_2, etc. is

$$\frac{N!}{n_1!n_2!\cdots}.$$

17. What is the entropy of the system in Question 16 in Joules/K? What is the entropic contribution to the free energy at $T = 300$ K?

18. What distribution of the nine particles in Question 16 would maximize entropy of the system? Why is this not the actual distribution observed when the system is in contact with a heat bath?

19. What is the thermal kinetic energy (in Joules) of 1 mole of an ideal gas at 300 K? What is the root-mean-square fluctuation in this energy in Joules? What is the thermal kinetic energy of one attomole of an ideal gas? What is the root-mean-square fluctuation in this energy in Joules?

20. Gaseous HI placed in a closed container at 425°C partially decomposes into H_2 and I_2. The equilibrium concentrations are 3.53×10^{-3} M for HI, and 4.8×10^{-4} M for H_2 and I_2. Calculate the equilibrium constant for this reaction (see Exercise 3.14).

21. If the ratio of products to reactants in equilibrium is 10:1, how much energy could be extracted on a per molecule basis at $T = 300$ K? Express your answer in eV and calories per mole.

c. Mathematical problems:

22. Molecules will posses, in addition to translational states (for which we calculated the partition function) vibrational, rotational, and electronic states. How would the partition functions corresponding to each of

these additional degrees of freedom modify the single particle partition function we derived? Why would the electronic partition function be close to unity at room temperature?

23. Derive Equation 3.58, the thermal average occupation function for a single state of a system of bosons.

24. Derive Equation 3.41 from 3.36. To take the derivative of the logarithm term, write it out as a sum in the form $\ln(\text{Constant}) + 3/2\ln(T)$, use $\frac{d}{dx}\ln(x) = \frac{1}{x}$ and also use the fact that $\ln(a/b) = \ln(a) - \ln(b)$ to pull the $\ln N$ term into the argument.

25. The quantum mechanical energy levels of a harmonic oscillator are given by $E_n = \hbar\omega(n + \frac{1}{2})$. Assume that a solid can be described as N oscillators all of frequency ω_E. Calculate the partition function for a single oscillator, and thus the Helmholtz free energy, F, for a single oscillator. Use $\langle E \rangle = -\frac{\partial \ln Z}{\partial \beta}$ and replace Z with F to show that

$$\langle E \rangle = \frac{1}{2}\hbar\omega_E + \frac{\hbar\omega_E}{\exp(\beta\hbar\omega_E) - 1}.$$

Thus, calculate the specific heat of an ensemble of N oscillators each with three modes of oscillation equal to ω_E in frequency. To sum the single particle partition function for $n = 0$ to $n = \infty$ you will need to use the result that $(1 - t)^{-1} = 1 + t + t^2 + \cdots$ for $|t| < 1$. This is the Einstein model for specific heats, and it explained why the specific heat of materials falls below the Dulong and Petit value as temperature is lowered. Note also how the introduction of quantized levels (with no restriction on occupation) has produced the Bose–Einstein result.

References

[1] Dauxois, T., M. Peyrard, and S. Ruffo, The Fermi–Pasta–Ulam 'numerical experiment': history and pedagogical perspectives. Eur. J. Phys., **26**: S3–S11 (2005).

[2] W.T. Coffey, Y.P. Kalmykov, and J.T. Waldron, *The Langevin Equation with Applications in Physics, Chemistry and Electrical Engineering*. 2nd ed. 2003, Singapore: World Scientific.

[3] Kramers, H.A., Brownian motion in a field of force and diffusion model of chemical reactions. Physica, **7**: 284–304 (1940).

[4] Hänggi, P., T.J. Mroczkowski, F. Moss, and P.V.E. McClintock, Bistability driven by colored noise: theory and experiment. Phys. Rev. A, **31**: 695–698 (1985).

[5] Hänggi, P., Reaction rate theory: 50 years after Kramers. Rev. Mod. Phys., **62**: 252–341 (1990).

[6] McCann, L.I., M. Dykman, and B. Golding, Thermally activated transitions in a bistable three-dimensional optical trap. Nature, **402**: 785–787 (1999).

[7] Howard, J., *Mechanics of Motor Proteins and the Cytoskeleton*. 2001, Sunderland, MA: Sinauer Associates.

[8] Fersht, A., *Enzyme Structure and Mechanism*. 1985, New York: W.H. Freeman.

[9]   McKane, A.J., J.D. Nagy, T.J. Newman, and M.O. Stefani, Amplified biochemical oscillations in cellular systems. J. Stat. Phys., **128**: 165–191 (2007).

[10]   Hill, T., *Thermodynamics of Small Systems*. 1994, New York: Dover.

[11]   Chamberlin, R.V., Mesoscopic mean-field theory for supercooled liquids and the glass transition. Phys. Rev. Lett., **82**: 2520–2523 (1999).

[12]   Feynman, R.P., *Statistical Mechanics*. 1972, Reading, MA: Addison-Wesley.

[13]   Elber, R., A. Ghosh, and A. Cardenas, Long time dynamics of complex systems. Acc. Chem. Res., **35**: 396–403 (2002).

[14]   Faradjian, A. and R. Elber, Computing time scales from reaction coordinates by milestoning. J. Chem. Phys., **120**: 10880 (2004).

[15]   Flores, S., N. Echols, D. Milburn, B. Hespenheide, K. Keating, J. Lu, S. Wells, E.Z. Yu, M.F. Thorpe, and M. Gerstein, The Database of Macromolecular Motions: new features added at the decade mark. Nucl. Acids Res., **34**: D296–D301 (2006).

[16]   Liphardt, J., S. Dumont, S.B. Smith, I. Tinoco, and C. Bustamante, Equilibrium information from nonequilibrium measurements in an experimental test of Jarzynski's equality. Science, **296**: 1832–1835 (2002).

[17]   Liboff, R.L., *Introductory Quantum Mechanics*. 1980, Oakland, CA: Holden-Day, Inc.

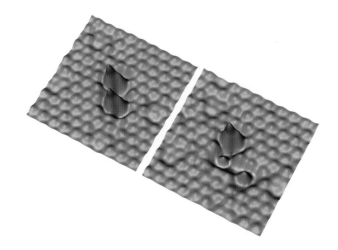

Plate 1 Chemical reaction induced by pushing two atoms together with a scanning tunneling microscope. See Fig. 1.5. (Courtesy of Professor Wilson Ho.)

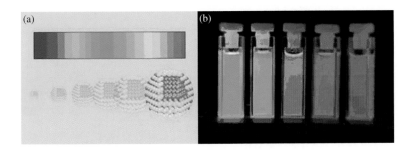

Plate 2 Fluorescence from quantum dots gets redder in color as the dots get larger. See Fig. 2.9. ((a) Courtesy of Dylan M. Spencer, (b) Source: Philips.)

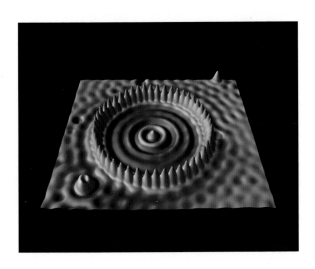

Plate 3 The "quantum corral." Iron atoms arranged in a circle trap electrons on the surface of nickel. See Fig. 2.13. (Photo courtesy of IBM.)

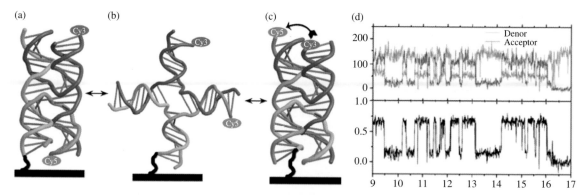

Plate 4 Mapping the dynamics of a DNA cross-structure using fluorescent labels attached to the DNA and recording their interaction using optical interactions between them. See Fig. 4.34. ((a), (b) and (c) Reprinted by permission from Macmillan Publishers Ltd: Nature Structural & Molecular Biology, Watching flipping junctions, 2003; (d) Reprinted by permission from Macmillan Publishers Ltd: Nature Structural and Molecular Biology, Structural dynamics of individual Holliday junctions, 2003.)

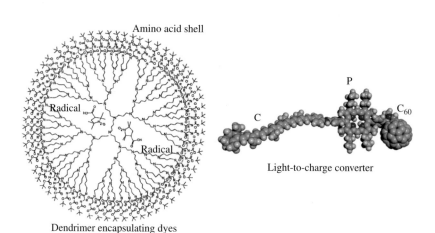

Plate 5 Structures synthesized by organic chemists. See Fig. 6.1. (Dendrimer reprinted with permission of John Wiley & Sons, Inc.; Light-to-charge converter courtesy of Professor Devens Gust of Arizona State University; Nano-muscle reprinted with permission from Linear artificial molecular muscles, Y. Liu, A.H. Flood, P.A. Bonvallet, S.A. Vignon, B.H. Northrop, H.-R. Tseng, J.O. Jeppesen, T.J. Huang, B. Brough, M. Baller, S. Magonov, S.D. Solares, W.A. Goddard, C.-M. Ho and J.F. Stoddart, J. Am. Chem. Soc., 2005. 127: 9745–9759, American Chemical Society.)

Plate 6 In ancient times, stained glass makers put small amounts of gold chloride into glass to make the brilliant reds and orange colors seen in this window. In the reducing environment, small colloidal gold particles formed, scattering light strongly at the plasma frequency (Chapter 9.6). (From http://en. wikipedia.org/wiki/Image:Heaton,_Butler_ and_Bayne01.png.)

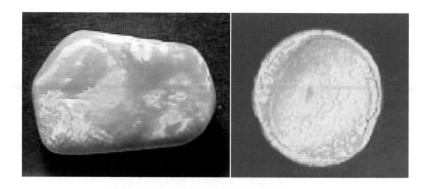

Plate 7 Bragg reflection from nanoscale periodic mineral structures in opal give rise to the colors characteristic of this gem (left image). The colloidal crystal shown in Fig. 9.15 produces colors by the same mechanism (right image). (Courtesy of Minnesota Materials Research Science and Engineering Center and Professor Andreas Stein. All rights reserved.)

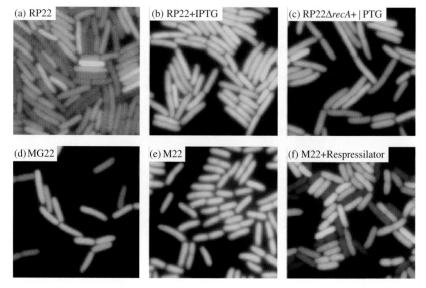

(a) RP22   (b) RP22+IPTG   (c) RP22Δ*recA*+ | PTG

(d) MG22   (e) M22   (f) M22+Respressilator

Plate 8 Significant fluctuations in gene expression occur from bacterium-to-bacterium because of the small number of molecules involved in the process. Here, a yellow color signifies a constant (high) level of gene expression whereas green and orange colors signify departures from the average. See Fig. 10.22. (Reprinted with permission from Stochastic Gene Expression in a Single Cell, Elowitz, M.B., A.J. Levine, E.D. Siggia, and P.S. Swain, Science, 297: 1183–1186 (2002). AAAS.)

Tools

Microscopy and manipulation tools

4

This chapter describes the important imaging and manipulation tools without which Nanoscience would be impossible. We begin with a description of the scanning tunneling microscope (STM; Section 4.1) because this was the first instrument that made the manipulation of single atoms feasible. Atomic scale imaging had been possible (using electron microscopes) for some time before the STM was invented, but the ease of operation and construction of STMs popularized Nanoscience and Nanotechnology. The STM has largely been superseded by its close relation, the atomic force microscope (AFM), because the AFM is capable of imaging insulating surfaces and therefore has a more general set of applications (Section 4.2). The AFM also enabled the measurement of the mechanical properties of individual molecules, and this application to single molecule force measurements is also reviewed here (Section 4.2.6). Despite these enormous advances in scanning probe microscopy, the electron microscope still offers the most accurate measurement of many nanoscale structures and we review some aspects of electron microscopy in Section 4.3. At around the same time that the AFM was becoming popular, the optics community began to develop techniques and dyes for fluorescent imaging of single molecules (Section 4.4). These methods broke the classical limits on optical resolution, allowing particles to be located with nanometer accuracy in optical images. The new optical methods also enabled new approaches to measuring motion at a single molecule level. The discovery that small particles could be manipulated with a light beam also occurred in this time frame (the mid-1980s), leading to the development of devices that manipulate single molecules attached to small particles (Section 4.5). Methods for the direct imaging of nanostructures and the measurement of their mechanical properties and their fluctuations have undergone a dramatic revolution in the past 20 years.

4.1 The scanning tunneling microscope

4.1.1 The scanning tunneling microscope—history

Nanoscience and nanotechnology burst onto the scene as a result of the invention of the scanning tunneling microscope[1] (STM) by Gerd Binnig and Heine Rohrer at IBM labs in Switzerland in 1979.[2] The electron microscope has long been the tool of choice for imaging nanoscale structures, and the AFM has largely superseded the STM as a probe of surface structure. But there is much to be gained by beginning with a description of the STM, despite the fact that its application is limited to the imaging of very clean (or chemically very well-defined) conducting surfaces. This is because many of the important features of the STM are shared by the AFM (discussed later in this chapter).

Binnig had suggested the idea of using tunneling to profile surfaces while still a graduate student in Frankfurt. His first thought was that he could use a very sharp metal whisker that would be scanned over the surface to record changes in tunnel current as the height of the surface changed. He soon realized that a fine whisker would be too susceptible to vibrations. Making such a microscope immune to vibrations requires the use of a short stubby probe, but Binnig also realized that even a stubby probe could give high resolution. Figure 4.1 reproduces a page from his laboratory notebook on which he calculated the contributions to tunnel current from a probe of 1000 Å radius as a function of position from the apex. Because the tunnel current decays exponentially with distance (see Equation 2.60) significant current is contributed only by the

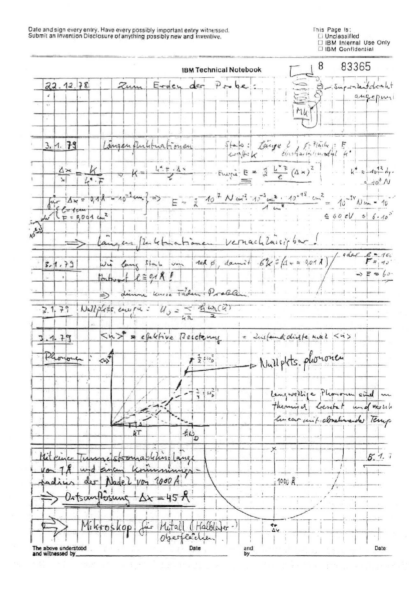

Fig. 4.1 Facsimile of a page from Binnig's lab notebook. At the bottom of the page he shows how most of the tunnel current from a probe of radius 1000 Å comes from a region of only 45 Å in radius. Thus, a short and stubby (hence vibration proof) probe could be used. (Reprinted with permission from C.F. Quate, Physics Today,[3] copyright 1986 by the American Institute of Physics, courtesy of Professor C.F. Quate.)

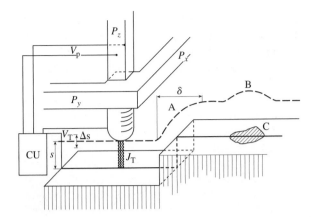

Fig. 4.2 Schematic layout of the first STM. The three piezoelectric rods are labeled P_X, P_Y, and P_Z. The region of highest tunnel current density is labeled J_T. The current is kept constant by raising the probe over steps (A) or (B) over regions where the chemistry of the surface causes the tunnel current to increase (C). (Reprinted from Figure 1 of Binnig et al.[2], Physical Review Letters, with permission from the American Physical Society and courtesy of Professor Christophe Gerber.)

portion of the probe that is closest to the surface. This results in a resolution of ~45 Å as shown (underlined) on the bottom left of Binnig's lab notebook (Fig. 4.1).

The first paper describing the operation and use of an STM appeared in 1982.[2] Figure 4.2 shows the arrangement of that first microscope. A metal probe was attached to a tripod consisting of three piezoelectric elements (shown in the figure as P_X, P_Y, and P_Z). A piezoelectric material is one that changes its dimensions when a voltage is applied across it. (The converse effect, the generation of a voltage by a strained piezoelectric material, is probably familiar as the basis of the percussion ignition systems found on gas appliances.) One of the elements, P_Z, is used to control the height of the probe above the surface. The other two (P_X and P_Y) are used together to scan the probe across the surface in a raster pattern. The probe is advanced toward the surface until a preset level of tunnel current is detected. At this point, the advance is terminated and the $x-y$ scan of the surface begins. During this scan, a feedback circuit is used to control the height of the probe above the surface so that the tunnel current remains constant. Changes in the height of the probe reflect either changes in the height of the surface (A in Fig. 4.2) or a change of the local work function (B in Fig. 4.2). (The work function is ϕ in Equation 2.60.) The signal that is used to correct the height of the probe is also sent to a computer display where the display intensity is used as a measure of height at a particular point on the raster scan.

A remarkable surprise emerged from the first application of the STM to study the surface of a gold crystal. Atomic scale steps (3 Å) were seen quite clearly in the scans across the gold surface. Clearly, the microscope was capable of imaging at a resolution much greater than the predicted 45 Å. This surprising result came about because, at the nanoscale, the probe is not really a sphere. It has an atomic scale structure, and if this includes one atom dangling from the very tip, then atomic scale resolution is possible. This remarkable effect only occurs if the surface being imaged is extremely flat. A rough surface will sample tunneling signal from around the edges of the tip, so resolution is indeed ultimately limited by the radius of the probe. However, many surfaces of interest can be prepared so as to be extremely flat and atomic resolution is regularly obtained on surfaces like this.

This unexpected atomic scale resolution was put to use to obtain images of the rather complex surface that results from annealing the silicon (111)* surface

*Crystal surfaces are specified by the direction of a vector perpendicular to the surface: here 111 means that the vector has a unit component along the x, y, and z axes. This particular surface is densely packed and has a low surface energy.

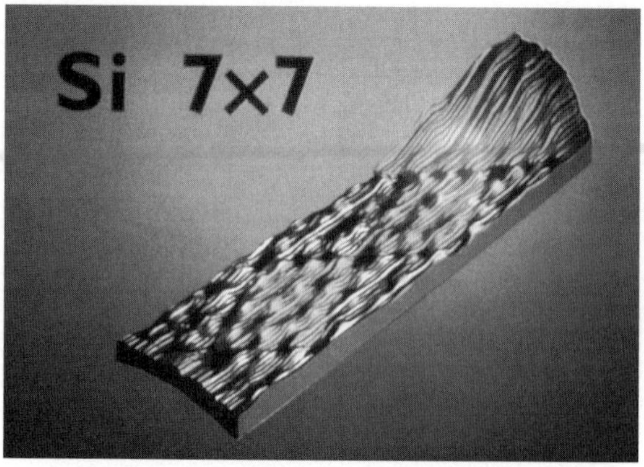

Fig. 4.3 Three-dimensional reconstruction of the silicon "7 × 7" surface. This model was made by cutting out and stacking traces from chart recorder paper. (Courtesy of IBM Zurich Research Laboratory. Unauthorized use not permitted.)

at high temperature in ultrahigh vacuum. The reconstructed silicon surface is called the silicon "7 × 7" surface because of the symmetry of the pattern obtained when low-energy electrons are diffracted from it. At the time (1983), the structure of this surface was a matter of considerable controversy, so a direct image, obtained by STM, had an enormous impact.[4] The first image of the silicon 7 × 7 surface reported in the 1983 paper is shown in Fig. 4.3. At this early stage even IBM Corp. had not successfully attached a computer to a scanning probe microscope, so this three-dimensional rendition was made by cutting up copies of traces made on an $x-y$ chart recorder, stacking them together, and gluing them!

The very first instrument was rather complicated. It relied on magnetic levitation of a superconductor to provide vibration isolation. Within a few generations of design the basic STM mechanism became so small, compact, and rigid that it was easily capable of atomic resolution when operated on a tabletop. In addition to giving atomic resolution images, the probe could be used to move things about on a surface at the nanometer scale. Unlike the electron microscope, the STM could be operated outside of a vacuum and atomic resolution was soon obtained with an STM operated in water.[5] Perhaps even more remarkably, this simple tool made it possible to touch atoms for the first time. All of this from an instrument that could be put together for a few hundred dollars by a graduate student over a period of just a few months! It is hardly surprising that the word nanotechnology burst into the popular lexicon at about this time.

4.1.2   The scanning tunneling microscope—how it works

Here we are going to take a fairly detailed look at the operation of an STM. A schematic form of the height control circuit is shown in Fig. 4.4. The triangular symbols represent operational amplifiers. Operational amplifiers are very high gain analog amplifiers usually available as integrated circuits. They have a very high input impedance (i.e., it takes essentially no current to drive them) and so when operated in negative feedback (some fraction of the output voltage fed back to an inverting input) the output goes to a level such that the voltage difference between the plus and minus terminals is zero. (For a fuller description see any modern text on the electronics.[6]) The schematic diagram in Fig. 4.4 shows a sample biased at V_{sample} volts with respect to ground. The STM tip

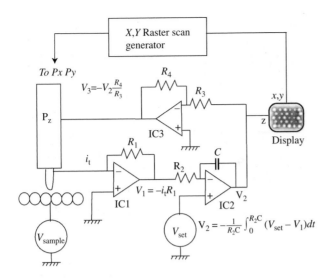

Fig. 4.4 Schematic circuit of the height control circuit of an STM. The triangles represent "operational amplifiers," very high gain amplifiers that are operated by feeding the output back to the input so as to keep the "+" and "−" input terminals at the same potential.

is connected to the inverting input of IC1. Because the noninverting input is connected to ground, the feedback acts to keep the voltage at the inverting terminal at the same potential as ground. Thus, the tip is effectively held at ground potential (this is called a virtual ground). To achieve this condition, the output voltage of IC1, V_1 goes to a value equal to minus the product of the tunnel current, I_t, with the feedback resistor, R_1. Thus, the voltage output from IC1 is directly proportional to the tunnel current (but of inverted sign). IC2 carries out the height control function. A programming voltage, V_{set}, is applied to the noninverting terminal and the capacitor in the feedback loop acts to integrate the difference between V_{set} and V_1. The output, V_2, is directly proportional to the difference between the measured tunnel current and the desired set-point tunnel current (represented by V_{set}). The error signal is sent to a high voltage amplifier, IC3, which drives the height control actuator, P_Z. The overall phase of the feedback is arranged to be negative; that is to say if the tunnel current increases, the probe is pulled away from the surface. Thus the signal, V_2, continues to change in a direction that increases or decreases the gap, as needed, until the difference between V_1 and V_{set} is zero. In this way, the control signal, V_2, also represents the height of the sample and it is used to modulate the brightness of a display. The x−y position of the display is synchronized to the x−y position of the probe by driving the probe and the display with the same waveforms used to generate a raster scanning pattern on the surface of the sample.

In early STMs, it was common to include a logarithmic amplifier in the circuit to compensate for the exponential dependence of current on tip height. In modern microscopes, the entire feedback arrangement is carried out digitally.

The piezoelectric scanning transducer is the single most important mechanical component of the STM. These transducers are made up of a material called lead zirconium titanate, abbreviated PZT (where the P stands for the chemical symbol for lead, Pb). These materials have an intrinsic expansion (or contraction) of ~3 Å per volt applied across them. In modern instruments, all three scanning functions, x, y, and z, are incorporated into one piezoelectric element as shown in Fig. 4.5. The scanner is formed from a long tube that is coated

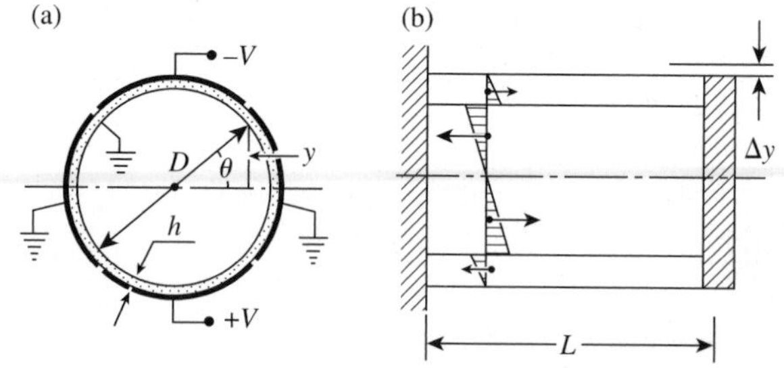

Fig. 4.5 Tube scanner for a scanning probe microscope. The inside of the tube is coated with one of the conductive electrodes that is grounded. The outside is coated with four electrodes (a). Opposite pairs are used for driving the tube in one direction. For example, if opposite and equal voltages $(+V, -V)$ are applied to the opposite y elements, then the tube expands on one side and contracts on the other, resulting in a bending motion that gives the y scan (B). A similar scheme is used with the other two electrodes for the x scan. Addition of a common bias to all four electrodes adds a component of expansion or contraction in the z direction. (Reprinted with permission from Electromechanical deflections of piezoelectric tubes with quartered electrodes, Applied Physics Letters, C. Julian Chen, Vol 60, Copyright 1992.)

on both the inside and the outside surface with metal electrodes. Application of a voltage across the wall of the tube causes the thickness of the wall to increase or decrease, depending on the direction of the field with respect to the polarization of the ceramic. Because the volume of the material remains approximately constant, there is also a change in the length of the tube. As a consequence of the geometry of this arrangement, the change in length of the tube is much larger than the change in the thickness of the wall. Typical tube scanners change their length by ~ 20 Å per volt. The tube scanner incorporates x and y motion as well by exploiting an outer electrode that is divided into four segments. As shown in Fig. 4.5, if one pair of opposing segments is driven in opposite senses (i.e., a voltage applied so as to expand on one side of the tube and contract on the other side) the overall effect is a bending of the tube, resulting in a lateral displacement of the probe along the direction of the bend. With the x scanning waveform applied to one pair of electrodes, the y scanning waveform applied to the other pair with the z control voltage applied uniformly to all four segments, three-dimensional control is achieved with one piezoelectric element.

Typical piezoelectric scanning elements have an intrinsic lowest mechanical resonant frequency of perhaps 50 kHz. This limits the fastest possible response of the microscope. Figure D.5 in Appendix D shows how the phase of the response of a mechanical system to a driving force changes with frequency. If driven at a frequency well above resonance, the response is precisely out of phase with the driving force, and the negative feedback required for current control becomes positive feedback, resulting in unstable operation of the microscope. Thus, the response time of the overall feedback loop must be made significantly longer than the reciprocal of the resonance frequency of the piezoelectric control system. This is done by choosing appropriate values of R_2 and C in the integrator part of the control circuit (Fig. 4.4). This is usually achieved empirically by adjusting the gain of the integrator (by changing R_2) so that it is set just below the point at which the control system becomes unstable (evident as spurious oscillations in the STM image). For a mechanical resonance of 50 kHz, the fastest stable response of the servo system will

Fig. 4.6 An STM for use in an ultrahigh vacuum chamber. Part of the tube scanner is visible in the middle of the picture. Wires on the top left and right carry scanning signals in and the tunnel-current signal out. The sample sits on a small carrier underneath the tube scanner. (Courtesy of Joost Frenken and G.J.C. van Baarle of Leiden Probe Microscopy BV.)

be longer than 20 μs. This will correspond to the shortest time in which one "pixel" of the image can be acquired (a pixel is the smallest element of the image displayed on the screen). Thus, a typical line scan in which 512 pixels are acquired will take at least 20 ms. A complete image consisting of 512 such lines would therefore require at least 5 seconds of scanning time. These numbers are rather typical of modern scanning probe microscopes, although special instruments have achieved image acquisition rates of hundreds of frames per second.

Building a system with a high intrinsic mechanical resonance has a second additional advantage. Referring again to Appendix D (Equation D.10), the response of a driven resonant system falls off as the square of the difference between the resonant frequency and the driving frequency. Specifically taking the limit of Equation D.10 in the case where $\omega \gg \omega_0$:

$$x \approx A_0 \left(\frac{f}{f_0} \right)^2 , \tag{4.1}$$

where $f = \omega/2\pi$ and $f_0 = \omega_0/2\pi$. Thus, mounting the microscope on a support with a very low intrinsic resonant frequency (e.g., 1 Hz) results in a suppression of a vibration that is approximately the ratio of this frequency to the mechanical resonance frequency of the microscope squared. For a 1 Hz resonance of the antivibration mounting and a 50 kHz resonance of the microscope, this is a suppression of vibrations of a factor 2.5×10^9. Thus all but the very largest building vibrations are reduced to subatomic levels at the probe of the microscope.

4.1.3 The scanning tunneling microscope—sensitivity and noise

In a well-designed microscope, the smallest detectable tunnel current is set by the Johnson noise in the resistor used in the current to voltage converter (R_1 in Fig. 4.4). Johnson noise reflects the thermal agitation of electrons in the resistor, and it can be thought of in terms of the equipartition theorem, with $1/2 k_B T$ of noise associated with the current and voltage variables. It is usual to describe noise in power per unit frequency (rather than just energy) and the noise power in a frequency interval Δf is given by

$$P = 4k_{B}T\Delta f. \tag{4.2}$$

Using the expression $P = i^2 R$ for the power dissipated in a resistor R when a current i flows through it yields the following expression for the root-mean-square Johnson noise in the current detector

$$\langle i \rangle_{RMS} = \sqrt{\frac{4k_{B}T\Delta f}{R}}. \tag{4.3}$$

For a 100 MΩ resistor at room temperature in a 5 kHz bandwidth (a reasonable response for an STM), the Johnson noise is $\sim$6 pA. We will learn in Chapter 7 that a small point contact has a resistance of 12.6 kΩ, so given that this resistance increases by a factor of 10 for every Å the probe is withdrawn from a metal surface (like gold—Chapter 2), a 6 pA current corresponds to a distance of $\sim$7 Å with a probe bias of a half volt. The Johnson noise may be reduced by increasing R, but this reduces the frequency response of the microscope because of the time constant associated with this resistor and stray input capacitance (typically several pF) to the current to voltage converter. A second physical limit is set by shot noise owing to the relatively small number of electrons per unit time in very small currents (this scales as $\sqrt{N}/N$) but this is not important in currents above $\sim$1 pA. Thus, the STM can be operated on currents as small as a few tens of picoamps. The bias between the tip of the sample is usually limited to $\sim$1 volt (to avoid field emission—see Section 4.4) so that tunneling may be sustained in junctions with conductances as small as a few tens of pico Siemens.

4.1.4   The scanning tunneling microscope—what the image means

The simple one-dimensional tunneling model discussed in Chapter 2 forms a reasonable starting point for understanding STM images. The tunnel current may be calculated using a version of Fermi's golden rule and a model of the tunnel barrier in which it is approximated as a series of square well potentials (this is called the Wentzel–Kramers–Brillouin—WKB—approximation). The resulting expression for the tunnel current is[7]

$$I = \frac{4\pi e}{\hbar} \int_{0}^{eV} \rho_{s}(r,E)\rho_{t}(r,(E-eV))T(E,eV,r)\,\mathrm{d}E, \tag{4.4}$$

where $\rho_{s}(r,E)$ and $\rho_{t}(r,E)$ are the density of states (the number of states per unit energy) of the sample and tip, respectively, at location r on the surface and energy E. In this expression, energy is referenced to the Fermi energy of the metal electrodes. (The Fermi energy of a metal corresponds to the highest occupied state in the metal—see Chapter 7.) For positive tip bias, most tunneling electrons are from sample to tip, whereas at negative tip bias electrons tunnel from tip to sample. The tunneling transmission probability $T(E, eV, r)$ is given by

$$T(E, eV, r) \propto \exp\left(-\frac{2S\sqrt{2m}}{\hbar}\sqrt{\frac{\phi_s + \phi_t}{2} + \frac{eV}{2} - E}\right), \quad (4.5)$$

where m is the mass of electron, ϕ_s and ϕ_t are the work functions of the sample and tip respectively, and S is the distance between the tip and the sample. The tunneling current measured in an experiment is a convolution of sample topography, the number of electronic states of the tip and the sample with energies close to the Fermi energy, and the tunneling bias applied between the tip and the sample. The average value of the work functions, $\phi = (\phi_s + \phi_t)/2$, is called the apparent barrier height. At a small tip–sample bias ($V \ll \phi$), the dependence of ϕ on V becomes negligible and if the tip material has a flat density of states in the region of the Fermi level, $\rho_t(r, E)$ can then be treated as a constant as well. With these conditions applied, Equation 4.4 becomes

$$I = \frac{4\pi e}{\hbar} \rho_t(0) \exp\left(-\frac{2S\sqrt{2m\phi}}{\hbar}\right) \int_0^{eV} \rho_s(r, E) \, dE. \quad (4.6)$$

Thus, operated at a constant tunnel current, the STM follows contours of constant density of states at the Fermi energy. This is not the same thing as the total density of electrons at the surface. STM images, even rather famous ones such as that of the silicon the 7×7 surface in Fig. 4.3, have "missing atoms" known to be present in the physical structure but not imaged in the STM image. A full understanding of STM images requires a detailed calculation of the surface density of states at the Fermi energy using, for example, the density functional techniques discussed in Chapter 2. STM images can be very difficult to interpret in the absence of a detailed theoretical model.

4.1.5 The scanning tunneling microscope—spectroscopy

Equation 4.6 states that the tunneling current, I, at a certain sample bias, V, is proportional to the total number of states between the Fermi level of the sample and the energy of the tip, $E_f + eV$. A negative sample bias gives the number of occupied states below the Fermi level, and a positive gives the number of unoccupied states above the Fermi level. An independent bulk measurement of the number of states per unit energy can be made using photoemission (measuring the number of electrons ejected from a sample as a function of the energy of incident radiation) or inverse photoemission (the surface is bombarded with energetic electrons and the spectrum of emitted light is measured). The STM can perform similar measurements on the atomic scale. The desired quantity is the number of states per unit energy (the density of states, DOS) rather than the integrated total number of states. For small bias, the density of states may be obtained as the derivative of Equation 4.6:

$$\frac{dI}{dV} \propto \rho_s(r, E) \equiv \text{DOS}(eV). \quad (4.7)$$

Therefore, the DOS function of a sample can be obtained by taking the derivative (dI/dV) of an I–V curve acquired at a constant tip position when a small bias

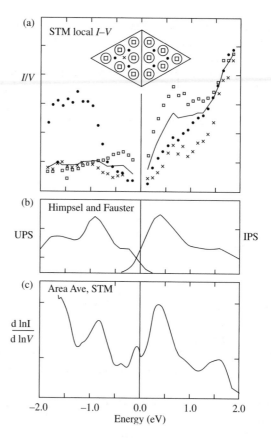

Fig. 4.7 Atomically resolved tunneling spectra, obtained on Si(111)-(7 × 7). (a) Plot of conductance vs. applied voltage measured at different locations within the Si(111)-(7 × 7) unit cell, depicted at top. (b) Density of states extracted from (large-area) photoemission and inverse photoemission data. (c) Normalized tunneling data averaged over several unit cells, showing correspondence with photoemission/inverse photoemission data. (Reprinted with permission from Physical Review Letters[8], copyright 1986 by the American Physical Society – courtesy of Professor R.J. Hamers.)

range is used. At larger bias, the intrinsic dependence of current on bias can be compensated for to some extent by forming the quantity

$$\frac{d \ln(I)}{d \ln(V)}. \tag{4.8}$$

An example of current–voltage curves captured at different points on the silicon 7×7 unit cell is shown in Fig. 4.7. The density of states obtained by averaging many such curves and forming the quantity defined in (4.8) is shown in the lower panel. This is compared to the results of photoemission and inverse photoemission which are shown in the middle panel.

Variations of this technique have proved very useful in molecular electronics and will be discussed in Chapter 8.

4.2   The atomic force microscope

4.2.1   The atomic force microscope—history

The AFM was invented by Binnig, Quate, and Gerber in 1986[9] with the goal of addressing the limitation of STM to imaging conducting samples. The first microscope used a scanning tunneling probe to detect very tiny deflections of a conducting cantilever pressed into contact with the surface. Much like the STM, the flexible cantilever (equipped with a sharp diamond point) was scanned over a surface while the height of the cantilever above the surface was adjusted so as to keep the tunnel current between the cantilever and a tunnel probe, placed above the cantilever, constant. This detection scheme proved unreliable and is no longer used, but the first paper captured the important physics of the AFM.[9] The first point made in that paper is that very soft cantilevers may be used as long as they are also made very small. The resonant frequency of a cantilever is given by the square root of the ratio of its spring constant, k, to its mass, m (see Appendix D):

$$f_0 = \frac{1}{2\pi}\sqrt{\frac{k}{m}}. \tag{4.9}$$

Thus, a weak spring (small k) can have a high resonant frequency if its mass, m, is small enough. A high resonant frequency will minimize the influence of building vibrations, as just discussed for the case of the STM (where the slowly responding element was the PZT tube scanner).

The second point made in that paper was that the ability to detect tiny deflections translates into an ability to detect tiny forces even when macroscopic springs are used. In that paper, they select the arbitrary distance of 0.16 Å and estimate that the force required to stretch an ionic bond by this amount is $\sim 10^{-8}$ N while stretching a weak van der Waals bond by the same amount would produce a force of $\sim 10^{-11}$ N. These are forces that are easily detectable using the cantilever spring and detection scheme that they proposed. This is because the bonds that hold atoms together have, near their equilibrium position, a spring constant that is quite similar to that of a macroscopic cantilever (see Exercises 4 and 5 at the end of this chapter).

The AFM has flourished, turning into a tool with many uses. It has several distinct imaging modes. It "feels" the surface, and has become the ultimate nanomanipulation tool, capable of pulling apart individual molecules or scratching or painting nanoscale structures. Equipped with a conductive probe, it has largely replaced the STM even for the task of nanoscale electronic measurements.

4.2.2 Detection scheme in a modern AFM

Modern AFMs almost all use an optical deflection scheme for monitoring the bending of the force sensing cantilever. This arrangement is shown schematically in Fig. 4.8. Light from a small diode laser is focused on to the back side

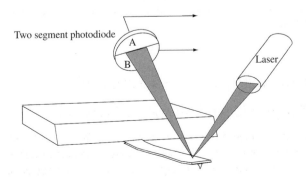

Fig. 4.8 Optical deflection scheme for monitoring the deflection of a force sensing cantilever. A beam from a small diode laser is reflected from the backside of the cantilever into a two-segment photodiode. In the arrangement shown here an upward movement of the beam results in more current from segment A and less current from segment B.

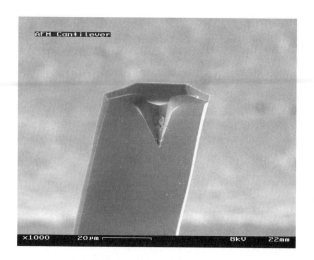

Fig. 4.9 Scanning electron microscope image of a cantilever showing the small probe fabricated onto the tip of the cantilever. The scale bar is 20 μm. The cantilever is 40 μm wide, 125 μm long, and 4 μm in thickness and has a spring constant of approximately 40 N/m. It is a stiffer type of cantilever, designed to pull away from a sticky surface (Fig. 4.13(b)). The radius of curvature of the force sensing probe can be as small as a few nm. (Image courtesy of Ami Chand, Applied Nanostructures, www.appnano.com.)

Fig. 4.10 A complete scanning probe microscope system (Agilent 5400). The microscope stands in the foreground with the scanner located on top of the microscope. The sample sits immediately under the scanner inside the glass chamber (used for control of the sample environment). The electronic controller is shown on the left and one of the screens displays the control software while the other displays the images that are collected. Switching between AFM and STM is achieved by changing the scanner. (Image courtesy of Agilent Technologies.)

of the force sensing cantilever (this is often coated with gold to enhance optical reflection) and reflected onto a two-segment photodiode which monitors the deflection of the reflected beam via the relative photocurrent produced in the two segments, i_A and i_B. A scanning electron micrograph of a typical force sensing cantilever is shown in Fig. 4.9. If the optical beam is aligned so that the two currents from the two photodiode segments are nearly equal when the cantilever is at the desired position, the deflection signal will be proportional to the difference between the two currents i_A and i_B. In practice, the signal that is used is normalized by the sum of the two currents in order to compensate for fluctuations in the laser power. Thus the deflection signal, δz, is proportional to

$$\delta z \propto \frac{i_A - i_B}{i_A + i_B}, \tag{4.10}$$

where δz means the deflection of the cantilever in units of nm. The signal measured by the controller is a voltage proportional to the current ratio defined in Equation 4.10. It is turned into a value in nm by calibrating the voltage signal using force-distance curves (Fig. 4.13). In a well-designed microscope, the detection limit is set by the shot noise on the laser signal. A 1 mW red laser emits $\sim 3 \times 10^{15}$ photons per second. The data acquisition time for the AFM is limited by the resonant frequency of the cantilever (or the scanning PZT, whichever is slower) but a typical period for the dwell time on each pixel is ~1 ms (cf. the STM response time discussed above). Therefore, $\sim 3 \times 10^{12}$ photons are collected in each pixel, so the shot noise ($\sqrt{N}$) is $\sim 2 \times 10^6$ photons per pixel. This results in a signal-to-noise ratio of better than one part in one million. *Thus, a deflection of the cantilever of somewhat smaller than one part in one million is readily detectable.* For a cantilever of 50 μm in length, this corresponds to a detection limit of ~50 pm (i.e., 0.5 Å).

A second limit on the smallest detectable signal lies in the thermal excitation of the force sensing cantilever. The equipartition theorem (Chapter 3, Section 3.6) states that the thermal energy in each degree of freedom of the cantilever is equal to $1/2 k_B T$. Thus, if the vertical deflection, z, of the cantilever corresponds to the motion of a harmonic oscillator of energy $\frac{1}{2} \kappa z^2$, where κ is the spring constant of the cantilever in the z direction, then the root-mean-square thermal deflection is given by

$$\sqrt{\langle z^2 \rangle} = \sqrt{\frac{k_B T}{\kappa}}. \tag{4.11a}$$

$k_B T$ at 300 K is ~4.2 pN·nm (see Appendix A). Thus for a cantilever of spring constant $\kappa = 1$ N/m (= 1 nN/nm), this noise is <1 Å.

The analysis given above is oversimplified. The noise spectrum of the cantilever is identical to its frequency response as a driven damped harmonic oscillator according to the fluctuation dissipation theorem (see Appendix D). Equation 4.11a applies to the response as integrated over all frequencies. In practice, the signal is detected in some narrow bandwidth, B (for 1 ms per pixel acquisition time B will be on the order of 1 kHz). Thus, the noise that enters the instrument electronics is somewhat less than what is predicted by Equation 4.11a. A fuller analysis[10] shows that for a cantilever that behaves like a harmonic oscillator with damping characterized by the quality factor Q (see Appendix D)

$$\delta z_{RMS} = \sqrt{\frac{4 k_B T B Q}{\kappa \omega_0}} \tag{4.11b}$$

when the cantilever is driven at its resonant angular frequency, ω_0, and

$$\delta z_{RMS} = \sqrt{\frac{4 k_B T B}{Q \kappa \omega_0}} \tag{4.11c}$$

when the cantilever is driven away from its resonant frequency. There is a significant advantage in using a very small cantilever with a very high resonant frequency, and collecting the signal in some small frequency window below the resonant frequency.[11] A higher ω_0 implies a smaller noise per $\sqrt{B}$ (noise is usually expressed as noise per $\sqrt{\text{Hz}}$). If the cantilever has a significant Q factor, then there is an additional benefit to driving it off resonance. (Cantilevers operated in water typically have a low Q of 2–5, while cantilevers operated in air may have a Q as high as 50.)

4.2.3   Surface forces and operation of the atomic force microscope

The interactions between atoms can be described by the Lennard Jones 6-12 potential given in Equation 2.107:

$$\phi(r) = 4\varepsilon \left[\left(\frac{\sigma}{r_{ij}} \right)^{12} - \left(\frac{\sigma}{r_{ij}} \right)^{6} \right].$$

The attractive term (in r^{-6}) decays quite slowly. This is a long-range interaction and it needs to be summed over all the atom pairs in the surface and the tip. The attractive interaction energy between some small volume of the sample and some small volume of the probe is given by

$$dW(r_{sp}) = -\frac{C\rho_s\rho_p}{r_{sp}^6} dV_s dV_p, \tag{4.12}$$

where ρ_s and ρ_p are the number densities of atoms in the sample and probe, respectively, and r_{sp} is the distance separating the elementary volumes dV_s and dV_p of sample and probe, respectively. The geometry for a sphere of radius, R, at a distance, D, from the surface of a plane is shown in Fig. 4.11. In the sphere each ring of radius x contains $2\pi x \rho_p \, dx dz$ atoms so the sum of the interaction energies between all the atoms in the probe and all the atoms in the surface is given by an integral over z from D to $D + 2R$ and over x from 0 to ∞. Taking the result from the book by Israelachvilli[12] (p. 157):

$$W(D) = -\frac{\pi^2 C \rho_s \rho_p R}{D}. \tag{4.13}$$

Note how the integration of a reasonably strongly decaying potential (r^{-6}) has become a very weakly decaying interaction (D^{-1}) when integrated out over macroscopic geometries. For this reason, the probe must come into hard contact with the surface being examined if high resolution is to be achieved (so that the interaction is dominated by the short-range repulsive term).

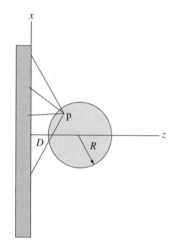

Fig. 4.11 Long-range interactions between a sphere and a plane. The van der Waals interaction is summed for each atom in the sphere with every atom in the surface.

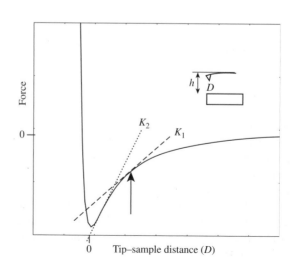

Fig. 4.12 The curve shows the relationship between force and distance for a spherical probe a height D above a plane. Contact occurs at $D = 0$ where strong repulsive forces dominate. The force vs. distance response of a spring is linear so that a probe a distance h above the sample is deflected by an amount $h - D$ yielding a restoring force $K(h - D)$ which, in equilibrium, is equal to the force between the probe and the surface. However, for a soft cantilever (spring constant K_1—long dashed line) the surface force increases more rapidly than the spring force at the D value marked by the arrow, pulling the tip into the surface. It is necessary to work with a cantilever of spring constant $\geq K_2$ (short dashed line), i.e., stiffer than the largest force gradient in the probe sample interaction.

A second consequence of the long-range interaction between the probe and the sample (Equation 4.13) is an instability that results when a weak cantilever is used. The attractive interaction force between the probe and the surface is given by

$$F(D) = \frac{\partial W(D)}{\partial D} = \frac{\pi^2 C \rho_s \rho_p R}{D^2} \qquad (4.14)$$

and a plot of this function is shown in Fig. 4.12 (the strong repulsive part has been added in at small values for $D < 0$ where the surface is contacted). In mechanical equilibrium, the surface force (Equation 4.14) is exactly balanced by the force generated by the displacement of the cantilever (shown as $h - D$ in the inset in Fig. 4.12). The surface force increases nonlinearly as the gap becomes smaller, so if a weak cantilever is used, a point is reached where the surface force increases more rapidly than the restoring force generated by the cantilever. The consequence is a mechanical instability at this point. The tip is pulled into the surface.

What happens once the tip touches the surface? Unless specially treated, most surfaces will adhere to one another with a force that is proportional to the surface energies, γ_i, of the materials. For a sphere of radius, R, in contact with a plane of the same material in vacuum, this force is

$$F_A = 4\pi R \gamma. \qquad (4.15)$$

For many materials, γ is on the order of 0.5 J/m^2 (see Table 11.5 in Israelachvilli[12]) resulting in an adhesive force of ~6 nN for a probe of radius 1 nm. Adhesion can be reduced by working in solvents where surface energies are greatly reduced, owing to interactions between the surfaces and the solvent.

These effects are manifested in so-called force–distance plots. The raw data is not a force–distance plot, but rather a plot of a signal proportional to cantilever displacement vs. the distance that the cantilever holder has been translated toward the sample. The displacement signal can be translated into a force using the slope of the line when the cantilever is in hard contact with the sample (this is unity because the surface moves with the cantilever in contact) and the known spring constant of the cantilever. The displacement of the cantilever holder can be translated into tip–sample distance by taking the obvious point of contact as the zero of separation, and subtracting the tip displacement from the displacement measured outward from this point.

Figure 4.13 shows some examples of uncorrected cantilever displacement vs. translation of the cantilever holder plots (these are what would appear on an instrument screen). The first plot in (a) shows both the effects of a jump-to-contact owing to a weak cantilever as well as the effect of a strong adhesion between the tip and surface. The second plot (b) is what might be expected with a stiffer cantilever. The third plot (c) shows an example of approach and retraction curves with an adequately stiff cantilever and minimal adhesion. This is the desired situation for high-quality imaging. A stiff cantilever, designed for "tapping" on the surface, is illustrated in Fig. 4.9.

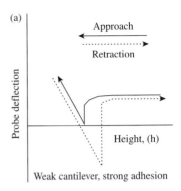

(a) Weak cantilever, strong adhesion

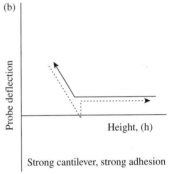

(b) Strong cantilever, strong adhesion

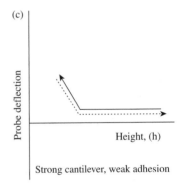

(c) Strong cantilever, weak adhesion

Fig. 4.13 Showing characteristic plots of the probe deflection vs. the height of the probe holder above the sample for (a) a weak cantilever and strong adhesion. The probe is pulled into the surface on the way in and sticks to it on the way out until the spring force of the cantilever exceeds the adhesion force. When in hard contact with the sample, the probe rides up with it so the slope of the plot after contact is unity. (b) Shows the same plot for a strong cantilever and strong adhesion (the deflection owing to adhesion is smaller because the cantilever is stronger). (c) Shows the same plot for a strong cantilever and weak adhesion. This latter behavior is required for high-quality images.

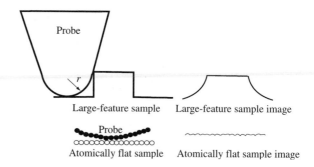

Fig. 4.14 Surfaces with a large-scale topography (top) interact with the large-scale geometry of the probe. Thus, the image of a sharp ridge takes on features at its edges that reflect the radius of curvature of the tip. However, atomic scale roughness on the same probe can yield atomic resolution on a very flat surface (as shown in the bottom example).

4.2.4   Resolution of the atomic force microscope

Because the resolution is dominated by contact between probe and sample, the shape of the probe limits the resolution that can be attained. Figure 4.14 shows how the image of a ridge with sharp edges is broadened by a probe of finite radius r. However, the same probe operated on a very flat surface might yield atomic resolution of the surface owing to small atomic scale asperities on the end of the probe, as shown in the bottom half of the figure.

If the shape of the probe is known, or can be measured by imaging, for example, a very sharp point on a sample, then methods exist for recovering the original contours of the sample. These are limited, however, because of the fact that concave features on the sample that are smaller in radius than the tip can never be reconstructed. This is because the image contains no information about them. Thus, referring to the top part of Fig. 4.14, if the radius, r, is known, the sharp ridge could, in principle, be reconstructed. However, if the sides of the ridge sloped inwards, this feature would be missed entirely by a reconstruction program. Simple formulae for reconstruction have been given by Keller[13] and a more comprehensive algorithm has been presented by Villarrubia.[14]

4.2.5   Imaging modes of the AFM

In its simplest mode of operation, known as contact mode, the probe is brought down to the surface until the desired contact force is measured. This force is usually set to be much larger than the noise discussed above, typically at least tens of piconewtons. A servo control (exactly analogous to that shown for the STM in Fig. 4.4) is used to maintain constant force as an image is generated from the height correction signal. This scheme of operation is illustrated in Fig. 4.15(a) and an example of a contact mode image is given in Fig. 4.16(a).[16] In the image in Fig. 4.16(a), internal features are clearly resolved in small proteins packed into an array. The contact force was adjusted from low to high as the probe was scanned from top to bottom to show how this changed the conformation of the protein. Achieving resolution like this in contact mode requires very careful sample preparation, and judicious choice of the solution in which the imaging is carried out in order to minimize attractive forces and adhesion.[18] The most popular mode of operation is known as dynamic force microscopy (illustrated in Fig. 4.15(b)). In this mode of operation, the probe is vibrated close to its resonance frequency and the electronics are set up so as to monitor the amplitude of the oscillating deflection signal. Typically, the oscillation amplitude is adjusted to some fixed amount, A_0, with the tip far

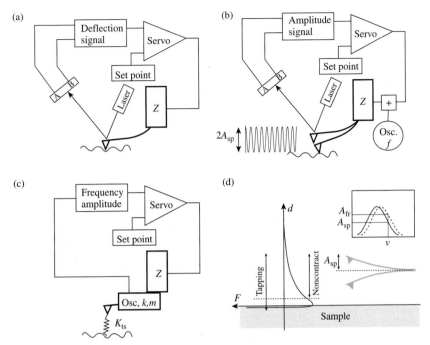

Fig. 4.15 Illustrating various imaging modes of the AFM. (a) Illustrates the control system for contact mode AFM where the deflection signal is maintained at a fixed level by adjusting the height of the sample as the sample is scanned. (b) Illustrates dynamic force microscopy. In this mode, the probe is oscillated at a frequency f by the application of a small mechanical vibration to the probe (or by direct application of a magnetic force to the probe). The amplitude of this oscillation signal, A_{sp}, is detected by a rectifying circuit, or a lock-in detector. The feedback adjusts the height of the tip so as to maintain constant amplitude of oscillation. (c) Illustrates frequency modulation mode operation. In this mode, the probe is attached to a high Q electromechanical oscillator and the frequency of oscillation converted to an amplitude signal. In this case, the set-point is set so that the mechanical oscillator operates at a slightly different frequency owing to interactions between the probe and the surface (K_{ts}). In this case, the height of the probe is adjusted so as to maintain the desired frequency of oscillation. (d) Shows how the dynamic force microscope mode (b) can be set-up so as to operate in noncontact or intermittent contact ("tapping") mode. (Reprinted with permission from Electro mechanical deflections of piezoelectric tubes with quartered electrodes, Applied Physics Letters, L. Julian Chen, Vol. 60, copyright 1992.)

from the surface, and the servo control set point adjusted so that advance of the probe is stopped when a set-point amplitude, Asp ($<A_0$), is attained. The probe is then scanned over the surface while the amplitude of oscillation is maintained constant by adjustment of the probe height. Usually, the microscope is operated in a regime in which the free amplitude is reduced as the probe hits the surface on its downward swing. This is called intermittent contact mode. It has the advantage of being easy to operate, and is quite gentle on soft samples. On very sticky samples, the amplitude is adjusted so as to be high enough that the probe will always pull away from the surface. It is quite difficult to drive oscillation of the probe when it is submersed in a fluid because of the volume of fluid the probe must displace as it oscillates. A high level of mechanical drive is required to oscillate a probe in fluid, and this can result in spurious responses of the microscope. One method for overcoming this problem utilizes cantilevers coated with a magnetic film, driving the cantilever into oscillation using a small solenoid close to the cantilever. An alternating current passed through this solenoid generates a time varying magnetic field that drives the cantilever into motion directly. This mode of operation is called magnetic AC mode ("MacMode").[19] An example of an image taken with this mode of operation is shown in Fig. 4.16(b). Dynamic force microscopy also

Fig. 4.16 Images obtained with the various imaging modes of operation of the AFM. (a) Shows a contact mode image of a layer of small proteins, demonstrating how the conformation of the proteins changes as the contact force is adjusted from 100 to 300 pN. This image was taken with a cantilever of a force constant $K = 0.1$ N/m in a salt solution that minimized adhesives and attractive interactions. (Courtesy of Professor Andreas Engel.)[16] (b) Shows a dynamic force microscope image of chromatin molecules (DNA plus the protein that folds it into chromosomes) taken in the intermittent contact mode using magnetic excitation of the cantilever ($K = 1$ N/m, $A_{sp} = 6$ nm). The image was obtained in a biological buffer solution. (c) Shows a dynamic force mode image taken in noncontact mode. These are 100 nm wide stripes of liquid octane condensed onto a nano-patterned surface. Any contact would disrupt the liquid surface. This image was made with a probe of $K = 40$ N/m, Asp <10 nm and the driving frequency 100 Hz above the resonance of the cantilever. (Reprinted from Checco et al.[15] Ultramicroscopy, 2006, with permission from Elsevier.) (d) Shows a very high-resolution image of the atomic structure of the silicon 7×7 surface obtained with FM mode AFM. Here, $K = 1800$ N/m, $A = 0.25$ nm, the free resonance of the oscillator was 14 MHz, and the set-point shift was +4 Hz. (F.J. Giessibl,[17] Reviews of Modern Physics, copyright 2003, American Physical Society. Courtesy of Professor Franz J. Giessibl.)

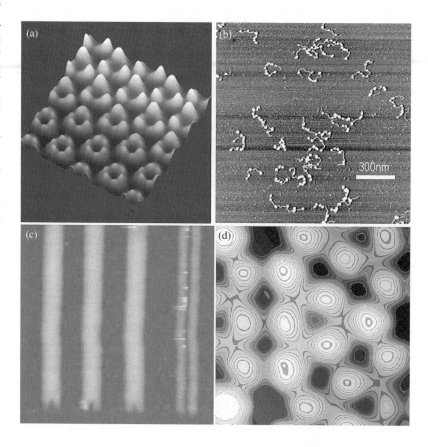

allows for the operation of the microscope in a noncontact mode. This operation is illustrated in Fig. 4.15(d). A force vs. distance plot is shown turned sideways to illustrate the point that if the probe oscillates in the region labeled "noncontact," it experiences an average attractive interaction and, if driven at a frequency somewhat above its free resonance, this results in a decrease of amplitude. This decrease is not because of the probe hitting the surface, but rather because of a change in its mechanical properties as a consequence of its interaction with the surface.[15] Noncontact operation results in lower resolution as discussed above, but it is very valuable when extremely soft samples must be imaged. Figure 4.16(c) shows an image of stripes of liquid octane assembled onto a nano-patterned surface. If the amplitude of oscillation of the probe was increased even a little bit, the resulting intermittent contact mode destroyed the liquid film on the surface.[15] A third mode of operation is shown in Fig. 4.15(c). This is called FM mode AFM. In FM mode, the cantilever is used as a self-driving mechanical oscillator. One popular arrangement is to use one arm of a quartz tuning fork (as used as the timing element in electronic watches). A sharp probe is formed on one of the arms of the tuning fork and the oscillation frequency of the fork measured as the probe is approached toward the surface. This technique is normally used on atomically flat surfaces in ultrahigh vacuum where the adhesion can be very high. Very stiff cantilevers are used to overcome this problem. Typically FM mode detects shifts in the resonant frequency of the tuning fork as small as a few parts in 10 million. The probe is scanned over the surface while the height of the probe is adjusted to keep this tiny shift in frequency owing to the tip–sample interactions (K_{ts} in Fig. 4.15(c)) constant. The resolution achieved by this technique now equals, or even exceeds, the best results of STM. For a comprehensive modern review, see the work of Giessibl.[17] A high-resolution image of the silicon 7×7 surface obtained with FM mode AFM is shown in Fig. 4.16(d).

4.2.6 Measuring forces

The deflection vs. distance curves shown in Fig. 4.13 can be exploited to measure interaction forces between the probe and the surface. Clearly, the measured adhesion force can be used to determine the local surface energy according to Equation 4.15. Similarly, the local elastic properties of the surface can be determined from small departures from a $45°$ slope in the hard contact part of the curve. To understand this, suppose that the surface was locally just as deformable as the cantilever, then the tip deflection when the tip is in contact with the surface would only be half the amount that the surface was translated toward the tip, resulting in a slope on the deflection vs. distance curves of less than $45°$. Most materials stiffen when they are compressed significantly, so these effects are only evident very near the point of first contact. In this region, the deformation of the surface can be extracted from a mechanical model of the deformation of a plane surface in contact with a cone of semi-angle α pushing on it with a force, F. The Hertzian model[20] gives the following result for the deformation of the surface, δ:

$$\delta = \sqrt{\frac{\pi(1-\sigma)F}{4G \cot \alpha}},$$
(4.16)

where G is the shear modulus of the material and σ is its Poisson's constant. Thus, the slope of the contact part of the deflection vs. distance curves when the probe first contacts a typical rubber is ~ 0.6 compared to the slope when it first contacts steel.[21]

In reality, studies of this sort are plagued by the unknown geometry of the AFM probe. To get around this problem, studies of adhesion and indentation often use cantilevers modified by the attachment of a well-characterized sphere.[22]

There have been many attempts to measure adhesion at the single molecule level using AFM, including attempts to stretch and break molecules and thus to record the forces required to do this. These present several challenges. The first is the general, unspecific, and poorly characterized interactions between an AFM probe and a surface. To overcome this problem, it is necessary to attach the system being studied to long tethers, one attached to the surface and the other attached to the probe, so that the probe is far from the surface when the interaction of interest is studied. The second difficulty comes from the notion of a distinct bond breaking force. As explained in Chapter 3, the making or breaking of chemical bonds is dominated by fluctuations that mediate passage over a rate-limiting transition state. Thus, bond breaking is a statistical event and must be characterized appropriately.

One approach to overcoming the nonspecific interaction between the probe and the surface is to engineer synthetic polymeric constructs in which the molecule under study is embedded in a repeating unit. Stuck at one end to

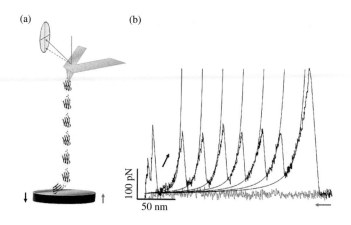

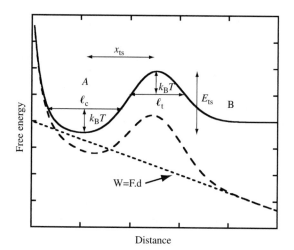

Fig. 4.17 (a) Protein with repeated globular domains (titin) trapped between a probe and a surface. (b) Force-distance curve showing sequential unfolding of each domain in turn. The curve is inverted relative to Fig. 4.13 so that peak adhesion forces point upwards. Each unfolding event is fitted (solid lines) with a "worm-like-chain" model of polymer extension. (Courtesy of Professor Piotr Marszalek, Duke University.)

Fig. 4.18 Showing how applying a force to a molecular system alters the rate of processes such as unfolding or unbinding. The top (solid) curve shows an energy landscape with the complex bound at A and unbound at B. The transition state is the maximum in potential at a distance x_{ts} from the minimum of the bound state. Applying a force F to this system does work on the system equal to $F \cdot d$, where d is the distance the system has moved under the application of the force (short dashed straight line). The sum of this work and the original potential gives a new tilted potential landscape, shown as the long dashed line. The energy of the barrier at the transition state has been lowered by the application of a force to the system. The parameters shown are used in the Evans version of Kramers' theory.

the surface by a chemical bond, and picked up by the probe somewhere along its length, such a polymeric repeat will show repeated structure in its force vs. extension curve. An example is shown in Fig. 4.17, using data obtained with a polyprotein (titin). Data are recorded only when the probe is well away from the surface and the force vs. extension curves show a distinct series of peaks where the force rises rapidly to a maximum and then falls again as one of the protein units in the polymeric construct unwinds. Polyproteins have even been engineered with alternating inserts of a well characterized protein as a calibration.[23] The response of molecular systems to the application of a force is described by a modification of Kramers theory (Section 3.15) owing to Evans.[24] The energy landscape for binding and unbinding of a complex is shown schematically in Fig. 4.18. Doing work on the system, by applying a force to it, adds energy that increases linearly with distance (the short dashed line in the figure) so that the overall energy landscape under the application of a force is tilted (long dashed line). This causes a lowering of the transition state energy and an increase in the rate of dissociation of the complex. Evans' formulation of the problem replaced the parameters ω_a and ω_b in the Kramers formula (Equation 3.91) with the widths, ℓ_c and ℓ_t, of the well and of the transition state at $1\,k_BT$ from the turning points (see Fig. 4.18)

$$\upsilon_{\text{off}} = \frac{D}{\ell_c \ell_t} \exp - \left(\frac{E_t}{k_B T} \right). \tag{4.17}$$

The molecular friction (γ in Equation 3.91) is replaced by a diffusion constant for the Brownian motion in the well, using the Einstein–Smoluchowski relation (Equation 3.86) and a factor k_BT that enters when the curvature parameters (ω_a and ω_b) are replaced by ℓ_c and ℓ_t. E_t is the height of the transition state (E_b in Equation 3.91). In the presence of an applied force, f, Equation 4.17 becomes

$$\upsilon_{\text{off}} = \frac{D}{\ell_c \ell_t} \exp - \left(\frac{E_{ts} - f x_{ts}}{k_B T} \right) \tag{4.18}$$

under the approximation that the distance to the transition state was not modified by the application of a force.

In a real AFM experiment, the force on the system is not constant, but rather continues to rise as the probe is pulled away from the surface. This modifies the picture shown in Fig. 4.18, and the result is an exponential dependence of the rate for breaking and unfolding on the rate at which the probe is pulled away from the surface. Specifically, if the loading rate is r_f (Newtons per second)

$$f_p = \frac{k_B T}{x_{ts}} \ln \left[\frac{r_f \upsilon_{\text{off}} x_{ts}}{k_B T} \right], \tag{4.19}$$

where x_{ts} is the distance to the transition state shown on Fig. 4.18 and f_p is the most probable force for bond breaking or unfolding. This is illustrated with experimental data taken by repeated unfolding of a small RNA molecule in Fig. 4.19.[25]

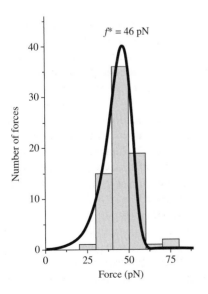

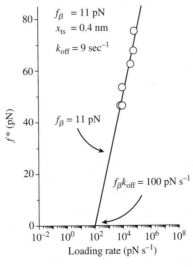

Fig. 4.19 The histogram on the left shows the distribution of forces measured when a small double helical RNA molecule was pulled apart at a loading rate of 10,000 pN/s. This yielded a most probable force of 46 pN for the breaking of the RNA strands. The experiment was then repeated for a variety of loading rates to produce the plot of most probable breaking force vs. loading rate shown on the right. This curve is related to the distribution of breaking forces at any one loading rate as shown by the predicted line overlaid on the histogram. (From the review by Williams[25] in *Scanning Probe Microscopies Beyond Imaging*, copyright Wiley-VCH: 2006 Verlag GmbH & Co. KgaA. Reproduced with permission.)

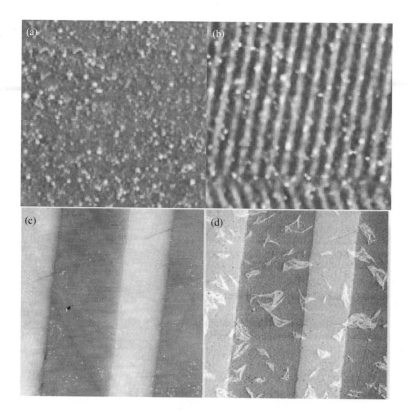

Fig. 4.20 Top: Showing a 10 μm square image of (a) the topography and (b) the magnetic domains in a piece of recording tape taken by magnetic force microscopy (Reprinted with permission from Pacific Nanotechnology.) The bottom images showed 40 μm scans of topography (c) and surface potential (d) over alternating TiO_2 and SiO_2 regions on the surface of a device. (Images courtesy of Veeco Instruments Inc.)

4.2.7   Some other force sensing modes

The AFM is capable of sensing the surface by means of a number of interactions. For example, the use of a magnetic probe will result in a deflection signal when a surface containing magnetized features is scanned. This is the basis of the magnetic force microscope used extensively to image features on magnetic disks as shown in Fig. 4.20(a) and 4.20(b). It is necessary to carry out the measurement in a way that separates topographical from magnetic information. One way of doing this is the so-called lift mode, in which a topographical scan-line is first recorded, the probe raised some distance above the surface, and then scanned again at a constant distance above the surface (over the same line across the sample) while the deflection owing to the long-range magnetic interaction is recorded.

A second type of sensing is electric. In electric force microscopy, a conducting probe is scanned out of contact with the surface while a bias is applied between the probe and the surface. The resulting Coulomb interactions cause deflections of the probe even though it is far from the surface. One sensitive way to record such data is to oscillate the voltage applied to the probe at a fairly high frequency and detect the component of the deflection at this frequency, f, and at the second harmonic ($2f$). The corresponding amplitude of the signal at the fundamental frequency in the deflection is proportional to the surface charge while the amplitude of the signal at $2f$ is proportional to the polarizability of the surface.[26] In a variant of this scheme, the bias applied to the probe is adjusted as it is scanned over the surface, in such a way as to keep the deflection owing to the electrostatic interaction constant. The voltage required to achieve this is a measure in the change of the local work function from point to point on the surface. This technique is called Kelvin force spectroscopy and it can be used to map the concentration of dopants in semiconductors because these change the work function locally. Kelvin probe mapping works even on surfaces of very low conductivity where STM would be impossible. An example of surface potential mapping is given in Fig. 4.20(d) (4.20(c) shows the topographic image of the same region).

4.2.8   Recognition imaging

Chemical information is hard to obtain from conventional AFM images. This problem is particularly troublesome when complicated samples are imaged by AFM (such as those containing many different biological molecules). One solution to this problem is to tag the probe with a chemical agent that recognizes specific targets on the surface. This approach is subject to the problem of non-specific interactions discussed above, but this problem can be overcome if the recognition element is tied to the probe by a long flexible, nonsticky, tether such as a polyethylene glycol molecule. Antibodies have been tethered to an AFM probe and used to detect individual antibody–antigen binding events.[27] This is not enough to produce a recognition image by itself. The waveform of an oscillating probe is indeed distorted when an antibody on a tether binds its target on the surface. However, the AFM imaging electronics operates to correct any change in amplitude owing to the binding of the antibody by adjusting the sample height, and this hides any evidence of a binding event in the signal amplitude. There is, however, a subtle indication of the binding event in a small displacement of the overall signal as illustrated in Fig. 4.21. Sensitive electronics can be used to detect this small shift in the waveform so that two images can

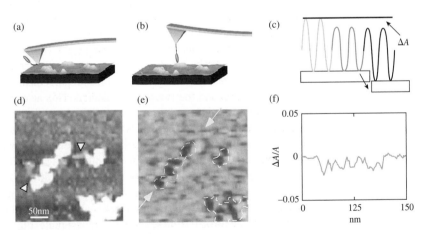

Fig. 4.21 The technique of recognition imaging. Recognition imaging detects the small shift in the peak value of the AFM cantilever displacement signal when an antibody tethered to an oscillating cantilever (a) binds to an antigen on the surface being scanned (b). In the absence of antibody binding, the tip displacement pattern is sinusoidal and cut off only at the bottom of the swing where the tip contacts the surface (lighter curve on left in panel (c), exaggerated for effect). When the attached antibody binds to an antigen, the stretched tether reduces the upward extent of the swing of the tip (middle part of curve in panel (c)), causing the microscope servo to restore the amplitude by pulling the tip away from the sample. As a result, the upper level of the signal falls by an amount ΔA (right part of curve in panel (c)). (d and e) Show a representative topographic image (d) and the corresponding recognition image (e) for a field of chromatin molecules scanned with an antibody to one of the proteins that comprise chromatin. The locations of this protein in the topographic scan are outlined on the recognition image as dashed white lines. (f) Shows a plot of the recognition signal as a fraction of the full amplitude ($\Delta A/A$) measured along the line joining the arrowheads in (e). The dips in the plot correspond to dark "recognition spots" in the recognition image. (Reprinted from Stroh et al.[28], published 2004, National Academy of Sciences.)

be displayed side-by-side: one is the regular topographic image, and the other is a map of regions where the overall signal was displaced owing to a binding event.[28] Excellent antibodies are needed because the force exerted by the oscillating tip acts to pull the antibody off the target molecule. The technique has been used to follow changes in the composition of the biological material as it is acted on by an enzyme.[29]

4.3   Electron microscopy

The first nanoscale images were obtained with an electron microscope, and electron microscopes remain to this day the most useful general purpose nanoscale imaging tool. The first demonstration of magnified images using electron beams occurred in Berlin in 1931 when Ernst Ruska showed that a projected image of a grid was magnified if the grid was placed immediately after the focal point of the converging electron beam and behind an imaging screen placed some distance in front of the grid. The magnification was exactly what one would expect from geometric optics; that is, the ratio of the distances between the focal point and the sample, and the sample and the screen. This was an exciting result for Ruska, because, believing electrons to be point particles, he thought he had the makings of a microscope that did not suffer from the limitations associated with the finite wavelength of light (which is what limits optical microscopes). In his Nobel Prize acceptance speech (Nobel Prize in physics, 1986), Ruska recalls his disappointment on hearing of de Broglie's theory of the wavelike nature of the electron, followed by his elation when he used de Broglie's formula (Equation 2.1) and calculated that the wavelength of electrons is very small indeed.

All electron microscopes have certain common features. The electron optics are contained in a high vacuum chamber ($\sim 10^{-8}$ Torr) and consist of a source of electrons (which may be a thermionic emission gun or a field emission gun), a condenser lens to collect electrons from the electron source and place them at one focus of an objective lens and the objective lens itself which forms a small spot on the sample. In the most commonly used form of electron microscopy, the small spot formed by the objective lens is scanned over the sample in a raster pattern by magnetic deflection coils as some signal indicative of the interaction between the beam and the sample is collected. This arrangement is called the scanning electron microscope, and an overall schematic representation of such a microscope is shown in Fig. 4.22. The layout of a commercial instrument is shown in Fig. 4.23. The vacuum column on the left of the instrument has an elaborate interlock system at its base for introducing samples in to the vacuum column without having to pump the whole system down. The darker cylinder behind the vacuum column holds cryogenic fluid for cooling an X-ray detector. The control electronics and display screens are shown on the right.

Images can be formed from several different signals (Fig. 4.22). If the sample is thin enough, the transmitted electron signal can be collected by a detector below the sample and an image formed from this transmitted current. This is called scanning transmission electron microscopy or STEM. Alternatively, elastically backscattered electrons can be collected from thicker samples using a detector placed above the sample. Yet another signal comes from the inelastically scattered electrons. These are excited from a sample by the incident beam and are distinguishable from the elastically scattered electrons by their lower

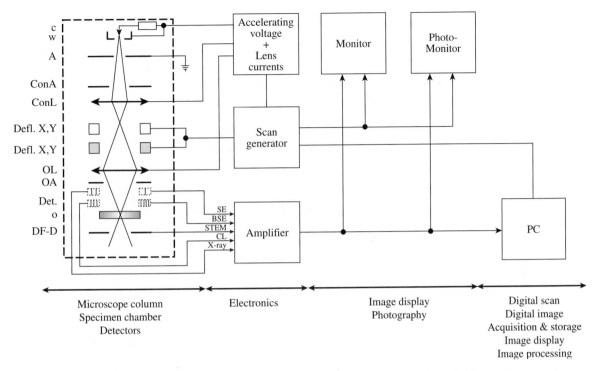

Fig. 4.22 Schematic layout of a scanning electron microscope. The vacuum column is shown inside the dashed box and it contains the electron gun, electromagnetic lenses, electromagnetic deflection coils, apertures, specimen stage, and detectors. The condenser lens, ConL, forms a bright image of the filament that is scanned (by the coils Defl X,Y) and then focused onto the sample (o) by the objective lens, OL. Backscattered electrons are collected by the plates above the sample (BSE), while electrons transmitted through thin samples are collected below the sample (STEM). Inelastically scattered secondary electrons can also be collected (SE) as can X-rays generated by the incident electrons. (Reprinted from Reichelt[30] with kind permission of Springer Science + Business Media. Courtesy of Professor Rudolf Reichelt.)

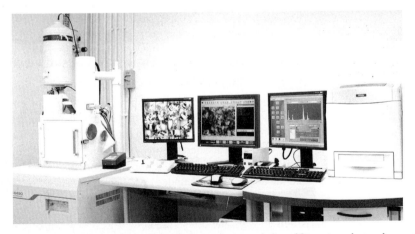

Fig. 4.23 The JEOL 6490 scanning electron microscope. This microscope has a resolution of 3 nm at an electron beam energy of 30,000 V. The sample chamber is differentially pumped which allows for pressures of up to 50 Pa (0.4 Torr) at the sample chamber while maintaining a high vacuum for the electron optics. This permits imaging in a humid atmosphere, maintaining the state of biological samples and avoiding electron beam charging for insulating samples. (Courtesy of JEOL USA Marketing.)

Fig. 4.24 SEM image of the eye of a house-fly. The scale bar marks 100 μm. The fly was coated with a thin film (15 nm) of gold before imaging to prevent charging. This required sample preparation is a drawback of SEM, but it is compensated for by the larger depth of field and more accurate dimensional analysis (there is no finite tip-size broadening) compared to scanning probe microscopes. (Image courtesy of the University of Wales Bioimaging Laboratory — http://www.aber.ac.uk/bioimage.)

energy. Inelastically scattered electrons are collected using an energy resolving detector to give a secondary electron image. Finally, if the incident electron beam is energetic enough, it can be used to excite X-ray emission from core electrons. By using an energy-sensitive X-ray detector, it is possible to make a map corresponding to X-rays from just one chosen element. In this way, the atomic composition of the sample can be mapped at very high resolution.

Scanning electron microscopes have some remarkable properties. Their resolution can approach 1 nm. The objective lens usually has a long focal length, and therefore also has a large depth of focus (5–30 mm). (This vastly exceeds the depth of focus of optical microscopes where the constraints of diffraction require a short working distance objective, and scanning probe microscopes where contact is required.) Thus, large objects can be imaged with very little loss of resolution. Figure 4.24 shows an SEM image of the eye of a house-fly. The sample in this case was coated with a thin gold film to minimize the accumulation of charge on its surface.

4.3.1   Electron microscopy—physics of electron beams

(a) *Beam generation:*
A thermionic emission gun is simply a heated filament, coated with a low work function metal, and heated to a temperature that is a significant fraction of the work function (i.e., >1000 K). In this way, electrons are thermally excited out of the metal and accelerated away from the filament by an anode held at a high positive potential (in Fig. 4.22 the filament is held at a high negative potential and the anode, A, is grounded). The dependence of thermionic current density, J, on temperature is given by the Richardson law

$$J = \frac{4\pi m e}{h^3} (k_B T)^2 \exp\left(-\frac{\phi}{k_B T}\right), \qquad (4.20)$$

where ϕ is the work function of the metal and the other symbols have their usual meaning. Thermionic emission guns are straightforward to manufacture, but the emitted electrons have a broad energy spectrum because they come from a hot source (cf. Fig. 3.5).

Field emission guns utilize extremely sharp probes to generate very high local electric fields. Treating the end of a very sharp probe as a sphere of radius a, the capacitance of the sphere is given by $4\pi\varepsilon_0 a$ so at a potential, V, the charge on the sphere, Q, as given by $Q = CV$, is $4\pi\varepsilon_0 aV$. Taken together with Coulomb's law for the electric field at the surface of a sphere of radius a containing charge Q this yields

$$E = \frac{1}{4\pi\varepsilon_0}\frac{Q}{a^2} = \frac{V}{a} \quad (4.21)$$

for the electric field at the surface of a sphere of radius a and electrostatic potential V relative to the surroundings. Thus, a point of radius 100 nm held at a potential of 1000 V experiences a field at its surface of 10^{10} V/m or 1 V/Å. Thus, if the work function of the metal, ϕ, is 2 eV, electrons at the Fermi energy can escape by tunneling through a barrier of just 2 Å distance (see Fig. 4.25). Calculation of the tunneling rate is a nice problem in one-dimensional quantum mechanics[31] and the result is the Fowler–Nordheim equation:

$$J \propto E^2 \exp - \left(\frac{4\sqrt{2m}}{3e\hbar}\frac{\phi^{3/2}}{E}\right). \quad (4.22)$$

Field emission guns are capable of producing quite large currents without any heating and consequently the electron beam is far more monoenergetic than the beam obtained from a thermionic source.

(b) *Beam focusing and resolution:*
Electron lenses suffer from aberrations that depend on the spread of electron energies (just like chromatic aberrations in optical lenses) so microscopes equipped with field emission guns are capable of producing much finer focal points and thus much higher resolution.

Electrons are deflected by electric and magnetic fields and their path can be predicted from Newton's law of motion:

$$m\ddot{\mathbf{x}} = -e\mathbf{V}\times\mathbf{B} - e\mathbf{E}, \quad (4.23)$$

where $\mathbf{V}$ is the velocity of the electron, $\mathbf{B}$ is the magnetic field, and $\mathbf{E}$ is the electric field. The first term is the Lorentz force and it acts perpendicular to both the velocity vector, $\mathbf{V}$, and the magnetic field vector, $\mathbf{B}$, according to the right-hand screw rule. Thus the electrons spiral around magnetic field lines. In practice, most of the accelerating potential is dropped between the cathode and the anode of the microscope above the magnetic focusing elements (see Fig. 4.22). The lens operates by generating a magnetic field along the axis of the microscope (see Fig. 4.26; the field lines run from P1 to P2). Electrons with a velocity component normal to this direction are forced to spiral around the field lines, and the action of the field on this rotational component of the velocity is such as to move the electrons back to the axis, resulting in focusing of the bundle.

Our discussion thus far has been limited to treating the electrons as point charges and this is a perfectly adequate description for understanding the operation of electron microscopes at resolutions of a few nanometers. Atomic scale

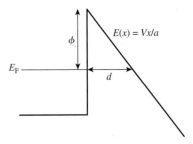

Fig. 4.25 Field emission: showing the variation of potential with distance perpendicular to the surface of a metal. Inside the metal (left) electrons at the Fermi energy are ϕ eV (the work function) below the energy of free electrons. The application of a large electric field, V/a, generates a small tunneling barrier (width d) for electrons at the Fermi energy to escape through.

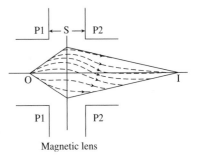

Fig. 4.26 Magnetic focusing of a diverging beam from O to a point on the image plane I. The dashed lines show the spiral paths of the electrons on the surface of the ray cones that connect the object to the image. (Based on Fig 1.4 from Agar et al.[32])

resolution requires an understanding of the wavelike nature of electrons as given by the de Broglie relation (Equation 2.1). Restated in terms of the electron energy in volts, V_0,

$$\lambda = \frac{h}{\sqrt{meV_0}}.$$
(4.24)

According to Equation 4.24 electrons accelerated by a potential as low as 5 kV have a wavelength of 0.17 Å, much smaller than the size of the atomic structure one would usually like to resolve by electron microscopy. Much higher accelerating voltages are used in practice because they simplify the design of the focusing lenses. These voltages are large enough that electrons move at a significant fraction of the speed of light. Correction for the relativistic change of the electron mass is usually achieved by replacing V_0 by a modified voltage V_r given by

$$V_r = V_0 + \left(\frac{e}{2m_0c^2}\right) V_0^2.$$
(4.25)

m_0 in Equation 4.25 is the rest of mass of the electron; e/m_0c^2 is $(511 \text{ keV})^{-1}$ in units of (electron volts)$^{-1}$. Thus, the size of the term in V_0^2 is proportional to the accelerating voltage divided by 1.022×10^4. Neglect of the relativistic correction amounts to about a 5% error in λ at 100 kV.

Despite the fact that the electron wavelength is considerably smaller than the size of atomic structure, wave effects are important in high-resolution imaging because of the way the contrast is generated. Atomic scale contrast arises because of interference between beams that have passed through rows of atoms and beams that have passed through the holes between rows of atoms, an arrangement that is only possible with extremely thin samples. In very thin samples, the phase difference between the beams that take these two paths is very small. The same problem occurs in thin optical samples, where the solution is a technique called differential interference contrast microscopy. Because of the small phase change across a thin sample, electrons focused precisely on the sample yield essentially no contrast. The resolution of individual atoms requires a precise defocusing of the microscope—one that is large enough to generate interference contrast (like the differential interference contrast in optical microscopes) but not so large that local atomic information is lost. For this reason, high-resolution TEM images are obtained with a beam fixed on one point on the sample. For a fuller description, see the book by Spence.[33] An example of an atomic resolution TEM image of a carbon nanotube is given in Fig. 4.27.

(c) *Sample preparation:*

Sample preparation and preservation of the sample in the electron beam are probably the two most demanding aspects of experimental high-resolution electron microscopy. This is particularly true for biological materials that are easily damaged in the incident electron beam. This has led to the development of the field of cryomicroscopy. In cryomicroscopy, the samples of interest are frozen into a thin layer of ice and exposed to tiny doses of electrons, doses so small that images cannot be generated from any one sample (i.e., molecule) but are

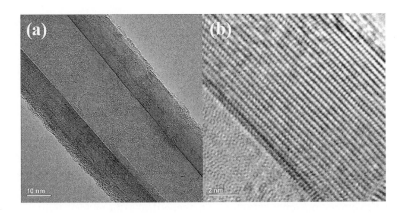

Fig. 4.27 High-resolution TEM image of a multiwalled carbon nanotube. The low resolution image on the left shows the bundle of walls while the high-resolution image on the right shows the atomic structure of one side of the nanotube.[35] (Reprinted with the permission of Cambridge University Press. Courtesy of Mario Miki Yoshida.)

rather assembled by averaging thousands of low contrast images using computer programs to align them for the averaging process. This averaging is easier to do if the signal can be collected in the form of an electron diffraction pattern taken from a thin two-dimensional crystal of the material being studied. Further details of the field of cryoelectron microscopy can be found in the review by Plitzko and Baumeister.[34]

4.4   Nano-measurement techniques based on fluorescence

The resolution of optical microscopy is limited by the wavelength of light. Specifically, using a microscope objective lens of numerical aperture, NA, given by (see Fig. 4.28)

$$NA = n \sin \theta \tag{4.26}$$

the resolution of an optical microscope is

$$r = \frac{\lambda}{2NA}. \tag{4.27}$$

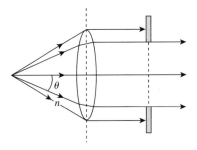

Fig. 4.28 The numerical aperture of a lens is determined by the collection angle, θ, and the refractive index of the medium, n.

This is the resolution determined by the Rayleigh criterion, which states that the resolution of a microscope is given by the distance at which two points are just resolved as separate features. In Equation 4.27, r is the distance separating two points for which the dip in light intensity at the midpoint between the images of the two points is 74% of the peak intensity (this is a dip that is barely discernible by eye). Thus, for a high-quality oil immersion objective lens (NA = 1.4) with blue illumination ($\lambda = 480$ nm) the resolution according to the Rayleigh criterion would be 171 nm.

The Rayleigh limit has been beaten using single fluorescent dye molecules as labels. In addition, these dye-label based techniques have enabled a number of new approaches for studying the dynamics of single molecules.[36]

In order to understand how the fluorescence from a single molecule can be measured, it is necessary to review some of the physics of optical fluorescence. Atoms and molecules absorb light because of the interaction of the

dipole moment of the atom, **D** (arising from the distribution of negative electrons around the positive nucleus), with the electric field, **E**, of the incident
light. This causes transitions from a ground state, ψ_g, to an excited state, ψ_e,
at a rate predicted by Fermi's Golden rule (Equation 2.94a) where the matrix
element is

$$H_{ge} = \langle \psi_g | e\mathbf{D} \cdot \mathbf{E} | \psi_e \rangle . \tag{4.28a}$$

The delta function in Fermi's Golden rule requires

$$\hbar\omega = E_e - E_g. \tag{4.28b}$$

These two equations amount to the requirements that (a) the incident light energy
exactly equals the energy difference between the ground and the excited state;
(b) if the molecule is aligned in space, then maximum absorption occurs when
the electric field is aligned with the dipole moment of the molecule; and (c) the
excited and ground state wavefunctions have opposite parity (i.e., change of
sign on inversion) because the parity of the dipole operator is odd so the integral
implied by Equation 4.28a is zero unless the two wavefunctions have opposite
parity. An isolated atom with just a single optical absorption will re-emit light
at the same frequency at which it was absorbed. However, the optical spectrum
of a dye molecule in a solvent is much more complex. The process is illustrated
by the so-called Jablonski diagram in Fig. 4.29. Each of the electronic states
of the molecule is split into a multiplicity of states by interactions with specific
molecular vibrations. The energy of a molecular vibration is a small fraction

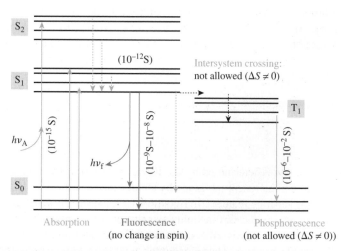

Fig. 4.29 Jablonski diagram showing the ground state (S_0) and two excited states (S_1 and S_2) of
a dye molecule. Each electronic level is split into a manifold of vibrational states owing to the
discrete quantized vibrations of the molecule. The initial rapid (10^{-15} s) absorption of a photon
from the ground state to one of the excited states is followed by vibrational relaxation (10^{-12} s)
to the lowest excited state where the electron remains for a period of time called the fluorescence
lifetime ($\sim 10^{-9}$ to 10^{-8} s) after which it falls into one of the vibrational levels associated with the
ground state, S_0, emitting a photon of fluorescence in the process. Optical transitions preserve the
spin of the electron so all the states labeled S are singlet states with the electron in the excited state
having a spin opposite to that of the remaining electron in the ground state. Interactions between
atoms can result in the excited electron becoming trapped in one of the triplet, T, states (both
electron spins in the same direction). (Courtesy of M. Levitus.)

of the energy associated with an electronic transition, and for this reason the energy spectrum of the molecule appears as sets of discrete series of states separated by large energies that correspond to electronic transitions, with a series of perturbed states separated by the small energy associated with molecular vibrations superimposed on the electronic states. The initial absorption of light occurs from one well-defined energy state (provided, of course, that the dye molecule is cold enough that its vibrational states are not excited thermally). However, the incident photon can drive the molecule into any one of a number of excited states depending on its wavelength. Thus even for a single electronic transition (say S_0 to S_1 in Fig. 4.29), a fairly broad spectrum of light wavelengths may be absorbed. In the gas phase, these different absorptions will correspond to a series of sharp lines, but for dyes dissolved in a solvent these transitions are smeared out into a single broad absorption spectrum. Vibrational relaxation of the molecule from one of these excited states down to the lowest vibrational state associated with the excited electronic state (e.g., S_1) occurs very rapidly, so that all of these possible absorptions lead to an accumulation of electrons in the lowest energy S_1 state. The molecule will remain in this state for a considerable time (the fluorescence lifetime, 10^{-9} to 10^{-8} s) until it decays spontaneously down to one of the manifold of vibrational states in the ground state. The emitted fluorescence is thus shifted to lower energy than the absorbed light, reflecting the average energy lost to vibrations in the process. This shift is called the Stokes shift of the molecule. An example of an absorption and emission spectrum is given in Fig. 4.30. The Stokes shift for this dye is $\sim$25 nm. Note how the absorption and emission spectrum appear to be mirror images of one another. If the distribution (i.e., manifold) of vibrational states is similar in the ground and excited states, then the same spread of energies that gave shape to the absorption spectrum also controls the shape of the emission spectrum.

The key to measuring single molecule fluorescence lies in the fact that the absorption rate will increase with the intensity of the electric field (cf. Equation 4.28a) up to the limit where the molecule is emitting one photon per fluorescence lifetime. Thus, one molecule irradiated by an intense laser can emit up to a billion photons per second, a number that is readily detectable. The development of practical single molecule detection had to await the development of suitable dyes. Interactions with the environment can result in excited state electrons becoming trapped in a triplet configuration (see Fig. 4.29). Floppy molecules can become trapped in nonradiating configurations. And, of course, the absorption cross section of the molecule must be large enough, the emitted fluorescence wavelength must be compatible with sensitive detectors, and the dye must be stable against photochemical reactions. These problems prevented a successful demonstration of single molecule fluorescence until the late 1980s.[37,38]

The ability to see single fluorescent molecules has enabled some remarkable new techniques in Nanoscience. Since the light source is known to be essentially a true point, its location can be determined to a remarkable degree of accuracy by fitting the intensity profile of the emitted light with a function that represents the point spread response of the microscope (i.e., the point spread function, PSF). This is illustrated in Fig. 4.31(a). Thus, by chemically attaching a suitable fluorescent dye to the particle or molecule to be tracked, and digitally recording images, the trajectory of a single molecule can be followed over time. The separation of two dyes that are very close to one another can be resolved to a high degree of accuracy if the dyes fluoresce at different wavelengths. This is

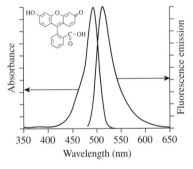

Fig. 4.30 Absorption and emission spectrum for fluorescein dye. The fluorescence maximum is shifted to longer wavelengths than the absorption maximum and the two spectra appear to be mirror images of one another. (Courtesy of M. Levitus.)

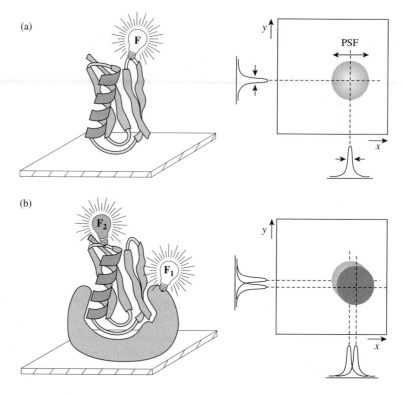

Fig. 4.31 Particle tracking using fluorescent dyes. The width of the point spread function of the microscope is no better than ~0.25 μm but the centroid of the spot can be located to within a few tens of nanometers by fitting the intensity distribution numerically (a). By using two different fluorescent dyes (b) and color filters so that the centroid of each spot may be determined separately, it becomes possible to locate two objects separated by just a few tens of nanometers. (From Weiss[36], Public Domain.) For color, see the images on CD.

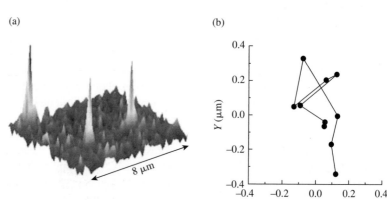

Fig. 4.32 Single particle tracking. The image (a) maps fluorescence count rate over the surface of the membrane of a cell containing fluorescently labeled proteins. The proteins are free to diffuse in the membrane and the position of any one protein can be tracked by following the fluorescence peaks as a function of time. The trajectory of one protein is shown at 10 ms intervals in (b). (Reproduced with permission, Lommerse et al.[39], Journal of Cell Science.)

done by recording two images through filters so that the point spread function of each dye can be fitted separately as illustrated in Fig. 4.31(b). An example of experimental data for single particle tracking is given in Fig. 4.32.

A second valuable application of single molecule fluorescence is single particle FRET (fluorescence resonant energy transfer). FRET utilizes a pair of dyes, the emission spectrum of one (the donor) overlapping the absorption spectrum of the other (the acceptor) as illustrated in Fig. 4.33. When the two dyes are close together, the dipole moment of one dye interacts with the dipole moment of the second, so that the dipole–dipole interaction mixes degenerate states of the two dyes (i.e., those associated with the overlap area in Fig. 4.33). The consequence is a delocalization of dipole energy between the two molecules in the energy range where emission and absorption profiles overlap. Thus if the donor is optically excited, it will transfer energy to the acceptor via a formation of this mixed state with an efficiency proportional to the strength of the dipole–dipole interaction between the two molecules. This interaction falls off as the sixth power of the distance between the two molecules (cf. the van der Waals interaction introduced in Chapter 2). It also depends on the relative orientation of the dipole moments of the two molecules, $\mathbf{D}_1 \bullet \mathbf{D}_2$. Specifically, the FRET efficiency is given by

$$E_{\text{FRET}} = \frac{1}{1 + \left(\frac{d}{R_0}\right)^6}, \qquad (4.29)$$

where d is the distance between the two molecules and R_0 is the *Förster radius* given by

$$R_0 = 9.78 \times 10^3 \left[\kappa^2 n^{-4} \varphi_D J(\lambda)\right]^{1/6}, \qquad (4.30)$$

where κ is a factor that depends on the relative orientation of the donor and acceptor ($0 \le \kappa^2 \le 4$), n is the refractive index of the medium, φ_D is the quantum yield of the donor, and $J(\lambda)$ is an overlap integral between the appropriate states of the donor and acceptor.[36] The value of R_0 depends on the dye pair that is chosen, but it lies typically in the range from 2 to 8 nm. Thus, in principle, the FRET efficiency is a molecular ruler, capable of reporting the distance between two dyes through the relative intensity of the donor and acceptor fluorescence signals. In practice, because the orientation of the two dyes is unknown in a single molecule experiment, the most straightforward use of the technique is simply to report on whether the two dyes are close to, or far from, one another. An example is shown in Fig. 4.34. This shows the fluctuations in FRET signal as a small DNA molecule (labeled with a donor and acceptor dye at two different positions) flips between different available conformations of the molecule.[41]

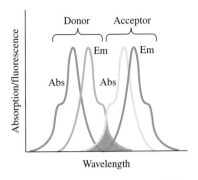

Fig. 4.33 Absorption (Abs) and emission (Em) spectra of two dyes (donor, acceptor) suitable for FRET. The donor emission overlaps the acceptor absorption in the region shown by the gray shaded area.

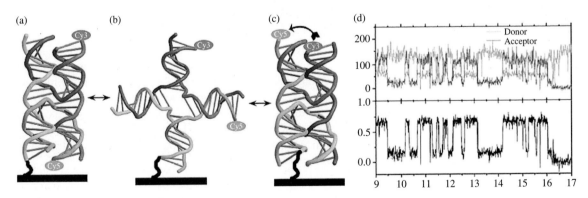

Fig. 4.34 Single molecule FRET experiment. A small DNA molecule, consisting of four interacting strands, was labeled by a donor dye (Cy3) at the end of one strand and an acceptor dye (Cy5) at the end of another strand. Three possible configurations of the molecule are shown (a, b, and c) but only one of the three (c) brings the dyes close enough to generate a FRET signal. A time trace of the green and red fluorescence signals is shown in the top part of (d) (darker trace is the red signal). The donor is excited and emits in the green unless the acceptor comes close in which case emission is in the red. Note how the green and red signals are anticorrelated. These raw data are translated into a FRET efficiency in the black trace at the bottom (the timescale units are seconds). (Reprinted by permission from Macmillan Publishers Ltd: Nature Structural & Molecular Biology, Watching flipping junctions, 2003.) For color, see Color Plate 4.

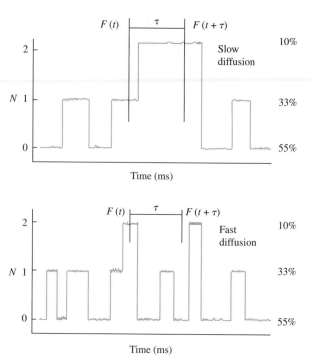

Fig. 4.35 Illustrating the principle of fluorescence correlation spectroscopy. A slowly diffusing fluorophore is more likely to emit more photons in greater number in a given time than a rapidly diffusing fluorophore. This is because the slow molecule remains in a favorable orientation for longer. (Courtesy Marcia Levitus.)

The types of data shown in Figs. 4.32 and 4.34 require rather long sampling times, on the order of 10 ms, and so cannot monitor rapid fluctuations. One way that more rapid fluctuations can be monitored is *fluorescence correlation spectroscopy* (FCS). In this technique, the number of photons arriving at the detector per unit time is recorded as a function of time. The autocorrelation function of this distribution of count rate vs. time is calculated according to

$$R(\tau) = \langle I(t) \cdot I(t+\tau) \rangle = \frac{1}{T} \int_0^T F(t) F(t+\tau) \, dt. \qquad (4.31)$$

This function decays more slowly for a slowly diffusing particle as illustrated in Fig. 4.35. This is because the fluorescence excitation geometry (cf. Equation 4.28a) remains in a favorable position for longer time if the molecule is rotating more slowly. Theoretical fits to the autocorrelation function permit the calculation of the diffusion constant of the molecule. With good enough data, it is possible to record motion on several timescales so that diffusive motion corresponding to both that of the particle through the beam, and internal motion of the particle, can be extracted.

4.5   Tweezers for grabbing molecules

In Section 4.2.6, we discussed the use of the AFM as a tool for manipulating single molecules. In fact, several such tools are now available and they are summarized in Table 4.1.

Table 4.1 Summary of the various tools for single molecule manipulation, showing the available range of forces (Fmin–max), spatial resolution (Xmin), and effective spring constant of the force transducer (Stiffness). (Adapted from the review by Bustamante et al.[42])

Methods	Fmin–max (N)	Xmin	Stiffness (N·m^{-1})	Applications	Practical advantages
Cantilevers (AFM)	10^{-11} to 10^{-2}	10^{-10}	0.001–100	Unfolding proteins/polysaccharides. Measuring bond strengths	High spatial resolution, commercially available
Microneedles (long, flexible needles not designed for imaging)	10^{-12} to 10^{-10}	10^{-9}	10^{-6} to 1	Molecular motors, strength of biopolymers	Good operator control, soft spring constant
Flow field (hydrodynamic forces from flow in channel)	10^{-13} to 10^{-9}	10^{-8}	n.a.	DNA dynamics, RNA polymerase	Rapid buffer exchange, simplicity
Magnetic field (pulling on magnetic particles)	10^{-14} to 10^{-11}	10^{-6}	n.a.	DNA elasticity, following processes involving twist	Simplicity, ability to apply torque
Photon field (laser tweezers)	10^{-13} to 10^{-10}	10^{-9}	10^{-10} to 10^{-3}	Molecular motors, protein unfolding, RNA folding	Well-defined anchor points, high force resolution

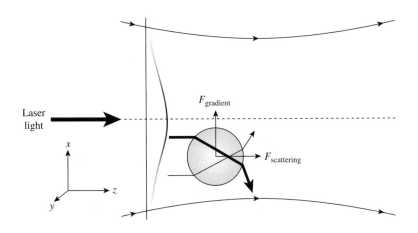

Fig. 4.36 Optical trapping of a dielectric particle in a focused laser beam. The arrows that cross the sphere show the refraction of light at each surface, indicating how the net sideways force on the particle is zero in the middle of the focused spot. The momentum imparted to the sphere by scattering of photons tends to push it along the z-axis, but scattering of the refracted rays can give rise to a compensating force that keeps the sphere at the point of highest intensity along the z-axis (i.e., at the focus). (Courtesy of Professor Carlos Bustamante, Professor of Molecular and Cell Biology, Physics and Chemistry at the University of California, Berkeley.)

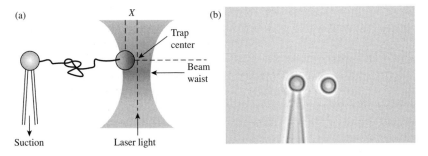

Fig. 4.37 Optical tweezers. The focused laser beam acts as a trap, drawing the small transparent sphere into the point of highest light intensity. (a) Shows a molecule (DNA) attached at one end to a small sphere that is trapped in the laser beam and at the other end to a small sphere that is attached to the end of a suction pipette. Moving the pipette away from the laser trap generates a stretching force on the DNA. (b) Shows a video micrograph of the arrangement shown schematically in (a). DNA is not visible in the optical image. (Courtesy of Professor Carlos Bustamante, Professor of Molecular and Cell Biology, Physics and Chemistry at the University of California, Berkeley.)

Here we review two of these techniques, optical tweezers and magnetic manipulation. We begin with a discussion of optical tweezers. Manipulation of small dielectric spheres in laser beams was first demonstrated by Ashkin and coworkers.[43] The geometry of an optical trap is illustrated in Fig. 4.36. The layout of a real experiment is illustrated in Fig. 4.37. Forces are exerted on the bead in the xy plane by refraction of the incident light. Clearly, these forces are a minimum at the center of the beam where the gradient of intensity is the same on each side of the sphere. What is not so obvious is that the net force on the bead from scattering of the incident light is also a zero along the z-axis when the bead is at the point of highest intensity, that is, the focal point.[43] This is because the focusing action of the bead itself results in a backward push that can balance the forward momentum imparted by scattering. Therefore, the particle becomes trapped at the focal point of the laser beam provided that it is large enough and has a large enough dielectric constant. Typically, plastic or glass beads of a diameter on the order of a micron are used. Typical laser sources for this application emit tens of mW or less, resulting in effective spring constants in the xy plane that range from 10^{-10} to 10^{-3} N/m, a 1000-fold or more less than the spring constants of levers used in the AFM. This greatly enhances force sensitivity (at the expense of spatial resolution). The lowered spatial resolution relative to the AFM comes as a consequence of the larger thermal fluctuations in the "softer" spring driven by $k_B T = 4.2$ pN·nm.

Once the optical trap is calibrated, measurement of the displacement of the bead in the trap provides a direct readout of the force on the bead. This displacement is commonly measured by one of the two techniques. The first technique relies on the measurement of the interference pattern between transmitted and scattered laser light from the bead. It is illustrated in Fig. 4.38. An instrument like this can resolve displacements as small as 3 nm, sampling data at kilohertz rates.[44] Displacement is calibrated by moving a bead fixed to a glass slide a known distance through the laser beam using a piezoelectric translation system. Force can be calibrated by a number of means. One method is to flow a solvent of known viscosity past a bead tethered to the surface by a flexible tether. The viscous drag on the bead will cause a displacement of the bead relative to its equilibrium position before the flow was started. Stokes' law can be utilized to calculate the viscous force (as corrected for the hydrodynamic effects of a nearby surface[45]).

A second technique for calibrating force exploits direct measurements of Brownian motion. Recalling the Langevin equation (3.65) for a free particle and adding a restoring force proportional to the displacement of the particle from

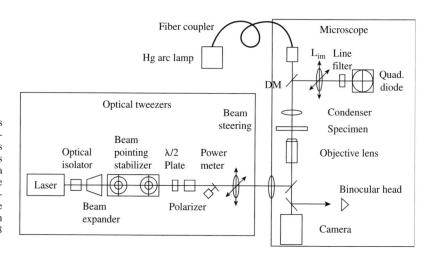

Fig. 4.38 Layout of an optical tweezers experiment incorporating backscatter detection of the position of the bead. DM indicates the location of a dichroic mirror that directs transmitted and scattered laser light onto a four quadrant photodiode (Quad Diode). The interference pattern between the transmitted and scattered light is used to determine the position of the bead. (Reprinted from Allersma et al.[44], Biophysical Journal 1998 with permission from Elsevier.)

the center of the trap to describe the case of a particle in an optical trap it yields

$$m\ddot{x} = -\alpha\dot{x} - \kappa x + F(t). \tag{4.32}$$

In this equation, we have replaced the velocity with the time derivative of displacement, κ is the spring constant of the optical trap, and $F(t)$ represents the random thermal driving force. In the limit of heavy viscous damping, the inertial term $m\ddot{x}$ can be ignored. The equation of motion is then

$$\dot{x} + 2\pi f_c x = \frac{F(t)}{\alpha}, \tag{4.33a}$$

where

$$f_c \equiv \frac{\kappa}{2\pi\alpha}. \tag{4.33b}$$

Thus, the response of the particle falls off for frequencies above this "corner frequency," f_c. Fourier analysis of the response to a random driving force yields the following result for the power spectrum (proportional to $|x(f)|^2$ of the random motions of the particle in the trap

$$P(f) = \frac{C}{f_c^2 + f^2}, \tag{4.34}$$

where C is a constant.[45] Thus, by recording the displacement of the particle as a function of time over some interval, calculating the power spectrum by Fourier transforming the displacement vs. time data, and fitting the resulting distribution by Equation 4.34, the corner frequency, f_c, is determined. The spring constant of the trap, κ, can then be obtained with Equation 4.33b (with α calculated from Stokes' law).

Optical traps have been used to measure the force generated by molecular motors and the size of the steps taken by molecular motors that generate muscle action. Here, we illustrate one of their first applications, which was

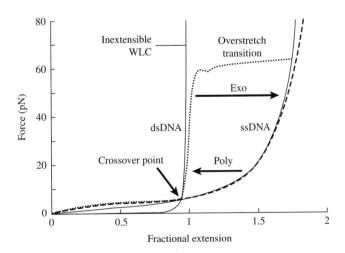

Fig. 4.39 Force vs. fractional extension for double stranded DNA (dsDNA, short dashed line) and single stranded DNA (ssDNA, long dashed line) obtained with optical tweezers measurements. The solid lines are theoretical fits. Double-stranded DNA gives less entropic restoring force at low extensions because its persistence length is longer than that of ssDNA. (Reprinted by permission from Macmillan Publishers Ltd: Nature, Nature Publishing Group, Ten Years of Tension: Single Molecule DNA Mechanics[46], copyright 2003.)

measurement of the elastic properties of a single DNA molecule. Using an experimental set-up like that illustrated in Fig. 4.37, Bustamante and coworkers have measured the force vs. extension curves for various different single- and double-stranded DNA molecules.[46] A summary of some of these data is given in Fig. 4.39. The entropic force produced by extending a polymer (cf. Chapter 3) dominates the force extension curve of both single- and double-stranded DNA at low extensions. The entropy of a given length of a double-stranded DNA is much smaller than that of the single-stranded DNA because of the constraint that the two polymers wrap around one another in a double helix. This gives rise to an effective flexible link that is quite long for double-stranded DNA. This effective flexible link length is called the persistence length, and it is ~50 nm for double-stranded DNA. It is <1 nm for single-stranded DNA. Consequently, at low extensions, a force of only 0.1 pN will extend double-stranded DNA significantly, whereas ~6 pN is required to extend single-stranded DNA a similar amount. The double helical structure of double-stranded DNA causes it to be relatively inextensible when pulled to its full length (fractional extension = 1) until, that is, the force applied to the double helix is large enough to disrupt the double helical structure, causing a transition to a new overstretched form of DNA. In contrast, single-stranded DNA continues to extend in a smooth way. This behavior is important for the biological processing of DNA because the molecular motors that process DNA impose forces of a few pN.[46]

Magnetic forces may also be used to manipulate molecules attached to small paramagnetic beads. A paramagnetic bead can be magnetized by an external field to produce a magnetic dipole that lies in the direction of the applied field. Such a bead feels no net attraction in a uniform field because it is pulled equally to the north and south poles of the magnet imposing the field. However in a field gradient, $\partial B / \partial z$ it experiences a force

$$F_z = -m \frac{\partial B}{\partial z},$$ (4.35)

where m is the magnetic moment of the bead induced by the applied field, B. Thus, inhomogeneous magnetic fields may be used to pull on paramagnetic beads, generating a force on molecules that are attached at one end to the bead and at the other end to a surface. The magnetic manipulation apparatus is not a molecular tweezer in the sense that the bead has an equilibrium position, but is rather a method for placing a constant force on a small bead.[47] One novel aspect of the magnetic force apparatus is that it can be used to apply a torque to an asymmetric bead, so it can be used to explore the effects of applying a mechanical twist to a single molecule. An experiment in which a single DNA molecules are overtwisted is illustrated in Fig. 4.40.

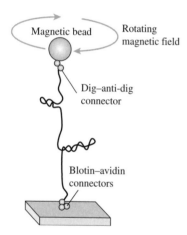

Fig. 4.40 Illustrating the use of magnetic forces for single molecule manipulation. A paramagnetic bead can be both pulled away from the surface, and if anisotropic, rotated by rotating the magnets that generate the field gradient near the bead. This illustrates an experiment in which a long DNA molecule was over twisted. (Reprinted by permission from Macmillan Publishers Ltd: Nature, Nature Publishing Group, Ten years of tension: Grabbing the cat by the tail, manipulating molecules one by one[42] copyright 2000.)

Labels in figure: Magnetic bead; Rotating magnetic field; Dig–anti-dig connector; Blotin–avidin connectors

4.6   Chemical kinetics and single molecule experiments

Single molecule experiments require dilution of the molecules to be studied. For example, there can only be one molecule in the optical trap system of a laser tweezer, or the illumination volume of a FRET or other type of fluorescence experiment. The most benign example is of a FRET or FCS experiment where the molecules to be studied diffuse through a volume illuminated by a microscope objective. Roughly speaking, the sample concentration cannot exceed one molecule per $(\mu m)^3$. This is 10^{-15} molecules/liter or <2 nM. As mentioned in Chapter 3, the dissociation constant, K_D, for many biological molecules lies in the micromolar range, characteristic of the concentration of proteins in cells. Thus, it would appear to be impossible to measure the properties of many biological complexes that consist of two or more molecules bound together, because, at the extreme dilution needed for single molecule experiments, all the molecules would be dissociated.

However, it is worth taking a look at the problem in a little more detail. As discussed in Chapter 3 (Section 3.19), the on rate, k_{on}, is strongly dependent on concentration, having a value in the range of 10^6 to 10^7 M^{-1}s^{-1}. But the off rate is determined entirely by the energy landscape of the bound pair and it does not depend on sample concentration. For $K_D = k_{off}/k_{on} = 1\,\mu$M, we get $k_{on} = 1$ to 10 s^{-1}. Thus, processes that occur on the subsecond timescale could be studied by forming complexes in a concentrated solution, and then diluting it rapidly to make measurements over a time comparable to k_{off}^{-1}.

4.7 Bibliography

C.J. Chen, Introduction to Scanning Tunneling Microscopy. 1993, NY: Oxford University Press. A comprehensive and clear text with a painless introduction to quantum mechanics. Good coverage of both theory and experimental techniques.

D. Sarid, Atomic Force Microscopy. 1992, New York: Oxford University Press. There are lots of formulae in this book (beam mechanics, stability, etc.).

E. Meyer, H.J. Hug and R. Bennewitz, Scanning Probe Microscopy: The Lab on a Tip. 2003, Berlin: Springer-Verlag. A good contemporary survey of the basics.

P.W. Hawkes and J.C.H. Spence (Editors), Science of Microscopy. 2007, New York: Springer. A comprehensive and contemporary survey of microscopy.

J.C.H. Spence, High Resolution Electron Microscopy, 3rd ed. 2003, Oxford: Oxford University Press. A standard reference for atomic resolution electron microscopy.

A. Smith, C. Gell and D. Brockwell, Handbook of Single Molecule Fluorescence Spectroscopy. 2006, Oxford: Oxford University Press. A balanced theoretical and experimental treatment of FRET and FCS.

At the time of writing, there does not appear to be a book on laser tweezer and magnetic pulling. The review by Bustamante, Macosko, and Wuite[42] is one of the best sources available at present.

4.8 Exercises

1. How many volts must be applied to a typical tube scanner (Section 4.1.2) to deflect it by 1 μm?
2. If building vibrations are a maximum of 100 μm at 0.1 Hz, what is the lowest resonant frequency that a control component of a scanning

probe microscope could have so as to be deflected by no more than 0.01 nm by these vibrations?

3. Sketch the shape you expect for a plot of $d \ln I / d \ln V$ vs. v for a semiconductor with a bandgap of 1 eV.

4. The O-H stretching vibration has a frequency of $\sim 10^{14}$ Hz. Assume that the hydrogen moves against a stationary oxygen, and calculate the "spring constant" associated with this vibrational mode using the simple "mass-on-a-spring" formula.

5. Silicon has a Young's modulus (E) of 179 GPa, and a density of 2330 kg/m^3.

   (a) What is the mass of a cantilever of length (l) 100 μm, width (w) 10 μm, and thickness (t) 1 μm?
   (b) From the formula $\kappa = 3EI/l^3$ where $I = wt^3/12$, calculate κ, the spring constant, in N/m.
   (c) What is the resonant frequency of the cantilever in Hz?

6. Calculate the total thermal RMS noise for cantilevers of spring constant 0.1 and 0.01 N/m. How do these numbers change for the 0.1 N/m cantilever if the noise is measured off resonance in a 1 kHz bandwidth? (Take the resonant frequency to be 50 kHz and the Q to be 20. $T = 300$ K.)

7. A protein of 10 nm diameter is imaged by an AFM probe of radius 20 nm. Estimate the full width of the image at its half-height.

8. Name the best AFM technique to:

   (a) Image a stiff protein bound well to a surface at very high resolution in solution?
   (b) Image a liquid film of lubricant on a hard drive?
   (c) Image trapped charges on the surface of a dielectric?

9. A certain complex is bound with an activation energy barrier of $2k_B T$ at room temperature. A force of 0.01 nN applied to the system increases the dissociation rate by a factor of 2. What is the distance from the bound state equilibrium to the transition state?

10. What type of electron microscopy technique would you use to

    (a) Map the elemental composition of a surface?
    (b) Image the atomic coordination in a thin film?

    What source is better for high-resolution microscopy, a field emission source or a thermionic source?

11. Calculate the relativistically correct wavelength of electrons accelerated through a potential difference of 100 kV.

12. How are the absorption and emission spectra of a dye related? Explain the relationship.

13. A photon detection system has a noise background of 100 counts per ms, and presents a solid angle to the source of 10^{-3} Sr. The overall efficiency of the optical microscope is 20%. For a red laser illumination ($\lambda = 630$ nm) of 1 mW in a parallel beam of 1 mm diameter, what must the absorption cross section of a dye be for a single molecule to be detected with a 2:1 signal-to-noise ratio in a 1 ms counting interval, assuming

the shortest fluorescence lifetime quoted in this chapter? To do this problem, first work out the number of photons crossing a unit area per second. The number adsorbed by the molecule per second will be this flux times the (unknown) cross section of the molecule. The adsorbed photons will be re-radiated as fluorescence into 4π Sr. So you can calculate the fraction collected by the microscope, and, knowing the efficiency of the optical system you can figure out what cross section you would need. The number is unrealistically huge, because this is a completely unsuitable geometry for single molecule measurements.

14. In a single molecule FRET experiment, the green channel (direct fluorescence) counts 1000 counts and the red channel (FRET signal) counts 200 counts in a given interval. The dye used has $R_0 = 6$ nm. What is the separation of donor and acceptor dyes, assuming that their alignment remains fixed during the experiment?

15. An optical trap has a spring constant of 10^{-3} N/m. By how much is a trapped bead of diameter 2 μm deflected in a flow of 0.005 m/s? Use Stokes law and the viscosity given for water in Appendix A.

16. What is the corner frequency for the trap and bead in Question 15 for free Brownian motion?

References

[1] Chen, C.J., *Introduction to Scanning Tunneling Microscopy*. 1993, NY: Oxford University Press.

[2] Binnig, G., H. Rohrer, C. Gerber, and E. Weibel, Surface studies by scanning tunneling microscopy. Phys. Rev. Lett., **49**(1): 57–61 (1982).

[3] Quate, C.F., Vacuum tunneling: A new technique for microscopy. Physics Today, 26–33 (1986).

[4] Binnig, G., H. Rohrer, C. Gerber, and E. Weibel, 7 × 7 Reconstruction on Si(111) resolved in real space. Phys. Rev. Lett., **50**: 120–123 (1983).

[5] Sonnenfeld, R. and P.K. Hansma, Atomic resolution microscopy in water. Science, **232**: 211–213 (1986).

[6] Horowitz, P. and W. Hill, *The Art of Electronics*. 2nd ed. 1989, Cambridge: Cambridge University Press.

[7] Drakova, D., Theoretical modelling of scanning tunnelling microscopy, scanning tunnelling spectroscopy and atomic force microscopy. Rep. Progr. Phys., **64**: 205–290 (2001).

[8] Hamers, R.J., R.M. Tromp, and J.E. Demuth, Surface electronic structure of Si (111)-(7 × 7) resolved in real space. Phys. Rev. Lett., **56**: 1972–1975 (1986).

[9] Binnig, G., C.F. Quate, and C. Gerber, Atomic force microscope. Phys. Rev. Lett., **56**(9): 930–933 (1986).

[10] Lindsay, S.M., The scanning probe microscope in biology, in *Scanning Probe Microscopy: Techniques and Applications*, 2nd ed., D. Bonnell, Editor. 2000, New York: John Wiley, p. 289–336.

[11] Viani, M.B., T.E. Schäffer, A. Chand, M. Rief, H.E. Gaub, and P.K. Hansma, Small cantilevers for force spectroscopy of single molecules. J. Appl. Phys., **86**: 2258–2262 (1999).

[12]   Israelachvilli, J.N., *Intermolecular and Surface Forces*. 2nd ed. 1991, New York: Academic Press.

[13]   Keller, D., Reconstruction of STM and AFM images distorted by finite-size tips. Surf. Sci., **253**: 353–364 (1991).

[14]   Villarrubia, J.S., Algorithms for scanned probe microscope image simulation, surface reconstruction, and tip estimation. J. Res. Natl. Inst. Stand. Technol., **102**: 425 (1997).

[15]   Checco, A., Y. Caib, O. Gangc, and B.M. Ocko, High resolution non-contact AFM imaging of liquids condensed onto chemically nanopatterned surfaces. Ultramicroscopy, **106**: 703–708 (2006).

[16]   Muller, D.J., G. Buldt, and A. Engel, Force-induced conformational change of bacteriorhodopsin. J. Mol. Biol., **249**: 239–243 (1995).

[17]   Giessibl, F.J., Advances in atomic force microscopy. Rev. Mod. Phys., **75**: 949–983 (2003).

[18]   Muller, D.J. and A. Engel, The height of biomolecules measured with the atomic force microscope depends on electrostatic interactions. Biophys. J., **73**: 1633–1644 (1997).

[19]   Han, W., S.M. Lindsay, and T. Jing, A magnetically-driven oscillating probe microscope for operation in liquids. Appl. Phys. Lett., **69**: 4111–4114 (1996).

[20]   Sneddon, I.N., The relationship between load and penetration in the axiosymmetric Bousinesq problem for a punch of arbitrary profile. Int. J. Eng. Sci., **3**: 47–56 (1965).

[21]   Tao, N.J., S.M. Lindsay, and S. Lees, Studies of microelastic properties of hydrated compact hard tissues. Biophys. J., **63**: 1165–1169 (1992).

[22]   Li, Y.Q., N.J. Tao, J. Pan, A.A. Garcia, and S.M. Lindsay, Direct measurement of interaction forces between colloidal particles using the scanning force microscope. Langmuir, **9**: 637–641 (1993).

[23]   Sarkar, A., S. Caamano, and J.M. Fernandez, The elasticity of individual titin PEVK exons measured by single molecule atomic force microscopy. J. Biol. Chem., **280**: 6261–6264 (2005).

[24]   Evans, E. and K. Ritchie, Dynamic strength of molecular adhesion bonds. Biophys. J., **72**: 1541–1555 (1997).

[25]   Williams, P.M., Force spectroscopy, in *Scanning Probe Microscopies Beyond Imaging*, P. Samori, Editor. 2006, Wiley VCH.

[26]   Terris, B.D., J.E. Sterns, D. Rugar, and H.J. Mamin, Contact electrification using force microscopy. Phys. Rev. Lett., **63**(24): 2669–2672 (1989).

[27]   Hinterdorfer, P., W. Baumgartner, H.J. Gruber, K. Schilcher, and H. Schindler, Detection and localization of individual antibody-antigen recognition events by atomic force microscopy. Proc. Natl. Acad. Sci. USA, **93**: 3477–3481 (1996).

[28]   Stroh, C., H. Wang, R. Bash, B. Ashcroft, J. Nelson, H. Gruber, D. Lohr, S. Lindsay, and P. Hinterdorfer, Single molecule recognition imaging microscopy. Proc. Natl. Acad. Sci. USA, **101**: 12503–12507 (2004).

[29]   Bash, R., H. Wang, C. Anderson, J. Yodh, G. Hager, S.M. Lindsay, and D. Lohr, AFM imaging of protein movements: Histone H2A–H2B release during nucleosome remodeling. FEBS Lett., **580**: 4757–4761 (2006).

[30] Reichelt, R., Scanning electron microscopy, in *Science of Microscopy*, P.W. Hawkes and J.C.H. Spence, Editors. 2007, New York: Springer. p. 133–272.

[31] Liboff, R.L., Introductory Quantum Mechanics. 1980, Oakland, CA: Holden-Day Inc.

[32] Agar, A.W., R.H. Alderson, and D. Chescoe, Principles and Practice of Electron Microscope Operation. 1974, Amsterdam: North Holland.

[33] Spence, J.C.H., *High Resolution Electron Microscopy*. 3rd ed. 2003, Oxford: Oxford University Press.

[34] Plitzko, J.M. and W. Baumeister, Cryo electron tomography (CET), in *Science of Microscopy*, P.W. Hawkes and J.C.H. Spence, Editors. 2007, New York: Springer. p. 535–604.

[35] Miki-Yoshida, M., J.L. Elechiguerra, W. Antúnez-Flores, A. Aguilar-Elguezabal, and M. José-Yacamán, Atomic resolution of multi-walled carbon nanotubes. Microsc. Microanal., **10**(Suppl 2): 370–371 (2004).

[36] Weiss, S., Fluorescence spectroscopy of single biomolecules. Science, **283**: 1676–1683 (1999).

[37] Shera, E.B., N.K. Seitzinger, L.M. Davis, R.A. Keller, and S.A. Soper, Detection of single fluorescent molecules. Chem. Phys. Lett., **174**: 553–557 (1990).

[38] Moerner, W.E. and L. Kador, Finding a single molecule in a haystack—detection and spectroscopy of single absorbers in solids. Anal. Chem., **61**: A1217–A1223 (1989).

[39] Lommerse, P.H.M., B.E. Snaar-Jagalska, H.P. Spaink, and T. Schmidt, Single-molecule diffusion measurements of H-Ras at the plasma membrane of live cells reveal microdomain localization upon activation. J. Cell Sci., **118**: 1799–1809 (2005).

[40] Churchill, M., Watching flipping junctions. Nat. Struct. Mol. Biol., **10**: 73–75 (2003).

[41] McKinney, S.A., A.C. Déclais, D.M.J. Lilley, and T. Ha, Structural dynamics of individual Holliday junctions. Nat. Struct. Mol. Biol., **10**: 93–97 (2003).

[42] Bustamante, C., J.C. Macosko, and G.J.L. Wuite, Grabbing the cat by the tail: Manipulating molecules one by one. Nature Rev. Mol. Cell Biol., **1**: 130–136 (2000).

[43] Ashkin, A., J. Dziedzic, J. Bjorkholm, and S. Chu, Observation of a single-beam gradient force optical trap for dielectric particles. Optical Lett., **11**: 288–290 (1986).

[44] Allersma, M.W., F. Gittes, M.J. deCastro, R.J. Stewart, and C.F. Schmidt, Two-dimensional tracking of ncd motility by back focal plane interferometry. Biophys. J., **74**: 1074–1085 (1998).

[45] Berg-Sørensena, K. and H. Flyvbjergb, Power spectrum analysis for optical tweezers. Rev. Sci. Instrum., **75**: 594–612 (2004).

[46] Bustamante, C., Z. Bryant, and S.B. Smith, Ten years of tension: single-molecule DNA mechanics. Nature, **421**: 423–427 (2003).

[47] Gosse, C. and V. Croquette, Magnetic tweezers: micromanipulation and force measurement at the molecular level. Biophys. J., **82**: 3314–3329 (2002).

5 Making nanostructures: top down

This chapter is about the conventional techniques used to produce nanostructures. The dominant use of nanostructures in manufacturing is, of course, in the electronics industry. This has driven the development of technology at a remarkable rate. In 1965, Gorden Moore, the founder of Intel, observed that the number of transistors in integrated circuits was doubling every 18 months. Plots of the growth of computer technology show this exponential growth over long periods and in any measure (number of transistors per unit area, calculations per second, etc.). For example, the author, Ray Kurzweil, has constructed a log-linear plot of the number of computations per second per $1000 of computer cost. It is an almost perfect straight line when plotted with available data from the year 1900 to the year 2000. This exponential growth is often referred to as Moore's Law. Though hardly a law of physics, it certainly captures a robust feature of rapid technology development. As a consequence of this breakneck pace of development, it is highly likely that this chapter about technology will become dated rapidly. The best that can be done is to give a snapshot of some of the tools available at the time of writing. It would be a mistake to think that many of the limitations discussed here will still apply in a few years.

We deal first with photolithography. This is often a required first step in making nanostructures, even if the features are no smaller than the micron scale, because the macro to nano interface requires many steps of structure of intermediate size. Nanolithography is made directly possible by some of the beam technologies such as electron beam lithography and focused ion beam (FIB) milling. We also review thin film technologies, both in the form of conventional deposition methods and the more exotic molecular beam epitaxy (MBE). Finally, we examine some novel polymer-based methods for making nanostructures such as block copolymer lithography masks and rubber stamp technologies. It is rather remarkable how many of these techniques are predicted quite accurately in Feynman's original talk (Appendix B and Chapter 1). I have chosen not to dwell on scanning probe–based lithography techniques[1] (and their near relatives like dip pen lithography[2]) because they are slow (though Moore's Law like growth may yet make this a mistake on my part). This chapter assumes some familiarity with basic semiconductor terminology—the physics of semiconductors is discussed in Chapter 7.

5.1 Overview of nanofabrication: top down

Top down means proceeding to build like a sculptor, chipping away at a block of marble to produce a statue. It is how most small structures are mass produced.

(The bottom-up approach is based on getting chemical components to assemble into the desired nanostrucures, much like living things do, and it is the subject of the next chapter.)

The classic approach to top-down assembly is photolithography (from the Greek words for "light," "stone," and "to write"). This technology is, in principle, limited to writing features that are no smaller than the diffraction limited resolution of optical lenses ($\sim$0.5 μm in the visible) but, as we shall see, devices with smallest features (called "critical dimensions") approaching 50 nm are in the pipeline! These most modern methods of lithography are beyond the reach of all bar a few of the leading semiconductor fabrication plants, and the start-up costs for producing devices with features this small are too large for the techniques to be widely applied in research (except at the best capitalized semiconductor companies). Nonetheless, photolithography with even micron-scale resolution is a useful precursor tool for generating nanostructures by other methods.

Other methods more suited to a research environment include electron beam lithography (e-beam lithography), which exposes lithography resists with very fine electron beams, and FIBs, which etch materials directly by bombarding them with energetic ions. e-Beam methods are slow, but one recent innovation may allow "stamping out" of nanostructures. A "master copy" is made by e-beam and then used to make rubber stamps that are molded on the nanoscale.

All these processes form but one part of the process of building a real device. In a semiconductor fabrication plant, the complete tool kit consists of at least the following steps: (a) *Oxidation*, to place a protective layer of silicon dioxide of precise thickness onto the silicon. (b) *Masking*, the photolithography step at which features are opened in the oxide window. (c) *Implantation*, the doping step in which the exposed silicon is implanted with ions to change its electronic properties in the regions exposed through the oxide window. Doping is usually carried out with an ion accelerator, and is combined with multiple steps of lithography to generate the needed regions of different types of doping. (d) *Etching*, to remove oxide and cut other features into regions of the silicon. This includes making holes right through parts of the structure for vertical wires called vias (enabling a three-dimensional layout of the devices). (e) *Metallization*, for contacts, transistor gates, and interconnects between the various components. This is performed either by evaporating or sputtering metal over an entire (masked) wafer. Metal can be deposited either by evaporation or by electrochemical means. (f) *Lift-off*, the complement of etching, is a process in which a number of layers are deposited on to a patterned photoresist which is then dissolved, pulling away the layers on top of the removed resist and leaving them on top of the bare semiconductor (see Fig. 5.7).

5.2 Photolithography

The standard approach to photolithograpy is illustrated in Fig. 5.1. The desired pattern is laid out either as opaque regions on a transparent mask (negative resist) or as transparent patterns on an otherwise opaque mask (positive resist). Only one type of mask is shown in the figure to illustrate the different outcomes with the different types of resist. The oxidized silicon, coated with a thin layer (100–2000 nm) of photoresist, is exposed to light of a wavelength that either

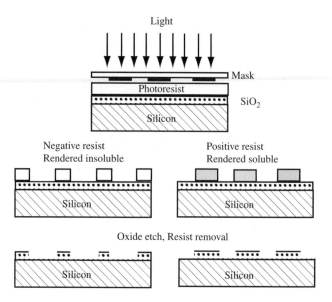

Fig. 5.1 Steps in photolithography, in this case showing the processes required to open windows in an oxide layer on silicon. Photoresist is either cross-linked by exposure to UV light (negative resist on the left) or cut into shorter chains (positive resist on the right). Windows are opened by rinsing off the cut (positive resist) or uncrosslinked (negative resist) resist material. The oxide on the Si is then etched away through these windows, leaving openings through which Si can be doped in the desired pattern. Similar processes allow metal layers and insulators to be applied as thin films in the desired patterns.

cross-links the resist (negative resist) or induces breaks to weaken it (positive resist). Resist is applied as a drop deposited at the center of a silicon wafer spinning at a few thousand RPM and spread by centrifugal force (which overcomes the surface tension of the resist). Once the thin layer is formed, it is stabilized by baking to drive off solvent. We will follow along the right-hand side of Fig. 5.1 to illustrate the next steps in lithography. This part of the figure illustrates the use of a positive resist, such as PMMA (poly(methylmethacrylate)). It is made soluble by exposure to UV radiation in air. This is because scission of the chains (via a photo-induced oxidation of the carbon backbone) lowers the molecular weight of the material in regions where it has been exposed to UV radiation, rendering it more soluble than neighboring (high molecular weight) material. After exposure to UV light through a mask, spraying the resist with a solvent dissolves the regions exposed to UV light and leaves behind the regions underneath the opaque parts of the mask. Figure 5.1 shows the mask being held in direct contact with the photoresist for exposure to light. This simple method is no longer used because it poses problems, both of obtaining good contact and of mechanical wear to the mask. More usually, the mask is placed in one focal plane of a projector system that demagnifies an image of the mask onto the photoresist.

After the exposed photoresist has been flushed away, some areas of the silicon dioxide are exposed, while other regions are protected from exposure to reactive chemicals by the remaining photoresist. The exposed silicon dioxide is readily removed by a solution of hydrofluoric acid (this is a very dangerous material and should not be used without proper training). Exposing the resist-patterned silicon wafer to an HF solution results in etching away of the oxide in the regions that were originally exposed to UV light (which is why the resist is called a positive resist). The bare silicon surface is highly reactive and will oxidize if left exposed to air for a long time. However, various processes, such as ion implantation, can be carried out soon after the original etching process,

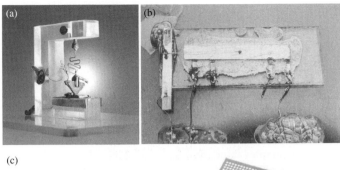

Fig. 5.2 From millimeter scale technology to nanometer scale technology in your laptop. (a) Shows the first transistor, developed at Bell Labs in 1947. (Image courtesy of Lucent Technologies, Public Domain.) (b) Shows the first integrated circuit developed at Texas Instruments, in 1959. (Image courtesy of Texas Instruments, Public Domain.) (c) This is the Intel Pentium processor, versions of which have critical dimensions as small as 65 nm with as many as half a billion transistors on a chip, operating at speeds of several gigahertz. (Image courtesy of Intel Corporation, Public Domain.)

resulting in the modification of the silicon in only those areas exposed through the photoresist (and hence through the protective oxide).

These basic steps, repeated with various types of deposition and removal of materials, constitute the basis of the semiconductor microfabrication (and now nanofabrication) of electronic components. The remarkable evolution of the electronics industry is illustrated in Fig. 5.2. Figure 5.2(a) shows the first transistor assembled by hand at Bell laboratories in 1947. Some idea of the scale is given by the paper clip used to apply force to the point contact junction. Figure 5.2(b) shows the first silicon integrated circuit built at Texas Instruments in 1959. A single slice of silicon was used to form both a transistor and its associated biasing resistors. This was the revolutionary step that led to the complex integrated circuits we use today. An example of this progress is shown in Fig. 5.2(c) which shows an Intel Corp. Pentium processor. Some of these processors have critical dimensions as small as 65 nm and some contain as many as half a billion transistors, operating at frequencies of gigahertz. Forming structures like this requires going well beyond conventional UV optics. Recalling Equation 4.27, the smallest feature that a conventional optical projection system can make is given by

$$r = \frac{k\lambda}{\text{NA}}, \tag{5.1}$$

where NA is the numerical aperture of the projection lens and k is a factor between 0.5 and 1.0 that depends on the resist as well as processing procedures. Producing features as small as 65 nm requires the use of very short wavelength excitation. This is achieved with excimer lasers, some of which can produce intense radiation of a wavelength <200 nm. Optics is usually so inefficient at these wavelengths that a great deal of the incident power is dissipated by absorption in the optical materials used to project the image of the photo mask. Limitations of the optics also restrict the area that can be imaged on the wafer, as off-axis aberrations lower the resolution of the projection system. Exposure systems that step the sample under the optics are being developed to overcome this problem. An example of such a system is shown in Fig. 5.3 and a schematic view of its operation is given in Fig. 5.4.

Figure 5.5 shows how complex patterns can be built up through successive steps of masking, etching, deposition, and lift-off. The figure caption describes the construction of a simple planar diode in detail, and shows the arrangement of a section of a complex three-dimensional CMOS chip in the lower panel of the figure.

Fig. 5.3 An excimer laser stepper lithography system. Excimer lasers emit radiation in the deep UV (248 nm to as short as 157 nm). The problem of imaging diffraction limited spots over large areas is overcome by using a stepper system that scans the wafer with nanometer scale accuracy under a fixed optical system. (Reprinted with permission of ASML Corporate Communications.)

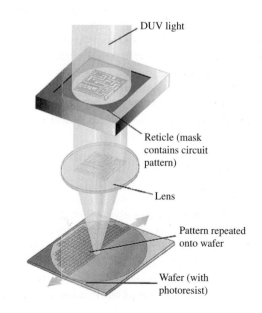

DUV light

Reticle (mask contains circuit pattern)

Lens

Pattern repeated onto wafer

Wafer (with photoresist)

Fig. 5.4 Guts of a UV stepper photolithography system. (Reprinted with permission of ASML Corporate Communications.)

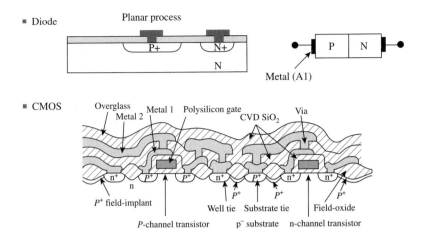

- Diode   Planar process
- CMOS

Metal (A1)

Overglass Metal 1 Polysilicon gate   Via
Metal 2   CVD SiO$_2$

P^+ field-implant

Well tie   Substrate tie   Field-oxide

P-channel transistor   p$^-$ substrate   n-channel transistor

Fig. 5.5 Electronic components made by photolithography. A simple planar diode is shown at the top. The PN junction is formed by diffusion of p-type dopants through a hole in the center of the mask. A heavy diffusion of p-type acceptors overcomes the pre-existing n-type doping of the silicon. A second hole on the right is used to guide the implantation of extra n-type donors to form a good electrical contact with the n-type substrate. An additional step of masking and metal coating forms the wired contacts for the diode. The much more complicated CMOS (complementary metal oxide on silicon), the structure below, is made by a more complex series of similar steps. (From David Johns and Ken Martin, *Analog Integrated Circuit Design*. Reprinted with permission, John Wiley & Sons Inc.)

5.3 Electron beam lithography

As discussed in Chapter 4, the scanning electron microscope (SEM) can form a spot as small as a few nanometers in diameter, and maintain this small spot size over a significant field of view. Energetic electrons may also be used to carry out scission processes in resists like PMMA so electron beam exposure can be used to generate very fine features in resists (in fact both positive and negative e-beam resists are available).[3] An e-beam lithography system (Fig. 5.6) is simply a SEM in which the beam can be turned on and off by computer and directed in a prearranged pattern over the surface of the resist-coated substrate. The most commonly used process makes metal features on the surface of the substrate using lift-off as illustrated in Fig. 5.7(a). A very thin layer of PMMA is spun onto a wafer which is then inserted into the sample stage of a SEM. The resist is irradiated by repeated scanning of an electron beam, the wafer removed and the damaged resist rinsed off. Thin polymer resists are quite transparent to kilovolt electrons, so heavy exposures are required. Once clean regions of the substrate are exposed, the entire structure is coated with a metal film, after which the remaining resist is dissolved to lift off the unwanted metal. Figure 5.7 also shows an example of a nanostructure fabricated with e-beam lithography (see also Fig. 1.1).

In practice, the resolution of e-beam lithography is much worse than one might expect, given the high resolution of SEMs. This is because secondary electrons scatter over a wide region and can cause damage to the photoresist even when their energy is as low as a few electron volts. This process is illustrated in Fig. 5.8. This illustrates the results of simulations of a 20 kV beam impinging on a PMMA resist. Secondary electrons are spread over a distance of nearly ± 2 μm. In practice, resolutions of $\sim$100 nm are readily obtained while features as small as 10 nm have been reported. Thus far, e-beam lithography does not compete with photolithography for rapid production of electronic circuitry. Its throughput is small because of the requirement of long electron exposures, and its resolution, in its simplest implementation, is barely better than that of the best photolithography. It is, however, an indispensable tool for making research structures for nanoscience in small quantities. Masks are

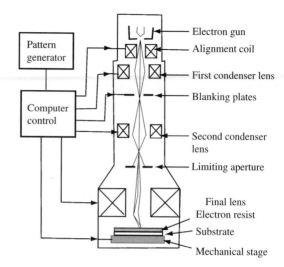

Fig. 5.6 SEM arranged for e-beam lithography.

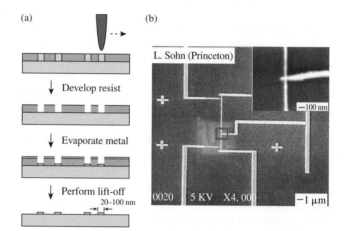

Fig. 5.7 (a) Shows the sequence of steps required to pattern a substrate by lift-off using e-beam exposure of a resist. An example of a structure made by e-beam lithography is shown on the right (b). It is a device for studying electron transport in nanowires. (Courtesy Dr. Lydia Sohn.) See also Fig. 1.1.

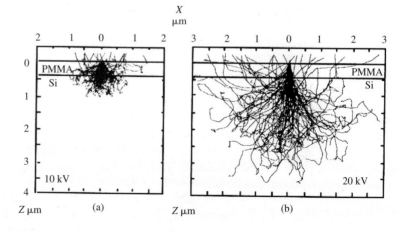

Fig. 5.8 Electron scattering trajectories calculated for a tightly focused spot on a PMMA layer on silicon for (a) a 10 kV beam and (b) a 20 kV beam. Secondary scattering of electrons results in a much broader feature in the exposed resist at the higher voltage. (Reprinted with permission from Kyser and Viswanathan[4], Journal of Vacuum Science & Technology, copyright 1975, American Institute of Physics.)

digitally generated and features are written directly without the need of elaborate stepper systems and expensive UV optics.

5.4   Micromechanical structures

The various device fabrication processes described above can be used to under-cut features on the wafer, thereby liberating small freestanding or cantilevered mechanical structures. This capability has led to the field of MEMS (micro-electro-mechanical systems). Silicon may be made resistant to etching by implanting certain dopants and can be selectively etched along particular crystal faces using reagents that dissolve silicon preferentially along certain directions. The possibilities for forming such structures are almost endless and we will limit our illustration here to the fabrication of probes for atomic force microscopy (AFM). Many of the processes and procedures used to make MEMS are described in texts such as the book by Madou.[5]

One process for making AFM probes with integrated tips is shown in Figs. 5.9 and 5.10. The first stage (Fig. 5.9a) consists of fabricating an array of sharp probes on a silicon wafer. A silicon {001} face is chosen for this purpose so that selective etching along the {110} face may be used to eventually remove silicon underneath the desired cantilever structure. Many of the steps in this fabrication are carried out using a technique called reactive ion etching (RIE). RIE uses a plasma of reactive gases (at low pressure in a vacuum chamber) to remove silicon or other materials selectively, depending on the chemical composition of the gases used in the plasma. The steps shown in Fig. 5.9 illustrate how an array of sharp points can be made under an array of small oxide circles or squares. Isotropic etching of the silicon results both in the removal of silicon directly downwards and also sideways, cutting into the material underneath the protective oxide masks. The result is that an array of sharp probes are formed, capped by plates of the unetched oxide. In this particular application, the goal was to produce sharp probes that stood on top of tall pillars in order to enable AFM imaging of deep features (see Section 4.2.4).[6] In order to form these tall pillars, the sharp walls of the probe were first selectively shaded with a polymer

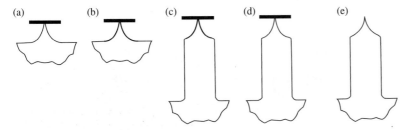

Fig. 5.9 Fabrication of high aspect ratio AFM probes. Oxide masks of a few microns in size are formed on a silicon {001} wafer which is then isotropically etched with SF_6 in a reactive ion etcher. The etching is terminated when a small pillar of $\sim$0.5 μm remains underneath the oxide mask (a). Next a polymer is selectively deposited onto the steep side walls in a CHF_3 plasma (b). An anisotropic etch (SF_6 and O_2) is used to cut preferentially down the {001} face of the silicon (c). The polymer is then removed in an oxygen plasma and the probe further undercut in SF_6 (d). The tip processing is completed by oxidation in a heated water oxygen mixture, followed by removal of excess oxide in HF. This final step results in a very sharp probe apex (e). (From Fig. 2 of Boisen et al.[6], reprinted with permission from IOP Publishing Ltd. and courtesy of Professor Anja Boisen.)

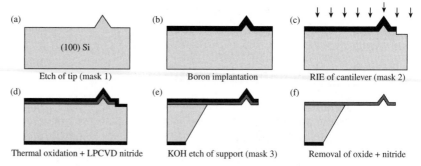

(a)	(b)	(c)
(100) Si		
Etch of tip (mask 1)	Boron implantation	RIE of cantilever (mask 2)
(d)	(e)	(f)
Thermal oxidation + LPCVD nitride	KOH etch of support (mask 3)	Removal of oxide + nitride

Fig. 5.10 Fabrication of the complete cantilever assembly. The silicon wafer containing an array of sharp probes (Fig. 5.9, but without the final oxidation step) (a) is implanted with boron to define an etch resistant layer with the desired thickness (b). The probe pattern is defined by a layer of photoresist and silicon outside of this protected area removed by a reactive ion etch in SF_6/O_2 plasma (c). A thick oxide is grown, both to finish off the tip and to place a protective layer on the sides of the probe to prevent loss of boron. In addition, an etch mask is defined on the back of the wafer by low pressure chemical vapor deposition of silicon nitride (d). Anisotropic etch in KOH cuts preferentially into the {110} face to remove the silicon that was not boron implanted (e). Finally, removal of the oxide layer in HF yields a single crystal boron doped silicon cantilever with a sharp probe at one end. (From Fig. 4 of Boisen et al.[6], reprinted with permission from IOP Publishing Ltd. and courtesy of Professor Anja Boisen.)

coating, and the wafer then etched to produce the pillars. The protective polymer was removed after this step, and the isotropic etching was continued to a point just before the protective plates of oxide fall off the end of the tip. The final step in the sharpening of the probe consisted of growing an oxide layer on the silicon and etching it away in HF, thinning the structure because of the removal of Si as part of the oxide. Fabrication of the underlying cantilever structure is illustrated in Figure 5.10. This process is carried out before the final oxide sharpening of the probes (the last step of probe preparation described above). Boron was implanted into the top of the wafer to a depth equal to the final desired thickness of the cantilever. The desired outline of the probe was then formed on the surface of the wafer using a photo resist. The photo resist protected the nascent cantilevers from a reactive ion etch that removed the surrounding boron doped silicon. At this point, the entire structure was oxidized as a precursor to probe sharpening and as a way of protecting the sides of the boron doped regions. Photolithography was used to pattern the back of the wafer for a deposition of silicon nitride in a way that left a series of open windows behind the cantilevers. An anisotropic etch using KOH selectively cut along the {110} face but left the boron doped silicon unscathed. This process results in an array of silicon chips with a cantilever protruding from each one. In a final step, the protective oxide was removed, to leave the desired thickness of boron doped single crystal silicon cantilever with a sharp probe at its end. Some idea of the three-dimensional geometry of AFM probes may be gleaned from the scanning electron micrograph shown in Fig. 4.9.

Many useful commercial products are made with technology like this. Figure 5.11 shows an optical switch made from a silicon mirror suspended from silicon springs, and oriented by the application of electric field applied between metal electrodes on the back of the mirror and electrodes on the surface of the silicon chip. Arrays of switched mirrors like this are used for high brightness video projection systems.

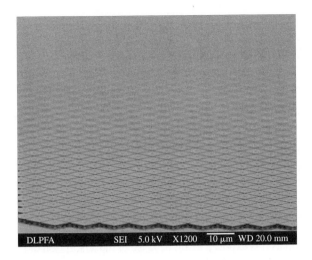

Fig. 5.11 Showing about 1000 of the 800,000 electronically tiltable mirrors on a MEMS-based mirror projection array. The hinges, transducers, and electronics are all located under each mirror. The gap between each mirror is 0.5 μm and devices have no defects because the human eye can catch even one defective pixel in a million. (Courtesy of Dr. Patrick I. Oden.)

5.5   Thin film technologies

Micro- and nanofabrication processes, and the attendant nanoscale imaging techniques, rely heavily on the formation of thin films on surfaces. These films are deposited by a number of methods that have in common the use of high or ultrahigh vacuum technology. We can gain some insight into the requirements for purity of the deposition system and quality of a vacuum from the kinetic theory of gases. Treating the gas as ideal, and taking the kinetic energy for translation along each of the x-, y-, and z-axes to be $\frac{1}{2}k_\mathrm{B}T$ yields the following result for the RMS speed of a molecule in centimeters per second at a temperature of T ($^\circ$K)

$$\sqrt{\langle v^2 \rangle} = 1.58 \times 10^4 \sqrt{\frac{T}{M}} \ \text{cm/s}, \qquad (5.2)$$

where M is the mass of the gas molecule in atomic mass units. This result can be used to estimate the number of molecules striking the surfaces of a container per unit time. Since

$$\langle v^2 \rangle = \langle v_x^2 + v_y^2 + v_z^2 \rangle$$

and half the molecules moving along a given axis will have a velocity vector oriented toward a given surface, the number of molecules striking a surface per unit time will be given by

$$N_\mathrm{s} = \frac{1}{2\sqrt{3}} \rho \sqrt{\langle v^2 \rangle}, \qquad (5.3)$$

where $\rho = N/V = P/k_\mathrm{B}T$ is the density of the gas (re-expressed here using the ideal gas law to rewrite the density in terms of pressure and temperature). Equations 5.2 and 5.3 may be used to estimate the number of molecules striking a surface per unit time at a given temperature. For example, if we assume that the molecule is oxygen (MW = 32) and the temperature is 300 K, we arrive at the conclusion that about 10^{15} molecules strike each centimeter of the surface per second at a pressure of 10^{-6} Torr. This is about equal to a monolayer of adsorbed molecule per second if every molecule striking the surface sticks to it. Thus, if a surface is prepared so as to be absolutely clean, a vacuum of 10^{-9} Torr is required for it to stay that way for times on the order of 1000 s. This is an unrealistic requirement for many vacuum deposition techniques, but standard practice for surface science research. In practice, the thin films are placed on surfaces at rather lower vacuums than this by first depositing a thin layer of a reactive species such as chromium or titanium that actively displaces contamination from the surface to be coated. It is important to realize that surfaces become contaminated immediately on exposure to laboratory air. For example, gold that is cleaned in an ultrahigh vacuum chamber, and then exposed to water under a clean nitrogen atmosphere, is hydrophilic (i.e., water spreads rapidly onto it). The gold surface becomes hydrophobic (i.e., not wetted by water) as soon as it is exposed to air, because of the adsorption of hydrocarbons onto its surface.[7]

The second issue in thin film deposition is the growth pattern of deposited material on the underlying surface. In a simple homogeneous system where element x is deposited onto a surface of a single crystal of element x, the growth is said to be epitaxial. That is, the crystal lattice grows smoothly out from the face of the crystal. Even in this rather special circumstance of epitaxial growth, the resulting surface is not necessarily smooth. Three distinct modes of crystal growth on a surface are illustrated in Fig. 5.12. In the simplest case (Frank–van der Merwe), the crystal grows layer by layer to extend uniformly. In Stranski–Krastanov growth, the coupling of a strong tendency to form a monolayer with a weaker tendency to form three-dimensional structures leads to an underlying uniform film with a surface that gets rougher as the film thickness is built up. In the case of Volmer–Weber growth, the tendency to form three-dimensional structures is so strong that monolayers are never formed. Which of these processes dominates depends on the relative strength of the interactions between the

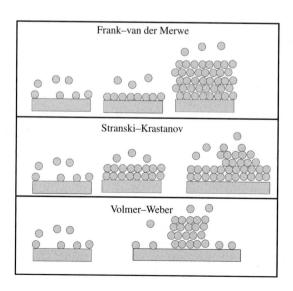

Fig. 5.12 Three distinct modes of epitaxial crystal growth. (Redrawn after Zangwill,[8] Fig. 16.6.)

impinging atoms and the surface, the interaction between neighbouring adatoms on the surface with each other, the temperature of the surface and the deposition rate and energy of the arriving atomic beam. Most thin films tend to deposit as polycrystalline layers composed of nanometer scale crystals that are randomly oriented with respect to each other, though recipes exist for growing quite crystalline metal films with careful control of the growth process.[9] In practice, the growth of epitaxial layers requires rather special circumstances (see below) so most deposited thin films are randomly oriented with respect to the underlying substrate.

Thermal evaporation is probably the simplest method for making a thin film. The material to be evaporated is placed in a tungsten "boat" (a container that also serves as a resistive filament) and is heated electrically. The boat is heated to above the melting temperature of the material to be evaporated, generally resulting in sublimation of the material out into the vacuum chamber. The sample to be coated is clamped in a temperature controlled sample stage at some distance from the heater. The incident flux of material onto the sample stage is usually measured with a quartz crystal thickness monitor. This is a quartz crystal oscillator which shifts its resonance frequency as material is deposited onto it. The sample may be placed at a considerable distance from the source in order to achieve a more uniform thickness of the deposited film. A more elaborate approach to thermal evaporation utilizes an energetic electron beam focused onto a crucible of the material to be evaporated. This arrangement— called e-beam deposition—can achieve the high temperature needed to deposit refractory metals such as titanium.

Sputtering is another approach for depositing thin films. It uses a beam of energetic ions to bombard the surface of the material to be deposited and literally chips it away so that chunks of the material fly from the source to the sample to be coated. The ion source is generated by passing a gas such as argon through a small orifice placed in a strong electric field so that the gas is ionized as it is passed into the vacuum chamber. It is then accelerated onto the sample (typically held at ground potential, while the ion gun is held at several kilovolts with respect to ground). The background ion pressure is kept small enough so as to permit rapid diffusion of sputtered materials. This technique is often useful for materials that are difficult to evaporate thermally, or for materials that have a complex composition and would be decomposed by thermal evaporation.

Yet another approach relies on the creation of reactive chemical species close to the surface to be coated. This is called chemical vapor deposition (CVD). The growth of an oxide layer in a heated steam/oxygen mixture described earlier is an example of CVD. Another example is the growth of polycrystalline silicon (polysilicon) using silane gas. Silane is introduced into the reaction chamber as a temperature of $\sim 600°C$ and a pressure of ~ 1 Torr resulting in the following reaction on the surface:

$$SiH_4 \rightarrow Si + 2H_2.$$

An example of a comprehensive research system for preparing clean surfaces, characterizing them, and depositing well-controlled layers onto them is shown in Fig. 5.13.

Variable temperature (STM/AFM, scanning Auger electron spectroscopy)

Gas exposure, conductivity and Hall effect measurements

Surface analysis (XPS, LEED ion profiling)

Thin film deposition (sputtering, evaporation, effusion cells, ion-beam)

Sample transfer arm

Sample introduction   Sample storage

Fig. 5.13 Thin film deposition and surface analysis system. (Courtesy of Professor Robert Lad, Laboratory for Surface Science and Technology, University of Maine.)

5.6   Molecular beam epitaxy

MBE[10] is a rather remarkable technology based on the epitaxial growth of atomic layers on a substrate. It represents a realization of Feynman's vision of the atom by atom growth of new materials. MBE instruments build up layered materials by sequential deposition of crystalline layers of particular atoms. Its application is limited to certain combinations of materials because of the constraints that one material must grow epitaxially (i.e., in exact atomic registration) on the crystal surface of another material. However, much research has established the required conditions for MBE growth of materials of interest to the electronics industry, and it is now used to produce whole wafers of layered materials of interest. Examples of layered materials made this way are GaAs/AlGaAs, Si/SiGe, InAs/AlSb, and InGaN/GaN.

The MBE instrument is, in principle, quite simple, resembling the simple thermal evaporator discussed above (as shown schematically in Fig. 5.14(a)). Key differences are that the sample is a carefully prepared and cleaned single crystal face placed in ultrahigh vacuum chamber. The atomic beam sources are placed far from the substrate to ensure uniform deposition and an elaborate control system allows precise timing of exposure to several different materials. Reflection high-energy electron diffraction is used to measure the thickness of deposited material via the oscillations in diffracted beam intensity caused by reflections from the interface and surface layers. This allows the thickness of the deposited material to be measured to better than monolayer precision (Fig. 5.14(b)). The growth of materials with lattice constants incompatible with the substrate is possible if the lowering of the surface energy caused by the adhesion of the deposited layer exceeds the strain energy required to distort the deposited layer. This can cause epitaxial registration with the substrate for even quite large mismatches in lattice constants (Fig. 5.15(a)). The associated strain energy accumulates with the number of added layers and therefore limits the thickness of "adlayers" that may be grown with a given degree of substrate–lattice mismatch. Some of the factors controlling the growth kinetics of the added layers are shown in Fig. 5.15(b). Growth rates depend on the fraction of incident atoms that stick, trapping of "adatoms" at special sites, the rate of two-dimensional diffusion on the surface, the association and dissociation rate of small clusters,

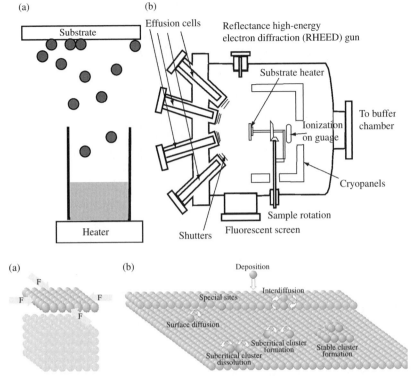

(a)

Substrate

Heater

(b)

Effusion cells

Reflectance high-energy
electron diffraction (RHEED) gun

Substrate heater

To buffer
chamber

Ionization
on guage

Cryopanels

Sample rotation

Shutters   Fluorescent screen

Fig. 5.14 Molecular beam epitaxy. (a) Atoms are evaporated from a heated effusion cell onto a substrate maintained at a controlled temperature. Effusion rate and temperature are adjusted for the best crystal growth. (b) Shows the layout of a practical apparatus. It is assembled in an ultrahigh vacuum chamber (base pressure better than 10^{-10} Torr), has several effusion sources (controlled by individual shutters) and cold baffles behind the sample to trap surplus material. A reflectance high-energy electron diffraction (RHEED) system is used to measure the thickness of the deposited material with atomic precision.

(a)

F
F   F
F

(b)

Deposition

Interdiffusion

Special sites

Surface diffusion

Subcritical cluster
formation

Subcritical cluster
dissolution

Stable cluster
formation

Fig. 5.15 Mechanisms of molecular beam epitaxy. (a) Shows the growth mechanism of a material of a lattice constant different from that of the underlying substrate. The driving force for adhesion of the added layer is lowering of surface energy of the original surface. This is opposed by a strained generated in stretching, or compressing (as shown here) the added layer so as to match it to the underlying substrate. This strain energy grows linearly with the number of layers added, while the lowering of surface energy is constant, and this limits the thickness of a strained layer that can be grown in this way. (b) Shows the various kinetic mechanisms that control the rate of layer growth (Courtesy of Dr. Jeff Drucker, Department of Physics and School of Materials, Arizona State University.)

and the formation rate of stable clusters. The substrate temperature and deposition rate are controlled carefully to achieve uniform growth similar to that shown in the upper panel of Fig. 5.12 (Frank-van der Merwe growth). This is difficult to do because elastic stress can be relieved through the formation of islands (Volmer–Weber or Stranski–Krastanov growth—Fig. 5.12).

By alternating the layers between, for example, small band gap semiconductors and wide band gap insulators, materials with remarkable electronic properties (stemming from the quantum confinement of electrons to the more conductive materials) can be made. This is discussed in more detail in Chapter 7. A transmission electron microscope (TEM) image of such a super lattice is shown in Chapter 1 (Fig. 1.4).

5.7   Self-assembled masks

An interesting approach for the nanolithography of a regular array of simple patterns lies in exploiting the self-assembly of block copolymers. Diblock

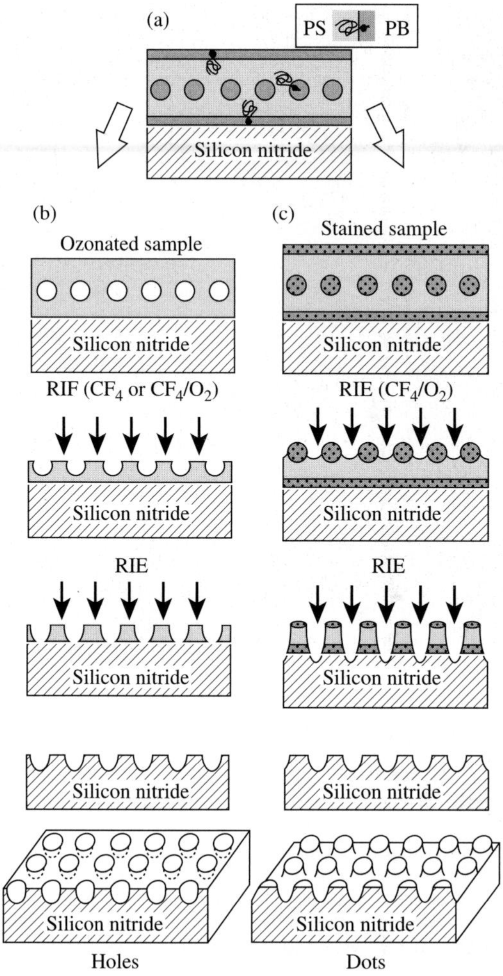

Fig. 5.16 A thin layer of a block copolymer composed of alternating chains of polystyrene and polybutadiene spontaneously forms nanometer scale phase-separated domains when spun as a thin layer on the surface of the silicon nitride wafer (a). The polybutadiene phase can be removed with an ozone treatment and the resulting masked substrate etched with RIE to produce holes under the regions previously occupied by the polybutadiene domains (b). Alternatively, the polybutadiene regions can be stained with a reagent containing a heavy metal which protects the underlying substrate against RIE, resulting in posts underneath the regions previously occupied by the polybutadiene domains (c). (From Fig. 1 of Park et al.[11], Science 1997, reprinted with permission of AAAS.)

copolymers consist of two different polymer chains, A and B, joined by a covalent bond to form a high molecular weight chain of alternating composition, ABABAB.... Immiscible polymers phase-separate into a domain structure that is often quite ordered. The size and the structure of the domains depend on the length of the individual blocks and on the nature of the polymers themselves. Figure 5.16 shows how a thin layer of a block copolymer containing polystyrene and polybutadiene blocks (in the ratio 36:11) was used as both a positive and a negative mask to form holes and pillars with a regular 30 nm spacing. The diblock copolymer was spun onto the wafer to form a 50 nm film (when cured). Removal of the polybutadiene component by exposure to ozone (Fig. 5.16(b)) resulted in a remarkably uniform array of small holes in the resist as shown in Fig. 5.17(a). RIE through these holes and into the silicon nitride substrate produces the array of holes shown in the TEM image in Fig. 5.17(b). Alternatively, the polybutadiene component could be selectively stained with a heavy metal that serves to protect the underlying substrate from subsequent etching (Fig. 5.16(c)). Structures made from cylindrical domains are illustrated in Fig. 5.17(c) (which shows the osmium stained polybutadiene

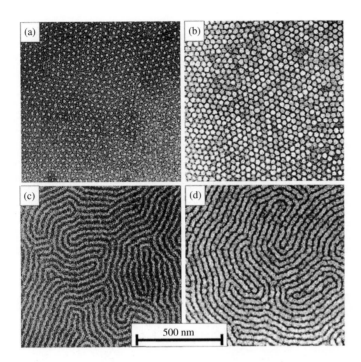

Fig. 5.17 TEM images showing (a) a spherical microdomain monolayer film after the removal of polybutadiene by ozone treatment, (b) the resulting array of holes in silicon nitride after RIE, (c) cylindrical microdomains in which the darker regions are osmium stained polybutadiene domains, and (d) the resulting cylindrical pattern etched into the silicon nitride surface. (From Fig. 2 of Park et al.[11], Science 1997, Reprinted with permission of AAAS.)

regions as dark) and in Fig. 5.17(d) (which shows the resulting cylindrical etch pattern of the silicon nitride). Though limited to the fabrication of regular patterns, this approach has found applications in the fabrication of magnetic nanostructures.[12]

5.8   Focused ion beam milling

The FIB mill[13] represents another realization of one of Feynman's fantastic predictions (Chapter 1). It is the "electron microscope operated backwards" to accelerate ions that chip materials away to form nanostructures. Other than the reversal of bias (the ion source is held at a high positive potential with respect to the grounded sample) and the use of an ion source rather than an electron gun, the FIB instrument is essentially identical to an SEM (and, with its programmable exposure, even more like an e-beam lithography machine). A comparison of the SEM and FIB is given in Fig. 5.18. Typical energies for ion milling lie in the range of 5–30 kV. Ions are many thousands of times heavier than electrons; therefore at a given energy, move much more slowly than electrons (in the ratio of the square root of the masses). This means that the Lorenz force owing to a magnetic field ($\mathbf{V} \times \mathbf{B}$ in Equation 4.23) is much smaller on an ion at a given energy than on an electron. For this reason, electrostatic focusing is used in ion beam instruments.

Liquid gallium is the preferred source for ion beam milling. Gallium metal has a low melting point so it is easy to construct a liquid source. The gallium is singly ionized at the gun (an example of which is shown in Fig. 5.19(a)). Gallium's atomic number (31) places it near the middle of the periodic table and, in consequence, momentum transfer is optimized for a wide range of targets

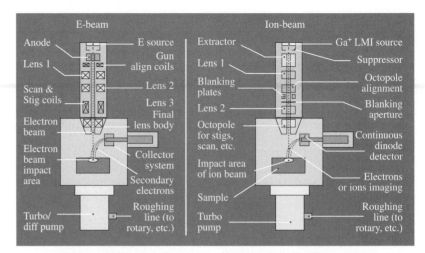

Fig. 5.18 Side-by-side comparison of a SEM and a FIB instrument illustrating the many similarities. (From Fig. 8 of focused ion beam technology capabilities and applications FEI Company, courtesy of FEI.)

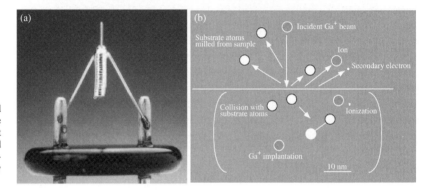

Fig. 5.19 Showing (a) the gallium liquid metal ion source and reservoir and (b) the various types of interactions between incident gallium ions and a surface. (From Figs. 4 and 5 of focused ion beam technology capabilities and applications FEI Company, courtesy of FEI.)

(maximum momentum transfer occurs in a collision when both particles have the same mass). Ions do not penetrate as far into a target as electrons do because of their strong momentum transfer to the target atoms. It is this momentum transfer that causes ejection of target atoms. The interactions of ions with the sample (see Fig. 5.19(b)) offer several different modes of imaging. Collection of the scattered ions permits a direct ion beam imaging mode, while collection of secondary electrons generated by impact ionization of the target enables another mode of imaging. In addition to scattering from the sample, gallium ions can become lodged within it, so a by-product of FIB milling is substantial implantation of gallium ions. Combined FIB/SEM instruments are available with dual columns, one a FIB mill and the other a high-quality SEM. The columns are tilted one with respect to the other, to enable direct SEM viewing of the area that is being FIB-milled.

Two examples of structures milled with a FIB are given in Fig. 5.20. The first example (5.20(a)) demonstrates the direct transfer of a bitmap image onto a sample at the nanoscale. The instrument comes with a software that will translate images directly into features etched into the target. The second example (5.20(b)) shows how the FIB can be used to prepare samples for electron microscopy using a technique called "lift out." In this method, the FIB is used

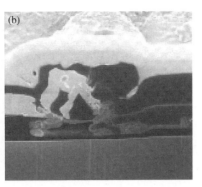

Fig. 5.20 Examples of ion beam milling. (a) Shows an image directly milled by ion beam irradiation of a gold film. The instrument includes computer control for direct transfer of a bitmap image into a structure. (b) SEM image of an insulator defect in a semiconductor device is revealed by cutting through the device using the FIB. FIB instruments can incorporate a second electron column for simultaneous imaging and milling. (From Figs. 13 and 14 of focused ion beam technology capabilities and applications FEI Company, courtesy of FEI.)

to cut straight through a section of a device on a semiconductor wafer, the region of interest plucked out of the wafer, and placed onto a TEM grid for high-resolution imaging. The FIB can also be used for selective deposition of materials using preloaded canisters of reagents for electron beam assisted CVD. Materials such as carbon and platinum may be "written" onto the surface as nanometer scale lines. This capability is the basis of the use of the FIB in the semiconductor industry as a tool for making repairs to semiconductor wafers and masks.

The resolution of the FIB depends on the scattering of the gallium ions within the target volume (Fig. 5.19(b)) and, like e-beam lithography (Fig. 5.8) feature sizes are considerably larger than the FIB spot. The smallest features that can be milled are typically on the order of a few tens of nanometers.

5.9 Stamp technology

Stamp technologies work by direct printing of nanoscale features, working just like a lithographic printing process, but at much higher resolution. It is a rather amazing fact that the contact printing technology invented by Gutenberg in the fifteenth century can be made to work down to the nanometer scale! It is yet another realization of Feynman's visionary predictions. Features down to ~10 nm in size have been printed by using a soft polydimethylsiloxane (PDMS) stamp as a printing press (Fig. 5.21(a)).[14] The desired shape of mold for the stamp is first cut directly into a hard substance like silicon using e-beam or FIB lithography. The mold is shaped to be complementary to the desired stamp shape. The PDMS is poured onto the mold and cured, either by chemical cross-linking or radiation cross-linking. The resulting silicone rubber is flexible and can be pulled away from the mold, yet retains the topographical features of the mold with nanometer scale precision. This rubber stamp can be used just like an office ink stamp, wetted with ink (the chemical reagent to be transferred) and pressed against the substrate to be patterned. If the chemical that is transferred is a resist, then subsequent etching will produce the desired structure in the substrate. In practice, the ability to transfer chemicals onto a surface may be the most important application of this technology. Selective printing of desired reagents in nanoscale patterns is an important part of building chemical sensors and diagnostic arrays. (Dip pen lithography is another method for patterning chemicals at the nanoscale.[2]) The great benefit of this approach is that the

Fig. 5.21 Illustrating the three approaches to nanoimprint lithography. (a) Soft imprint lithography using a polydimethylsiloxane stamp to transfer chemicals to the surface of a wafer to be etched. In this case, the materials act as an etch resist, but the material to be imprinted could also be a functional chemical such as a DNA molecule in a gene chip. (b) Nanoimprint lithography with a hard stamp. The stamp is pressed into a thermoplastic on the surface of the wafer at a temperature above the glass transition temperature of the plastic. The plastic is then cooled and the stamp withdrawn to leave a patterned resist. (c) In step and flash imprint lithography, the resist is a UV-curable low molecular weight resin. The hard stamp is pressed into the resin which is then cured with UV light.

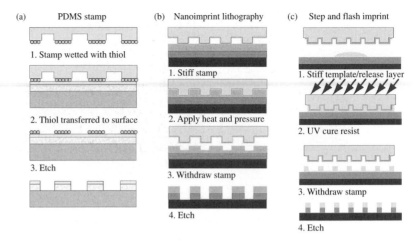

Fig. 5.22 MOSFET with a 60 nm gate. All four levels of lithography were carried out by nanoimprint methods. (Reprinted with permission from Zhang and Chou[18], Applied Physics Letters, copyright 2003 American Physical Society.)

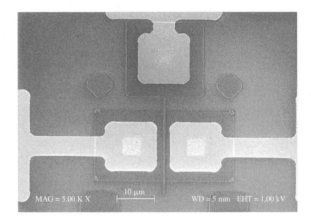

stamps are very easy to produce once the mold is made, so damage to the stamp is not a serious problem because a new stamp is readily produced.

Another type of stamp lithography uses a hard stamp and a soft resist.[15] Two versions of this nanoimprint lithography are shown in Fig. 5.21(b) and (c). In the original form of nanoimprint lithography,[16] the wafer to be patterned is covered with a thin layer of a thermoplastic resist. The resist is heated to above its glass transition temperature (a temperature at which it becomes soft) and the hard mold pressed into the soft plastic. The plastic is then cooled and the mold pulled away. The thinner regions of plastic may be removed to allow etching of the underlying substrate. A somewhat easier version of this technology is the step and flash imprint lithography shown in Fig. 5.21(c).[17] This is similar to nanoimprint lithography, but the resist is a liquid resin that is cross-linked by exposure to UV light after the mold is pressed into the liquid. The surface of the mold is covered with a release agent that allows the mold to be removed from the imprinted surface after the resin is cured. Commercial systems are now available for nanoimprint lithography and Fig. 5.22 shows an example of a MOSFET with a 60 nm gate fabricated entirely by nanoimprint lithography.[18] The same technology has been used to make fluidic devices with liquid channels of nanometer scale dimensions.[19]

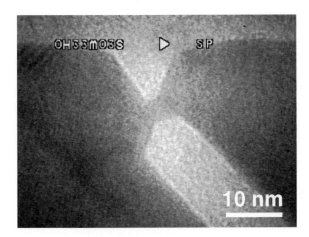

Fig. 5.23 Feedback-controlled electromigration was used to make this gap in a gold nanowire. The gold atoms are removed as atomic layers, resulting in a gap terminated almost perfectly along crystal symmetry directions. (Reprinted with permission from Strachan et al.[21], Physical Review Letters, copyright 2008, American Physical Society. Courtesy of Professor Douglas R. Strachan, University of Kentucky.)

5.10 Nanoscale junctions

All of the lithographic and nanofabrication techniques described thus far are limited to making features down to a few tens of nanometers in size. This is still too large to act as an interface to the atomic and molecular world (cf. molecular electronics, Chapter 8). One way to make very tiny junctions is to break open a thin wire made by conventional nanolithography. This can be performed by a process called electromigration, essentially blowing the wire open as though it were a fuse. The process works in soft metals like gold where very high current densities lead to substantial local heating and movement of the nearly liquid metal by the current. As might be imagined, the technique is quite hard to control at the nanometer scale and there is some controversy about the local atomic structure of the junction. The formation of these junctions has been studied in situ by SEM in an ultrahigh vacuum instrument, resulting in improved understanding of the mechanism whereby nanoscale junctions are formed.[20] Realizing the uncontrolled nature of the "fuse-blowing" approach to making these nanogaps, Strachan and colleagues[21] have devised a control circuit that reduces the current through the junction as it begins to break. The result is that the gold atoms appear to flow away from the hot spot in atomic layers, forming beautiful crystal facets in junctions that can now be controlled to almost atom precision. An example of this work is shown in Fig. 5.23.

Electrochemistry can also be used to either grow or dissolve very small wires in nanometer scale gaps. Small gaps are fabricated by e-beam lithography, and are subsequently filled in by depositing metal into the junction electrochemically. Alternately, metal can be stripped from a very thin wire to open it up in a controlled way. This controlled growth and dissolution has been used to study quantum conductance in nanowires[22] a subject we will discuss in Chapter 7.

5.11 Bibliography

M. Madou, Fundamentals of Microfabrication. 1997, Boca Raton, FL: CRC Press. An elementary survey of current techniques and some that might come into use in the future.

H. Levinson, Principles of Lithography. 2001, Bellingham, WA: SPIE. A good engineering reference for engineers in the field.

D.M. Mattox, The Foundations of Vacuum Coating Technology. 2003, Norwich: Noyes/William Andrew. A survey of current coating techniques, including CVD, but not MBE.

J. Orloff, L. Swanson and M. Utlaut, High Resolution Focused Ion Beams: FIB and Applications. 2002, Berlin: Springer. A comprehensive introduction by pioneers in the field.

5.12   Exercises

1. Design and sketch masks to be used with negative photoresist for making an array of five wires of smallest width 2 μm on a silicon wafer. The wires are to be connected at each end to contact pads (1 mm^2) for wire bonding. The wires are to be broken to have 20 nm gaps. What technique would you use to do this? The wire to be broken must be thin (no more than 50 nm) but the contact pads need to be thick (1 μm) for wire bonding. Outline the steps you might take to make the final device. You may have to research resist treatments for lift-off.

2. An eximer laser produces radiation at 157 nm, and optics with a numerical aperture of 1.4 is available. What is the smallest size of feature you could make in a resist? How might smaller final features be made in the underlying silicon with the same optics?

3. One way to deposit wires on a substrate is to make a thin metal mask with apertures in it, and to then evaporate the desired metal so that it is deposited only where wanted. Why is the lift-off process better? (List as many reasons as you can think of.)

4. Estimate the resolution that could be obtained in e-beam lithography when a 10 kV beam is used with a PMMA resist.

5. Estimate how much lattice mismatch might be tolerated when one monolayer of a mismatched material is epitaxially deposited onto a different substrate. To do this, use the value of spring constant, κ, for a covalent bond estimated in Question 4 of Chapter 4 to calculate the energy of a distortion δx according to $\Delta E = \frac{1}{2}\kappa(\delta x)^2$. Assume that this is compensated by loss of the surface energy of the substrate, using the value 0.5 J/m^2 given in Chapter 4 (Section 4.2.3). Assume that atoms are packed at a distance equal to the Si lattice constant of 2.3 Å.

6. Conducting AFM probes do not usually make a good electrical contact to gold in ambient air unless a significant force is applied. Explain.

7. What process would you use to make regions of silicon resistant to etching by KOH?

8. Using a Web search, make a list of as many materials as you can that can be deposited as thin layers by CVD.

9. What, of the following conditions, might favor epitaxial growth (and why)? High or low substrate temperature? High or low deposition rate?

10. Can you think of a real-world application for an array of magnetic nanostructures like the ones referred to at the end of Section 5.7?

11. FIB would be one way to cut the wires in Question 1. What effect might the Ga deposition have on the conductivity of the gaps?

12. A PDMS stamp is to be used to transfer thiolated (functionalized with a sulfur) DNA onto a gold surface, exploiting the fact that the gold–sulfur bond is quite strong. What treatments might the gold need for fine features to be printed with good uniform printing of the DNA spots? You will have to do some research or asking around.

References

[1] Tseng, A.A., A. Notargiacomo, and T.P. Chen, Nanofabrication by scanning probe microscope lithography: a review. J. Vac. Sci. Technol. B, **23**: 877–894 (2005).

[2] Piner, R.D., J. Zhu, F. Xu, S. Hong, and C.A. Mirkin, "Dip Pen" nanolithography. Science, **283**: 661–663 (1999).

[3] Howard, R.E. and D.E. Prober, in *VLSI Electronics: Microstructure Science*, N.G. Einspruch (ed.). 1982, New York: Academic Press. p. 146.

[4] Kyser, D.F. and N.S. Viswanathan, Monte Carlo simulation of spatially distributed beams in electron-beam lithography. J. Vac. Sci. Technol., **12**: 1305–1308 (1975).

[5] Madou, M., *Fundamentals of Microfabrication*. 1997, Boca Raton, FL: CRC Press.

[6] Boisen, A., O. Hansen, and S. Bouswstra, AFM probes with directly fabricated tips. J. Micromech. Microeng., **6**: 58–62 (1996).

[7] Smith, T., The hydrophilic nature of a clean gold surface. J. Colloid Interface Sci., **75**: 51–53 (1980).

[8] Zangwill, A., *Physics at Surfaces*. 1988, Cambridge: Cambridge University Press.

[9] DeRose, J.A., T. Thundat, L.A. Nagahara, and S.M. Lindsay, Gold grown epitaxially on mica: conditions for large area flat faces. Surf. Sci., **256**: 102–108 (1991).

[10] Tsao, J.Y., *Materials Fundamentals of Molecular Beam Epitaxy*. 1993, San Diego: Academic Press.

[11] Park, M., C. Harrison, P.M. Chaikin, R.A. Register, and D.H. Adamson, Block copolymer lithography: periodic arrays of 10^{11} holes in 1 square centimeter. Science, **276**: 1401–1404 (1997).

[12] Cheng, J.Y., C.A. Ross, E.L. Thomas, H.I. Smith, R.G.H. Lammertink, and G.J. Vansco, Magnetic properties of large area particle arrays fabricated using block copolymer lithography. IEEE Trans. Magnetics, **38**: 2541–2543 (2002).

[13] Giannuzzi, L.A. and F.A. Stevie, *Introduction to Focused Ion Beams*. 2005, New York: Springer.

[14] Xia, Y. and G.M. Whitesides, Soft lithography. Angew. Chem. Int. Ed., **37**: 550–575 (1998).

[15] Guo, L.J., Recent progress in nano imprint technology and its applications. J. Phys. D, **37**: R123–R141 (2004).

[16] Chou, S.Y., P.R. Krauss, and P.J. Renstrom, Imprint lithography with 25-nanometer resolution. Science, **272**: 85–87 (1996).

[17] Coiburn, M., T. Bailey, B.I. Choi, J.G. Ekerdt, S.V. Sreenivasan, and C.C. Willson, Development and advantages of step-and-flash lithography. Solid State Technol., **44**: 67–68 (2001).

[18] Zhang, W. and S.Y. Chou, Fabrication of 60-nm transistors on 4-in. wafer using nanoimprint at all lithography levels, Appl. Phys. Lett., **83**: 1632–1634 (2003).

[19] Cao, H., Z.N. Yu, J. Wang, J.O. Tegenfeldt, R.H. Austin, E. Chen, W. Wu, and S.Y. Chou, Fabrication of 10 nm enclosed nanofluidic channels. Appl. Phys. Lett., **81**(1): 174–176 (2002).

[20] Taychatanapat, T., K.I. Bolotin, F. Kuemmeth, and D.C. Ralph, Imaging electromigration during the formation of break junctions. Nano Lett., **7**: 652–656 (2007).

[21] Strachan, D.R., D.E. Johnston, B.S. Guiton, S.S. Datta, P.K. Davies, D.A. Bonnell, and A.T. Johnson, Real-time TEM imaging of the formation of crystalline nanoscale gaps. Phys. Rev. Lett., **100**: 056805 (2008).

[22] Li, C.Z. and N.J. Tao, Quantum transport in metallic nanowires fabricated by electrochemical deposition/dissolution. Appl. Phys. Lett., **72**: 894–896 (1998).

Making nanostructures: bottom up

6

Chemical synthesis has been the primary route to self-assembly of complex materials since the time of the alchemists. Its importance is even more today, and skill in making new compounds and adding functional groups to existing compounds is a key tool in nanoscience. This chapter begins with a brief outline of the synthetic methods, tools, and analytical techniques used by organic chemists. We then turn to review the many interactions, weaker than a covalent bond, but somewhat stronger (in binding energy) than thermal fluctuations, that permit the self-assembly of complex systems in biology, and ever increasingly in nanotechnology. We consider the formation of lipid vesicles (fatty spheres that self-assemble in water) as our example of equilibrium self-assembly. Clever synthesis techniques exploit chemical kinetics to allow certain kinds of structures to be made with a narrow size range not possible in thermodynamic equilibrium (examples are the synthesis of quantum dots and nanowires). We end with a look at a new strategy for self-assembly of very complex systems based on the remarkable properties of the DNA that carries the genetic code in biology. DNA nanotechnology is sowing the seeds of a revolution in self-assembly.

6.1 Common aspects of all bottom-up assembly methods

Bottom-up assembly generally means making complex nanostructures starting from the random collisions of the components dissolved in a solvent. Biology provides the most striking example of this process; complex macroscopic organisms emerge from the chaotic soup of the fertilized egg, and it is this self-assembly of living systems that provides much of the inspiration for current research in this field. Complex electronic circuitry that assembles itself from a chemical soup is presently the stuff of science fiction, but, as we shall see at the end of this chapter, tools have been invented that bring this science fiction closer to becoming science fact.

Entropy plays a central role in self-assembling systems made by equilibrium methods. The final structure must be reasonably stable at room temperature, but it also must be weakly enough bound so that erroneous structures dissociate, allowing the system to explore the large number of configurations needed to find the desired configuration of lowest free energy. Self-assembly of complex structures requires achieving just the right balance between entropy at a given temperature and the binding enthalpies of the various components. Other methods are kinetic, relying on trapping the system in some nonequilibrium configuration by rapidly changing the concentration of a reactant, changing the solvent, or changing the temperature of the system. All of these considerations apply to any form of bottom-up assembly, including the synthetic chemistry of even rather simple compounds. Later in the chapter, we will examine "equilibrium" nanostructures that spontaneously assemble, as well as nanostructures that are made by kinetically trapping metastable intermediate structures.

We are going to start with a description of chemical synthesis that ignores the complications of thermodynamics and chemical kinetics. This simple view works pretty well because the enthalpies involved in the formation of strong covalent bonds are so large ($\sim 100 k_B T$ at room temperature) that they allow us to think in simple mechanistic terms, at least as a first approximation.

Traditional chemical synthesis is an important route for assembling nanostructures (chemists rightly insist that they have been building nanostructures for centuries). The goal is to produce the desired product through strong and very specific interactions. Chemical reactions occur far from true thermodynamic equilibrium (because the process of reactants turning into products is not readily reversible) but certain thermodynamic ideas are valuable. For example, we know that we can drive a reaction further in the forward direction by increasing the free energy of the reactants relative to the products simply by increasing the concentration of the reactants. If the reaction enthalpy from a given type of bonding greatly exceeds thermal energy, then the product is stable, so that the reaction proceeds in one direction only, at a rate determined both by diffusion of the reactants in solution and the energy of the transition state for a reaction to proceed. Even with highly specific chemical reactions that form stable bonds, entropy can still complicate the picture, and few chemical synthesis procedures result in a pure product. Thus, characterization and purification of the resulting compounds are as much a part of organic synthesis as the carrying out of reactions. We will give an overview of some aspects of organic synthesis in the next section.

Some knowledge of synthetic chemistry is essential for budding nano scientists. Small structures have large surface to volume ratios and the surfaces are covered with atoms, atoms that cannot satisfy chemical bonding requirements with the neighboring atom of the same material. For this reason, surface chemistry considerations can dominate the properties of small particles. As another example, experiments frequently require that nanoparticles, whether they be biological molecules or solid-state particles, be placed at specific locations on a solid surface. This linking together of components of the system requires chemistry. One may want to place labels or tags at specific locations (such as fluorescent dyes in the case of FRET discussed in Chapter 4). Specific chemical attachment is required to do this. There are many reasons (over and above the direct synthesis of nanostructures) why experimental nanoscience requires a good chemistry laboratory.

6.2   Organic synthesis

Organic chemistry dates back to before the discovery of the structure of benzene, C_6H_6, a ring of six carbon atoms (see Chapter 8 and Appendix K). That carbon could form such structures was completely unexpected. Much of the richness of organic chemistry derives from the remarkable chemistry of carbon. Carbon has four valence electrons ($2s^2$, $2p^2$) but is always found in nature bonded to itself (or other elements) and these bonding interactions perturb the valence orbitals of carbon. The new perturbed electronic structures are simple combinations (hybrids) of the atomic orbitals of the isolated atom. The process is called hybridization (we will discuss bonding in organic materials in Chapter 8). Now, we need to know that carbon atoms involved in bonding interactions can form four sets of orbitals composed of the 2s orbital and the three 2p orbitals (called sp^3 hybridization) or alternatively three orbitals composed of the 2s orbital and two of the 2p orbitals (called sp^2 hybridization) with one p orbital left over. One s orbital can hybridize with one p orbital (with 2p orbitals left over) in sp hybridization. These combinations of orbitals can form bonding states containing two electrons each (single bonds), four electrons each (double bonds), or six electrons each (triple bonds). In consequence, the possible structures that carbon can form are essentially unlimited. Figure 6.1 shows some contemporary examples of remarkable nanomachines assembled by synthetic methods.

The first example is a nanoscale box called a dendrimer, assembled by repeated addition of branched structures to smaller branched structures, much in the way that tree limbs divide to form smaller branches. Carried out in three dimensions, this series of reactions forms a particle that is relatively less dense at the center and becomes progressively more dense as one proceeds toward the exterior of the dendrimer. The hollow interior was used as a box to contain

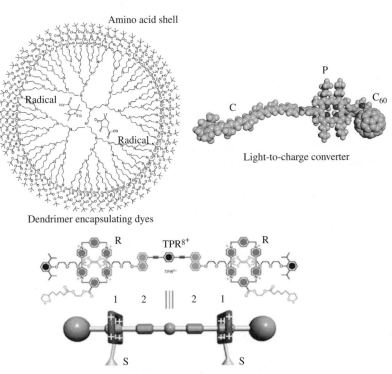

Dendrimer encapsulating dyes

Light-to-charge converter

"Nano-muscle" actuator based on rotaxanes

Fig. 6.1 Examples of complex and beautiful nanostructures synthesized chemically. Clockwise: The dendrimer (top left) is synthesized by repeated addition of branched structures to the ends of existing branched structures. This structure was synthesized in the presence of reactive radicals (compounds with unpaired electrons) which became trapped when the outer layer was capped by the addition of a dense amino acid structure. Thus, it serves as a chemical box to trap and preserve these normally unstable compounds.[1] (Reprinted with permission from John Wiley & Sons Inc., courtesy E.W. Meijer.) The light-to-charge converter (top right) absorbs light at the porphyrin chromophore (P) which then donates an electron from its excited state to the C_{60} molecule, and accepts an electron into its ground state from a carotene molecule (C). The result is a long-lived charge-separated state that can be used to do chemical work. (Courtesy of Professor Devens Gust of Arizona State University.) The nano-muscle actuator (bottom) is based on a pair of rotaxane molecules (molecules free to slide on a shaft) that switch from position 1 to position 2 as the oxidation state of the central molecule (TPR) is changed from 8^+ to 12^+. The sulfur "feet" (S) cause it to expand or contract the gold surface to which it is attached. (Reprinted with permission from Liu et al.[2] J. Am. Chem. Soc. Published 2005 by the American Chemical Society.)

radicals (reactive compounds with unpaired electrons) that would not have been stable if left free in solution.[1]

The second example is a nanomachine that converts light into a long-lived charge-separated state that can be used to provide chemical energy. It is a molecule that mimics part of the process of photosynthesis carried out by plants. The porphyrin in the middle of the molecule is a chromophore that absorbs light. An isolated porphyrin molecule would rapidly decay to its ground state by fluorescence emission of a photon (Chapter 4). However, in this molecule the porphyrin is coupled to a buckyball (the spherical C_{60} molecule) that rapidly grabs the electron from its excited state, and to a carotene molecule that rapidly injects an electron back into the ground state of the porphyrin. Thus, the porphyrin returns to its ground state leaving a positive charge on the carotene and a negative charge on the C_{60}. There is no easy way for these charges to recombine, so the light energy is now stored in a charge-separated state that can be used as a source of chemical energy.[3] This molecule is analyzed in Chapter 8.

The third example makes use of a wonderful structure called a rotaxane. This is an assembly in which one molecule slides freely on another, but cannot come off because of covalent "stoppers" (rather like a sliding bead on the rail of an abacus). The example shown in Fig. 6.1 utilizes two rotaxanes, one placed on each side of a highly charged group in the middle. Changing the charge of the group in the middle causes the sliding components to change position. The sliding rings have chemically sticky feet attached to them (sulfur atoms) so changing the charge state of the central group moves the "feet" in or out. The structure was christened a "nano muscle" for its ability to convert chemical energy into linear mechanical actuation.[2]

Synthesis of exotic nanomachines like these is very challenging, but it consists of a number of steps, each one of which can be quite straightforward: (a) specific coupling reactions between simpler components, (b) protection and deprotection of reactive groups as needed, followed by (c) separation of the desired component and (d) analysis of the structure to make sure that the desired product was synthesized. We will outline these basic components of organic synthesis below.

A chemical reaction proceeds when a compound with valence electrons in a high-energy state collides with a compound that can rearrange (usually by adding or losing atoms in the case of the chemical coupling reaction) to accept those electrons into a state of lower energy. (Chapter 8 deals with the topic of electron transfer reactions which, to one degree or another, underlie all chemical reactions.) Since there are many components (including the solvent) participating in these reactions, entropy plays a role, so the process is best described by the transfer of electrons from states of higher free energy in the reactants to states of lower free energy in the products. Compounds containing unpaired electrons (radicals) tend to be unstable, so the usual pathway for a reaction is that a compound of higher free energy supplies additional electrons to form a more stable product of lower free energy. It is usual to call one reactant the attacking reagent and the other the substrate. If the bond breaks in such a way that the electrons remain with one fragment the reaction is said to be heterolytic. An attacking reagent that brings an electron pair is called a nucleophile and the corresponding heterolytic reaction is called nucleophilic. An attacking reagent that takes an electron pair is called an electrophile and the reaction is said to be electrophilic. Nonheterolytic reactions are those in which the bond breaks in such a way as to leave a single electron on each fragment, generating a pair of radicals. Such "homolytic" processes are referred to as "free radical mechanisms" and usually involve subsequent reactions between the free radicals and other components of the system. Yet another type of electron transfer occurs in organic rings (like benzene) when electrons transfer around the ring (referred to as a pericyclic mechanism). Finally, bonds can be broken with the simultaneous transfer of a proton (e.g., HCl gas dissolving in water to form hydrochloric acid). These reactions are called acid–base reactions.

This very brief overview of organic synthesis gives a small taste of a vast and rich field (described with great clarity in the comprehensive textbook *March's Organic Chemistry*[4]). As an illustration, we will describe two coupling reactions of great utility in many nanoscience experiments. These are (a) the coupling of one compound to another containing a primary amine ($-NH_2$) through the use of activated carboxylate group called an NHS ester and (2) the coupling of a compound containing a thiol ($-SH$) to another containing a maleimide (see Fig. 6.2). These reactions are relatively straightforward to carry out and hundreds of compounds are available with NHS esters or maleimides attached to them. These coupling reactions are routine enough that they can be carried out by following the instructions provided by the manufacturers of the reagents. NHS ester coupling is an example of a nucleophilic attack where the primary amine is the nucleophile and the carboxylate group is the substrate. The resulting bond (see Fig. 6.2(a)) is called an amide bond. In the figure, R_1 and R_2 represent the compounds to be coupled together by the amide bond. In the case of maleimide reagents (Fig. 6.2(b)), the thiol serves as the nucleophile, though the reaction is accompanied by the transfer of a proton to an oxygen on the modified maleimide ring (the intermediate structure in Fig. 6.2(b)) which then goes on to form the thioether linkage in the final structure on the right. Both of these reactions are robust if carried out in the appropriate conditions.

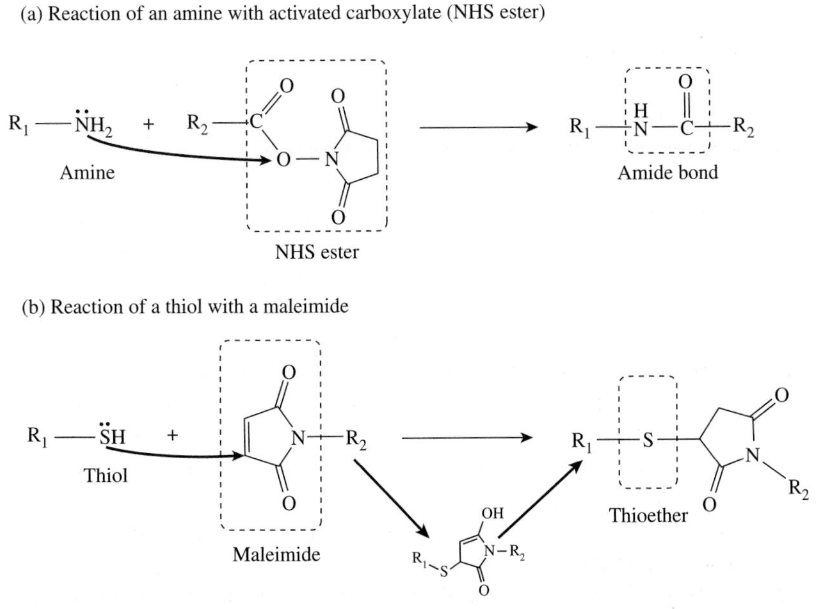

(a) Reaction of an amine with activated carboxylate (NHS ester)

(b) Reaction of a thiol with a maleimide

Fig. 6.2 Coupling reagents that rely on nucleophilic attack on an electrophile. (a) Shows the reaction of a primary amine with an activated carboxylate. The lone pair of electrons on the nitrogen (the nucleophile) are donated to the C–O bond, breaking it. The reaction is carried out in buffered solution which takes up the proton released from the amine. The end result is a stable amide bond linking R1 and R2. (b) Shows another common coupling reaction in which two electrons are transferred from the sulfur to the maleimide ring. The reaction passes through an intermediate in which the proton is captured by the ring. The intermediate decomposes to form the stable thioester that links R1 and R2 (courtesy of Professor Peiming Zhang, Arizona State University).

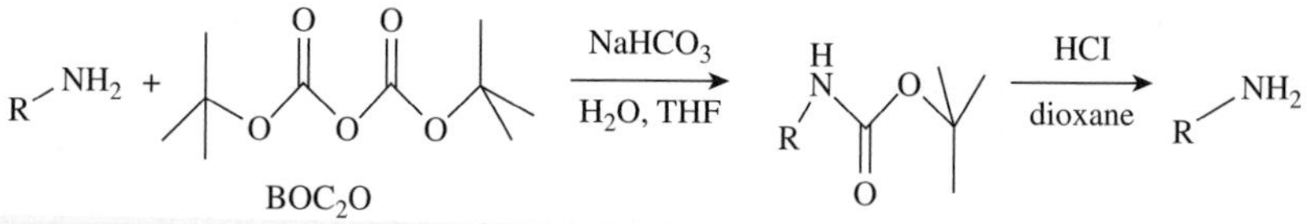

Fig. 6.3 Example of the use of a removable protecting group. BOC groups (t−butoxycarbonyl) are readily attached to amines in aqueous conditions producing a BOC-amine that is stable against nucleophilic attack and bases. The BOC group is readily removed when desired using acid conditions (http://www.organic-chemistry.org/frames.htm and http://www. organic-chemistry.org/protectivegroups/amino/boc-amino.htm).

For example, the nucleophilic attack on the carboxylate, shown in Fig. 6.2(a), requires that the solution take up the proton lost from the amine, so a buffered (i.e., pH controlled—see Section 3.18) solvent is required. The NHS ester itself is subject to attack by water, so solutions cannot be stored indefinitely.

There are a large number of fairly robust coupling reactions of the type just described but it can become an impossible challenge to make a precursor with different and stable coupling groups at every place to which one wishes to link a new component. Thus a second, and important, set of tools is based on the use of protecting groups that serve to shield reactive components from attack until they are required for the next step in the synthesis. The protecting group is removed at the point when access to the reactive group is needed. One type of protecting group (t-butoxycarbonyl, also called BOC) is shown in Fig. 6.3. It reacts readily with amines in aqueous solution, and prevents them from attacking electrophiles until such point as it is removed by decreasing the pH. It functions well as a protecting group, so long as the reactions preceding its removal do not involve the use of acids.

The foregoing discussion has shown how it is possible to map a route for making a complex product (similar to the examples shown in Fig. 6.1) via a series of reactions, protections, and deprotections. It is rare indeed that the desired product purifies itself by precipitating out of solution as a beautiful and pure crystal. Therefore, purification of the desired product is an essential component of organic synthesis. This purification is usually coupled with structural analysis so that the desired component can be identified and the purification of just one product carried out on a larger scale.

Many forms of purification are based on chromatography. Chromatography utilizes a remarkable phenomenon whereby different dissolved species will separate if traveling through a solvent moving through a porous aggregate of fixed material to which some solvent sticks (the "stationary phase"). Its name comes from the circumstances of its original discovery: liquid extracted from plants and poured over a column packed with fine particles separates into bands of distinct colors. Nowadays, many different methods are available for detecting the different bands so they do not have to be colored per se. Most common is measurement of the optical absorbance at one or more wavelengths, or measurement of the refractive index of the solution. The solution is usually pushed all the way through a long column packed with small beads using a high pressure pump. The beads are typically made from a silica gel capped with small hydrophobic carbon chains. Many other materials are available, and many of the beads come with a large variety of chemical groups attached to them in order to enhance the separation of particular target compounds.

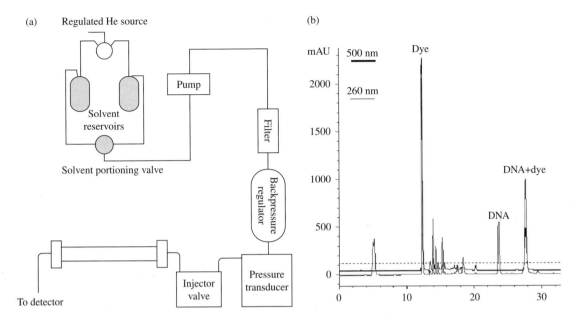

Fig. 6.4 High-performance liquid chromatography (HPLC). (a) Shows the layout of a modern HPLC. Solutions to be analyzed/purified (maintained in a helium atmosphere) are pumped at high pressure through a column containing small beads. Material eluted from the column is detected via its optical absorbance (for example). (b) Shows the output from two detectors, one operating at 500 nm and the other at 260 nm as a mixture containing a dye and DNA, and DNA coupled to the dye passes through the column. The desired product (DNA plus dye) comes out last. (Data courtesy Quinn Spadola.)

The physical layout of a high-performance liquid chromatography (HPLC) apparatus is shown in Fig. 6.4(a). An example of an experimental readout from the optical detectors is shown in Fig. 6.4(b). In this case, the reactants included fluorescein (a dye) and DNA, and the desired product was a compound containing both fluorescein and DNA. The pure DNA peak is readily identified because DNA absorbs light at 260 nm but not at 500 nm. Both the fluorescein and the desired product absorb light at both wavelengths, so identifying the peak that contains the correct product requires that each of the separate components be analyzed by other techniques (see below). Once the desired product is verified, the HPLC separation can be carried out on larger columns using larger volumes so that the (now calibrated) correct fraction can be collected.

Once a method is available for making something that contains the desired product, and once the technique can be found for separating the various products, the final essential ingredient is a method for identifying the desired product and showing that it has the required structure. One of the first methods for doing this was X-ray crystallography, a technique that is applicable only to materials that form high-quality crystals. New techniques have been developed that do not require crystals. Two of the most useful analytical techniques in chemistry are time-of-flight mass spectrometry and nuclear magnetic resonance (NMR) analysis.

Time-of-flight mass spectroscopy measures the time taken by ions of a fixed energy to travel a fixed distance, as illustrated in Fig. 6.5(a). This figure illustrates the technique of matrix-assisted laser desorption/ionization (MALDI). A small amount of the sample is absorbed into a matrix that is readily ablated by an intense laser beam. It is then placed on an electrically isolated sample

(a) 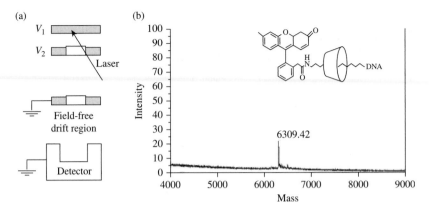 (b)

Fig. 6.5 Matrix-assisted laser desorption/ionization (MALDI) time-of-flight mass spectroscopy. (a) Schematic diagram of the instrument. Ions are generated by an intense laser beam incident on a matrix containing the sample and held at a negative potential V_1 with respect to ground. A second gating electrode, held at a potential V_2, can be used to gate the flow of ions from the sample, while a third grounded electrode accelerates positive ions into a field-free region where they drift with the velocity acquired under the accelerating potential. A detector measures the arrival of positive charge as a function of time after a burst of the ions was passed into the system by the gating electrode. Heavier ions take longer to reach the detector, so the readout is a series of peaks corresponding to the charge to mass ratio of the detected ions. (b) Shows a plot of data obtained for a molecule consisting of a rotaxane formed by the threading of an alkane chain by a cyclic sugar molecule. One end of the alkane chain is stoppered with a dye molecule while the other end is stoppered with a short DNA molecule. The peak occurs at the calculated mass of the complex for a singly charged ion, confirming the purification procedure used to isolate this complex substance. (Data from Quinn Spadola.)

stage in an ultrahigh vacuum chamber and held at a large negative potential so that, following ablation, positive ions are accelerated into the body of the chamber. An intermediate gating electrode is a used to pulse the emerging beam of ions, which then enter a field-free drift region to be collected by a detector at the far end. Equating the kinetic energy of the accelerated ions ($\frac{1}{2}mv^2$) to the energy gained by being accelerated by a potential difference V (ZeV, where Z is the number of charges on the ion), the time of flight over a distance s is given by

$$t = s\sqrt{\frac{m}{2ZeV}}. \tag{6.1}$$

Modern timing electronics are accurate enough to determine a difference in time of flight owing to just one atomic mass unit for even large proteins with masses up to 300 kDa. Detectors respond to even just a few ions so that, given efficient ionization of the sample and optimized collection geometry, samples as small as a few hundred to a few thousand molecules can be detected. An example of data collected by MALDI mass spectroscopy is shown in Fig. 6.5(b). The axis is calibrated in units of mass to charge ratio, m/Z. Lines are also present owing to the presence of various ionic species (different charge numbers). These can usually be separated from features that are mass dependent because they occur at exactly integer multiples of delay time. In the example shown in Fig. 6.5(b), the peak corresponds to the singly ionized version of the rotaxane complex (shown in the inset). This reagent was separated from unwanted products by

(a)

(b)

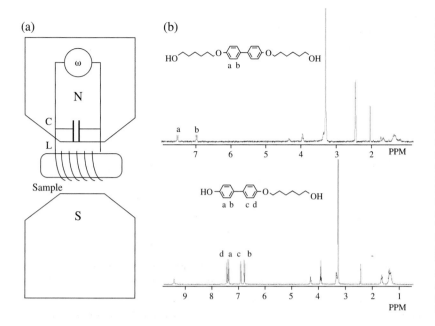

Fig. 6.6 Nuclear magnetic resonance (NMR). A schematic layout of the apparatus is shown in (a). The sample is enclosed by the inductor, L, part of an LC resonant circuit tuned to the magnetic resonance frequency of the nuclei in the sample (when exposed to a magnetic field imposed by the poles of the magnet (N,S)). A small loss of energy from the applied field at resonance gives rise to a peak in the NMR spectrum, examples of which are shown in (b). Identical nuclei (e.g., the protons attached to each of the carbons) have resonance frequencies that differ (chemical shifts) depending on the location of the carbon in the compound. The two spectra show the chemical shifts for carbon atoms in the aromatic rings. The carbons labeled "a" and "b" in the symmetric structure (top) have only two chemically distinct environments and so give rise to only two peaks (labeled "a" and "b"). In contrast, the asymmetric compound (bottom) shows four peaks for these carbon atoms reflecting the four chemically distinct environments for these carbons, a, b, c, and d. (Data from Quinn Spadola.)

HPLC and this mass spectroscopy (Fig. 6.5(b)) verified that the correct fraction had been collected from the HPLC instrument.

Mass spectroscopy will verify that the product contains the desired types and number of atoms, but it cannot determine the spatial structure of the compound. The most versatile technique for doing this is NMR illustrated in Fig. 6.6(a). NMR detects the interaction of a radio frequency electric field with the spin of a nucleus. Protons have a spin of $\frac{1}{2}$ (in units of $\hbar$) while more complex nuclei have spins that depend on the number of protons and neutrons. The nuclear spin is zero if the number of neutrons and the number of protons are both even, half integer if the number of neutrons plus the number of protons is odd, and integer if the number of neutrons and the number of protons are both odd. A nucleus has an intrinsic magnetic moment given by

$$\mu = \gamma I,\tag{6.2}$$

where I is the nuclear spin in units of $\hbar/2$ and γ is a fundamental constant that is different for each nucleus (called the gyromagnetic ratio). A classical way to think of this is that a spinning charge is the equivalent of a circulating current that induces a magnetic field. In a magnetic field, B, the energy of the nucleus splits into levels labeled by the magnetic quantum number m (see Section 2.16 in Chapter 2 for the definition of the magnetic quantum number for a hydrogen atom).

$$E_m = m\gamma IB.\tag{6.3}$$

The energy difference between two levels (γIB) is generally very small, even for large fields, and the Boltzmann distribution at room temperature implies that the various spin levels are essentially equally occupied. The very tiny difference in occupation between two levels leads to a minute amount of absorption at an angular frequency

$$\omega_{mn} = \frac{E_m - E_n}{\hbar} = \frac{(m-n)\gamma IB}{\hbar}. \tag{6.4}$$

For most nuclei and reasonable values of B (which can be up to several Tesla if a superconducting magnet is used), this frequency is in the MHz to GHz range. For example, the proton resonance occurs at 42.58 MHz in a field of 1 T. A classical way to think of the process is absorption of radiation as a magnet of moment, μ, is made to precess around a magnetic field, B. Energy is passed from one spin to another (spin–spin relaxation) or to heating up the environment as the system of spins returns to the equilibrium Boltzmann distribution (spin–lattice relaxation).

NMR has become a powerful analytical tool because the local magnetic field felt by the nucleus depends on the local distribution of electrons. In the case of proton NMR, the electron is in an s-state and its precession generates a magnetic field that opposes the applied field. This lowers the field felt by the proton. The consequence is a lower resonant frequency for the proton NMR, an effect called a "diamagnetic chemical shift." This shift is small (parts per million) but quite characteristic of the environment of the proton and therefore useful in identifying local structure. Examples of proton NMR spectra are given in Fig. 6.6(b).

Other nuclei can be targeted by choosing the radio frequency appropriately. The all-important carbon atom does not generate an NMR signal because ^{12}C has six neutrons and six protons, but the isotope ^{13}C is abundant enough to facilitate NMR. ^{13}C NMR shows a much larger range of chemical shifts than proton NMR, because signals are both upshifted and downshifted depending on the symmetry of the local occupied orbitals.

Pulsed techniques make NMR yet more powerful. It is relatively easy to saturate the occupation of spins at one chemical shift (i.e., put all the spins in a high-energy state) because of the small energy gap between the ground and excited states. Once this is done, the NMR spectrum tracks the way that the magnetization is transferred to neighboring atoms. Experiments like this are capable of producing two-dimensional maps that show the relative location of pairs of nuclei. These data can be used to assemble a complete structure for the compound under study.

6.3　Weak interactions between molecules

We were able to give a simplified discussion of synthetic chemistry because of the strength ($\sim 100 k_B T$) of covalent bonds. However, most self-assembly involves units that interact with energies of just a few times $k_B T$. We call a bond "weak" if the interaction free energy is on the order of $k_B T$ at room temperature, so that the bond rapidly breaks and reforms as a result of thermal fluctuations. Entropy plays a large role in the formation of self-assembled structures put together with weak bonds. Erroneous high-energy structures can dissociate and then recombine to form the desired structure with the lowest free energy. All interatomic interactions are quantum mechanical in origin, but it is very convenient to classify relatively weak interactions in terms of classical, usually electrostatic, bonding mechanisms.

Coulomb interactions between bare charges and covalent bonds are both strong—on the order of electron volts per atom pair. Weaker interactions exist between neutral molecules that nonetheless have an electric dipole moment (polar molecules), between polarizable molecules, and between the combinations of such atoms and molecules, for example, between charged molecules and polariable molecules. There are, in addition, quantum mechanical interactions that involve protons that are shared by groups of electronegative atoms. These are called hydrogen bonds and they are extremely important in biomolecular assemblies. A whole new class of weak interactions arises between molecules dissolved in a solvent, where molecule–molecule bonding competes with molecule–solvent and solvent–solvent bonding. Of all solvents, water is the most important. Molecules that interact weakly with water but strongly with each other are called hydrophobic and one speaks of the "hydrophobic interaction" as pushing such molecules together in a solution. Molecules that interact more strongly with water are called hydrophilic. We describe these important types of weaker interaction in more detail below.

The various types of interactions among atoms, molecules, and ions are illustrated in Fig. 6.7, on the basis of a table in Jacob Israelachvili's classic book *Intermolecular and Surface Forces*.[5] The approach to calculating these interactions is (a) to calculate the Coulomb interaction between all the point charges or induced charges and (b) average the interactions if they are weak enough that thermal fluctuations drive a range of molecule–molecule orientations (i.e., if the bond strength is on the order of k_BT). Many of the weaker interactions involve polar molecules. These are molecules with a dipole moment, for example, molecules in which bonding is accompanied by asymmetric charge transfer (see the discussion of HCN in Chapter 8). Dipole moments characterize the amount of charge displacement: charges of $\pm q$ separated by a distance x have a dipole moment $\mu = qx$. These moments are expressed in units of Debyes, 3.336×10^{-30} Cm. Charges of $\pm e$ separated by 1 Å correspond to a dipole moment of 4.8 Debyes. A symmetric molecule like benzene (C_6H_6) has no dipole moment, whereas a highly polar molecule, like water, has a dipole moment of several Debyes (1.85 Debyes for water). Some molecules (so-called zwitterions) spontaneously charge in water through the transfer of a proton from one group to another, producing a very large dipole moment. An example is the amino acid glycine:

The Coulomb interaction between a fixed dipole and a point charge (Fig. 6.7) results in an interaction energy that falls off as r^{-2} (i.e., more rapidly than the Coulomb interaction between point charges which falls off as r^{-1}). If the interaction is weak enough so as not to orient the dipole in the field of the point charge and the dipole is free to rotate (in solution, for example), then the interaction falls off even more rapidly (as r^{-4}) and becomes temperature dependent (proportional to T^{-1}) because of the Boltzmann weighted averaging of the angular orientation.[5] Two fixed dipoles (Fig. 6.7) contain both negative

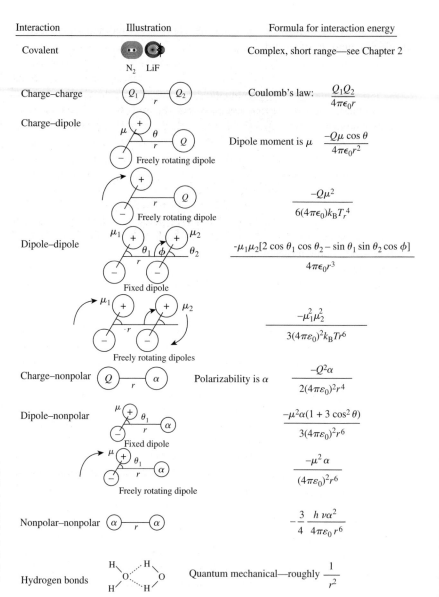

Interaction	Illustration	Formula for interaction energy
Covalent	N_2 LiF	Complex, short range—see Chapter 2
Charge–charge	$Q_1 \quad r \quad Q_2$	Coulomb's law: $\dfrac{Q_1Q_2}{4\pi\epsilon_0 r}$
Charge–dipole	Freely rotating dipole	Dipole moment is μ $\dfrac{-Q\mu\cos\theta}{4\pi\epsilon_0 r^2}$
	Freely rotating dipole	$\dfrac{-Q\mu^2}{6(4\pi\epsilon_0)k_B T r^4}$
Dipole–dipole	Fixed dipole	$\dfrac{-\mu_1\mu_2[2\cos\theta_1\cos\theta_2 - \sin\theta_1\sin\theta_2\cos\phi]}{4\pi\epsilon_0 r^3}$
	Freely rotating dipoles	$\dfrac{-\mu_1^2\mu_2^2}{3(4\pi\varepsilon_0)^2 k_B T r^6}$
Charge–nonpolar	Polarizability is α	$\dfrac{-Q^2\alpha}{2(4\pi\varepsilon_0)^2 r^4}$
Dipole–nonpolar	Fixed dipole	$\dfrac{-\mu^2\alpha(1 + 3\cos^2\theta)}{3(4\pi\varepsilon_0)^2 r^6}$
	Freely rotating dipole	$\dfrac{-\mu^2\alpha}{(4\pi\varepsilon_0)^2 r^6}$
Nonpolar–nonpolar	$\alpha \quad r \quad \alpha$	$-\dfrac{3}{4}\dfrac{h\,\nu\alpha^2}{4\pi\varepsilon_0\,r^6}$
Hydrogen bonds		Quantum mechanical—roughly $\dfrac{1}{r^2}$

Fig. 6.7 Types of interactions among atoms, molecules, and ions. The expressions given here (after Fig. 2.2 from Israelachvili[5]) are for interactions in vacuum. Outside of a vacuum, the electrostatic interactions (all but the first and last) are reduced by an amount equal to the inverse of the dielectric constant of the medium they are immersed in. Interaction free energies are generally lower than the interaction energies given here because of the reduced entropy when one molecule aligns with another.

(i.e., attractive + charge to − charge) and positive (i.e., repulsive + to + and − to −) interaction energy terms that sum to fall off as r^{-3}. Dipoles aligned in the same direction repel one another (positive interaction energy) whereas dipoles aligned in the opposite direction attract one another (negative interaction energy).

One might think that freely rotating dipoles would have no interaction (because the time-averaged field they produce is zero) but the Boltzmann thermal average favors attractive interactions, and this results in a weak attractive interaction that falls off inversely with temperature and has a distance dependence of r^{-6}. This is the form of the long-range attractive part of the van der Waals interaction (cf. the Lennard-Jones potential in Chapter 2) and it arises because of fluctuations in dipole moment.

Molecules with no intrinsic dipole moment may have a moment μ, induced by the electric field, E, of a nearby charge or dipole according to $\mu = \alpha E$, where α is the polarizability of the molecule. Polarizability is proportional to molecular volume (specifically, $4\pi\varepsilon_0 \times$ volume). The volume is traditionally given in units of Å^3 (so a volume of 1 Å^3 corresponds to a polarizability of 1.11 $\text{C}^2\text{m}^2\text{J}^{-1}$ in MKS units). The volume in question is that occupied by electronic states of the molecule (see Chapter 8). So benzene, while having no dipole moment, is highly polarizable with $\alpha = 10.3 \times 4\pi\varepsilon_0 \text{Å}^3$ (compared to $0.814\pi\varepsilon_0 \text{Å}^3$ for a helium molecule). This is because of the delocalized states associated with the six-carbon-atom ring (discussed in Chapter 8). The interaction of a polariable molecule with a fixed charge falls off as r^{-4} (Fig. 6.7). Interactions of both fixed and rotating dipoles with polariable molecules fall off as r^{-6}.

Even completely symmetrical atoms (like the noble gases) undergo fluctuations in charge distribution, giving rise to a transient dipole moment, and thus a dipole—polariable molecule interaction that falls off as r^{-6} (London dispersion energy—Fig. 6.7). This is true even at absolute zero because of quantum fluctuations. Thus, even noble gas atoms feel significant van der Waals attraction. For example, the melting point of solid xenon is $161°\text{K}$. This solid is held together entirely by London dispersion forces.

Hydrogen bonding (Fig. 6.7) is a special case of a charge—dipole interaction. It is shown here as the interaction of the $O(-)$–$H(+)$ dipole associated with the OH group of water with the residual negative charge on the oxygen atom. There is a significant quantum mechanical element to this interaction. The light proton responds to the electric field by moving away from the oxygen to which it is bonded toward the second oxygen involved in the hydrogen bond. This has two consequences. First, the hydrogen bond is stronger than one would calculate from a point dipole approximation and, second, bond lengths are changed significantly from what they are in the absence of a hydrogen bond. Although this may seem like an obscure effect, its consequences are enormously important. Water is the most abundant liquid on this planet and an essential component of life, and its properties are dominated in many respects by hydrogen bonding. For example, solid ice is less dense than water as a consequence of the hydrogen bonding structure in ice. Ice is also more polariable than water, probably because of the way in which protons can hop between equivalent sites in the hydrogen bonded network. The crystal structure of ice shows that the distance between the proton and the oxygen involved in hydrogen bonding is less (0.176 nm) than the sum of the van der Waals radii (0.26 nm) of the oxygen and hydrogen, but still considerably more than the length of the covalent bond O–H (0.1 nm). The strength of hydrogen bonds lies in the range of 10–40 kJ/mole (0.1–0.4 eV per bond). Thus in a population of hydrogen bonded molecules at room temperature ($k_B T = 25$ meV at 300 K), a fraction $\exp(-0.1/0.025)(\sim 2\%)$ of the bonds are broken at any one time. Specific pairs of hydrogen bonds are responsible for the precise chemical pairing of the genetic codes in opposite strands of DNA molecule, as we shall discuss shortly.

Liquid water owes its remarkable density (relative to ice) to the many transient networks of hydrogen bonds that are constantly forming and breaking in the liquid. Consequently, introducing a molecule that does not form hydrogen bonds into water raises the energy of the system as the water reorganizes to hydrogen bond around the guest molecule (as shown in Fig. 6.8). In consequence, molecules that do not hydrogen bond (alkanes, hydrocarbons,

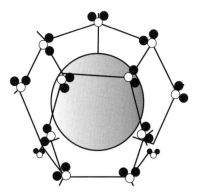

Fig. 6.8 Cage structures formed by water molecules around non-hydrogen bonding species in order to preserve the hydrogen bonding structure of the water network.

fluorocarbons, inert atoms) tend to be pushed together inside these water cage structures. This apparent attraction between these molecules gives rise to the hydrophobic interaction.

6.4 Vesicles and micelles

One of the most studied self-assembling systems is the association of so-called amphiphiles in aqueous solutions. Amphiphiles are molecules that contain both a hydrophilic and a hydrophobic component. Some of the most familiar amphiphiles find common application as detergents. In a detergent, a hydrophobic component binds oily dirt while the hydrophilic part helps to pull the complex into solution. By far the most widespread use of amphiphiles by humans (and most other life-forms) is in the construction of the cells from which we are made: the walls of cells are self-assembled from amphiphilic molecules.

Amphiphiles will spontaneously form small vesicles that are impervious to water when placed in aqueous solutions. Phospholipids are a common type of amphiphile found in cell walls and the structure of one phospholipid is shown in Fig. 6.9. The hydrophobic part of the molecule consists of a pair of long alkane chains (repeated units of methylenes, $-CH_2-$) connected together by a head group that contains a number of charged atoms, most notably the negative oxygens on the phosphate group. Amphiphiles like this tend to associate into lipid bilayers, consisting of two layers that come together to form a sandwich where the hydrophilic groups are the "bread" and the hydrophobic tails are the "filling" (see Fig. 6.10). In addition to forming lipid bilayers, amphiphiles can also form small micelles or vesicles. Micelles are clusters that consist of a single layer of amphiphiles in which all the hydrophobic tails point inwards to exclude water completely while the hydrophobic heads point outwards. Vesicles (i.e.,

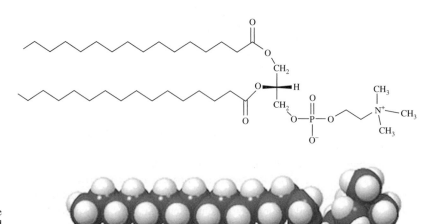

Fig. 6.9 Chemical structure (top) and space filling model (bottom) of a phospholipid molecule. The long alkane chains on the left are hydrophobic while the charged phosphate is hydrophilic. In water, these molecules spontaneously organize into vesicles where the hydrophilic groups point out into the water and the hydrophobic groups come together to exclude water.

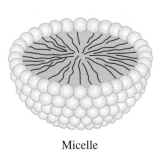

Micelle

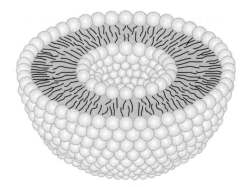

Liposome

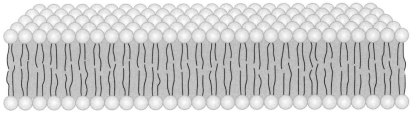

Bilayer sheet

Fig. 6.10 Structures of a micelle, liposome (=vesicle), and bilayer. (From *Molecular Cell Biology*, 4th edn, by H. Lodish, A. Berk, S.L. Zipursky, P. Matsudara, D. Baltimore, J. Darnell. © 2000 W.H. Freeman and Company. Used with permission.)

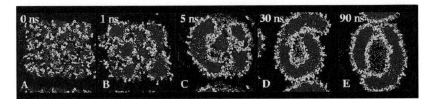

Fig. 6.11 Molecular dynamics simulation of the spontaneous formation of a vesicle as dipalmitoylphosphatidylcholine (DPPC) molecules organize spontaneously so that the hydrophilic heads (light) are in contact with water and the hydrophobic tails (darker) are in contact with one another. The plots show a cross section through the structure. A circular slice of the vesicle is almost completely formed after 90 ns (two defects remain). (Reprinted with permission from de Vries et al.[6], J. Am. Chem. Soc. Published 2004 by the American Chemical Society.)

liposomes) are bilayer spheres that enclose water. These structures are shown in Fig. 6.10. Which type of vesicle forms depends to a large extent on the geometry of the individual molecules.

The driving force for the assembly of the structures is the hydrophobic interaction. Very large-scale molecular dynamics simulations show how the structuring of water segregates dissolved lipids into a self-organized aggregate, provided that the initial concentration is high enough. The spontaneous formation of a bilayer vesicle is illustrated in a series of structures generated with a molecular dynamics simulation of dipalmitoylphosphatidylcholine (DPPC) molecules in water in Fig. 6.11. A total of 1017 DPPC molecules were distributed randomly in part of a box containing 106,563 water molecules and molecular dynamics simulations ran for 90 ns.[6] The DPPC molecules can be seen forming an almost complete vesicle over the time of the simulation (the final structure is not quite a sphere, being cut by two defects). We will consider the thermodynamic factors that drive these processes in the next section.

6.5　Thermodynamic aspects of self-assembling nanostructures

Self-assembly is a topic treated naturally in the context of nano-thermodynamics,[7] where the size of the self-assembled aggregate is one of the parameters that determine the free energy. The nano-thermodynamics formalism of Hill was touched on in Chapter 3, but true to our word we are going to give a simplified treatment based on the law of mass action and parameterization of the chemical potential. We focus on the self-assembly of amphiphiles, but the general principles applies to many other kinds of self-assembly.

Consider a system in equilibrium that consists of a mole fraction X_1 of monomers and a mole fraction X_N of aggregates each composed of N monomers. The total concentration of solute molecules is given by

$$C = X_1 + X_N, \tag{6.5}$$

where C is a total mole fraction of solute (equal to the moles of solute per liter divided by 55.5, the number of moles of water in one liter of water plus the number of solute molecules). Defined this way, C can never exceed unity.

One way to form an aggregate of N monomers is for N molecules to associate simultaneously, giving an association rate

$$k_A = k_1 X_1^N, \tag{6.6}$$

where k_1 is the association rate constant. The aggregates will dissociate into monomers at a rate

$$k_D = \frac{k_2 X_N}{N}, \tag{6.7}$$

where k_2 is the dissociation rate constant for the aggregate and k_D gives the equivalent dissociation rate into monomers. In equilibrium, association rates (6.6) and dissociation rates are equal and

$$K = \frac{k_1}{k_2} = \frac{X_N}{N X_1^N}, \tag{6.8}$$

where K is the equilibrium constant for the aggregation process. Solving for X_1 in terms of C and K yields

$$X_1 = \left[\frac{(C - X_1)}{NK} \right]^{1/N}. \tag{6.9}$$

Since the maximum possible value of $(C - X_1)$ is unity, we arrive at the conclusion that

$$X_1 \leq \left[\frac{1}{NK} \right]^{1/N}. \tag{6.10}$$

Thus, the concentration of monomer cannot be increased above a critical value, $[NK]^{1/N}$, called the *critical micelle concentration*. Above this value, all additional added monomer is converted completely to aggregates.

The equilibrium constant for the monomer-to-aggregate interconversion must also be equal to the ratio of the Gibbs distributions for the two forms. In terms of the chemical potentials for monomer and aggregate this is

$$K = \frac{k_1}{k_2} = \frac{\exp(N\mu_1^0/k_BT)}{\exp(N\mu_N^0/k_BT)} = \exp\left(\frac{N}{k_BT}(\mu_1^0 - \mu_N^0)\right), \qquad (6.11)$$

where $N\mu_1^0$ is the chemical potential of N monomers and $N\mu_N^0$ is the chemical potential of N molecules in an aggregate in which the chemical potential per molecule is μ_N^0. This can be written in terms of mole fractions using Equation 6.8:

$$X_N = N\left\{X_1 \exp\left(\left[\mu_1^0 - \mu_N^0\right]\Big/ k_BT\right)\right\}^N. \qquad (6.12)$$

It is clear from Equation 6.12 that the formation of aggregates will be driven only if the chemical potential of a molecule in an aggregate is lower than that of an isolated monomer. If this were not the case, then $\mu_1^0 = \mu_N^0$, so $X_N = NX_1^N$ and, since $X_1 < 1$, monomers will dominate. So the formation of aggregates requires a size-dependent chemical potential.

To illustrate the notion of a size-dependent chemical potential, consider the case where molecules accumulate together into a spherical aggregate. The total number of molecules in the sphere is proportional to the volume $\frac{4}{3}\pi r^3$, while the number of bonded molecules on the surface is proportional to the surface area of the sphere, $4\pi r^2$. Thus, the chemical potential per particle must be of the form

$$\mu_N^0 = \mu_\infty^0 + \text{constant} \times \frac{N^{2/3}}{N} = \mu_\infty^0 + \frac{\alpha k_BT}{N^{1/3}}, \qquad (6.13)$$

where α is a "bond energy" expressed in units of k_BT for convenience, and μ_∞^0 is the chemical potential in an infinite aggregate. For a spherical aggregate[8]

$$\alpha = \frac{4\pi r^2 \gamma}{k_BT}, \qquad (6.14)$$

where r is the effective radius of a monomer and γ is the interfacial energy (between the monomer and the solvent).

The expression given in Equation 6.13 applies to a number of other shapes with different exponents of N :[5]

$$\mu_N^0 = \mu_\infty^0 + \frac{\alpha k_BT}{N^p}, \qquad (6.15)$$

where p is a number that depends on the dimensionality and/or the shape of the particle. Inserting the relation given by Equation 6.15 into Equation 6.12 yields

$$X_N = N\left\{X_1 \exp\left[\alpha(1 - 1/N^p)\right]\right\}^N \approx N\left[X_1 \exp(\alpha)\right]^N. \qquad (6.16)$$

This expression gives us another way to express the critical micelle concentration. Since X_N can never exceed unity, X_1 cannot exceed $e^{-\alpha}$. That is, for spheres,

$$(X_1)_{\text{crit}} \approx \exp\left\{-\frac{4\pi r^2 \gamma}{k_B T}\right\}. \qquad (6.17)$$

For a water–hydrocarbon interface $\gamma \approx 50 \, \text{mJ/m}^2$ and $r \approx 0.2$ nm for a methane molecule. Thus, $4\pi r^2 \gamma \approx 2.5 \times 10^{-20}$ J and $\alpha = 6$. Accordingly, the critical micelle concentration for a small hydrocarbon in water corresponds to a mole fraction of $\sim 2 \times 10^{-3}$. Above this concentration, all of the added hydrocarbon goes into an infinite aggregate (as an oil phase floating on top of the water phase).

Equation 6.16 may also be used to gain some insights into the distribution of aggregate sizes. Taking the case where $p = 1$ (p is the shape factor in Equation 6.15) allows us to take the second term in the exponential ($1/N$) outside the curly brackets, giving

$$X_N = N \, [X_1 \exp \alpha]^N \exp(-\alpha). \qquad (6.18)$$

Above the critical micelle concentration $X_1 \exp(\alpha) \leq 1$ and $\exp(-\alpha)$ is a constant leading to the result that $X_N \propto N$ for small N. Thus, for $p = 1$, the concentration of aggregates grows linearly with the aggregate size, until N is so large that the term $[X_1 \exp \alpha]^N$ begins to dominate, bringing X_N down to zero as N goes to infinity. (This is in contrast to the cases where $p < 1$, where the aggregate size grows infinitely large at the critical micelle concentration) Thus, the distribution of aggregate sizes is sensitive to the geometry of the aggregates.

So far, we have paid no attention to molecular geometry. This plays an important role in the shape of the aggregate that can form. For an amphiphile like the DPPC molecule shown in Fig. 6.9, the critical parameters are the length of the hydrocarbon chain, ℓ_c, the volume occupied by the hydrocarbon chain, v, and the area, a_0, of the head group. (These are all semiempirical parameters somewhat related to the geometry of the molecule but determined in practice by fitting data.) A dimensionless shape factor, $v/a_0\ell_c$ plays an important role in determining the aggregate geometry:[8]

$$\frac{v}{a_0\ell_c} < \frac{1}{3} \text{ implies that spherical micelles are favored,} \qquad (6.19a)$$

$$\frac{1}{3} < \frac{v}{a_0\ell_c} < \frac{1}{2} \text{ implies that nonspherical micelles are favored,} \qquad (6.19b)$$

$$\frac{1}{2} < \frac{v}{a_0\ell_c} < 1 \text{ implies that vesicles or bilayers are favored,} \qquad (6.19c)$$

and

$$\frac{v}{a_0\ell_c} > 1 \text{ favors "inverted" cone-like structures.}[8] \qquad (6.19d)$$

Amphiphiles with small hydrocarbon chains, therefore, tend to form micelles while amphiphiles with larger hydrocarbon chains (or double hydrocarbon chains—such as DPPC, Fig. 6.9) tend to form vesicles or bilayers. The final folding of a planar bilayer into a spherical vessel is determined by a balance between the excess surface energy associated with the edges of a bilayer and the elastic energy required to fold a planar layer into a sphere.

6.6 A self-assembled nanochemistry machine—the mitochondrion

As a reward for wading through the thermodynamics and kinetics of amphiphiles, we will divert to describe a remarkable nanomachine that is largely assembled from phospholipids. The mitochondria occupy a significant fraction of the cytoplasm (part of a cell outside the nucleus) of all eukaryotic cells (the cells of complex multicelled organisms like you and me). They are the machines that allow us to use atmospheric oxygen to complete the oxidation of sugars to produce biological energy. They have a fascinating evolutionary history: mitochondria carry with them their own DNA (though they are not self-sufficient, requiring gene products from the host cell). Presumably, they became incorporated into anaerobic organisms as a symbiotic partner that helped these cells survive the increase in atmospheric oxygen that occurred a few hundred million years ago. Anaerobic organisms are only capable of oxidizing glucose to a compound called pyruvate, releasing enough chemical energy to make two ATP molecules (ATP is adenosine triphosphate, the energy source of all living things). In contrast, eukaryotic cells (incorporating mitochondria) oxidize glucose all the way to carbon dioxide, releasing enough chemical energy to make at least 30 molecules of ATP. Hence, mitochondria are a major part of our energy metabolism.

Figure 6.12 shows an electron micrograph of a mitochondrion accompanied by a schematic illustration of a cross section through its two sets of lipid bilayer membranes. The outer membrane acts as a kind of filter, only allowing small molecules into the intermembrane space. Pyruvate and oxygen are transported into the matrix inside the inner membrane where a series of enzymes reduce the oxygen to water and oxidize the pyruvate to carbon dioxide. The net result of these reactions is that protons are pumped into the intermediate space to make it acidic. The gradient of protons (between the intermembrane space and the inner matrix) drives a remarkable molecular motor called ATP synthase (see Chapter 10). ATP synthase uses this proton gradient to synthesize energetic ATP

(a)

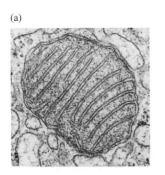

(b)

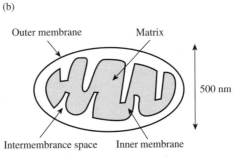

Outer membrane   Matrix

500 nm

Intermembrance space   Inner membrane

Fig. 6.12 Showing (a) an electron microscope image of a mitochondrion and (b) a schematic illustration showing the arrangement of an outer membrane made from a lipid bilayer and a convoluted inner membrane also made from a lipid bilayer. (EM image is reproduced with permission from Chapter 4 of *The Genetic Basis of Human Disease* by G. Wallis published by the Biochemical Society 1999. Copyrighted by the Biochemical Society. http://www.biochemj.org.)

from less energetic ADP. The ATP is then transported out of the mitochondrion to other places in the cell where it is used to drive chemical processes that require energy.

6.7   Self-assembled molecular monolayers

A very useful type of self-assembly occurs when molecules attach to a surface to which they are chemically attracted. The chemical attraction will eventually lead to a dense layer of molecules on the surface, and if the head groups of molecules on the surface are not reactive, then further accretion of molecules ceases, leaving a monolayer on the surface. Such molecular monolayers are described extensively in the classic book by Ulman.[9]

One of the most studied and readily assembled molecular monolayers consists of alkane chains ($-CH_2-$ repeats) modified at one end with a thiol ($-SH$) group that attaches chemically to a gold surface.[10] Monolayers can be formed by alkane chains that vary in length from just a few carbon atoms up to chains

(a)

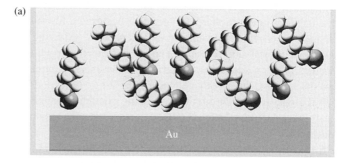

(b)

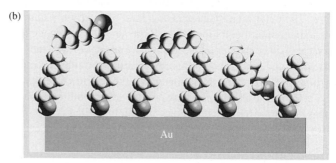

(c)

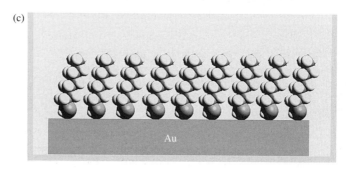

Fig. 6.13 Preparation of alkanethiol self-assembled monolayers. A flat gold substrate (usually presenting predominantly 111 faces) is immersed in a solution (a) containing alkane molecules terminated in a thiol (SH) group. The sulfur binds the gold quite strongly pulling the molecules down on to the surface (b) so that after diffusion on the surface, they assemble into an organized monolayer (c). Using a mixture of alkanethiols with various head groups makes it possible to prepare a mixed monolayer.

Fig. 6.14 High-resolution STM image of an dodecanethiol monolayer on an Au(111) surface. The image size is 40 nm × 40 nm. The hexagonal packing of the individual molecules is clearly visible. (Reproduced with permission from P.S. Weiss,[11] published by Accounts of Chemical Research, 2008. Courtesy of Professor Paul Weiss.)

that are so long that they are difficult to dissolve (>20 methylene groups). Longer chains (12 or more methylene units) pack together with stronger van der Waals interactions and therefore tend to form more ordered films, though reasonably well-ordered films can be formed by chains with as few as six or eight methylene units. The assembly of an alkanethiol monolayer is illustrated in Fig. 6.13. Typically, an ethanol solution of a few mM concentration of alkanethiol is kept in contact with a clean gold surface for several hours. If the head group is not reactive, the film remains stable after the sample is pulled out of the solution and rinsed. A scanning tunneling microscope (STM) image of two highly ordered regions of such a monolayer of octanethiol (CH_3-$(CH_2)_7$-SH) is shown in Fig. 6.14. In structures like this, the separation of individual molecules on the surface is ∼5 Å. The van der Waals radius of the alkane chain is somewhat less than this (4.6 Å), so the chains tend to tilt at an angle of ∼30° to fill the available space.

These self-assembled monolayers have great utility as a way of modifying the chemical properties of a surface. Simple alkane chains terminate in a methyl group (–CH_3), an extremely unreactive entity that bestows a low surface energy to the modified film. Alkanethiols are available with a wide variety of different end groups and these can be used to change the characteristics of the surface. For example, a surface covered with methyl terminated alkanes is extremely hydrophobic, whereas a surface covered in carboxylate groups (–COOH) is extremely hydrophilic.

At the time of writing, one of the unresolved mysteries of the formation of these films is the mobility of the molecules on the surface and the final resting place of the gold–sulfur bond. The sulfur–gold interaction is very strong (184 kJ/mole, ∼$300k_BT$) so it would not seem possible in principle that absorbed molecules could move around to pack into an ordered structure. Electronic structure calculations indicate that the weak link is the gold–gold bond in

Fig. 6.15 A series of STM images showing a single octanedithiol molecule (HS-$(CH_2)_8$-SH) inserted into an octanethiol (CH_3-$(CH_2)_7$-SH) monolayer on gold. The bright spot (a) is the protruding sulfur on the top of the inserted molecule. The inserted molecule has disappeared on a second scan (b) only to reappear on a third scan (c). This on-off switching reflects the instability of the contact between the molecule and the underlying gold.[13] (From Ramachandran et al.[13] Science, 2003. Reprinted with permission of AAAS.)

(a)   (b)   (c)

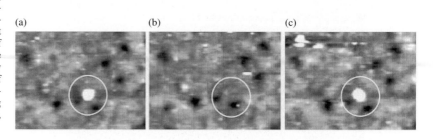

clusters where a sulfur is attached to one of the gold atoms. The interaction becomes so weak that sulfur, with the gold atom attached to it, moves over the surface in almost liquid-like manner.[12] The consequent fluctuation in the molecule metal contact can be seen directly in STM images. Figure 6.15 shows a series of STM images taken over a region of an alkanethiol monolayer into which an alkane dithiol molecule (i.e., containing a thiol headgroup in place of CH_3) has been inserted. The thiol headgroup appears as a bright spot in the image (Fig. 6.15(a)). The spot spontaneously disappears (Fig. 6.15(b)) only to reappear again (Fig. 6.15(c)), indicating that the electronic contact between the molecule and the underlying gold broke and then reformed again.[13] One consequence of this mobility of the sulfur–gold bond is a substantial restructuring of the gold surface resulting in the formation of pits on the surface that are one gold atom in depth.[14]

6.8 Kinetic control of growth: nanowires and quantum dots

The synthesis of small particles as precipitates from solution phase reactions is a familiar process. For example, when a solution of lead nitrate is added to a solution of potassium iodide, a brilliant yellow precipitate of (insoluble) lead iodide is formed:

$$2KI(aq) + Pb(NO_3)_2(aq) \rightarrow PbI_2(s) + 2KNO_3(aq)$$

Liquid or vapor phase synthesis of nanostructures relies on a similar process in which the solid product is either separated into a second liquid phase as a colloid (suspension of small solid particles) or grown on a solid support.

The first example (in 1857!) of the synthesis of solid nanoparticles by a precipitation reaction in a mixed solvent (carbon disulfide/water) was Michael Faraday's reduction of an aqueous gold salt with phosphorus in carbon disulfide to produce a ruby-colored aqueous solution of dispersed gold particles.[15] Carried out in a mixed solvent, the reduction reaction that produces gold metal particles from the gold salt stops short of producing large solid precipitates because the gold nanoparticles become suspended as colloids in the second component of the solution.

This 1857 strategy is still used today to produce metal nanoparticles[16] and quantum dots.[17] As an illustration we describe the synthesis of quantum dots of

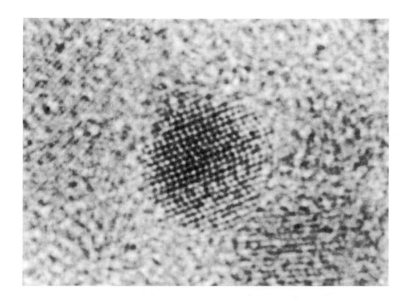

Fig. 6.16 Transmission electron micrograph of an 8 nm diameter CdSe nanoparticle. (Reprinted with permission from Murray et al.[16], J. Am. Chem. Soc. Published 2004 by American Chemical Society.)

cadmium compounds such as CdSe. A cold solution of one of the components is injected into a hot solution of the other component. A precipitation reaction begins as soon as the two components intermingle, and because of the rapid cooling caused by the injection of the cold solution, further growth of crystallites is halted. The solution is subsequently reheated, causing small crystallites (which have a larger surface to volume ratio and are therefore less stable) to dissolve and recrystalize onto more stable existing crystallites to produce a much more uniform size distribution of crystallites (a process called Ostwald ripening). Figure 6.16 shows an atomic resolution image of a CdSe nanoparticle synthesized using this method.

A second illustration of this kinetic control of growth of a precipitate is the vapor phase deposition of silicon that grows as a nanowire underneath a liquid formed from a gold–silicon eutectic (Fig. 6.17). This figure shows a gold nanoparticle used as a seed for the growth of a silicon nanowire. The gold nanoparticles are deposited onto a surface and heated in the presence of the precursors that generate silicon (chemical vapor deposition). The silicon is rapidly incorporated into the gold to form a liquid alloy, which, when saturated with silicon, produces a silicon precipitate that grows under the nano droplet of the eutectic alloy. In the right conditions, it is possible to grow nanowires of a few nanometers in diameter and up to several microns in length. In contrast to carbon nanotubes (which can be either semiconductors or metals—Chapter 9), these silicon nanowires all have well-determined electronic properties. An example of a field effect transistor made from a silicon nanowire is shown in Fig. 6.18.

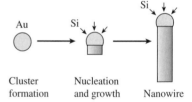

Cluster formation Nucleation and growth Nanowire

Fig. 6.17 Growth of silicon nanowires. A gold nanoparticle serves as the nucleation point for the growth of a nanowire. The gold is heated to a temperature such that when bombarded with silicon (by a chemical vapor deposition process) it forms a gold–silicon eutectic (low melting point alloy of silicon gold) that, saturated with silicon, results in the nucleation of a solid silicon nanoparticle that grows underneath the gold–silicon eutectic as a nanowire. (Reproduced with permission from Lu and Lieber[18], J. Phys. D: Applied Physics 2006, IOP Publishing and courtesy Wei Lu.)

6.9 DNA nanotechnology

DNA nanotechnology represents the most stunning confluence of biology, nanoscience, and bottom-up self-assembly. It was invented by Ned Seeman in 1986.[19] Seeman realized that, in addition to having very selective "sticky ends,"

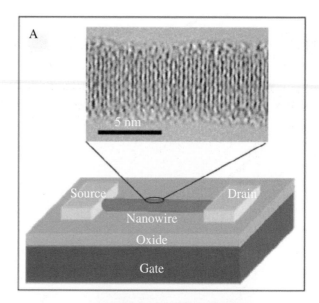

Fig. 6.18 (A) Schematic arrangement of a silicon nanowire field effect transistor showing the metal source and drain electrodes with an inset showing an atomic resolution image of the silicon nanowire. (b) Scanning electron micrograph of the assembled silicon nanowire transistor. (Reproduced with permission from Lu and Lieber[18], J. Phys. D: Applied Physics 2006, IOP Publishing and courtesy Wei Lu.)

DNA could also be programmed to form junctions at predetermined points along the polymer. With programmable sticky ends and three-way junctions, DNA is a marvelous engineering material for self-assembly of nanostructures. The age of DNA nanotechnology has come into being because the technology for the synthesis, purification, and characterization of synthetic DNA is now widespread and cheap as a consequence of the revolution in biotechnology.

To appreciate DNA nanotechnology, we must first examine the structure of the DNA double helix (Fig. 6.19). The discovery of this structure is surely the most important scientific event of the twentieth century (a story chronicled in the excellent book *The Eighth Day of Creation*[20]). The structure exemplifies many of the principles of self-assembly discussed earlier in this chapter. The repeating backbone of the polymer consists of sugar rings (called deoxyribose

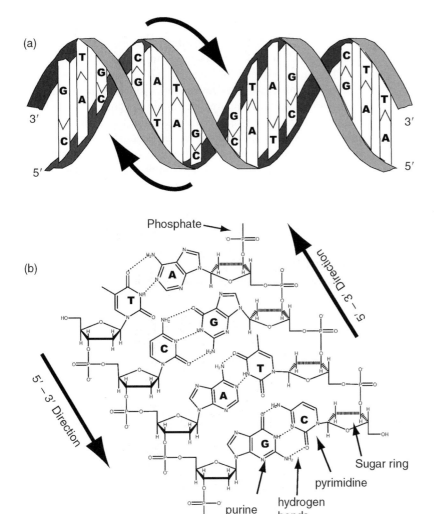

Fig. 6.19 Structure of the DNA double helix. The overall arrangement is shown in (a). The double helix consists of two phosphate-sugar polymers that run in opposite directions and are coupled together by pairs of organic rings (the bases) which hydrogen bond together in two distinct pairing patterns (A with T, G with C) that precisely fill the gap between the two polymers. The overall arrangement is twisted into a helix with a repeat distance of 3.4 nm, corresponding to 10 of the base pairs. The structure is illustrated in more detail in (b). The backbone consists of a five-atom sugar ring coupled to a phosphate group that carries a negative charge. The sugar ring is not symmetric and this gives rise to a directionality in the backbone, specified according to the number of the carbon atoms on the sugar ring at which the phosphate enters and leaves. Thus, the strand on the left is called the $5'$–$3'$ strand and the strand on the right is called the $3'$–$5'$ strand. The hydrogen bonds between the base pairs are illustrated as dotted lines. Adenine (A) and thymine (T) are connected by two hydrogen bonds, while guanine (G) and cytosine (C) are connected by three hydrogen bonds.

because they are composed of ribose rings missing an oxygen) linked together by phosphate groups that are singly ionized in water. (The dry material is a salt, carrying with it one positive ion for each phosphate.) These negatively charged groups account for the solubility of DNA and point outwards into the solution (and are the reason that dissolved DNA is an acid). They also give DNA many of its unusual properties: a long piece of DNA is, in effect, a single ion of enormous charge, called a polyelectrolyte. The two phosphate sugar backbones in the double helix have an interesting property of directionality: one phosphate attaches to the sugar ring at its $3'$ carbon (the labeling is arbitrary and fixed by convention) while another attaches to the sugar ring at its $5'$ carbon. Thus, one can specify a direction for the backbone as, for example, $5'$–$3'$. The two chains run in opposite directions, evidenced by the fact that X-ray diffraction patterns show that the molecule has a pseudo-inversion symmetry (which it could not have if the chains were oriented in the same direction). Organic groups, called bases, are attached to each of the sugars. There are four types of bases, adenine (A), thymine (T), guanine (G), and cytosine (C). The composition of DNA is somewhat random in terms of the average amount of each base in DNA from a given species. However, adenine and thymine are always present in equimolar proportions as are guanine and cytosine. The reason for this is that the larger bases containing two organic rings (adenine and guanine—called purines) pair specifically via two or three hydrogen bonds (see Fig. 6.19) to each of the two smaller bases (cytosine and thymine—called pyrimidines). The purine guanine pairs specifically with the pyrimidine cytosine and the purine adenine pairs specifically with the pyrimidine thymine. (In the face of all this biochemical nomenclature, it may come as some comfort to know that nature is quite sparing with its chemical inventions, and we will see that many of these components are used for other functions, as reviewed in Chapter 10.)

The bases both donate and accept protons. (They are called bases because the nitrogens can act as a nucleophile, donating electron pairs, a behavior that is the charge equivalent of accepting protons, making these structures act as "Lewis bases".) Hydrogen bonds form between the nitrogens on one base and the oxygens on another with a proton shared by both. Guanine and cytosine share three hydrogen bonds, while adenine and thymine are more weakly paired by two hydrogen bonds. These bases are rather insoluble in water and as a result hydrophobic interactions are important in keeping them in the interior of the DNA double helix.

The genetic code is carried in the sequence of bases as read along one of the chains. The fact that the second chain must carry a complementary sequence (A for T, G for C, etc.) is the basis of the preservation of genetic information on cell division—each strand acts as a template for the synthesis of a complementary strand. Starting from an initiation sequence, the genetic code can be read off in blocks of three bases, each block of three bases coding for one amino acid residue in a protein.

To summarize, the DNA double helix will spontaneously assemble from single strands if the sequence of bases on one strand is complementary to that on the other. The two strands are held together by specific hydrogen bonds between the complementary bases and by hydrophobic interactions, as well as direct electronic attractions between the bases, keeping water out of the interior and forcing the bases to stack on top of one another like a pile of dinner plates. These forces are opposed by the hydrophilic phosphate backbone and by electrostatic repulsion between one negatively charged helix and the other negatively charged helix.

The thermodynamic stability of a double helix is determined by the sequence of bases and the length of specifically paired double helix. It also depends on salt concentration, because at very low salt concentration the charge on one phosphate backbone is not screened from the charge on the other, giving rise to a Coulomb repulsion that destabilizes the double helix.

The free energy that stabilizes paired purine–pyrimidine bases is on the order of $k_B T$ at room temperature,[21] so small lengths of double-helical DNA (small polymers are called oligomers) are unstable. About eight bases must be paired specifically for a double helix to be remain bound together at room temperature.

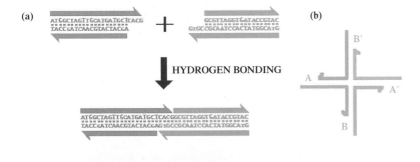

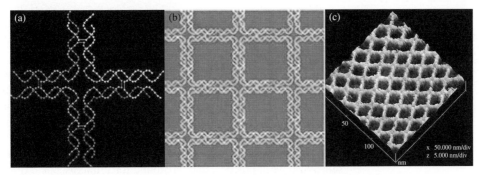

Fig. 6.20 The structural basis of DNA nanotechnology. (a) Shows how two linear DNA molecules will join at specific points by the base pairing of complementary overhanging sequences. Strands that are complementary to two other strands form crossover structures as illustrated in (b). For example, strand A is complementary at its 3′ end (arrowhead) to the 5′ end of strand B and complementary at its 5′ end to the 3′ end of strand B′. Thus, it forms a double helix with strand B initially, crossing over to form a double helix with strand B′ (courtesy of Professor Hao Yan, Arizona State University).

Fig. 6.21 Design of a four-way crossover structure that produces a rigid planar tile (a) which, with appropriate sticky ends, will spontaneously assemble into the lattice structure shown in (b). The distance between adjacent tiles in this design is 20 nm. (c) Shows an atomic force microscope image of this structure, self-assembled by annealing a mixture of DNA strands of the appropriate sequences (courtesy of Professor Hao Yan, Arizona State University).

DNA nanotechnology is based on specific base pairing used to join DNA at specific locations as shown in Fig. 6.20(a). The second key component is the formation of junctions at specific locations.[19] Such junctions occur naturally in genetic recombination, a process in which genes are shuffled between mother and father. The insertion point, called a Holliday Junction, consists of a crossover between two double helices with one strand remaining on one double helix and the other crossing over to the second double helix (and vice versa for the second double helix). In the case of genetic recombination, long runs of sequence on one helix are complementary to the sequence on the other, so the Holliday junction moves as DNA is transferred from one double helix to the other. Seeman's invention was to design a Holliday junction in which the sequence fixed the junction at one place on the double helices as shown in Fig. 6.20(b).

An example of a complex structure made by DNA nanotechnology is shown in Fig. 6.21. This is an array of crosses that self-assemble to form a tiled pattern with a repeat distance of 20 nm. Each cross consists of a pair of double helices that are bound together at the junction with a common strand (light blue in Fig. 6.21(a) – see the color image on CD) to form a four-way junction. Each end is terminated in two double helices, each of which has a sticky end (single-stranded overhang) so that the tile will spontaneously pair with tiles that have complementary sticky ends. The design of a lattice of crosses is shown in Fig. 6.21(b). The lattice is made by spontaneous assembly of the individual

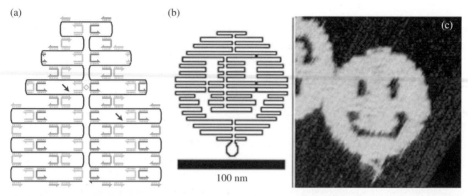

Fig. 6.22 Assembly of large and complex structures using DNA origami. The general principle is illustrated in (a). A long template strand (of unique and nonrepeating sequence—shown by the continuous black line) is annealed with a number of short strands (shown as the small loops and lines) that either form cross-links between different points on the template strand (loops) or fill in other regions to make them double helical (lines). (b) Shows how a template strand would have to be folded (no small "helper strands" shown) to make a nanoscale "smiley face." (c) Shows an AFM image of a nanoscale smiley face made in exactly this way. (Reprinted by permission from MacMillan Publishers Ltd.: Nature Publishing Group, P. Rothmunde, Nature[22] 2006.)

crosses as the complementary sticky ends join together. Such arrays readily self-assemble in practice, and an atomic force microscope (AFM) image of a self-assembled tile array is shown in Fig. 6.21(c).

The great power of this technology lies in its simplicity. The DNA sequences required to form a tile are first designed. On the basis of these designs, a set of synthetic oligomers are synthesized (at a cost of the few cents per base per nanomole), put into a solution containing a little salt, heated to 90°C (so that all base pairing is broken) and then cooled slowly so that the complex structure can self-assemble.

Structures formed this way can be remarkably complex yet produced with a yield that can approach 100%.

Figure 6.22 illustrates a technique called DNA origami.[22] In this approach, one very long strand is used as the master template for the final structure, and small oligomers (called helper strands) are added to cross-link the long template by forming double-helical regions. The principle of the approach is illustrated in Fig. 6.22(a). The long strand is shown schematically as a continuous black line (no bases shown). In this case (Fig. 6.22(a)), the desired final structure is the Christmas tree shape. The template strand has a unique and nonrepeating sequence and it is shown folded as desired. The next step in the design is to add cross-links by working out the sequence of small loops that are needed to form a double helix with any two remote parts of the template strand to pull it into the desired shape. These are shown by the small loop-like structures in Fig. 6.22(a). In general, not all regions can be made double helical by such loops. These remaining regions of single-stranded DNA are filled in using small oligomers that complete the double helix but do not add a cross-link (shown as the small linear pieces in Fig. 6.22(a)). Figure 6.22(b) shows the way in which a long template strand would need to be folded to make a "smiley face." This particular structure was designed to utilize a 7000 base length of viral DNA, and for simplicity, the small helper strands are not shown in this figure. Figure 6.22(c) shows an AFM image of a real nanoscale smiley face self-assembled in this way.

At the time of writing, DNA nanotechnology has yet to find widespread commercial application but it is hard to believe that it will be long in coming.

6.10 Bibliography

M.B. Smith and J. March, March's Organic Chemistry, 5th ed. 2001, New York: Wiley. The standard reference for organic synthesis.

J.N. Israelachvili, Intermolecular and Surface Forces. 1985, New York: Academic Press. Straightforward treatment of intermolecular interactions and many aspects of self-assembly.

A. Ulman, Ultrathin Organic Films. 1991, San Diego: Academic Press. The standard reference on all aspects of organic thin films.

B. Alberts, D. Bray, J. Lewis, M. Raff, K. Roberts and J.D. Watson, Molecular Biology of the Cell. 3rd ed. 1994, New York: Garland Press. A great primer on biochemistry and molecular biology.

6.11 Exercises

1. How many nearest neighbors does carbon have in (a) graphite and (b) diamond?

2. A small DNA molecule of precisely 10 bases was found to fly on a mass spectrometer with mass multiple peaks (assuming a charge, Z, of 1) each separated by 23 atomic mass units. Explain. (*Hint*: Look for an element of mass 23 on the periodic table.)

3. The yield of a reaction utilizing an NHS ester was observed to fall after the reagent was left dissolved in water for a week. Why?

4. A feature is observed in a mass spectrograph at a m/Z value of 1000 Daltons (assuming $Z = 1$). Where would the $Z = 2$ feature show up, if wrongly interpreted as a $Z = 1$ feature?

5. The magnetogyric ratio for C^{13} (spin $\frac{\hbar}{2}$) is a quarter of that of a proton. What is the magnetic resonance frequency of C^{13} at a field of 2 T?

6. Predict the number of lines in ^{13}C NMR spectra of benzene (below left) and toluene (below right).

(Note that the bonds between carbon in the rings are in fact all equivalent despite the alternating pattern of double and single bonds shown here.)

7. A molecule with a dipole moment of 1 Debye interacts with a point charge of $1e$ at a distance of 1 nm. Assume the molecule rotates freely and that the temperature is 300 K and calculate the interaction energy. Was the assumption of free-rotation justified?

8. What is likely difference in micelle structure between an amphiphile with one hydrophobic chain and one with two hydrophobic chains?

9. Calculate the critical micelle concentration for an equilibrium constant, K, of 10^{80} for aggregates of 20 monomers ($N = 20$).
10. In Question 9, the aggregates have a radius of 1 nm. Assume that the temperature is 300 K and calculate the surface energy, γ, in J/m^2.
11. Surface X-ray scattering (SXS) can produce reflections from surface films. Using the theory for electron diffraction from a slit in Chapter 2 as a guide, calculate the angle of the first diffraction peak in an SXS experiment on an alkanethiol monolayer for an X-ray wavelength of 1 Å.
12. Oxidizing a methyl group produces a carboxylate group. How will the wetting properties of an alkanethiol SAM change if it is oxidized?
13. Design DNA sequences for the shortest self-assembled system of three strands that could form a stable "Y" shape.

References

[1] Jansen, J.F.G.A., R.A.J. Janssen, E.M.M. Brabander-van den Berg, and E.W. Meijer, Triplet radical pairs of 3-carboxyproxyl encapsulated in a dendritic box. Adv. Mater., **7**: 561–564 (1995).
[2] Liu, Y., A.H. Flood, P.A. Bonvallet, S.A. Vignon, B.H. Northrop, H.-R. Tseng, J.O. Jeppesen, T.J. Huang, B. Brough, M. Baller, et al., Linear artificial molecular muscles. J. Am. Chem. Soc., **127**: 9745–9759 (2005).
[3] Kodis, G., P.A. Liddell, A.L. Moore, T.A. Moore, and D. Gust, Synthesis and photochemistry of a carotene-porphyrin-fullerene model photosynthetic reaction center. J. Phys. Org. Chem., **17**: 724–734 (2004).
[4] Smith, M.B. and J. March, *March's Organic Chemistry, 5th ed.* 2001, New York: Wiley.
[5] Israelachvili, J.N., *Intermolecular and Surface Forces, 2nd ed.* 1991, New York: Academic Press.
[6] de Vries, A.H., A.E. Mark, and S.J. Marrink, Molecular dynamics simulation of the spontaneous formation of a small DPPC vesicle in water in atomistic detail. J. Am. Chem. Soc., **126**: 4488–4489 (2004).
[7] Hill, T.L., *Thermodynamics of Small Systems.* 1994, New York: Dover.
[8] Israelachvili, J.N., *Intermolecular and Surface Forces.* 1985, New York: Academic Press.
[9] Ulman, A., *Ultrathin Organic Films.* 1991, San Diego: Academic Press. p. 443.
[10] Nuzzo, R.G. and D.L. Allara, Adsorption of bifunctional organic disulfides on gold. J. Am. Chem. Soc., **105**: 4481–4483 (1983).
[11] Weiss, P.S., Functional molecules and assemblies in controlled environments: formation and measurements. Acc. Chem. Res., **41**: 1772–1781 (2008).
[12] Beardmore, K.M., J.D. Kress, N. Gronbech-Jensen, and A.R. Bishop, Determination of the headgroup-gold(111) potential surface for alkanethiol self-assembled monolayers by ab initio calculation. Chem. Phys. Lett., **286**: 40 (1998).
[13] Ramachandran, G.K., T.J. Hopson, A.M. Rawlett, L.A. Nagahara, A. Primak, and S.M. Lindsay, A bond-fluctuation mechanism for stochastic switching in wired molecules. Science, **300**: 1413–1415 (2003).

[14] McDermott, C.A., M.T. McDermott, J.B. Green, and M.D. Porter, Structural origin the surface depressions at alkanethiolate monolayers on Au(111): a scanning tunneling and atomic force microscopic investigation. J. Phys. Chem., **99**: 13257–13267 (1995).

[15] Faraday, M., The Bakerian lecture: experimental relations of gold (and other metals) to light. Philos. Trans. R. Soc. London, **147**: 145–181 (1857).

[16] Brust, M., M. Walker, D. Bethell, D.J. Schiffrin, and R. Whyman, Synthesis of thiol-derivatised gold nanoparticles in a two-phase liquid-liquid system. J. Chem. Soc., Chem. Commun., **17**: 801–802 (1994).

[17] Murray, C.B., D.J. Noms, and M.G. Bawendi, Synthesis and characterization of nearly monodisperse CdE (E = S, Se, Te) semiconductor nanocrystallites. J. Am. Chem. Soc., **115**: 8706–8715 (1993).

[18] Lu, W. and C.M. Lieber, Semiconductor nanowires. J. Phys. D.: Appl. Phys., **39**: R387–R406 (2006).

[19] Seeman, N.C., Nucleic acid junctions and lattices. J. Theoret. Biol., **99**: 237–247 (1982).

[20] Judson, H., *The Eighth Day of Creation*. 1979, New York: Simon & Schuster.

[21] SantaLucia, J., A unified view of polymer, dumbbell and oligonucleotide DNA nearest neighbor thermodynamics. Proc. Natl. Acad. Sci. USA, **95**: 1460–1465 (1998).

[22] Rothmunde, P., Folding DNA to create nanoscale shapes and patterns. Nature, **440**: 297–302 (2006).

Applications

Electrons in nanostructures

The electronic properties of materials change when electrons are confined to structures that are smaller than the distance between scattering events (i.e., the mean free path) of electrons in normal solids. In this chapter, we will discuss what happens to electrons that are confined to one-dimensional structures (i.e., constrictions or "wires") and zero-dimensional structures (i.e., small particles). Two-dimensional confinement will be dealt with in a discussion of semiconductor heterostructures in Chapter 9. As a prerequisite for this material, we begin with a broad overview of conduction in normal solids, including the "free electron" model of metals and the band structure theory of the electronic states in periodic solids.

7.1 The vast variation in the electronic properties of materials

The electrical properties of materials vary vastly. We think of electrical properties in terms of resistance: the copper wires that carry electrical power have a low resistance and the glass insulators that support power lines have a very high resistance. Resistance depends on geometry and a more intrinsic quantity is resistivity, ρ. For example, a rod of material of length, L, and cross-sectional area, A, has a resistance

$$R = \frac{\rho L}{A}. \tag{7.1}$$

ρ is purely a material property, having units of Ω-m. The resitivities of some common materials are shown in Table 7.1. (The units here are Ω-m, but Ω-cm are more commonly used in the semiconductor industry.) Few physical quantities vary as much as resitivity: the range between copper and rubber is nearly 24 orders of magnitude!

Only a fraction of the electrons in a given material are involved in conducting electricity. For example, only one of the 29 electrons in each copper atom in a copper wire is free to carry a current. We shall see that these electrons move very quickly – about 10^6 m/s. However, they are also scattered very frequently – on average about every 40 nm in copper. The net current is carried by a slow drift of this randomly scattered cloud of electrons. The drift velocity depends upon the voltage drop across the copper wire, but for the small voltages dropped across typical appliance leads (a fraction of a volt per meter at high current) it is

Table 7.1 Resistivities of various
materials at 20° C.

Materials	ρ (Ω-m)
Silver	1.6×10^{-8}
Copper	1.7×10^{-8}
Aluminum	2.8×10^{-8}
Tungsten	5.5×10^{-8}
Iron	10×10^{-8}
Lead	22×10^{-8}
Mercury	96×10^{-8}
Nichrome	100×10^{-8}
Carbon (graphite)	3500×10^{-8}
Germanium	0.45
Silicon	640
Wood	10^{8} to 10^{14}
Glass	10^{10} to 10^{14}
Rubber	10^{13} to 10^{16}
Amber	5×10^{14}
Sulfur (yellow S_8)	1×10^{15}

a fraction of a mm per second. Despite the fact that they are in a minority, these "free electrons" give metals remarkable properties. In addition to their ability to pass electric currents, incident optical fields set free electrons into a motion that opposes the incident field, reradiating the light as a reflection, accounting for the optical reflectivity of most metals. Most of the elements in the periodic table are metals, with only those few on or to the right of the diagonal B-Si-As-Te-At being semiconductors or insulators. On the other hand, most compounds are insulators. Thus, metal oxides are insulators (e.g., Al_2O_3) or semiconductors (such as Cu_2O—the reason why it is possible to make good electrical contacts to partly oxidized copper wires).

We will see that this incredible variation in the electronic properties of materials has its origin in their quantum mechanical *band structure*.

7.2   Electrons in nanostructures and quantum effects

The electronic properties of bulk materials are dominated by electron scattering. This acts like a frictional force, so that the electron travels at a "drift velocity" such that the force owing to an applied field (field = voltage drop per unit distance, force = charge × field) is just equal to the friction force. Since the current is proportional to the drift velocity of the electrons, we can see why current is proportional to voltage in most conductors (i.e., Ohm's law is obeyed). The scattering events that contribute to resistance occur with mean free paths that are typically tens of nm in many metals at reasonable temperatures. *Thus, if the size of a structure is of the same scale as the mean free path of an electron, Ohm's law may not apply.* The transport can be entirely quantum at the nanoscale.

Another nanoscale phenomenon occurs if the structure is so small that adding an electron to it causes the energy to shift significantly (compared to $k_B T$)

because of Coulomb charging. Further charging can be impeded, because the structure now has a higher electrostatic potential than before a charge was added, leading to the phenomenon of the *Coulomb blockade*. Before we can consider this phenomenon, we must review the basics of electron transport: (1) the free electron model and (2) electronic band structure.

7.3 Fermi liquids and the free electron model

The free electron model treats conduction electrons as a gas of free, noninteracting particles, introducing interactions only as a means for particles to exchange energy by scattering. It was introduced by Drude, who justified it solely on the basis of the results produced by a simple model based on this assumption. It is remarkable that the Drude model works. To begin with, it ignores the long-range Coulomb repulsion between electrons. Even more seriously, we now know that the Pauli principle means that most electrons in a material are forbidden from moving anywhere (Drude's model predated quantum mechanics).

The explanation of why the free electron model works is subtle. It was first proposed, as a hypothesis, by Landau who called it the "Fermi liquid" model of metals, a model that has been remarkably successful in explaining the electrodynamics of metals.[1] The Landau hypothesis has been proved rigorously, but here will only sketch the basic idea. Figure 7.1 is a plot of the thermal average occupation number as a function of temperature for Fermions reproduced from Fig. 3.5, with plots for zero temperature and a temperature corresponding to 0.05μ (μ is the chemical potential). The chemical potential at $T = 0$ is called the Fermi energy, the energy of the highest occupied state at zero temperature. Mobile particles are produced *only* because thermal fluctuations promote electrons from below the Fermi energy to above it. Thus carriers are not produced alone, but in pairs, corresponding to a net positive charge in a state below the Fermi level, and a net negative charge in a state above it. This association between an electron and its "correlation hole" is part of the reason

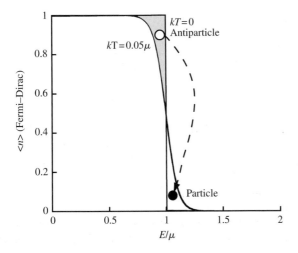

Fig. 7.1 Fermi liquid theory of a metal. The carriers are not the electrons themselves (which are immobile at low temperature) but "quasiparticles" formed when an electron is excited above the Fermi energy by a thermal fluctuation, leaving behind a "hole" or antiparticle.

that Coulomb interactions may be ignored. Another, less obvious, consequence of this process of excitation of carriers is that the "quasiparticles" (see Chapter 2.21) that result from thermal fluctuations of the original "Fermi gas" really are stationary states of the system, if they are close in energy to the Fermi energy. This is a consequence of the fact that, very close to the Fermi energy, the number of states that an excited electron can scatter into becomes vanishingly small. This is because it becomes impossible to satisfy the constraints of energy conservation, momentum conservation, and the Pauli principle when the energy difference between the quasiparticle and the Fermi energy goes to zero. (For a very readable derivation of this result, see the book by Abrikosov.[1]) Second, the quasiparticles themselves follow the Fermi–Dirac distribution for an ideal gas (Equation 3.57). Finally, for many situations, the energy of the quasiparticles (which is an energy relative to the Fermi energy) is quadratic in their momentum,

$$E = \frac{\hbar^2 k^2}{2m^*} \tag{7.2}$$

just as is the case for free electrons (Equation 2.36b) save that the mass is replaced by a parameter, m^*, called the *effective mass*.[1] We shall see below that a typical Fermi energy is 200 times the thermal energy, so the constraints of the Fermi liquid model hold at all reasonable temperatures. Thus, we can proceed to use a remarkably simple model that, at first sight, must appear to be absurd!

The Drude model of a free electron metal: Suppose that the mean time between collisions that scatter electrons is $\langle \tau \rangle$ and that the mean drift velocity (also averaged in the same way) is $\langle \mathbf{v} \rangle$, then Newton's second law gives

$$m \frac{\langle \mathbf{v} \rangle}{\langle \tau \rangle} = -e\mathbf{E}, \tag{7.3}$$

where $\mathbf{E}$ is the applied electric field and m is the mass of the electron. Ohm's law is

$$\mathbf{J} = \sigma \mathbf{E}, \tag{7.4}$$

where σ is the conductivity (resistivity $\rho = 1/\sigma$) of the material and $\mathbf{J}$, the current density, is

$$\mathbf{J} = ne \langle \mathbf{v} \rangle, \tag{7.5}$$

where n is the electron density. The above three equations yield the Drude result for conductivity

$$\sigma = \frac{1}{\rho} = \frac{ne^2 \langle \tau \rangle}{m}. \tag{7.6}$$

Copper has a resistivity of 1.6 $\mu\Omega$-cm at 273 K so that, according to Equation 7.6, the mean time between scattering events is 2.7×10^{-14} s. This is in good agreement with measurements. If we (incorrectly!) take the velocity of electrons to be thermal speed ($m\langle v \rangle^2 = 3k_B T$), then we find that the mean free path

($\ell = \langle v \rangle \langle \tau \rangle$) is only on the order of 1 nm, tens to hundreds of times smaller than actually observed. Thus, while the Drude model predicts many transport phenomena correctly (in addition to Ohm's law), it is fundamentally flawed, as might be guessed from the discussion of Fermi liquid theory above.

The Sommerfeld model: Treating the electron as a free quantum particle and putting two electrons (spin up and spin down) into each of the dn states per unit wave vector given by Equation 2.54 (Chapter 2), the total number of electrons, N, in a volume, V, must satisfy

$$N = 2 \times \int_0^{k_F} \frac{V k^2 dk}{2\pi^2} = \frac{k_F^3}{3\pi^2} V, \qquad (7.7)$$

where k_F is the wave vector of the highest filled state (we have not considered the possibility of excitations so we are limited to zero temperature at the moment). k_F is called the *Fermi* wave vector. Electrons with this wave vector have an energy

$$E_F = \frac{\hbar^2 k_F^2}{2m}. \qquad (7.8)$$

E_F is called the Fermi energy, the chemical potential at $T = 0$. A "Fermi velocity" can be calculated from the ratio of the momentum, $\hbar k_F$, to the mass as

$$V_F = \frac{\hbar k_F}{m}. \qquad (7.9)$$

The result we just obtained is for N electrons in volume V, so in terms of electron density, $n = N/V$, Equation 7.7 becomes

$$n = \frac{k_F^3}{3\pi^2}. \qquad (7.10)$$

Thus, all the properties of the free electron gas may be calculated from the electron density, n (this is the starting point for density functional theory— recall Section 2.22). The electron density is calculated using the volume of a unit cell of the crystal and the number of valence electrons per unit cell.

Taking a concrete example, Li has a lattice constant of 3.49 Å with a body-centered cubic lattice. The conventional unit cell contains two atoms and has a volume of $(3.49 \text{ Å})^3$. Thus, the volume per atom (and hence per electron, because Li is monovalent) is 2.17×10^{-23} cm^3. Equations (7.10), (7.9), and (7.8) are used to calculate k_F, V_F, and E_F, respectively, and the values obtained are listed in Table 7.2.

Table 7.2 Free electron properties of lithium, calculated using the free electron mass, m.

n	k_F	v_F	E_F
4.6×10^{22} cm^{-3}	1.12 Å^{-1}	1.23×10^8 cm/s	4.74 eV

7.4   Transport in free electron metals

Thus far, we have pictured the electrons as an ideal "Fermi gas" filling up all available energy levels up to the Fermi energy. This zero temperature picture will not help us understand transport (of heat or current). This must, of necessity, involve excited states (the quasiparticles of Fermi liquid theory) because electrons must have empty states to move into under the influence of thermal or electric potential gradients. If we assume that the applied electric fields are small, then the main source of excitation is thermal, leading to a distribution of occupation numbers around the Fermi level similar to that shown in Fig. 7.1. The energy spread (i.e., the energy range over which the occupation number differs significantly from 0 or 1) is on the order of $k_{\mathrm{B}}T$ (as can be seen by estimating the energy range where $0.25 < \bar{n}_i < 0.75$ from the plots of $\bar{n}_i$ vs. E in Fig. 3.5).

The result is that transport involves only a fraction of carriers, $f(n)$, on the order of

$$f(n) \approx \frac{k_{\mathrm{B}}T}{E_{\mathrm{F}}}. \tag{7.11}$$

At 300 K, only about 1 in 200 electrons in Li (4.74/0.025—Table 7.2) participate in transport. The corrected Drude expression for conductivity (Equation 7.6) becomes

$$\sigma = \frac{1}{\rho} \approx \frac{ne^2 \langle \tau \rangle}{m} \frac{k_{\mathrm{B}}T}{E_{\mathrm{F}}}. \tag{7.12}$$

Likewise, following from the fact that the specific heat of an ideal gas is $\frac{3}{2}nk_{\mathrm{B}}$, and using Equation 7.11, the specific heat of a Fermi gas is

$$C_{\mathrm{V}} \approx \frac{3}{2}nk_{\mathrm{B}}\left(\frac{k_{\mathrm{B}}T}{E_{\mathrm{F}}}\right). \tag{7.13}$$

Thus, the specific heat of a metal depends linearly on temperature, a result that is quite different from the T^3 low-temperature dependence of specific heat that arises from lattice vibrations (e.g., see Chapter 5 of Kittel's book on solid state physics[2]).

7.5   Electrons in crystalline solids: Bloch's theorem

It is possible to make exact calculations of the properties of crystalline solids containing billions of atoms as a consequence of *Bloch's theorem*. This famous theorem reduces the problem from one involving the entire solid to a much smaller one involving only the atoms in the fundamental repeat unit of the solid.

The physically measurable properties of a periodic lattice must, themselves, be periodic; that is, in terms of a lattice translation vector, **R** (a vector that is precisely an integer number of repeat units long), some property described by a function $U(\mathbf{r})$ must obey

$$U(\mathbf{r}) = U(\mathbf{r} + \mathbf{R}). \tag{7.14}$$

These measurable quantities will involve terms in the *square* of the wavefunctions, $\psi(r)$. Thus, the constraint (7.14) must be generalized to include the possibility of a *phase factor* multiplying the wavefunction. This can have any value so long as its square (in $\psi^*\psi$) is unity. This is satisfied by wavefunctions of the form

$$\psi_{n,k}(\mathbf{r}) = \exp[i\mathbf{k} \cdot \mathbf{r}]U_{n,k}(\mathbf{r}), \tag{7.15}$$

where $U_{nk}(\mathbf{r})$ is a function that is periodic in the lattice repeat distances $\mathbf{R}$ (i.e., $U_{nk}(\mathbf{r}) = U_{nk}(\mathbf{r} + \mathbf{R})$). This satisfies the constraints discussed above because $\exp[i\mathbf{k} \cdot \mathbf{r}] \times \exp[-i\mathbf{k} \cdot \mathbf{r}] = 1$. Note that replacing $\mathbf{r}$ by $\mathbf{r}+\mathbf{R}$ in 7.15 leads to

$$\psi_{n,k}(\mathbf{r} + \mathbf{R}) = \exp[i\mathbf{k} \cdot \mathbf{R}]\psi_{n,k}(\mathbf{r}). \tag{7.15a}$$

Wavefunctions of the type described by 7.15 and 7.15a are known as *Bloch states*. The index k in Equation 7.15 refers to the value of the wave vector $\mathbf{k}$, and the index n refers to the *band* with which the wavefunction is associated (see below). The important consequence of Bloch's theorem is that:

All the properties of an infinite crystal can be described in terms of the basic symmetries and properties of a unit cell of the lattice together with an extra parameter $\mathbf{k}$.

$\mathbf{k}$ is called the "crystal momentum." It is not a real mechanical momentum, but rather a mathematical embodiment of how the "phase" of the wavefunction can change in a real crystal. For an infinite crystal, the magnitude of $\mathbf{k}$ can go to zero (i.e., infinite wavelength). At the other end of the length scale, wavelengths less than a lattice constant are not physically meaningful, because measurable quantities must always have the periodicity* of the lattice. Features that vary more rapidly can always be described in terms of longer wavelength features, because there is nothing in the lattice to be displaced or altered at spacings smaller than those of the atoms that make up the lattice. The shortest wavelength of any excitation in the lattice must correspond to twice the spacing between atoms (twice because a wave has a $+$ peak and a $-$ peak). This smallest distance in terms of wavelength corresponds to a maximum value for $\mathbf{k}$ of $2\pi/a$ (in a direction where the lattice constant is a). Thus, the constraint of a smallest possible spatial wavelength corresponds to a maximum allowed value of $\mathbf{k}$. So, for an infinite crystal, the allowed range of $\mathbf{k}$ lies between 0 and $2\pi/a$. This range of $\mathbf{k}$ is called the *first Brillouin zone*. Since $-\mathbf{k}$ and $+\mathbf{k}$ directions are equivalent (the square of the phase factor is unity either way) and the properties of the lattice repeat every time $\mathbf{k}$ increases by $2\pi/a$ in some direction, it is conventional to plot quantities that depend on $\mathbf{k}$ in the range $-\frac{\pi}{a} < |\mathbf{k}| < \frac{\pi}{a}$. This is called the *reduced zone scheme* (see the next section).

*This is true only in the noninteracting electron picture. There exist incommensurate excitations of crystal lattices such as charge density and spin density waves that do not have the period of the lattice.

7.6   Electrons in crystalline solids: band structure

We will limit our discussion to one dimension for simplicity at this point (though we shall see that living in a three-dimensional world really does matter). We use the Bloch states described above as a trial solution of the Schrödinger equation with a periodic potential to represent the chain of atoms. Starting from an atomic wavefunction, $\psi_s(r)$, one can construct a trial Bloch function, ψ_T, using Bloch's theorem in the form of 7.15a as follows:

$$\psi_T = \sum_n \exp[ikna]\psi_s(r - na), \qquad (7.16)$$

where n is an integer and $\psi_s(r - na)$ is a wavefunction with the periodicity of the lattice.

In perturbation theory (Section 2.18, Chapter 2), the change in energy between the atomic case ($E = E_s$ for isolated atoms, no interactions) and the crystal lattice case ($E = E_k$) for electrons from one atom interacting with a neighboring atom via an interaction Hamiltonian (ΔU^{op}) is given to first order by

$$E_k - E_s = \langle \psi_T | \Delta U^{op} | \psi_T \rangle, \qquad (7.17)$$

where ψ_T for a crystal is given by Equation 7.16 and we have used the Dirac notation introduced in Chapter 2 (i.e., Equation 7.17 implies integrating the product of the complex conjugate of ψ_T with $\Delta U^{op}\psi_T$ over all space). If we assume that matrix elements of the interaction Hamiltonian are nonzero only for nearest neighbors ($n = \pm 1$) and zero otherwise, then the only terms that remain when (7.16) is put into (7.17) are

$$E_k - E_s = \langle \psi_s | \Delta U^{op} | \psi_s \rangle + \langle \psi_s | \Delta U^{op} | \psi_s(r - a) \rangle \exp ika$$
$$+ \langle \psi_s | \Delta U^{op} | \psi_s(r + a) \rangle \exp -ika. \qquad (7.18)$$

Defining

$$E_s + \langle \psi_s | \Delta U^{op} | \psi_s \rangle = \varepsilon_0 \qquad (7.19)$$

(the "on-site energy") and

$$\langle \psi_s | \Delta U^{op} | \psi_s(r \pm a) \rangle = \tau \qquad (7.20)$$

(the "hopping matrix element"), Equation 7.18 becomes

$$E_k = \varepsilon_0 + 2\tau \cos ka. \qquad (7.21)$$

(Note that $\tau < 0$.) This wave vector dependence of the energy is compared with the free electron result ($E_k = (\hbar^2 k^2)/2m$) in Fig. 7.2.

For very small wave vectors near the Brillouin zone center, $E(k)$ is quadratic, as is the case for free electrons (this is because the expansion of $\cos\theta$ for small θ is $1 - \frac{\theta^2}{2}$). However, as $ka \to \pm\pi$, the curve turns over with $\frac{\mathrm{d}E(k)}{\mathrm{d}k} \to 0$.

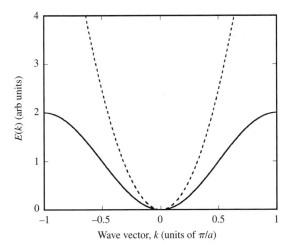

This turnover has an important physical consequence. The gradient of the energy with respect to wave number is the "group velocity" of a wave (for a plane wave this is just the velocity as given by speed = frequency × wavelength). In the case of a "wave packet" like an electron, the group velocity describes the velocity of the associated "particle." Thus

$$v_g = \frac{d\omega}{dk} = \frac{1}{\hbar}\frac{dE(k)}{dk} \tag{7.22}$$

is the group velocity of the electron. So, at $ka = \pm\pi$, *the electron's group velocity falls to zero.* This value of k corresponds exactly to $\lambda = \frac{2\pi}{k} = 2a$, which is the condition that the electron is *Bragg-reflected* by the crystal potential. The combination of forward and (Bragg) reflected waves results in *standing wave*. The electron does not propagate at these values of k. The flattening of the function $E(k)$ causes an increase in the density of states near $ka = \pm\pi$. The number of states per unit k in one dimension is, of course, constant (depending only on the size of the material), so the number of states per unit energy must increase as the dependence of energy on k gets weaker near the point of Bragg reflection.

We have considered only one atomic state, ψ_s. Had we considered a series of atomic states, we would have obtained a series of bands, one for each atomic state (as perturbed by the lattice). In this view, it is easy to see why "band gaps" must exist between allowed states: they correspond to the gaps between allowed atomic eigenstates (but which have been broadened into bands of energy width 4τ by interactions between neighboring atoms in the crystal lattice). It is quite complicated to discuss the size of the energy gap between one allowed band and the next with the model used here (the "tight-binding" model) because this is a function both of ε_n (the on-site energy for the nth band) and τ.

Band gaps are easier to introduce in the so-called nearly free electron model of band structure. This takes the opposite view, starting with free electrons that are weakly perturbed by a periodic potential of magnitude ΔU. The result of doing this (e.g., see Ashcroft and Mermin,[3] Chapter 9) is a parabolic band in k with energy gaps of $2\Delta U$ at values of $ka = \pm n\pi$. In either case, the bands

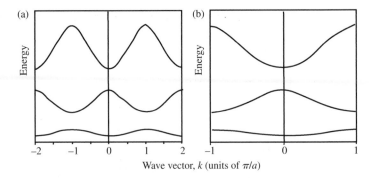

Fig. 7.3 Showing (a) the *extended zone* presentation of $E(k)$ for three bands and (b) the *reduced zone* scheme. The gaps between the highest and lowest energies in each band correspond to sets of propagation vectors that are not allowed inside the periodic potential.

(each band corresponding to a particular atomic level) can be "folded" back into the first Brillouin zone as shown in Fig. 7.3.

Band structure accounts for the electrical properties of solids. States are always filled up from the lowest energy in a manner similar to what was done for free electrons in Equation 7.7. To do this, for some complicated dependence of E on k requires that the integral over k is replaced with an integral over $E(k)$ that is carried out over the allowed bands. These bands are filled up by the requirement that the integral yield the total number of electrons in a volume V (just as was done for the simple case of the free electron case in Equation 7.7).

Some possible outcomes are as follows:

1. *The Fermi level lies inside an allowed band.* In the one-dimensional model discussed here, this would occur for a material with one valence electron per unit cell, where the band would be half full. More complex possibilities arise in three dimensions (see the next section). Materials with partially filled bands have unoccupied states available immediately adjacent in energy to the Fermi level, so electrons may move freely. These materials are the metals.
2. *The Fermi level lies at the top of a band.* Taking ΔU as the perturbing potential in the nearly free electron model (analogous to τ in the tight-binding model) when

$$2\Delta U \gg k_{\mathrm{B}}T, \tag{7.23}$$

no states exist within $k_{\mathrm{B}}T$ of the Fermi level and the material is an *insulator*. Electrons could be excited to the next band by a large enough electric field, and this is the process of dielectric breakdown that can turn even very good insulators into conductors.
3. If

$$2\Delta U \approx k_{\mathrm{B}}T \tag{7.24}$$

electrons can be thermally excited into an empty band and the material is a *semiconductor* (or *semimetal* if the gap is very small). Thermal excitation of an electron from the previously full band (*valence band*) into the previously empty band (*conduction band*) also opens up a second channel for current in semiconductors. Empty states now exist in the (previously full at $T = 0$ K) valence band, allowing current to flow in this band also.

Electrons in semiconductors near the empty states at the top of the valence band have an interesting property. Accelerating them toward the band edge causes them to *slow down* (since their group velocity goes to zero at the band edge). Their effective mass, given by analogy with classical mechanics,[†]

$$m^* = \left[\frac{1}{\hbar^2} \frac{\partial^2 E(k)}{\partial k^2} \right]^{-1} \tag{7.25}$$

[†] From $E = \frac{1}{2}mv^2 = \frac{p^2}{2m}$, $m = \left[\frac{\partial^2 E}{\partial p^2} \right]^{-1}$, so the quantum analog, Equation 7.25 follows using $p = \hbar k$. The curvature of $E(k)$ is different in different directions in the crystal, so the value of m^* is different in different directions.

is *negative* (i.e., the slope *decreases* as $ka \to \pm\pi$). An alternative description is to assign the particles a *positive effective mass* but call their charge positive. In that case, these electrons do indeed accelerate as their energy is increased (i.e., they act as though they have positive mass) but they behave as though their charge is *positive* because the acceleration is in the opposite direction to that for free electrons. These "quasiparticles" (the antiparticles obtained on promoting electrons from full bands) are called *holes*. The conductivity of semiconductors can be increased by doping them with impurities with states near the edges of one of the bands. If the difference in energy between states on the impurity is comparable to $k_B T$, the impurities can donate (or gain) electrons by thermal excitation. Dopants that donate electrons to the conduction band (such as the five valent P, As, or Sb in the case of silicon) make the semiconductor electron rich and the material is referred to as "n-type." Dopants that take electrons from the valence band (such as the three valent Al and B in the case of Si) make the semiconductor hole rich (the charge carriers are positive) and the material is referred to as "p-type."

7.7   Electrons in 3D—why copper conducts; Fermi surfaces and Brillouin zones

The simplified discussion just given for the one-dimensional problem is unsatisfactory. It tells us why the univalent alkali metals are metals, but appears to predict that the vast majority of metals in the periodic table (i.e., those with two valence electrons) should be insulators, with two electrons in a filled valence band. The problem lies with our oversimplistic one-dimensional treatment. To understand how electrons interact with a crystal lattice in three dimensions, we have to introduce a representation of the crystal structure in wave vector space. We have already done this for the one-dimensional case, where it was clear that a Bragg reflection occurs when the wave vector of the electron reaches out to the points $\pm\frac{\pi}{a}$. A solid constructed out of all the planes that correspond to Bragg reflections in wave vector space is called the reciprocal lattice. It is quite simple to construct this for a cubic lattice, because the reciprocal lattice is also a cube of sides $\frac{2\pi}{a}$ (dimensions are 1/length in reciprocal space). A unit-cell in the reciprocal lattice provides a beautiful geometric tool for working out the directions in which X-rays would be scattered by a crystal lattice. These are all the directions for which the difference between the incident and scattered wave vectors is a vector that touches a point on the reciprocal lattice (e.g., see Kittel,[2] Chapter 2). A unit-cell in the reciprocal lattice of a simple cubic lattice is shown schematically in Fig. 7.4(a).

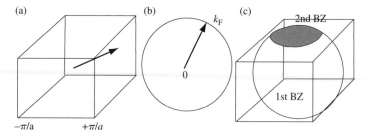

Fig. 7.4 Fermi surfaces in three dimensions. (a) Shows a unit cell in the reciprocal lattice of a simple cubic lattice. This is the representation of the real space lattice in wave vector space. Wave vectors that span the gap from the origin (in the middle of the reciprocal lattice) to touch one of the planes (at $\pm\frac{\pi}{a}$) correspond to wavelengths that are Bragg reflected in that particular direction. (b) Shows a Fermi sphere of radius k_F corresponding to a completely filled band (two electrons per unit cell) for noninteracting electrons. The free electron Fermi sphere must intersect Bragg reflection planes. The region of the Fermi sphere lying outside the first Brillouin zone (BZ) is shown (for one plane only) as the shaded region in (c).

To describe a simple gas of noninteracting electrons in three dimensions, it is necessary only to extend the idea of a Fermi wave vector to make a sphere that encloses all the wave vectors of occupied states. This Fermi sphere is shown schematically in Fig. 7.4(b).

Only wave vector values that will fit inside the reciprocal lattice will not be Bragg reflected. The volume of the Fermi sphere gives the number of occupied electronic states, but, because of the difference in the volumes contained by a sphere and a cube, the Fermi sphere *must* intersect Bragg planes if the volume (and hence the number of allowed k states) of the unit cell in reciprocal space and in the Fermi sphere are to be equal. This is shown schematically in Fig. 7.4(c).

Recalling that a band gap is introduced whenever the electron's wave vector touches a Bragg plane, the Fermi sphere must become truncated when interacting with the potential of a cubic lattice. The result is a full first *Brillouin zone* (1st BZ in Fig. 7.4(c)) and a partially full second Brillouin zone (2nd BZ in Fig. 7.4(c)). It is this second partially full Brillouin zone that provides the free electrons to divalent metals.

Copper forms a close-packed structure known as a face-centered cubic lattice for which the reciprocal lattice is a body-centered cubic lattice (see Kittell,[2] Chapter 2). The residue of the Fermi sphere that lies outside the first Brillouin zone falls into several partly filled bands corresponding to higher-order Brillouin zones.

7.8 Electrons passing through tiny structures: the Landauer resistance

We know that confining electrons to small structures leads to discrete eigenstates as discussed in Section 2.15 (Chapter 2). But what about the problem of "free" electrons passing from one macroscopic conductor to another through a nanoscale constriction? This is, for example, what happens in the scanning tunneling microscope (STM). We learn here that *if the restriction is smaller than the scattering length of the electrons, it cannot be described as a resistance in the "Ohm's law" sense.* It is important to point out at the outset that the result we are about to discuss does not apply to a long "wire," because the Fermi liquid model does not hold (see Chapter 8). Rather it applies to electrons that tunnel from one bulk electrode to another by means of a small connecting constriction.

This problem is quite simple to analyze by means of Fermi's Golden Rule and the result is quite remarkable. If the source and sink of electrons are connected by N channels (i.e., N types of electronic wavefunctions can occupy the gap), then the resistance of the gap, R_g, is just

$$R_g = \frac{h}{2Ne^2} = \frac{1}{N}R_L, \tag{7.26}$$

where R_L is the Landauer resistance ($h/2e^2$) which is approximately equal to 12.9 kΩ. This is a powerful result. It appears that the materials used in a "quantum point contact" do not contribute to its resistance. Thus, any STM probe that just touches a conducting surface (i.e., the constriction is of the same size as the de Broglie wavelength of the electron so that only one quantum channel exists) will form a junction of resistance 12.9 kΩ. Given this relationship, we can now use Equation 2.60 in Chapter 2 to write the resistance of a tunnel junction of gap L as

$$R = R_L \exp\left[1.02\sqrt{\phi}L\right] \approx 12.9 \exp\left[1.02\sqrt{\phi}L\right] \text{ k}\Omega. \tag{7.27}$$

This powerful approximate relationship yields a useful experimental method for knowing when a quantum contact is just made.

The Landauer resistance was introduced by Landauer using a rather indirect argument[4] because, at the time, time-dependent perturbation theory had not been extended to the case where one reservoir of particles is flowing into another, out of equilibrium, because of the application of a potential difference between the two reservoirs. This generalization has now been made (see Chapter 17 of the book by Nitzan[5]). The end result is that the Fermi Golden Rule (Equation 2.94b) is valid for this situation too:

$$P(m, k) = \frac{2\pi}{\hbar} \left|\langle\psi_m| \hat{H}' |\psi_k\rangle\right|^2 \rho(E_k).$$

Thus, to the extent that the matrix elements for each of the electronic transitions from a left electrode to a right electrode are the same, the current that flows between the electrodes will be proportional to the number of states available for the electrons to flow into at them applied bias, V.

Referring to Fig. 7.5(b), a bias, V, applied across the two macroscopic electrodes will shift the Fermi levels of one relative to the other by an amount eV so that the net current will be proportional to the number of states in this energy range. Specifically, per unit area $\mathbf{i} = ne\langle\mathbf{v}\rangle$ and

$$n = \frac{dn}{dE}eV = \frac{dn}{dk}\frac{dk}{dE}eV. \tag{7.28}$$

In the quantum structure, $\langle\mathbf{v}\rangle = v_g$, the group velocity as given by Equation 7.22. Furthermore, the interval between allowed k states is $2\pi/d$ for a

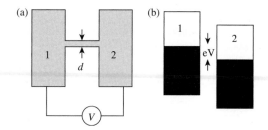

Fig. 7.5 (a) Arrangement of two bulk conductors connected by a small channel with bias V applied across the gap. (b) Corresponding energy diagram where the filled states are shown in black.

one-dimensional structure (Equation 2.53b in Chapter 2) so that the density of states (states per unit length) is

$$\frac{dn}{dk} = \frac{1}{2\pi}. \tag{7.29}$$

Thus,

$$i = ne\,v_g = 2 \times N \times \frac{1}{2\pi}\frac{1}{\hbar v_g} \times eV \times ev_g, \tag{7.30}$$

where the factor 2 accounts for the two allowed spin states per k state and N is the allowed number of quantum states in the channel. Equation 7.30 yields the following result for the quantum of conductance (conductance is the reciprocal of resistance) per channel ($N = 1$)

$$G_0 = \frac{2e^2}{h} = 77.5 \; \mu S \; \text{or, equivalently } R_0 = 12.9 \; \text{k}\Omega. \tag{7.31}$$

The material properties (in the form of v_g or $\frac{dE}{dk}$) are canceled out of this expression, so the expression is universal, independent of the material lying between the source and sink of electrons. The Landauer conductance or resistance is a fundamental constant associated with quantum transport. It is important to realize that the Landauer resistance is *not* a resistance in a conventional sense: *no power is dissipated in the quantum channel.* Rather, it reflects how the probability of transmission changes as the bias is changed. Dissipation (and thus Ohm's law) requires scattering. This occurs in the bulk electrodes, but not in the nanochannel. This is the reason why current densities in the STM can be enormous (10^9 A/m^2) with no damage to probe or sample.

This universal nanoscale behavior is readily demonstrated experimentally by measuring the current that passes through a fine gold wire as it is stretched to a breaking point. As the wire narrows down to dimensions of a few Fermi wavelengths, quantum jumps are observed in the current (current is carried by a finite number of states in the constriction—Fig. 7.6(a)). The plateaus between each jump correspond to stretching of the gold filament. It yields until another layer of atoms is peeled away, causing the number of allowed states to fall. Data for the final three states are shown in Fig. 7.6(b). Steps occur at almost exact multiples of G_0. This is, in fact, a rather special property of gold. The steps are observed close to, but not exactly, at multiples of G_0 for other metals owing to the details of the matrix element that we ignored in the simple arguments leading to Equation 7.31.[6]

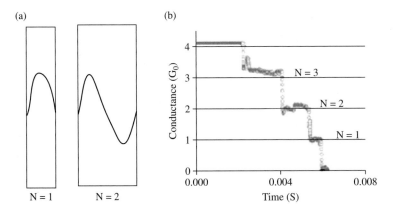

Fig. 7.6 (a) Illustrating a very narrow channel in which just one ($N = 1$) or two ($N = 2$) quantum modes can propagate. (b) Measured Landauer steps in the conductance of a gold break junction. (Courtesy of Professor Jin He, Arizona State University.)

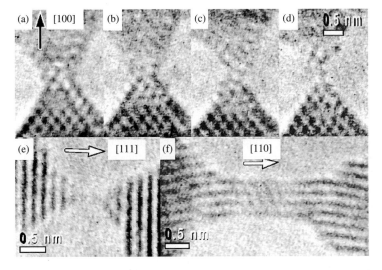

Fig. 7.7 Atomic structure of quantum wire constrictions (Reproduced with permission from Rego et al.[7], Phys. Rev. B, Copyright 2003 American Physical Society. Courtesy of Luis G.C. Rego, Universidade Federale de Santa Catarina.)

In general, an exact conductance may be calculated from the Landauer–Buttiker formula:

$$G = \frac{2e^2}{h} \sum_{ij} |T_{ij}|^2, \qquad (7.32)$$

where T_{ij} are the matrix elements that connect states i which correspond to electrons on one side of the junction to states j on the other side.

At what size scale do quantum events dominate? The upper limit occurs when the size of the structure approaches the mean free path for scattering: hundreds to thousands of Å at room temperature (and even larger lengths at lower temperatures). The lower limit occurs when only one mode is available in the channel, i.e., when the channel is on the order of a Fermi wavelength in diameter (i.e., $2\pi/k_F$). For lithium (Table 7.2), this would be $\sim$6 Å, i.e., atomic dimensions.

Transport measurements have been carried out on atomic scale junctions at the same time as their structure was imaged, using a mechanical "break junction" placed into a high-resolution transmission electron microscope (TEM).[7] TEM images for some quantum point contacts are shown in Fig. 7.7.

7.9   Charging nanostructures: the Coulomb blockade

If a small conducting particle is placed in the middle of a tunneling gap between two electrodes, electrons can tunnel from the first electrode to the second by hopping from one electrode to the small particle and then from the small particle to the second electrode. Such an arrangement is shown in Fig. 7.8(a). Two electrodes are separated by a tiny gap, in which sits a small metal particle of radius a. This simple geometry opens up a number of interesting pathways for electron transfer. One possibility is that the electron tunnels straight through the whole structure, a process that can happen with high efficiency if energy levels are in just the right place. Such a process is called resonant tunneling and it will be discussed in Section 7.10. Here, we consider the situation where the electron hops on to the center particle and then hops off to the second electrode. When we discuss resonant tunneling, we will see that the condition for the Coulomb blockade to occur is that the tunneling rate between the dot and the other electrodes must be sufficiently small. This hopping process is very sensitive to the potential of the center island because the charging energy of a small particle, even for one electron, can be quite significant. If the charging energy is greater than the thermally available energy, further hopping is inhibited, leading to a region of suppressed current in the current–voltage characteristic. Once the applied bias exceeds this "Coulomb blockade barrier" current can flow again. When the potential is increased to the point that the particle becomes charged with two electrons, a second blockade occurs. This process is repeated for each integer (number of electrons) occupation of the particle and the resulting series of steps in the current–voltage characteristic is called a Coulomb staircase.

The energy to charge a capacitor (C Farads) with an electronic charge is $e^2/2C$ ($e/2C$ in units of electron volts), and the capacitance of a sphere of radius, a, is $4\pi\varepsilon\varepsilon_0 a$. Thus, the voltage required to charge an island is given by

$$\Delta V = \frac{e^2}{8\pi\varepsilon\varepsilon_0 a}\text{volts.} \tag{7.33}$$

Taking $\varepsilon = 1$, with $\varepsilon_0 = 8.85 \times 10^{-12}$ F/m and $a = 1$ nm, we obtain $\Delta V = 0.7$ V. With $a = 100$ nm this falls to $\Delta V = 7$ mV. Thus, the Coulomb blockade should be observable in a very small particle at room temperature, whereas

Fig. 7.8 (a) Arrangement of an experiment to measure the Coulomb blockade. Two electrodes are supported on an insulating substrate separated by a small gap containing a conducting island of nm to hundreds of nm size. (b) The single electron transistor has similar geometry but a gate electrode close to the island can alter its potential to overcome the blockade.

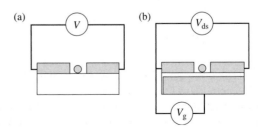

temperatures well below 70 K would be required to observe the effect in a 100 nm particle.

The situation is actually a little more complicated than this. The two tunnel junctions can be modeled by an electronic circuit as shown in Fig. 7.9. The two "resistors" (R_1 and R_2) model the tunnel current that flows across each of the two gaps when a bias V is applied across the whole structure. The two capacitors represent the charge stored by the central island; the capacitances of the junctions between the first electrode and the island (C_1) and the second electrode and the island (C_2) will help to determine the potential of the central island relative to each electrode.

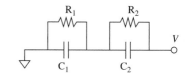

Fig. 7.9 Electrical circuit model of the two-junction Coulomb blockade experiment. (After Fig. 1 of Hanna and Tinkham.[8])

Such a circuit analysis has been carried out by Hanna and Tinkham.[8] They used conventional circuit theory to determine the charge on the central particle and its potential relative to the left and right electrodes, subject to the constraint that charge transfers onto and off the particle occur as integer multiples of the electronic charge. They give the following expressions for the current as a function of voltage:

$$I(V) = e \sum_{n=-\infty}^{\infty} \sigma(n) \left[\Gamma_2^+(n) - \Gamma_2^-(n) \right] = e \sum_{n=-\infty}^{\infty} \sigma(n) \left[\Gamma_1^-(n) - \Gamma_1^+(n) \right].$$

$$(7.34)$$

Here,

$$\Gamma_j^{\pm}(n) = \frac{1}{R_j e^2} \left(\frac{-\Delta E_j^{\pm}}{1 - \exp\left(\frac{\Delta E_j^{\pm}}{k_B T} \right)} \right), \quad j = 1, 2,$$

$$(7.35)$$

and

$$\Delta E_1^{\pm} = \frac{e}{C_1 + C_2} \left(\frac{e}{2} \pm (ne + Q_0) \pm C_2 V \right)$$

$$(7.36)$$

$$\Delta E_2^{\pm} = \frac{e}{C_1 + C_2} \left(\frac{e}{2} \pm (ne + Q_0) \pm C_1 V \right).$$

$$(7.37)$$

In (7.34), $\sigma(n)$ is the normalized distribution of charges on the central particle (i.e., the fraction of the total with charge n). It is determined from the relations:

$$\frac{\sigma(n)}{\sigma(n+1)} = \frac{\Gamma_1^-(n+1) + \Gamma_2^-(n+1)}{\Gamma_1^-(n) + \Gamma_2^+(n)}$$

and

$$\sum_{n=-\infty}^{\infty} \sigma(n) = 1.$$

The Γ_j s in 7.35 are the thermally activated hopping rates between the particle and the left and right electrodes, and the $\Delta E^{\pm}$ terms represent the charging energy of the central particle, as given by the expressions (7.36) and (7.37). The term in Q_0 represents the possibility of a residual floating charge on the

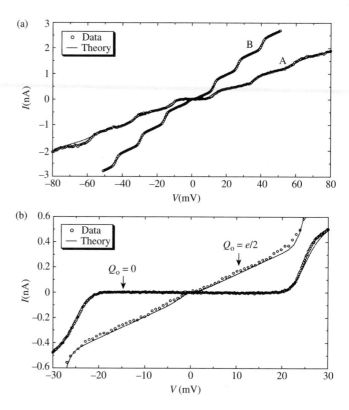

Fig. 7.10 Experimental measurements of *I–V* curves from nanoscale double junctions (dots) fitted with Coulomb blockade theory (lines). (Reprinted with permission from Hanna and Tinkham[8], Phys. Rev. B. Copyright 1991 American Physical Society. Courtesy of Professor M. Tinkham.)

central island. These expressions were used to fit current–voltage curves taken by a STM probe held over a grainy gold film at low temperature. As shown in Fig. 7.10, the agreement between the experimental data and this theoretical model was excellent.

7.10   The single electron transistor

The existence of a potential-dependent blockaded region in the *I–V* curve of double junction devices is the basis of a new type of transistor called the single electron transistor (SET). The SET consists of an isolated metal particle coupled by tunnel junctions to two microscopic electrodes, similar to the Coulomb blockade experiment. However, in the case of the SET, the isolated metal particle is capacitively coupled with a gate electrode that is used to control the potential of the metal particle independently. When the potential is adjusted to values that correspond to high conductance through the small metal particle, the transistor is on. This arrangement is shown schematically in Fig. 7.8(b).

Here, we show data for another arrangement, illustrated in Fig. 7.11(a). The "particle" was constructed as circular disc made from conductive *n*-type GaAs, insulated from source and drain contacts by thin films of the insulating alloy AlGaAs. The potential of the dot was controlled by a gate insulated from the dot by a layer of InGaAs. An energy diagram (Fig. 7.11(b)) shows how a finite source–drain voltage (V_{sd}) opens a "window" of potential for tunneling via the quantum dot. The alignment of the levels on the dot (i.e., energies for

(a)

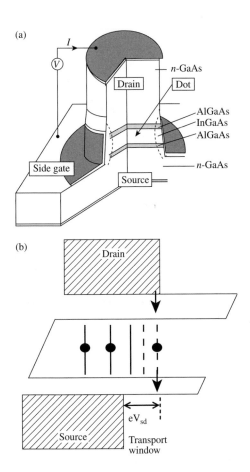

(b)

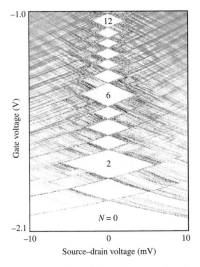

Fig. 7.11 (a) Structure of an SET using a circular quantum dot made from *n*-GaAs sandwiched between two electrodes (also *n*-GaAs) and isolated with two insulating films (AlGaAs) through which electrons can tunnel. InGaAs insulates the dot from the gate electrode. (b) Shows an energy diagram for the dot trapped between the electrodes. Electrons tunnel in an energy window determined by the source–drain bias (eV$_{sd}$). States of the dot in this window empty into the drain (dashed lines) and are filled by the source. (From Kouwenhoven et al.[9] Science 1997, reprinted with permission from AAAS.)

electronic occupation of the dot) relative to the Fermi energy is controlled by the gate potential.

The data obtained are three dimensional, because the current is measured as a function of both source–drain voltage and gate potential. It is conventional to plot not current, but its derivative with respect to V_{sd}. This is the source–drain differential conductance, so a plot of $\partial i/\partial V_{sd}$ shows the regions where the SET is "on" (as areas of darker color in Fig. 7.12). The center of Fig. 7.12 is occupied by white "diamonds" where the differential conductance is zero. To understand this, take a look at the "Coulomb staircase" in the upper part of Fig. 7.10. Between each step in current, the conductance is constant, so the differential conductance is zero. At each step, the current jumps sharply, so the differential conductance is very large (just at these points). Thus, the *I–V* curve in Fig. 7.10 corresponds to a line up the middle of the differential conductance plot in Fig. 7.12 (for $V_{sd} = 0$). The *I–V* curve in Fig. 7.9 would come from integrating the conductance along this line.

The levels of the dots are separated in energy just as they would be for an atom (shown by the lines in the middle of Fig. 7.11(b)). When V_{sd} exceeds the level spacing, the current is not blockaded at all, as there is always an unfilled level within the dot to transport an electron. This value of bias corresponds to the edges of the diamonds (at V_{sd} is ± a few mV, depending on the level). The circular dot was chosen as a model of a two-dimensional atom, and it

Fig. 7.12 Differential conductance, di/dV_{sd} (darker color = higher conductance) plotted as a function of both source–drain voltage and gate voltage. Fine structure outside the central diamonds reflects excited states. (From Kouwenhoven et al.[9], Science 1997, reprinted with permission from AAAS.)

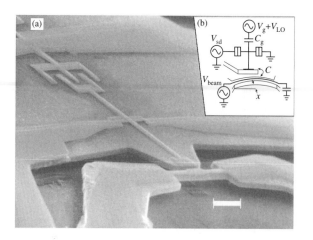

Fig. 7.13 A SET used as a nanomechanical sensor. (a) Shows an SEM image of a device shown schematically in (b). The scale bar is one micron. The vibrating beam at the bottom is biased and alters the potential of the gate electrode in the SET (upper left) to a degree that depends on the capacitance, C, between the beam and the gate electrode. (Reprinted from Knobel and Cleland[10], Nature 2003 with permission from MacMillan Publishers Ltd.)

shows atomic features. For example, the particularly stable diamonds for 2, 6, and 12 electrons are consequences of atomic "shell" structure.[9] The ability of the SET to sense charge at the single electron level opens up new possibilities for sensors, and Fig. 7.13 shows an electromechanical sensor based on the capacitative coupling of a vibrating beam to the gate of an SET.[10]

7.11   Resonant tunneling

In the preceding description of the Coulomb blockade, we treated charging of the central particle classically, requiring only that the electronic charge itself be quantized. It may appear obvious that charge should be quantized, but, as we have seen electrons can become delocalized, as in the case of atomic bonds, or band structure in semiconductors and metals. Thus, the condition for Coulomb blockading is that the electron localize on the quantum dot. The requirement for this to happen is that the tunneling "resistance" of the contacts exceed h/e^2, equal to twice the Landauer resistance.[11] If the coupling between the electrodes and the "dot" is stronger than this, the whole system must be treated quantum mechanically. This is quite straightforward to do in one dimension following the methods introduced in Chapter 2 where we solved tunneling problems by matching boundary conditions at points where the potential changed. A model for the double gap junction is shown in Fig. 7.14. In this diagram, the central particle has been modeled by introducing a square well of width $2R$ with a first bound state of energy E_0. The figure is taken from a paper that analyzed a spherical well based on this one-dimensional cross section,[12] but the result for the one-dimensional and three-dimensional case is in fact the same. Denoting the tunneling rate from the left of the barrier onto the localized state as Γ_L and the tunneling rate from the localized state to the right of the barrier as Γ_R, the following expression is found for the zero bias conductance of the junction (ignoring electron spins)[††][11]:

[††]But note that in the absence of an applied bias there is no net tunnel current because tunneling in either direction, localized state to electrode or electrode to localized state, occurs at the same rate.

$$G = \frac{4e^2}{h} \frac{\Gamma_L \Gamma_R}{(E - E_0)^2 + (\Gamma_L + \Gamma_R)^2}. \tag{7.38}$$

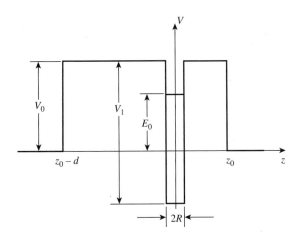

Fig. 7.14 One-dimensional potential energy barrier model for resonant tunneling. Electrons incident from the left face a barrier of height V_0 containing a localized state at energy E_0. (Reprinted with permission from Kalmeyer and Laughlin[12], Phys. Rev. B, Copyright 1987 American Physical Society. Courtesy of Professor R.B. Laughlin.)

This is a remarkable result. In the case where the incident electron has the same energy as the localized state in the barrier, $E = E_0$ and

$$G = \frac{4e^2}{h}\frac{\Gamma_L\Gamma_R}{(\Gamma_L + \Gamma_R)^2}. \tag{7.39}$$

Now let us further suppose that the structure is symmetric (i.e., the localized state is in the middle of the barrier so that $\Gamma_L = \Gamma_R$). In this case, Equation 7.39 states that the conductance becomes equal to e^2/h, half the Landauer conductance as given in Equation 7.31 (the appropriate Landauer conductance for this case because spins were ignored which accounts for the factor of 2).

Thus, at resonance ($E = E_0$), the localized state is acting like a metallic channel that connects the left and right electrodes. Note that in arriving at this result we did not specify values for Γ_L and Γ_R, merely requiring that they be equal. Apparently, this formula (Equation 7.38, first derived for nuclear scattering by Breit and Wigner) says that a resonant state half way between the Earth and Moon could mediate a metallic tunneling connection on astronomical scales! This implausible result is an illusion arising from the fact that we have ignored charging of the localized state. If the tunneling rates between the electrodes and the localized state are small enough, charge accumulation on the localized state becomes significant (see below) and, if the energy of the state depends on its charge, a Coulomb blockade results. The requirement for resonant tunneling to hold is that the conductance between any one electrode and the central state must be equal to or greater than half the Landauer conductance ([77.5/2] μS). Thus, a metallic point contact is required between the left and right electrodes and the center localized state for resonant tunneling to occur.

Viewed this way, the observation of unity transmission of electrons through a resonant structure is not at all surprising. One way to make a resonant structure between two electrodes would be to place a metal atom (being the same element as the atoms in the electrodes) in close contact with metal atoms at the apex of each electrode. In this case, resonant tunneling is nothing more than one description of electron delocalization on a lattice of metal atoms. In the limit of very tight coupling, it is possible to consider a molecule in a metallic gap as mediating resonant tunneling if states of the molecule lie close to the Fermi energy of the metal. A band structure model (cf. Section 7.6) has been solved for

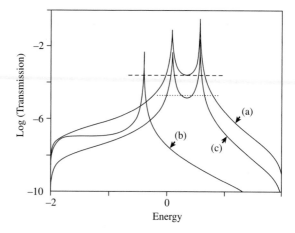

Fig. 7.15 Transmission vs. energy for a tight-binding model of resonant tunneling through a molecule bound into a gap in a one-dimensional wire. Curves A and C are for a molecule with two states lying in the conduction band of the wire with $\Gamma_L = \Gamma_R$ in curve A and $\Gamma_L = 4\Gamma_R$ in curve C. Curve B is for a molecule with just one state in the conduction band. The dashed lines show the transmission expected through the gaps to the left and the right of the molecules if no molecule is present, so the portion of the curves that lie above the dashed lines represent the enhancement in vacuum tunneling owing to resonant tunneling. (Reprinted from Lindsay et al.[13] J. Phys. Chem. 1990 with permission from the American Chemical Society.)

this problem and curves of transmission vs. energy are shown in Fig. 7.15. The peak transmissions are much larger than the vacuum tunneling rate through a gap equal to the size of the actual gap minus the length of the molecule, showing how resonance enhances tunneling.

7.12   Coulomb blockade or resonant tunneling?

We have alluded to the fact that a strong coupling between electrodes and a central particle can lead to resonant tunneling if the particle has a state at the Fermi energy, but that the process becomes more like the Coulomb blockade as the coupling between the electrodes and the central particle is made weaker. Specifically, the Coulomb blockade requires that the tunneling resistance of the contacts to the central particle exceeds twice the Landauer resistance. Why is this? Some insights can be gained by solving the time-dependent Schrödinger equation for a potential like that shown in Fig. 7.14. The results of such a simulation are shown in Fig. 7.16. In order to show the time development of the charge distribution, the electron is modeled as a Gaussian wave packet launched from the left of the resonant tunneling barriers in Fig. 7.16(a) and (e). The series of plots of charge density in Fig. 7.16(a) through (d) show the oscillations that build up on the left of the barrier as the incoming wave packet interferes with its own reflection from the barrier (7.16(b)). As time progresses (7.16(c)), the charge density on the resonant state at the center increases, continuing to increase until (7.16(c)) the amount that leaks out of the second barrier to the right reaches an equilibrium value (if the input flux of electrons continues from the left). The calculation is repeated for larger barriers in Fig. 7.16(e) through (h). A similar pattern is seen, but the oscillations in the charge density on the left and the build-up of charge on the central resonant state are much larger. The charge density on the central state continues to increase until it compensates for the larger barrier for tunneling to the right. This simulation shows how weaker coupling leads to increased charge build-up on the central particle. In these simulations, the energy of the state on the central particle was not allowed to change as that particle charged, so these simulations do not show a Coulomb blockade effect. They do show how resonant tunneling corresponds

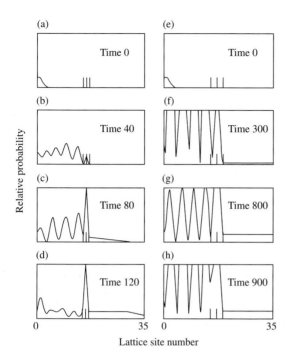

Fig. 7.16 Time development of the charge density for a wave packet incident (from the left) on a pair of barriers containing a localized resonant state. (a–d) Shows the time development for small barriers (strong coupling). (e–h) Shows the time development for larger barriers (weak coupling). (Redrawn after Lindsay et al.[14])

to the limit in which the central state mediating resonant transport does not charge significantly. In the Coulomb blockade limit, the central state acquires an entire electronic charge.

The Coulomb blockade picture represents a starting point for considering electron transfer between molecules, but the picture is significantly complicated by the role the environment plays in the case of molecular electron transfer. Electron acceptors often have their Coulomb energy lowered by significant polarization of the environment (corresponding to a large dielectric constant as the environment screens the local charge). This polarization normally arises as a consequence of slow nuclear motions of the atoms surrounding the charge center. This is a problem that we will discuss in Chapter 8 when we consider the oxidation and reduction of molecules.

7.13 Electron localization and system size

Disorder plays an important role in the electronic properties of nanostructures. This is because the impact of disorder on electron transport becomes more significant in systems of low dimensionality. This occurs in structures with nanoscale dimensions in at least one direction. Examples are two-dimensional electron gases trapped in thin sheets of semiconductor (made by molecular beam epitaxy), one-dimensional wires such as silicon nanowires, carbon nanotubes, and molecular wires (Chapter 8), or zero-dimensional quantum dots.

Experimental evidence for this behavior is found in the temperature dependence of the resistivity of various types of nanostructure. (A comprehensive description of this field is given in the book by Imry.[15]) In the free electron

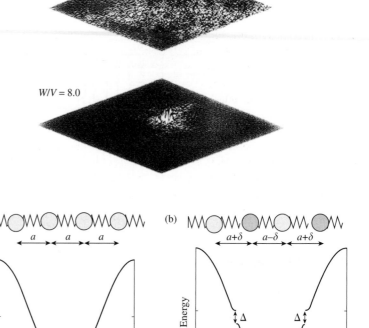

Fig. 7.17 Charge density distributions calculated for electrons in random potentials. The ratio W/V measures the width of the distribution of potential values in relation to the mean value of the potential. At $W/V = 5.5$, the electrons already show some signs of clumping, becoming strongly localized at $W/V = 8$. (Reprinted from Yoshino and Okazaki[16], J. Phys. Soc. Japan with permission from the Journal of the Physical Society of Japan.)

Fig. 7.18 Peierls distortion. (a) A linear chain with a half-filled band (one electron per unit cell of size a) can always distort (b) to double the unit cell size (every other atom shifted by an amount δ) to introduce a new Brillouin zone boundary halfway through the original Brillouin zone. The resulting gap at the new Brillouin zone boundary (Δ) lowers the electronic energy at the top part of the occupied band by $\Delta/2$. If this exceeds the elastic energy required to introduce the distortion ($\frac{1}{2}k\delta^2$), the structure spontaneously distorts into an insulator with a full band.

model, the resistivity of a material increases with increasing temperature as electron scattering increases. It is, however, experimentally observed that increasing the disorder in a system can lead to a resistivity that *decreases* with temperature. This effect is particularly pronounced in systems that are essentially one or two dimensional. The effect is a consequence of "Anderson localization." Anderson examined the problem of electronic states in a highly disordered system, carrying out perturbation theory for the limit where the width of the distribution of atomic potentials (W) is greater than the mean value of the potential itself (V), showing that this leads to localization of electrons. Computer simulations demonstrating this are shown in Fig. 7.17.[16] Results are shown for random distributions of potentials in which the ratio of the width to the distribution (W/V) is 5.5 and 8. For $W/V = 8$, the electrons are almost completely localized. To understand this result, one can view the formation of bands in periodic structures as resulting from the constructive interference of all possible scattering events in a periodic potential. The scattering that leads to constructive interference of waves propagating in allowed bands will be destroyed when the crystal is disordered.

A special type of localization occurs in one-dimensional systems, and this is important in the field of conducting polymers. The effect is called the Peierls distortion (Fig. 7.18). Figure 7.18(a) shows a one-dimensional array of atoms that have one valence electron each, so that this array should form a one-dimensional metal in which the conduction band is half filled. Now, consider a distortion of a linear uniform row of atoms that pulls every other atom together (Fig. 7.18(b)). This distortion results in halving of the Brillouin zone in wave vector space because the real space lattice is now doubled in size. Thus, the electrons that were part of a half full band are now a part of a full band (two electrons per unit cell because there are two atoms) and electrons at the top of this band belong to states distorted from a parabola by scattering at the new Brillouin zone boundary. This distortion results in a gap that lowers the energy of the electrons at the top of the new band. If the electronic energy is lowered more than the energy required to drive the distortion of the lattice, then the lattice will spontaneously distort. Thus, a chain of atoms with uniform bonds will make a transition to a chain of atoms with alternating long and short bonds, driving a transition from a metal (half-filled band) to an insulator (full band). This is the band structure explanation of the distortion that leads to bond alternation in unsaturated carbon chains.

7.14   Bibliography

C. Kittel, Introduction to Solid State Physics, 8th ed. 2004, New York: Wiley. The standard undergraduate introduction to solid state physics.

N.W. Ashcroft and N.D. Mermin, Solid State Physics. 1976, New York: Holt, Rinehart and Winston. The standard graduate solid state physics text.

Y. Imry, Introduction to Mesoscopic Physics. 1997, Oxford: Oxford University Press. A comprehensive survey of electrons in mesoscale devices. Written at a physics graduate level.

A. Nitzan, Chemical Dynamics in Condensed Phases. 2006, Oxford: Oxford University Press. Comprehensive, clear, and very mathematical account of electrons in solids from a chemical perspective.

7.15   Exercises

1. What is the resistance of a wire of 1 mm diameter and 1 m length for (a) silver, (b) copper, and (c) nichrome?
2. If the copper wire (1 mm cross diameter, 1 m length) was, instead, made up of a material with a mean time between electron collisions of 10^{-13} s at room temperature, what would its resistance be?
3. The free electron density in Na is $2.65 \times 10^{22}/cm^3$. Calculate k_F, V_F, and E_F.
4. Excitations of a free electron gas: plasmons. Consider a slab of metal of unit area that is polarized to produce an electric field, E. All the polarization occurs at the surface where we will assume that electrons protrude a distance, d, on one side (and the positive nuclei protrude a distance, d, on the other side). Assume that the charge density is n per unit volume. Gauss's law says that the flux of E ($E \cdot A$) is equal to the net charge (divided by ε_0) on the surface through which the field passes. Set up Newton's law of motion for this displaced charge and

determine the frequency of simple harmonic motion. From this, show that the frequency of plasmons (plasma oscillations) in the metal is given by

$$\omega = \sqrt{\frac{ne^2}{m\varepsilon_0}}.$$

Estimate the plasma resonance of lithium (in Hz).

5. Show that electrons at the Brillouin zone boundary have the correct wavelength to be Bragg reflected by the lattice.
6. Plot E vs. k for one dimension for which the lattice constant is 0.25 nm, with $\varepsilon_0 = 10$ eV and $\tau = -1$eV. Plot units of eV on the energy axis and nm^{-1} along the momentum axis.
7. An STM is operated with a 1 Å tunnel gap using metals of 5 eV work function. What is the resistance of the tunnel gap?
8. Calculate the charging energy of

   a. An iron atom (assuming the radius is 1.25 Å).
   b. A 100 nm diameter quantum dot.

   How are these energies altered if the atom or dot is placed in water?

9. What is the resistance of a quantum point contact that allows three electron channels to connect the electrodes on each side?
10. What is the resonant tunneling transmission of an electrode-dot-electrode system with coupling constants such that the tunneling rate on one side (left electrode to dot) is 10 times the tunneling rate on the other (right electrode to dot)? What is the resistance of this system at resonance?
11. The length of a carbon–carbon single bond is 0.154 nm and the length of a carbon–carbon double bond is 0.134 nm. Assume that the structure of polyacetylene:

is a consequence of a Peierls distortion from an average bond length of 0.144 nm and take the stiffness of the C–C bond to be 200 N/m. Estimate a lower bound on the band gap caused by the Peierls distortion.

References

[1] Abrikosov, A.A., *Fundamentals of the Theory of Metals*. 1988, Amsterdam: North Holland.
[2] Kittel, C., *Introduction to Solid State Physics*. 8th ed. 2004, New York: Wiley.
[3] Ashcroft, N.W. and N.D. Mermin, *Solid State Physics*. 1976, New York: Holt, Rinehart and Winston.
[4] Landauer, R., Electrical resistance of disordered one-dimensional lattices. Philos. Mag., **21**: 863–867 (1970).

[5] Nitzan, A., *Chemical Dynamics in Condensed Phases*. 2006, Oxford: Oxford University Press.

[6] Ruitenbeek, J.M.v., Quantum point contacts between metals, in *Mesoscopic Electron Transport*, L.L. Sohn, L.P. Kouwenhoven, and G. Schön, Eds. 1997, Amsterdam: Kluwer Academic Publishers, pp. 549–579.

[7] Rego, L.G.C., A.R. Rocha, V. Rodrigues, and D. Ugarte, Role of structural evolution in the quantum conductance behavior of gold nanowires during stretching. Phys. Rev. B, **67**: 045412-1-10 (2003).

[8] Hanna, A.E. and M. Tinkham, Variation of the Coulomb staircase in a two-junction system by fractional electron charge. Phys. Rev. B, **44**: 5919–5922 (1991).

[9] Kouwenhoven, L.P., T.H. Oosterkamp, M.W.S. Danoesastro, M. Eto, D.G. Austing, T. Honda, and S. Tarucha, Excitation spectra of circular, few-electron quantum dots. Science, **278**: 1788–1792 (1997).

[10] Knobel, R.G. and A.N. Cleland, Nanometre-scale displacement sensing using a single electron transistor. Nature, **424**: 291–292 (2003).

[11] Kastner, M.A., The single electron transistor and artificial atoms. Ann. Phys. (Leipzig), **9**: 885–894 (2000).

[12] Kalmeyer, V. and R.B. Laughlin, Differential conductance in three-dimensional resonant tunneling. Phys. Rev. B, **35**: 9805–9808 (1987).

[13] Lindsay, S.M., O.F. Sankey, Y. Li, and C. Herbst, Pressure and resonance effects in scanning tunneling microscopy of molecular adsorbates. J. Phys. Chem., **94**: 4655–4660 (1990).

[14] Lindsay, S.M., O.F. Sankey, and K.E. Schmidt, How does the scanning tunneling microscope image biopolymers? Comments Mol. Cell. Biophys., **A7**: 109–129 (1991).

[15] Imry, Y., *Introduction to Mesoscopic Physics*. 1997, Oxford: Oxford University Press.

[16] Yoshino, S. and M. Okazaki, Numerical study of electron localization in Anderson model for disordered systems: spatial extension of wavefunction. J. Phys. Soc. Japan, **43**: 415–423 (1977).

8 Molecular electronics

Molecular electronics lies at the intersection of chemistry with nanoelectronics, and many of the concepts learned in Chapter 7 will be combined with chemical concepts in this chapter. We begin by considering two examples of molecular electronic devices. The first is a molecular diode, proposed in the 1970s, though still not a commercial device. The second is the photovoltaic molecule that converts light to chemical energy, introduced as an example of chemical synthesis in Fig. 6.1. It has been successfully synthesized and could be producing electricity commercially one day. We use these as examples in this chapter.

Chemical bonding builds molecules, and we introduce the simple model of bonds developed by G.N. Lewis as a first step. We will learn something of the way that orbitals may be constructed for even quite complex molecules using the orbitals of the component atoms as a starting point. We improve on the treatment of the hydrogen molecular ion given in Chapter 2 to include the possibility that basis wavefunctions are not orthogonal and go on to give an outline of the useful *secular determinant* that comes from an energy minimization based on perturbation theory. This treatment is simplified even further using approximations that constitute the Hückel model.

The electronic properties of large molecules, especially in real environments (like solvents), at finite temperatures are still beyond the reach of computer codes in most cases, so we will learn about experimental probes of electron transfer in the form of the beautiful experimental science of electrochemistry. It is absolutely essential for both understanding and quantifying the charge-transfer properties of molecules.

The underlying theory of charge transfer in molecular systems—Marcus theory—is one of the most powerful applications of what we have already learned about the role of fluctuations in small systems in driving escape from a metastable state, and a direct extension of the Kramers theory discussed in Chapter 3. We will examine this theory in some detail and derive the Marcus expression for the activation energy for charge-transfer processes. We will use some of the ideas of Marcus theory in a discussion of charge transfer in organic solids, particularly the concept of hopping conductance. Charge trapping is particularly interesting in one-dimensional (1-D) systems, and we briefly discuss conducting polymers. We refer to "charge" transfer because the ideas we will discuss can usually be applied to both electrons and holes. However, when we use the word "donor" we mean a donor of electrons. Likewise, the word "acceptor" means an acceptor of electrons.*

We end with a brief survey of *single molecule electronics*. This is probably what is most often thought of as "molecular electronics," though it is a relatively undeveloped field compared to bulk organic electronic materials.

*In discussing hydrogen bonds the convention is reversed: a "donor" in hydrogen-bonding donates *protons*.

8.1 Why molecular electronics?

Molecular electronic components are perhaps the ultimate designed functional nanostructures. As the feature sizes in conventional electronic circuitry based on silicon chips become smaller and smaller, the atomic structure and atomic scale defects of silicon become an issue. For example, the probability of finding even one dopant atom in a nanometer-sized cube becomes so small that no two nanometer-sized cubes would be the same. The goal of molecular electronics is to approach electronic component design from the atomic scale upward, using molecules that are designed to carry out an electronic circuit function in their own right. The component elements of electronic circuits would be built at the smallest possible scale: that of individual molecules.[1,2]

Organic conductors have a long history, dating at least as far back as a 1911 paper on "organic amalgams," in which McCoy and Moore stated that "It is possible to prepare composite metallic substances from non-metallic constituent elements."[3] Current interest in the field was sparked by the publication of a seminal paper by Aviram and Ratner.[4] They proposed a structure for a molecular diode (Fig. 8.1). This proposal helped create molecular electronics as we think of it today—single molecules acting as active devices in a circuit. The molecule invented by Aviram and Ratner consists of an electron accepting group (TCNQ) coupled to an electron donating group (TTF) via a molecular bridge that isolates the two groups, but through which electrons can tunnel. An electron donor is a molecular group with a highest occupied electronic state that lies at a high energy (closer to the vacuum). An electron acceptor is a molecular group that is easily charged, having a low lying (further below the vacuum) unoccupied state. Thus the "downhill" direction of energy transfer is always from donor to acceptor (for electrons—an alternative view is to think of holes flowing uphill from acceptor to donor). If a metal electrode injects a charge into the donor, it will flow naturally to the acceptor, where it can pass into a second electrode of lower potential. Charge is not transported in the opposite direction, because of the probability of transfer from an acceptor to a donor is tiny at room temperature. Thus the structure should act as a molecular diode, passing current in one direction, but not the other.

Proposals like this have fueled the notion of synthesizing computers by chemistry, with billion dollar fabrication facilities for computer chips replaced by test tubes. We shall see in this chapter that we are a long way from this vision.

(a) (b)

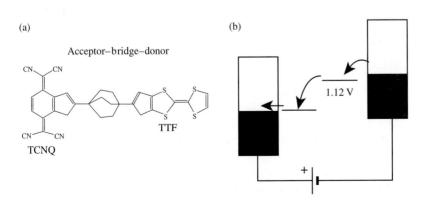

Fig. 8.1 (a) Chemical structure of an acceptor–bridge–donor molecule proposed as a molecular rectifier. The electron acceptor is tetracyanoquinodimethane (TCNQ) and the donor is tetrathiofulvalene (TTF). The bridge consists of three parallel chains of methylene ($-CH_2-$) groups. If sandwiched between electrodes (b) added electrons would prefer to move from the donor to the acceptor, resulting in a diode-like response. The energy difference between donor and acceptor levels is discussed in Section 8.10. (Figure (a) is adapted from Aviram and Ratner.[3])

That said, molecular electronic materials are already widely used in electronics. Examples are liquid crystals in display panels, organic conducting materials in light emitting diodes, and in transistors, particularly in applications where the electronics are printed onto flexible plastic surfaces. At the other end of the scale, there has been tremendous progress in understanding charge transport in individual molecules tethered to electrodes. To discuss this field, we first need to address the following questions:

- What are the electronic states occupied by electrons on complex, many atom molecules?
- How do molecules charge to form ions?
- What makes a "donor" and what makes an "acceptor" of electrons?
- How do charges move through a solid molecular material?
- Can a linear molecule make a wire (a "molecular wire")?
- How are electrons transferred between metal electrodes and molecules, and how do the energies of available states on molecules align with the Fermi level of metals or semiconductors?

Central to these questions is the issue of how a molecule becomes charged and what makes a particular molecule a "donor" or "acceptor." Even without introducing electrodes and conventional circuitry, some amazing devices can be constructed by coupling donors and acceptors together. One such example is the "Light-to-Charge" converter shown in Fig. 6.1. This is a molecule designed to simulate part of the process of photosynthesis.[5] The middle part of the molecule is a porphyrin pigment that absorbs incident light. The light is absorbed because the energy of the incident photons matches the energy required to promote an electron from the ground state of the molecule to an excited state. This excited state on the molecule can be thought of as a "molecular exciton," consisting of a bound electron–hole pair with the electron orbiting the positive charge left behind when it was promoted into the excited state. This state does not live for long because the electron falls back into the ground state, emitting a photon (fluorescent emission) and leaving the molecule back in its ground state. However, the molecular machine shown in Fig. 6.1 is designed to prevent this recombination. The C_{60} (Bucky ball) on the right is an excellent electron acceptor, whereas the long chain molecule on the left (a carotene molecule) is an excellent electron donor. Thus, when the excited state of the porphyrin pigment is created by the absorption of a photon, the electron in the excited state transfers very rapidly onto the Bucky ball while the remaining hole on the porphyrin molecule is rapidly filled with an electron donated by the carotene. In consequence, this molecular device serves to create a long-lived charge-separated state when illuminated with visible light. One day synthetic devices like this might power our world with clean, renewable energy, a goal for the nanodesigner that is at least as worthy as test tube fabrication of computers.

8.2   Lewis structures as a simple guide to chemical bonding

We understand, as a result of our simple discussion of bonding in Chapter 2, that atoms share electrons because of the lowering of quantum mechanical kinetic energy that comes about when the electron is placed in a larger "quantum box" consisting of the pair of bonded atoms. This net attractive interaction is balanced by the Coulomb repulsion between the atomic nuclei, and, when many electrons are present, the constraints imposed by the Pauli exclusion principle. This is a useful start, but it will not get us far in understanding the electronic properties of molecules. A remarkably simple qualitative model was invented by G.N. Lewis, based on some rules that most chemical bonds appear to obey. In elaborating these rules, we will take as given facts like the number of valance electrons available to a particular atom (a nonobvious number discussed further in Section 8.5).

Multielectron atoms can have two electrons in a given orbital, and Lewis observed that molecules tend to populate the states associated with bonds so as to give each atom participating in the bond a "closed shell" electronic count. For bonds involving hydrogen atoms, this is two electrons, corresponding to the $2s^2$ state of He. For bonds involving p-states, this is eight electrons (six for the three p-states and two for the s-states in the outer shell). Much like the lowering of electronic energy that comes from filling atomic states (to make the noble gasses), molecular states tend to have the lowest energy when the count of the "shared" electrons corresponds to a full shell for each atom. Lewis structures tell us little about the shape of a molecule, but they do predict how electrons are shared in bonds.

Each "full" bonding state is shown as a line connecting a pair of atoms:

$$\text{H}\text{---}\text{H}$$

Molecules often form more than one type of bonding state (i.e., a new occupied electronic state arising from the interactions and lowered in energy relative to the unperturbed atomic states). Thus, atoms may be shown with two (a "double bond") or even three (a "triple bond") bond-lines connecting them:

$$\text{H}_2\text{C}\!=\!=\!\text{CH}_2 \qquad \text{HC}\!\equiv\!\equiv\!\text{CH}$$

Double bonds hold four electrons and triple bonds hold six electrons. The remaining valence electrons, not involved in forming bonds, are shown as dots in completed Lewis structures.

Armed with these rules, we can begin to construct Lewis structures. Let us start with the simplest cases of H_2 and Cl_2 molecules:

$$\text{H}\!\cdot\!+\!\cdot\!\text{H} \longrightarrow \text{H}\!:\!\text{H} \longrightarrow \text{H-H}$$

$$:\!\ddot{\text{Cl}}\!\cdot\!+\!\cdot\!\ddot{\text{Cl}}\!: \longrightarrow :\!\ddot{\text{Cl}}\!:\!\ddot{\text{Cl}}\!: \longrightarrow :\!\ddot{\text{Cl}}\text{-}\ddot{\text{Cl}}\!:$$

Each atom now has a "full" shell, counting the two shared electrons in each bond. Moving on to the slightly more complicated ammonia molecule, the Lewis structure is

$$\ddot{\underset{\cdot\cdot}{\text{N}}}\!\cdot \; + \; 3\text{H}\!\cdot \; \longrightarrow \; \text{H -} \overset{\cdot\cdot}{\text{N}} \text{- H} \atop \text{H}$$

Three of the valence electrons are shared with the three hydrogen atoms to make three single bonds. The remaining electrons on the nitrogen form a localized concentration of negative charge referred to as a "lone pair." Lone pairs tend to be reactive and much of the interesting chemistry of nitrogen compounds originates with this lone pair of electrons.

Let us now turn to double bonds, taking CO_2 as an example. There is no way to put two oxygen atoms together with one carbon using single bonds. The solution is a pair of double bonds:

$$:\ddot{O}: + \cdot\dot{C}\cdot + :\ddot{O}: \longrightarrow \ddot{O}{=}C{=}\ddot{O}$$

The two double bonds hold eight electrons, contributing the carbon octet, while each double bond contributes half of each of the oxygen's quartets. We could have achieved octets with one single bond to one oxygen atom and one double bond to the other, but this would have left the oxygen atoms with more or less than their electron "share" (of six valence electrons each). This notion of how electrons are shared is quantified by the concept of a "formal charge." The formal charge is given by the difference between the number of valence electrons of an atom and the number of electrons actually allocated to it in a Lewis structure, counting half the bonding electrons. Each of the oxygen atoms in CO_2 has four electrons in lone pairs plus $\frac{1}{2}$ of the four electrons in the double bond. This is six electrons, exactly equal to the number of valence electrons in oxygen, so the formal charge is zero. Lewis structures are drawn so as to minimize formal charge. When a formal charge is nonzero, the extra electrons are allocated to the more electronegative atom. The existence of a formal charge is a good predictor of the formation of a polar bond.

Finally, we will take a look at a triple bond. The only way to generate an octet on the carbon in hydrogen cyanide (HCN) is to invoke a C-N triple bond:

$$H{-}C{\equiv}N:$$

The dangerous reactivity of this molecule arises with the lone pair on the nitrogen and the high charge in the triple bond. This makes HCN bind the hemoglobin molecule in blood much more strongly than oxygen, with tragic consequences for the cyanide inhaler.

So here again is a summary of the steps taken to draw Lewis structures:

1. Find the sum of all the valence electrons in each of the atoms in the molecule. If it is anion, subtract one, and if it is a cation, add one. For example:

$$PCl_3 \qquad 5 + 3 \times 7 = 26$$

2. Choose the *least* electronegative atom and put it at the center of the structure, connecting the other atoms by single bonds:

$$Cl{-}\underset{\underset{\displaystyle Cl}{|}}{P}{-}Cl$$

3. Complete octets for the outer atoms—remember that the bonds count for two electrons each:

$$
\begin{array}{c}
\ddot{\text{:Cl:}} \\
| \\
\text{:Cl} \!-\! \text{P} \!-\! \text{Cl:}
\end{array}
$$

4. Complete the octet for the central atom:

$$
\begin{array}{c}
\ddot{\text{:Cl:}} \\
| \\
\text{:Cl} \!-\! \text{P} \!-\! \text{Cl:}
\end{array}
$$

5. If you run out of electrons before you form an octet on the central atom, keep forming multiple bonds until you do:

$$
\text{H} \!-\! \text{C} \!-\! \text{N:} \quad \longrightarrow \quad \text{H} \!-\! \text{C} \!\equiv\! \text{N:}
$$

6. Now assign formal charges. The formal charge is the valence number minus (the number of electrons in lone pairs plus half the number of electrons in bonds). For example, the formal charges on each atom in

$$
\text{: O} \!-\! \text{C} \!\equiv\! \text{O:}
$$

are $6 - 7 = -1$, $4 - 4 = 0$, and $6 - 5 = +1$.

The formal charges in

$$
\text{O} \!=\! \text{C} \!=\! \text{O}
$$

are $6 - 6 = 0$, $4 - 4 = 0$, and $6 - 6 = 0$. The correct structure is the one that minimizes formal charges, which is why the second structure was shown earlier as an example of double bonds. If a nonzero formal charge is required, make sure that the negative charge is placed on the most electronegative atom.

These simple rules give a reasonably good account of bonding in a wide variety of compounds. The approach even works with ions. Positive ions are represented with one electron less than the valence number and negative ions with one electron more. Once again, the goal is to complete octets on all the atoms in the molecule. The Lewis structures are not iron-clad predictors of molecular bonding, because some elements do not follow the octet rule, but they are a much easier guide to molecular structure than solving the Schrödinger equation!

Linus Pauling took Lewis structures to another level of sophistication by introducing the idea of *resonance*. Lewis structures can be ambiguous simply because the molecule has a high symmetry, so there is no "right" way to draw the structure. Benzene, C_6H_6, is a good example:

Which structure is correct? The one on the left or the one on the right? Each carbon satisfies the octet rule and each hydrogen shares two electrons. Pauling realized that these are examples of *degenerate states* in the quantum mechanical sense, so that new states must be calculated using the degenerate states as a basis in degenerate perturbation theory (see Chapter 2.20). In this case, the new states are delocalized over the entire benzene ring. The bonding is described as *aromatic* and not shown as alternating double and single bonds.[†] In organic chemistry, carbons are not usually shown (just the bonds). Hydrogens are also omitted, on the understanding that an undercoordinated carbon makes up its bonding with hydrogen atoms. With these conventions, the two Lewis structures for benzene and the resultant aromatic structure are drawn as follows:

[†] Systems with alternating double and single bonds that result in delocalization are referred to in general as "conjugated".

Aromatic bonds are an important stabilizing factor in organic molecules, and we will discuss them further in more mathematical terms. But a significant amount of chemical knowledge comes from the simple quantum-chemistry-in-pictures approach given by Lewis structures combined with the idea of resonance.

8.3   The variational approach to calculating molecular orbitals

In this section, we will review the variational approach to building up molecular orbitals from atomic orbitals. At the end of the process, we will draw diagrams of the energy levels associated with the various types of orbitals. Knowing how these energy levels are filled with electrons will often be all that we need in order to understand the electronic properties of the molecules.

The variational formulae are derived as follows: Multiplying the time-independent Schrödinger equation (Equation 2.31) from the left by ψ^* and integrating both sides yields the following result for the energy of a stationary state:

$$E_0 = \frac{\langle \psi \,|H|\, \psi \rangle}{\langle \psi | \psi \rangle},\qquad(8.1)$$

where we have deliberately *not* used the fact that eigenstates of the Schrödinger equation satisfy $\langle \psi_n | \psi_m \rangle = \delta_{nm}$. If we did not know the eigenstates for a system, we might, nonetheless, guess a trial function for a particular state, ϕ.

If we use Equation 8.1 to calculate an "energy," E_ϕ, we would always get a value greater than (or, at best, equal to) E_0 because E_0 is the ground state energy, the lowest allowed by quantum mechanics, given the intrinsic quantum mechanical kinetic energy of the electron. This self-evident statement can be proved by expanding ϕ in a basis in terms of which the correct wavefunction, ψ, can also be expressed. All the terms that differ from those that make up ψ contribute positive corrections to the total energy calculated this way. Only when $\phi = \psi$ is $E_\phi = E_0$. Of course, in evaluating (8.1) for the trial wavefunction, we have to keep the denominator because for the trial wavefunctions $\langle \phi_n | \phi_m \rangle \neq \delta_{nm}$ in general.

Starting with a trial wavefunction

$$\phi = \sum_i c_i f_i, \tag{8.2}$$

the optimum values for the coefficients, c_i, of the functions, f_i, will be found from minimizing the calculated energy with respect to the coefficients, i.e.,

$$\frac{dE_\varphi}{dc_i} = \frac{\langle \sum_i c_i f_i | H | \sum_i c_i f_i \rangle}{\langle \sum_i c_i f_i | \sum_i c_i f_i \rangle} = 0. \tag{8.3}$$

To keep it simple, we will consider a trial wavefunction consisting of just two functions, f_1 and f_2:

$$\phi = c_1 f_1 + c_2 f_2. \tag{8.4}$$

Inserting this into (8.1) and using the definition $\langle f_1 | H | f_2 \rangle = H_{12} = H_{21} = \langle f_2 | H | f_1 \rangle$

$$E_\phi = \frac{c_1^2 \langle f_1 | H | f_1 \rangle + 2c_1 c_2 H_{12} + c_2^2 \langle f_2 | H | f_2 \rangle}{c_1^2 \langle f_1 | f_1 \rangle + 2c_1 c_2 \langle f_1 | f_2 \rangle + c_2^2 \langle f_2 | f_2 \rangle}. \tag{8.5}$$

In Chapter 2, we exploited the symmetry of the problem to invoke the relation $H_{12} = H_{21}$, but it is, in fact, a general property of functions that satisfy an eigenvalue equation. In order to have real eigenvalues, the matrix elements of the Hamiltonian must be Hermitian (i.e., $H_{ij} = H_{ji}$). Because f_1 and f_2 are functions (not operators), it is trivially true that $\langle f_1 | f_2 \rangle = \langle f_2 | f_1 \rangle$. The integrals in the denominator will have values other than 1 or 0 if f_1 and f_2 are not orthogonal functions. Defining these *overlap integrals* by

$$S_{ij} = \langle f_i | f_j \rangle \tag{8.6}$$

and using a similar notation for the matrix elements of the Hamiltonian, Equation 8.5 becomes

$$E_\phi = \frac{c_1^2 H_{11} + 2c_1 c_2 H_{12} + c_2^2 H_{22}}{c_1^2 S_{11} + 2c_1 c_2 S_{12} + c_2^2 S_{22}}. \tag{8.7}$$

The process of taking the derivative with respect to c_1 or c_2 is simplified by multiplying both sides by the denominator and using the chain rule to take the derivative of the left-hand side. For example, differentiating with respect to c_1:

$$(2c_1 S_{11} + 2c_2 S_{12})E_\phi + \frac{dE_\phi}{dc_1}(c_1^2 S_{11} + 2c_1 c_2 S_{12} + c_2^2 S_{22}) = 2c_1 H_{11} + 2c_2 H_{12}.$$

For a minimum, we require that $\frac{dE_\phi}{dc_1} = 0$, leading to

$$(c_1 S_{11} + c_2 S_{12})E_\phi = c_1 H_{11} + c_2 H_{12}.$$

Repeating the process but differentiating with respect to c_2 produces a second equation. The pair of linear equations to be solved for c_1 and c_2 are

$$\begin{aligned}
\left(H_{11} - E_\phi S_{11}\right)c_1 + (H_{12} - E_\phi S_{12})c_2 = 0 \\
(H_{12} - E_\phi S_{12})c_1 + (H_{22} - E_\phi S_{22})c_2 = 0
\end{aligned} \tag{8.8}$$

These equations have a nontrivial solution only when the secular determinant is zero:

$$\begin{vmatrix} H_{11} - E_\phi S_{11} & H_{12} - E_\phi S_{12} \\ H_{12} - E_\phi S_{12} & H_{22} - E_\phi S_{22} \end{vmatrix} = 0. \tag{8.9}$$

The variational approach for two basis states has been turned into a problem of solving the quadratic equation for E_ϕ produced by multiplying out the elements of the secular determinant (8.9). There are, in consequence, two values for E_ϕ corresponding to the bonding and antibonding states. Inserting these two values into the coupled equations, 8.8, for c_1 and c_2 produces the two (approximate) wavefunctions corresponding to the bonding and antibonding states. At this point, the approximations used in Section 2.20 can be seen as equivalent to setting $S_{ij} = \delta_{ij}$.

More complex problems, involving a larger number of basis states, follow this pattern. With N basis states, the result is an $N \times N$ secular determinant that produces a polynomial of degree N for E_ϕ. Consequently, there are N roots giving N values of E_ϕ. When each of these roots is put into the N coupled equations for the N coefficients c_i, the wavefunction corresponding to the particular root is found.

8.4   The hydrogen molecular ion revisited

We will illustrate the use of linear combinations of atomic orbitals (LCAO) for molecular bonds by working through the problem of the hydrogen molecular ion (first treated in Section 2.20) again. To simplify what follows we work in *atomic units*. This is a self-consistent system of units that sets each of the following quantities equal to unity by definition:

- e (the charge on the electron),
- m_e (the mass of the electron),
- a_0 (the Bohr radius of the hydrogen atom), and
- $1/4\pi\varepsilon_0$.

Doing this makes the equations much simpler (but keeping track of dimensions can be tricky). Atomic units are described in Appendix J.

A minimal basis for the molecular orbital of H_2^+ is the two hydrogen 1s atomic orbitals, which, in atomic units are

$$\phi_a = C \exp(-r_a)$$
$$\phi_b = C \exp(-r_b),$$

(8.10)

where C is a normalization factor and r_a and r_b are the distances of the electron from the two nuclei, expressed in units of the Bohr radius.

A trial wavefunction is

$$\psi = c_a \phi_a + c_b \phi_b$$

(8.11)

with a corresponding probability density

$$|\psi|^2 = c_a^2 \phi_a^2 + c_b^2 \phi_b^2 + 2 c_a c_b \phi_a \phi_b.$$

(8.12)

We could proceed by evaluating the secular determinant (8.9) but symmetry can often be used to determine a relationship between the coefficients in the trial wavefunction, and this is especially straightforward here. From the inversion symmetry of the H_2^+ ion, we expect that $c_a^2 = c_b^2$, which gives $c_a = \pm c_b$, and thus we have two molecular orbitals, given by

$$\psi_g = c_a(g)(\phi_a + \phi_b) \equiv \sigma_{1s}$$
$$\psi_u = c_a(u)(\phi_a - \phi_b) \equiv \sigma_{1s}^*$$

(8.13)

where the suffixes g and u are used to denote symmetric and antisymmetric orbitals, respectively. By symmetric and antisymmetric, we mean that changing the labels of the atoms (a for b and vice versa) either does not change the sign of the wavefunction (symmetric or even case) or does change the sign of the wavefunction (antisymmetric or odd case). (The labels g and u come from the German words gerade and ungerade meaning "even" and "odd".) In modern molecular orbital notation, these two orbitals are called σ_{1s} and σ_{1s}^*.

For the wavefunctions to be normalized, we require

$$\int |\psi_g|^2 \, dv = c_a(g)^2 \int (\phi_a^2 + \phi_b^2 + 2\phi_a\phi_b) \, dv = 1$$

(8.14)

and

$$\int |\psi_u|^2 \, dv = c_a(u)^2 \int (\phi_a^2 + \phi_b^2 - 2\phi_a\phi_b) \, dv = 1.$$

(8.15)

If the basis functions, ϕ_a and ϕ_b, are normalized, then

$$\int |\phi_a|^2 \, dv = \int |\phi_b|^2 \, dv = 1.$$

(8.16)

Using the definition of the overlap integral, S_{ab}, (8.6)

$$S_{ab} = \int \phi_a^* \phi_b \, d\mathbf{v}, \tag{8.17}$$

the Equations 8.15 and 8.16 become

$$c_a(g)^2(2 + 2S_{ab}) = 1 \Rightarrow c_a(g) = (2 + S_{ab})^{-1/2} \tag{8.18}$$

and

$$c_a(u)^2(2 - 2S_{ab}) = 1 \Rightarrow c_a(u) = (2 - S_{ab})^{-1/2}. \tag{8.19}$$

Now we know the constants in Equation 8.13, we can write down normalized LCAO approximations to the molecular orbitals:

$$\sigma_{1s} = (2 + S_{ab})^{-1/2}(\phi_a + \phi_b), \tag{8.20}$$

$$\sigma_{1s}^* = (2 - S_{ab})^{-1/2}(\phi_a - \phi_b). \tag{8.21}$$

These functions are plotted in Fig. 8.2. The results of exact calculations using the two-proton potential in the Schrödinger equation are shown as dashed lines for comparison.

The σ notation has a wider application because the molecular orbitals found in this simple case share features found in more complex molecules:

- There are as many molecular orbitals for the combined system as there are atomic orbitals for the atomic components. In this case, two 1s oribtals combine to make a σ and a $\sigma*$ orbital.
- The two orbitals combine because their symmetry allowed them to: being both s orbitals, overlap occurs as the atoms are brought together. In some situations, the phase of the wavefunctions is such that the amount of negative wavefunction in the overlap exactly cancels the amount of overlap with a wavefunction of positive sign. Such pairs of orbitals do not contribute to bonding; they form nonbonding molecular orbitals.
- The molecule has a different symmetry from the atoms. Atomic s-states do not alter their properties for rotations about the atom center because they

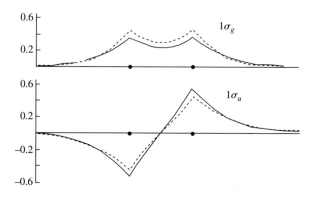

Fig. 8.2 The LCAO wavefunctions (solid lines) and exact wavefunctions (dashed lines) of H_2^+. In modern notation, the molecular orbitals would be labeled σ_{1s} (labeled $1\sigma_g$ here) and σ_{1s}^* (labeled $1\sigma_u$ here). (Redrawn from Slater.[6])

are spherically symmetric. The molecule made from the two 1s atoms has cylindrical symmetry. It does not alter its properties when rotated about the axis connecting the two atoms (Fig. 8.2 shows only a slice through this cylinder). Even very complex molecular orbitals are called σ orbitals if they have this property of being invariant under rotations about the axis connecting two bonded atoms.

- The bonding (σ) orbital concentrates charge between the atoms. The antibonding ($\sigma*$) orbital places a node between the atoms.

Whether or not a particular combination of atomic orbitals form bonding or antibonding orbitals depends on the nature of the orbitals. For example, we shall see that the symmetry of the molecular orbital that forms bonds from p orbitals is different from the σ orbitals because of the sign change of the p orbital on inversion.

We turn next to the energies of the states associated with the σ orbitals. These are found by evaluating the Hamiltonian matrix element of each of the states:

$$E_{g,u} = \int \psi_{g,u}^* \hat{H} \psi_{g,u}\, \mathrm{d}\mathbf{v} = (2 \pm 2S_{ab})^{-1} \int (\phi_a^* \pm \phi_b^*)\hat{H}(\phi_a \pm \phi_b)\, \mathrm{d}\mathbf{v}. \tag{8.22}$$

where H is the Hamiltonian operator for the molecule (the Hamiltonian in atomic units is given in Appendix J). The matrix elements are just energies and defining $H_{aa} = \int \phi_a^* \hat{H} \phi_a\, \mathrm{d}\mathbf{v}$ and $H_{ab} = \int \phi_a^* \hat{H} \phi_b\, \mathrm{d}\mathbf{v}$, Equation 8.22 becomes

$$E_{g,u} = 2(2 \pm 2S_{ab})^{-1}(H_{aa} \pm H_{ab}). \tag{8.23}$$

For H_2^+, the Hamiltonian is relatively simple, consisting of the kinetic energy operator of the electron, electrostatic attraction of the electron to the two nuclei, and electrostatic repulsion between the two nuclei. In atomic units, this is

$$\hat{H} = -\frac{1}{2}\nabla^2 - \frac{1}{r_a} - \frac{1}{r_b} + \frac{1}{R}, \tag{8.24}$$

where ∇^2, the "Laplacian operator," means the second derivative taken in polar coordinates. The integrals H_{aa}, H_{ab}, and S_{ab} can be evaluated analytically as functions of R, the distance between the two nuclei. This is the subject of Problem 4. Carrying out the integrals (following Problem 4) gives

$$H_{aa} = -\frac{1}{2} + e^{-2R}\left(1 + \frac{1}{R}\right), \tag{8.25}$$

$$H_{ab} = e^{-R}\left(\frac{1}{R} - \frac{1}{2} - \frac{7R}{6} - \frac{R^2}{6}\right), \tag{8.26}$$

$$S_{ab} = e^{-R}\left(1 + R + \frac{R^2}{3}\right). \tag{8.27}$$

By placing each of these terms into Equation 8.23, the interaction energy between the two atoms is obtained as a function of the distance between them.

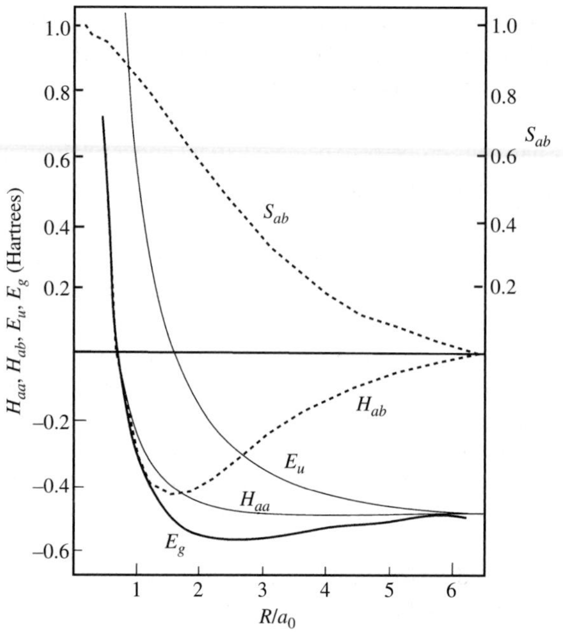

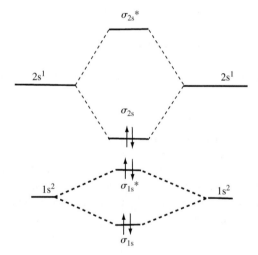

Fig. 8.3 The functions H_{aa}, S_{ab}, and H_{ab} and LCAO MO energies plotted against internuclear distance. The energies are in Hartrees and the distances are in Bohr radii. The overlap is dimensionless. (Redrawn from Slater.[6])

Fig. 8.4 Bonding for Li_2. The occupied σ_1 orbitals contribute nothing to bonding energy because both bonding and antibonding states are equally occupied. Only states where the occupation of bonding orbitals exceeds the occupation of antibonding orbitals contribute to bonding. This is why valence states are important and core electrons can usually be ignored.

A plot of the contributions to the total electronic energies in the two orbitals is shown in Fig. 8.3. The energy of the σ_{1s}^* state (E_u on the figure) increases as the two atoms approach. The energy of the σ_{1s} state (E_g on the figure) has a distinct minimum at the equilibrium bonding distance near $2a_0$. The form of the plot is similar to the Lennard-Jones 6-12 potential introduced in Chapter 2.

The bonding of other atoms with s orbital valence electrons is similar, except that an exchange interaction adds to the repulsion in molecules composed of multielectron atoms. A diagram of the energies of the molecular orbitals for Li_2 is shown in Fig. 8.4. Li has the electronic configuration $1s^2 2s^1$. The corresponding molecular configuration is $\sigma_{1s}(2)\,\sigma_{1s}^*(2)\,\sigma_{2s}(2)$, where the number of electrons in each orbital is shown in parentheses. Note that only the $\sigma 2$ orbitals contribute to bonding. There are as many electrons in the (increased energy) σ_1^* states as in the (decreased energy) $\sigma 1$ states.

8.5 Hybridization of atomic orbitals

Our discussions thus far have given us no insight into the spatial pattern of bonding in a molecule more complex than one consisting of two s-state atomic orbitals. Nor have they given precise definition to what we mean by a "valence electron." In order to illustrate the problem, consider the carbon atom, with electronic configuration $1s^2\, 2s^2 2p_x^{\,1} 2p_y^{\,1}$. The two highest occupied orbitals are p-states. The 2s electrons reside in a full s-shell. Therefore, one might expect carbon to be two valent. But it is not. The carbon atom usually forms *four* bonds, tetrahedrally coordinated about the carbon atom. We are misled when we think of the carbon in a molecule (like methane, CH_4, for example) as "atomic" carbon. It is not. The solution of the Schrödinger equation for a carbon atom surrounded by four hydrogen atoms is different from the solution of the Schrödinger equation for an isolated carbon. While this statement is trivially true, it does not give much insight into the formation of molecular bonds. It is much more constructive to think of the way in which atomic states morph into molecular states as the interaction between the atoms is "turned on." This process is referred to as *hybridization* and it can be described by perturbation theory, much as described above. Put another way, methane, CH_4, is a compound formed when carbon reacts with hydrogen. The four bonds must all be the same, so they cannot come from three p orbitals and one s orbital. The states must mix. The way in which states contribute to the final wavefunction depends on energy, as described by Equation 2.91. The closer in energy that atomic orbitals are, the more they will need to be mixed together to describe the state of a perturbed atom. As made clear by Fig. 2.18, states corresponding to a given principal quantum number are closer in energy than states associated with a different shell. Accordingly, the valence states of an atom are generally quite well described by mixtures of the states in the outermost shell. Furthermore, the lower lying shells will not contribute to the bond energies if they are fully occupied (Fig. 8.4). In the case of carbon, we are concerned with the 2s and 2p states. The states of molecules comprising carbon that contribute to the bonding energy generally correspond to the four states made from a 2s and three 2p states. The resulting mixed states are referred to as sp^3 hybrids. The appropriate LCAO for carbon is

$$\psi_1 = \frac{1}{2}(2s + 2p_x + 2p_y + 2p_z)$$

$$\psi_2 = \frac{1}{2}(2s - 2p_x - 2p_y + 2p_z)$$

$$\psi_3 = \frac{1}{2}(2s + 2p_x - 2p_y - 2p_z) \tag{8.28}$$

$$\psi_4 = \frac{1}{2}(2s - 2p_x + 2p_y - 2p_z)$$

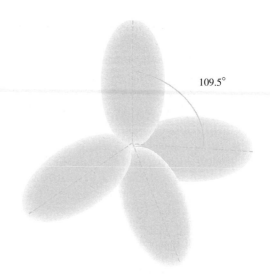

109.5°

Fig. 8.5 Illustrating sp³ hybrid orbitals in carbon. These are not states in isolated carbon but rather represent the way the atomic orbitals recombine to form molecular orbitals when carbon bonds with other atoms.

Drawing unit vectors along the $x, y,$ and z directions shows that these four orbitals point to the corners of a tetrahedron, as illustrated in Fig. 8.5. These orbitals are not those of the final molecule. In the case of methane, they would serve as a basis for calculating the mixture with four hydrogen 1s states. The notion of hybrid orbitals is useful though, because it describes what the atom is likely to do when it forms a molecule. sp³ Hybridization occurs in diamond and many molecules. Carbon in graphite hybridizes quite differently. In graphite, the atoms are threefold coordinated as described by the three states that arise from hybridizing an s- and 2p states—sp² hybridization.

The difference between atomic orbitals and molecular orbitals can be considerably more complicated than in carbon. For example, the electronic configuration of oxygen is $1s^2 2s^2 2p_x^2\, 2p_y^1 2p_z^1$ so it would seem obvious that the $2p_y$ and $2p_z$ orbitals would combine with the 1s states of hydrogen to form a water molecule. If this were really the case, the angle H-O-H would be 90°, rather than the observed 104.5°. This lies somewhere between the 90° expected for 2p orbitals and the 120° expected for sp² hybrid orbitals. The reason for the discrepancy lies with the extra charge density associated with the two remaining lone pairs from the second shell, introducing an interaction that requires sp hybridization.

8.6   Making diatomic molecules from atoms with both s- and p-states

p-states have different symmetry from s-states. They have a node at the atomic center and lobes at each side with the sign of the wavefunction changing from one lobe to the other. We consider first states that are made from overlapping p_z states chosen so that the z-axis is the molecular axis as shown in Fig. 8.6. The resulting states have cylindrical symmetry and are thus referred to as σ_{2p} and σ_{2p}^*.

The p_x and p_y states add to give charge density that lies above and below the molecular axis, and so they do not have cylindrical symmetry. The result

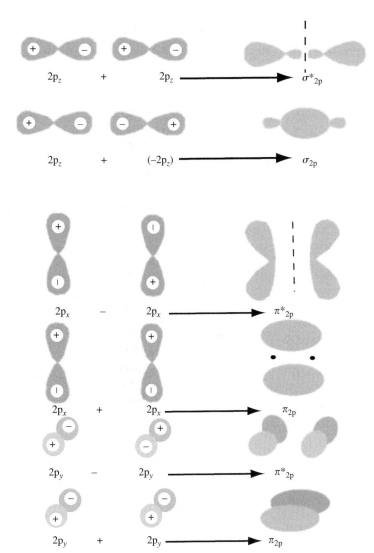

Fig. 8.6 Addition of p-states along the molecular axis. The sign change across lobes leads to the $2p_{za} - 2p_{zb}$ state being the bonding orbital. Both bonding and antibonding orbitals have σ symmetry.

Fig. 8.7 Addition of p_x and p_y states to form π and π^* states.

is molecular orbitals of different symmetry, denoted as π states. The formation of these states is illustrated in Fig. 8.7.

The approximate ordering of the energies of these states is as follows: The σ_{1s} are the lowest lying (Fig. 8.4). The strong end-on interaction of the p_z states results in a large split between the σ_{2p} and σ_{2p}^* energies, with the result that the σ_{2p} is the lowest energy state formed from p orbitals. The π_{2p} states are doubly degenerate (there are equivalent energies from the p_x and p_y contributions) and come next, followed by the π_{2p}^* states, with the σ_{2p}^* being highest in energy. This ordering turns out to be correct for O_2 and F_2. A molecular orbital energy diagram (Fig. 8.8) shows how the atomic states evolve into molecular energy levels.

The splitting in Fig. 8.8 neglects interactions between s- and p-states, so is not correct for hybridized orbitals. s-p Hybridization results in significant changes to the energy level scheme just discussed, so the energy levels of

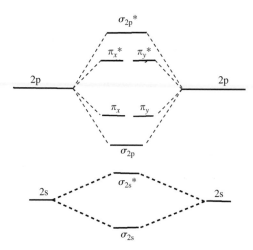

Fig. 8.8 Molecular orbital energy level diagram for a diatomic molecule composed of atoms with occupied 2p valence orbitals, and no s–p hybridization (e.g., O_2 and F_2).

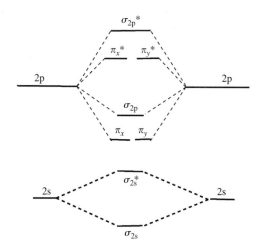

Fig. 8.9 Molecular orbital energy level diagram for a 2p diatomic molecule with s–p hybridization (e.g., B_2, C_2, and N_2).

B_2, C_2, and N_2 are ordered as shown in Fig. 8.9. Nitrogen has five valence electrons, so, according to Fig. 8.9, the valence states of N_2 are filled as follows: $\sigma_{2s}(\uparrow\downarrow), \sigma_{2s}^*(\uparrow\downarrow), \pi_{2p}(\uparrow\downarrow, \uparrow\downarrow), \sigma_{2p}(\uparrow\downarrow)$ to accommodate all 10 electrons from the valence states of the two atoms.

The oxygen molecule (Fig. 8.8) is interesting, because its 16 electrons (4 in the σ_{1s} and σ_{1s}^* states) do not fill up the π^* states (these can hold up to four electrons, but oxygen contributes only two). Do they go into the $\pi_{2p_x}^*$ or the $\pi_{2p_y}^*$ states? Hund's rule is the tiebreaker here: to minimize electron–electron interactions, each of these states is singly occupied, so the electronic configuration is $\sigma_{2s}(\uparrow\downarrow), \sigma_{2s}^*(\uparrow\downarrow), \sigma_{2p}(\uparrow\downarrow), \pi_{2p}(\uparrow\downarrow, \uparrow\downarrow), \pi_{2p_x}^*(\uparrow), \pi_{2p_y}^*(\uparrow)$ yielding a total of 12 electrons from the $n = 2$ states on the two oxygen atoms. The unpaired spins give molecular oxygen the property of being *paramagnetic*. This is vividly demonstrated when liquid oxygen is poured on to the pole pieces of a powerful magnet. It clings to the pole pieces as the magnetic field gradient (see Equation 4.35) traps the magnetized molecules.

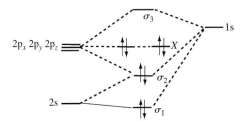

Fig. 8.10 MO energy level diagram for HF. Two of the occupied atomic orbitals have energies lowered by interaction with the hydrogen, so though only a single bond in a Lewis structure, four electrons are involved in bonding interactions (in the σ_1 and σ_2 states). The states marked "X" do not change in energy and correspond to two lone pairs in the remaining 2p orbitals. The σ_3 state is an antibonding state. The occupied states are shifted toward the fluorine levels, a consequence of substantial charge transfer in this very polar bond.

Broadly similar considerations apply to heteronuclear diatomic molecules, save that the energies of the orbitals involved in bonding may be very different. We will use HF as an example. The total of eight valence electrons occupy four molecular orbitals. The two highest energy states are degenerate π orbitals with no overlap with the hydrogen 1s state. They are nonbonding orbitals that contain four electrons corresponding to two of the lone pairs evident in a Lewis structure.

$$\text{H}-\ddot{\underset{\displaystyle{}}{\text{F}}}\colon$$

The molecular orbital energy level diagram is shown in Fig. 8.10. An MO analysis shows that four electrons in the molecule are involved in bonding. Two occupy two lone pairs, and the energy of the occupied states is closer to F than H, reflecting substantial charge transfer onto the electronegative fluorine.

8.7 Molecular levels in organic compounds: the Hückel model

Energy level diagrams for unsaturated carbon compounds, for example, ethene, C_2H_4, $H_2C{=}CH_2$ show that the highest occupied molecular orbital (called the *HOMO*) has a π character, and the lowest unoccupied molecular orbital (called the *LUMO*) has a π^* character. This is in contrast to a saturated compound like methane, CH_4, where the highest occupied state is a σ orbital, with the lowest unoccupied state having σ^* character. Thus, many of the properties of unsaturated organic molecules can be understood from a model that incorporates only the π orbitals.

A particularly simple model, originally devised by Hückel, makes the following assumptions:

(1) The orbitals are nearly orthogonal, so the overlap integrals are approximated by $S_{ij} = \delta_{ij}$.
(2) The Hamiltonian matrix elements evaluated on each carbon, H_{aa}, are same for each carbon, and have a constant value α (approximately the energy of a 2p electron in carbon, but taken as the zero reference for

energy in this model). The quantity α is exactly analogous to the on-site energy, ε_0, introduced in the last chapter (Equation 7.19).

(3) The Hamiltonian matrix elements are nonzero between nearest carbon neighbors only, and have a constant value β (determined by fitting data to experiment as having a value of -75 kJ/mole). The quantity β is exactly analogous to the hopping matrix element, τ, introduced in the last chapter (Equation 7.20).

Taking the example of ethene, C_2H_4, each carbon atom contributes one electron in a $2p_z$ state. Accordingly, the trial wavefunction for the π molecular orbital is

$$\psi_\pi = c_1 2p_{zA} + c_2 2p_{zB}. \qquad (8.29)$$

c_1 and c_2 are found using the secular determinant (8.9) which, with the Hückel approximations, reduces to

$$\begin{vmatrix} \alpha - E & \beta \\ \beta & \alpha - E \end{vmatrix} = 0. \qquad (8.30)$$

This is essentially identical to the hydrogen molecular ion problem we solved in Section 2.20. The secular equation yields $(\alpha - E)^2 = \beta^2$ giving $E = \alpha \pm \beta$. Because $\beta < 0$ (for an attractive interaction between bonded carbons), the lowest energy state is $E = \alpha + \beta$. Two electrons occupy this state, so the energy of the π state in ethene is $E_\pi = 2(\alpha + \beta)$. The equations for c_1 and c_2 are given by (8.8) with the Hückel approximations:

$$\begin{aligned} c_1(\alpha - E) + c_2\beta = 0 \\ c_1\beta + c_2(\alpha - E) = 0 \end{aligned} \qquad (8.31)$$

For $E = \alpha + \beta$ either equation yields $c_1 = c_2$ so $\psi_\pi = c_1(2p_{zA} + 2p_{zB})$. With the Hückel approximation $S_{ij} = \delta_{ij}$, the requirement that $\langle \psi_\pi \mid \psi_\pi \rangle = 1$ yields $c_1 = \frac{1}{\sqrt{2}}$. Thus

$$\psi_\pi = \frac{1}{\sqrt{2}}(2p_{zA} + 2p_{zB}). \qquad (8.32)$$

8.8   Delocalization energy

It is instructive to compare the molecular orbital energies of ethene with those of butadiene, a molecule that looks like two ethene molecules joined by a single bond, shown below in the three forms we have used for visualizing carbon compounds:

$$CH_2{=}CH{-}CH{=}CH_2$$

Each double bond keeps the carbons in a plane defined by the two CH bonds in the CH_2 groups at the ends. Defining this as the *x-y* plane, the hydrogen

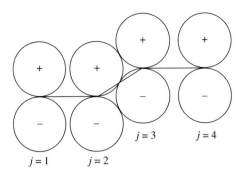

$j = 3$ $j = 4$

$j = 1$ $j = 2$

Fig. 8.11 Butadiene as a row of four carbon p_z states.

1s orbital and the carbon 2s, $2p_x$, and $2p_y$ orbitals have a reflection symmetry with respect to this plane. These orbitals combine to give molecular orbitals that have the same reflection symmetry and are thus referred to as σ orbitals because their spatial characteristics are similar to the σ orbitals of diatomic molecules. The σ orbitals are localized on the bond, and the electrons in them are generally tightly bound, having ionization energies on the order of 10 eV. The $2p_z$ orbitals are antisymmetric for reflections in the molecular plane (yz) as are the molecular orbitals formed by combining them, and they have the symmetry of π orbitals. We shall ignore the obvious structure of butadiene (it exists in two forms corresponding to rotations about the middle σ bond) and treat it as simply a row of four carbon atoms (Fig. 8.11). In this diagram, there will be four different π orbitals corresponding to four different linear combinations of the carbon $2p_z$ states. We can express all four compactly according to

$$\psi_{\pi i} = \sum_{j=1}^{4} c_{ij} 2p_{zj}. \tag{8.33}$$

The index j refers to the carbon atom (Fig. 8.11) and the index i refers to each of the four π orbitals. The secular determinant in the Hückel approximation is

$$\begin{vmatrix} \alpha - E & \beta & 0 & 0 \\ \beta & \alpha - E & \beta & 0 \\ 0 & \beta & \alpha - E & \beta \\ 0 & 0 & \beta & \alpha - E \end{vmatrix} = 0. \tag{8.34}$$

We can factor β out of each column and let $x = \frac{\alpha - E}{\beta}$ (you can demonstrate this to yourself by expanding a two by two version) to get

$$\beta^4 \begin{vmatrix} x & 1 & 0 & 0 \\ 1 & x & 1 & 0 \\ 0 & 1 & x & 1 \\ 0 & 0 & 1 & x \end{vmatrix} = 0 \tag{8.35}$$

for which the secular equation is

$$x^4 - 3x^2 + 1 = 0. \tag{8.36}$$

This can be solved for x^2, giving

$$x^2 = \frac{3 \pm \sqrt{5}}{2}. \tag{8.37a}$$

The four roots are $x = \pm 1.618, \pm 0.618$. The resulting energy diagram for butadiene is shown in Fig. 8.12. The total energy of the occupied π orbitals in butadiene is

$$E_\pi = 2(\alpha + 1.618\beta) + 2(\alpha + 0.618\beta) = 4\alpha + 4.472\beta. \tag{8.37b}$$

This is to be contrasted with the π orbital energy of ethene, $E_\pi = 2(\alpha + \beta)$. If we take the view that a butadiene molecule is just two ethene molecules (as far as the π-electron energy is concerned), we would conclude that

$$E_\pi = 4\alpha + 4\beta \text{ (for the two ethene model).}$$

Butadiene has a π-orbital bonding energy that is 0.472β, or 35 kcal/mole lower (β is negative) than expected. This more stable bond arises from the greater *delocalization* energy of the π electrons on the *conjugated chain* of the butadiene molecule. The larger confinement volume has lowered the kinetic energy of the electrons. In the case of an infinite chain (Chapter 7), the delocalization energy is related to the energy bandwidth for bonding states, 2τ.

The wavefunctions are found by substitution into the four linear equations in c_i obtained from the secular equations for each value of E. For example, taking the solution $x = -1.62$, we have

$$-1.62c_1 + c_2 = 0 \tag{8.38}$$

$$c_1 - 1.62c_2 + c_3 = 0, \tag{8.39}$$

$$c_2 - 1.62c_3 + c_4 = 0, \tag{8.40}$$

$$c_3 - 1.62c_4 = 0. \tag{8.41}$$

With the normalization condition,

$$c_1^2 + c_2^2 + c_3^2 + c_4^2 = 1,$$

we obtain $c_1 = 0.37$, $c_2 = 0.60$, $c_3 = 0.60$, and $c_4 = 0.37$. The corresponding molecule orbital is

$$\psi_1 = 0.37 \, 2p_{z1} + 0.60 \, 2p_{z2} + 0.60 \, 2p_{z3} + 0.37 \, 2p_{z4}. \tag{8.42}$$

A similar analysis gives the other three molecular orbitals as

$$\psi_2 = 0.60 \, 2p_{z1} + 0.37 \, 2p_{z2} - 0.37 \, 2p_{z3} - 0.60 \, 2p_{z4} \quad (E_2 = \alpha + 0.62),$$

$$\tag{8.43}$$

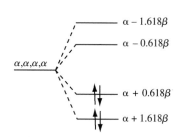

$$\alpha - 1.618\beta$$
$$\alpha - 0.618\beta$$
$$\alpha, \alpha, \alpha, \alpha$$
$$\alpha + 0.618\beta$$
$$\alpha + 1.618\beta$$

Fig. 8.12 MO level diagram for butadiene from Hückel theory (recall that $\beta < 0$).

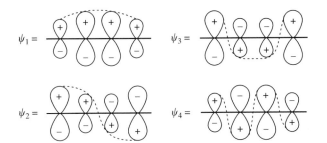

Fig. 8.13 Hückel MOs for butadiene. ψ_2 is the HOMO. ψ_3 (the LUMO) and ψ_4 are antibonding orbitals. (From Fig. 10.23 of McQuarrie and Simon[7] with permission from University Science Books.)

$$\psi_3 = 0.60\, 2p_{z1} - 0.37\, 2p_{z2} - 0.37\, 2p_{z3} + 0.60\, 2p_{z4}, (E_4 = \alpha - 0.62\beta),$$

$$(8.44)$$

$$\psi_4 = 0.37\, 2p_{z1} - 0.60\, 2p_{z2} + 0.60\, 2p_{z3} - 0.37\, 2p_{z4}, (E_4 = \alpha - 1.62\beta).$$

$$(8.45)$$

These are sketched in Fig. 8.13. Note how higher energy orbitals change more rapidly, and particularly how the phase of the orbitals changes at every site for the orbital of highest energy.

Benzene (see the end of Section 8.2 for a Lewis structure) consists of a ring of six carbons, so Hückel theory results in a 6×6 secular determinant that looks exactly like an extension of Equation 8.34 to 36 entries. The results of the analysis are given in Appendix K. Most importantly, the delocalization energy for this ring molecule is even greater than what was found for butadiene—a total of some 150 kJ/mole. The "resonance stabilization" of the structure proposed by Pauling in his extension of Lewis theory (Section 8.2) is dramatic.

The carotene molecule,

represents one of the largest conjugated structures found in nature. Its electronic states are so closely spaced that it readily loses an electron, acting as a donor in photosynthesis and it is used as the donor in the light-to-charge converter molecule (Fig. 6.1). The gap between the π and π^* states is small, so the molecule absorbs visible light, giving carrots (which contain a lot of carotene) their color. Carotene is about as extended as conjugated molecules get. It turns out that, for much larger structures, the added delocalization energy for adding an extra butadiene-like unit falls below $k_B T$ and the simple quantum mechanical picture fails to apply as a consequence of thermal fluctuations.

As a final note, some molecules can be made to switch their degree of conjugated bonding in response to optical illumination. Bonds are broken and rearranged as the molecule relaxes down from its optically excited state. An example of such a molecule is shown in Fig. 8.14. This molecule is transparent until it is exposed to UV light. This causes it to switch to a structure in which the extent of the conjugated bonding (double–single–double, etc.) is increased.

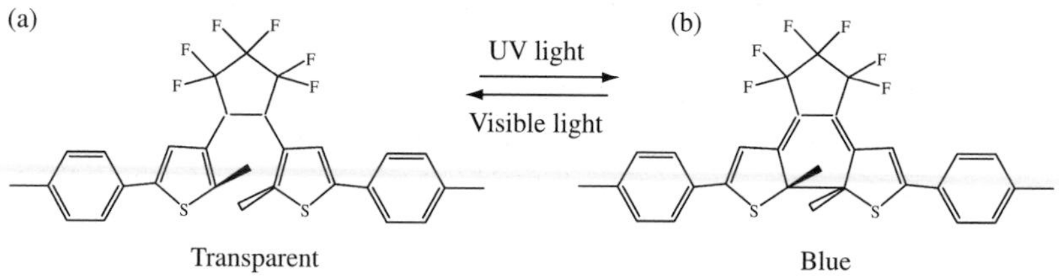

Fig. 8.14 A photochromic molecule. When the central ring is open (a) the molecule absorbs in the UV, relaxing back from the optically excited state into a new geometry; and (b) with a closed, fully conjugated central ring. This ring now absorbs in the visible, so the molecule is blue. Excitation by visible radiation drives a transition back to the open form of the molecule.

The molecule then absorbs in the blue. Exposure to visible light will eventually cause it to switch back to its transparent form. Molecules like this are called photochromic, and they are used as active tints in eyeglasses.

8.9   Quantifying donor and acceptor properties with electrochemistry

The theoretical discussion above should have given us a better idea of what we mean when we talk about molecules as electron donors or acceptors. Molecules with HOMOs that are high in energy are likely to be donors relative to molecules that have LUMOs that are low in energy. Our outline was basic, but even with powerful codes (such as the DFT discussed at the end of Chapter 2) locating the energies of states and their change on charging (i.e., gaining or losing an electron) is not an easy task. Molecular electronic structure can be measured by techniques like photoelectron spectroscopy (Chapter 9 in McQuarrie and Simon[7]). But these are measurements made in the gas phase, and chemistry is usually carried out in solution.

Electrochemistry is precisely the right tool for characterizing the propensity of particular compounds to act as donors or acceptors. Electrochemistry is so powerful, and so central to molecular electronics, that a self-contained introduction to the subject is an essential prerequisite for the rest of this chapter.

8.9.1   What electrochemistry can do

Electrochemistry is the study of charge transfer between a molecule (dissolved in a conducting electrolyte) and an electrode. The electrochemical community has worked hard to establish the conditions under which such reactions are reproducible, and to establish useful and reproducible potential scales with which to quantify the ease of driving an electron transfer reaction.[8] The conditions in which electrochemical measurements are made are almost certainly not conditions compatible with electronics manufacturing, but electrochemical data represent a starting point in designing molecular devices.

8.9.2   Ions in solution

Ions are dissolved charged species (atoms or molecules) that can transfer electrons to or from an electrode by virtue of their charge and electron transfer reactions with the electrode. It costs energy to charge an atom or molecule

(Problem 2.23) so why do ions form if the charged atom is more energetic than the neutral atom?

The energy required to create an ionic species in water may be calculated from the energy obtained by discharging an ion in a vacuum and subtracting from that the energy required to charge the ion in solvent. The resulting energy is called the Born solvation energy[9]:

$$\Delta \mu_i = -\frac{z^2 e^2}{8\pi \varepsilon_0 a}\left[1 - \frac{1}{\varepsilon}\right] \tag{8.46}$$

given here for an ion of charge ze and radius a in a medium of dielectric constant, ε.

Equation 8.46 shows that there is a substantial *enthalpic* driving force for ions to move into solvents of high dielectric constant. But ions in a salt are arranged so as to minimize the Coulomb energy, so the Born solvation energy still represents a net energy cost. We really need to look at the *free energy*, taking account of the *entropy change* associated with the formation of an ionic solution from a salt. Many salts will readily dissolve in solvents because of the increased entropy of ions in solution.

Consider a salt in contact with an electrolyte, originally containing few ions. Ions will pass from the solid into the liquid until the chemical potential in solid and liquid is equalized. Equating the entropy gain with the enthalpy cost, the equilibrium concentration of ions, X_s, will be reached when

$$k_B T \ln(X_s) = \Delta \mu_i, \tag{8.47}$$

where $\Delta \mu_i$ is the enthalpy change required to dissolve the ions, and we have used the ideal gas expression for the entropy of a solution in terms of concentration of the dissolved species (Chapter 3). $\Delta \mu_i$ is given approximately by

$$\Delta \mu_i \approx -\frac{e^2}{4\pi \varepsilon \varepsilon_0 (a_+ + a_-)} \tag{8.48}$$

for a monovalent salt (such as sodium chloride). The enthalpy change on dissolution is negative. Here a_- and a_+ are the ionic radii of the anion and cation. Solving 8.47 and 8.48, and using $a_+ + a_- = 2.76\,\text{Å}$ (for sodium chloride) yields a value for X_s of ~0.09 mol/mol in water ($\varepsilon = 80$). This is roughly equal to the experimental value of 0.11 mol/mol for the solubility of sodium chloride. The agreement shows how dissolution of a salt is driven by entropy.

This description applies to a block of salt (which is made of ions) in contact with water. But what about a metal, made of neutral atoms, in contact with water?

8.9.3 Electrodes and ions in equilibrium

Consider a magnesium electrode in contact with water. Magnesium metal is not soluble in water but magnesium ions are. There is an entropic driving force for the formation of magnesium ions. However, each magnesium ion that leaves the electrode must leave two electrons behind it in the metal, causing the potential of the electrode to increase. The process will continue until the potential of the

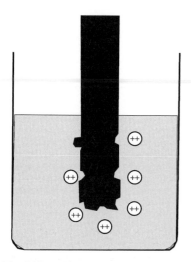

Fig. 8.15 Equilibrium between metal atoms in an electrode (black) and ions (white circles) dissolved in electrolyte.

electrode is raised to the point that electrons can no longer be donated to the metal. This point is reached when the Fermi level of the metal is exactly in equilibrium with the chemical potential of the solution. Another way to say this is that the free energy of ions in solution increases with the logarithm of their concentration until the point that the back reaction rate (ion + electrons → metal) becomes equal to the forward reaction rate (metal → ion + electrons). This process is illustrated in Fig. 8.15.

The potential to which an electrode moves when it is in equilibrium with a given concentration of ions in solution is a well-defined quantity. The energy needed to remove an electron from the metal to far from the metal is equal to the average of the work done in creating a cation and the work gained in discharging one, another way of saying that the Fermi level of the metal and the chemical potential of the solution are aligned. The Fermi level shifts, relative to the vacuum, because of the surface charge that accumulates on the metal, and this is related to the potential shift by Poisson's equation. The surface charge is, in turn, compensated by an accumulation of ions near the electrode surface, resulting in a *double layer* at the electrode surface. The Debye model of a double layer was presented in Section 3.5. It is difficult to measure the potential of the metal electrodes relative to the vacuum (but it has been done[10]). This problem has been solved in electrochemistry with the invention of the *reference electrode*. The reference electrode consists of a "standard" metal in contact with a "standard" solution of ions via some "standard" interface. One example is the *standard hydrogen electrode*. This consists of platinum in contact with a saturated aqueous solution of hydrogen gas, formed by bubbling hydrogen at atmospheric pressure through the solution. Platinum adsorbs hydrogen, forming a reservoir of hydrogen molecules in equilibrium with protons in the water. If negligible current is drawn through this platinum electrode, then the potential drop between the electrode and the solution is fixed purely by the concentrations of H_2 and H^+ at the surface of the electrode. This arrangement is shown in Fig. 8.16. The system on the left is the reference electrode (1 M H^+ in equilibrium with 1 atmosphere of H_2 gas at an electrode surface) and the system on

Fig. 8.16 Measuring the potential of a metal electrode ("Metal") in equilibrium with a salt solution with respect to a hydrogen reference electrode. The salt bridge allows the chemical potential of the two solutions to equilibrate without significant cross contamination on the time scale of a measurement. If the meter is replaced with a load, the system operates as battery with a potential difference somewhat less than the equilibrium potential (depending on the current drawn by the load).

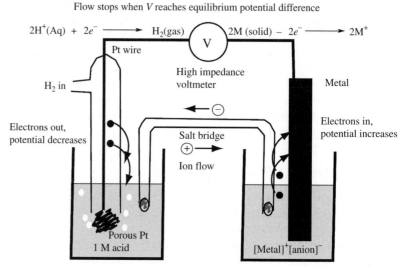

the right is, in the case shown in the figure, a monovalent metal in equilibrium with its salt in solution. The voltage difference between the two electrodes then establishes the potential for the reaction

$$2M^+ + 2e^- \rightarrow 2M$$

relative to the potential for the reference reaction

$$2H^+ + 2e^- \rightarrow H_2$$

by means of the potential difference read from the voltmeter. The voltmeter must be high impedance, meaning that it draws essentially no current, so as to leave the equilibrium distribution of ions unperturbed. The ion flow stops, of course, at the equilibrium potential difference between the two electrodes. If, however, the voltmeter is replaced with a load, the system acts as a battery, driving current through the load (at a voltage drop less than the equilibrium potential) until the reagents in the cell are exhausted.

This ingenious scheme circumvents the issue of absolute potentials relative to the vacuum. However, potentials measured relative to the standard hydrogen electrode can be related to vacuum potentials. In a series of clever experiments, an electrode, together with its associated double layer, was transferred from an electrochemical cell and into an ultrahigh vacuum chamber where ultraviolet photoemission measurements were used to determine the work function. This experiment yielded the result that the Fermi level of the standard hydrogen electrode lies ~4.4 V below the vacuum.[10]

There are many other types of reference electrode. One example is a silver wire in contact with solid silver chloride which is, in turn, in equilibrium with a standard silver chloride solution. If the amount of solid AgCl on the surface of the wire does not change significantly, and the concentration of the AgCl solution is kept constant, then the potential drop between metal and liquid is kept fixed by this interface of "constant polarization" (electrochemists call such interfaces "ideal nonpolarizable electrodes").

Reference electrodes allow the potential of the experimental electrode to be determined with respect to the (constant) potential of the reference electrode using a high impedance voltmeter between the two electrodes. It is by these means that standard scales are established that quantify the ease of electron transfer and identify donors and acceptors. For the electrode shown in Fig. 8.15, this was for equilibrium between magnesium metal and magnesium ions. The equilibrium could also be between ions of different oxidation value, for example, Fe^{2+} and Fe^{3+}. The electrode itself does not have to (and generally does not) participate in the reactions directly. For example, it could be a noble metal electrode (gold, platinum) in contact with a solution containing Fe^{2+} and Fe^{3+} ions. In this situation, the potential does not depend on the specific metal of the experimental electrode (so long as it does not react with ions in the solution) because a polarization layer will form on the metal to bring the Fermi level into equilibrium with the chemical potential of the solution. Thus, equilibrium potentials measured on these reference scales give a true measure of the ease of forming an ion or the ease with which the oxidation state of a given ion is changed. The list of some standard reduction potentials on the normal hydrogen electrode scale is given in Table 8.1 for some reduction reactions involving metal ions, water and hydrogen.

Table 8.1 Standard potentials for some electron transfer (reduction) reactions. The symbol $\Leftrightarrow$ means that the rate for the forward reaction is the same as the rate for the backward reaction so that the reactants and products shown are in equilibrium with an electrode held at the potential E^0 with respect to the standard hydrogen electrode (the scale is defined by the reduction of protons at 0 V). At lower (more positive) potentials, the equilibrium shifts toward more of the ions and less of the metals. Reactants are shown on the left and products on the right. Metals at the top of the list will easily lose an electron to positive ions low on the list. Gold ions are very easy to reduce (at $+1.83$ V). Conversely, gold metal is very hard to oxidize (at -1.83 V). This resistance of gold to oxidation is why it is known as a noble metal

Reduction reaction	E^0 (V)
$Li^+ + e^- \Leftrightarrow Li$	-3.04
$K^+ + e^- \Leftrightarrow K$	-2.92
$Ba^{2+} + 2e^- \Leftrightarrow Ba$	-2.90
$Ca^{2+} + 2e^- \Leftrightarrow Ca$	-2.87
$Na^+ + e^- \Leftrightarrow Na$	-2.71
$Mg^{2+} + 2e^- \Leftrightarrow Mg$	-2.37
$Al^{3+} + 3e^- \Leftrightarrow Al$	-1.66
$Mn^{2+} + 2e^- \Leftrightarrow Mn$	-1.18
$2H_2O + 2e^- \Leftrightarrow H_2 + 2OH^-$	-0.83
$Zn^{2+} + 2e^- \Leftrightarrow Zn$	-0.76
$Cr^{2+} + 2e^- \Leftrightarrow Cr$	-0.74
$Fe^{2+} + 2e^- \Leftrightarrow Fe$	-0.44
$Cr^{3+} + 3e^- \Leftrightarrow Cr$	-0.41
$Cd^{2+} + 2e^- \Leftrightarrow Cd$	-0.40
$Co^{2+} + 2e^- \Leftrightarrow Co$	-0.28
$Ni^{2+} + 2e^- \Leftrightarrow Ni$	-0.25
$Sn^{2+} + 2e^- \Leftrightarrow Sn$	-0.14
$Pb^{2+} + 2e^- \Leftrightarrow Pb$	-0.13
$Fe^{3+} + 3e^- \Leftrightarrow Fe$	-0.04
$2H^+ + 2e^- \Leftrightarrow H_2$	0.00
$Au^+ + e^- \Leftrightarrow Au$	1.83

All the reactions in Table 8.1 are called "half reactions" in the sense that they only show one-half of the reactions undergone in the complete cell (consisting of the experimental—or *working* electrode and the reference electrode). In a complete circuit, a reaction at the second electrode must be taking up electrons to maintain electrical neutrality. For the standard reaction potentials listed in Table 8.2 that second electrode is the standard hydrogen reference electrode.

Standard potentials are defined for exactly equal concentrations of the reduced and oxidized species at a given electrode. At other concentration ratios, the standard potential must be adjusted using the concentration dependence of the free energy (Chapter 3). Equation 3.45, when applied to the concentration dependence of electrochemical reactions, is called the Nernst equation.

$$E = E^0 + \frac{RT}{nF} \ln \frac{C_O}{C_R}, \tag{8.49}$$

where the factor RT/nF reflects the use of volts as the measure of energy; n is the number of electrons transferred in each reaction; and F is the Faraday constant, or the total charge per mole of electrons. The concentrations (or activities) of the oxidized and reduced species are C_O and C_R and $\frac{(C_O)^n}{C_R}$ is the reaction quotient, Q, in Equation 3.45.

In practice, measurement of the standard potential is not usually carried out in equilibrium conditions, but rather by using a technique called *linear sweep voltammetry* (or "cyclic voltammetry" if the sweep is repeated). An electronic circuit is used to sweep the potential of the working electrode ("Metal" in Fig. 8.16) with respect to the reference electrode, while measuring the current through the working electrode. All other things being equal, the current through the electrode is a maximum when exactly half the population of ions is reduced and exactly half is oxidized. The potential at which this occurs is called the *formal potential*. At more negative potentials, the solution contains fewer oxidized ions, and more reduced ions, so the current rises as the number of ions becoming oxidized increases. But above the formal potential, though the rate of the oxidation process is increasing, the number of ions still available for oxidation is falling. Thus, the oxidation current falls with increasing positive potential above the formal potential. Such a plot of current vs. potential (a cyclic voltammogram) is shown in Fig. 8.17. This particular cyclic voltammogram is for a thin film of molecules containing iron atoms in the Fe^{2+} or the F^{3+} oxidation states. Because the ions are contained in a film, problems of diffusive transport of ions to the surface are eliminated, and the peak on the way up corresponds to a current that is maximum when exactly half the ions are oxidized to Fe^{3+}, falling as the remainder become oxidized. The negative peak on the return sweep occurs when exactly half the population of Fe^{3+} ions has been reduced to Fe^{2+}. The current then falls as there are fewer ions to reduce. The process is reversible (the peak on the way up is in almost the same place as the peak on the way down). Such a symmetric voltammogram is a consequence of tethering the molecules containing the Fe^{2+}/Fe^{3+} to the electrode surface, avoiding time delays associated with mass transport of ions from the bulk solution. The interpretation of more complex cyclic voltammograms, such as those obtained from dissolved ions, is described in detail in texts such as the book by Bard and Faulkner.[8]

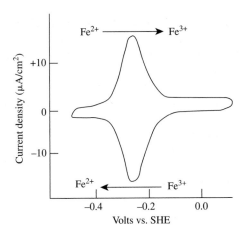

Fig. 8.17 A linear sweep voltammogram (called a "cyclic" voltammogram if repeated). The current flowing through an electrode in contact with a film of molecules containing Fe^{2+} (or Fe^{3+}) is plotted as a function of the potential of the electrode with respect to the Standard Hydrogen Electrode (SHE). The current peaks when exactly half the ions are reduced (or oxidized). That is, in this case, when half the ions are Fe^{2+} and the other half are Fe^{3+}. No current flows when essentially all the population are in the same state well above (all Fe^{3+}) and well below the peak (all Fe^{2+}). The residual current away from the peak is caused by capacitative charging of the electrolyte–metal interface and is proportional to the sweep rate. Note that the potential for an equilibrium between Fe^{2+} and Fe^{3+} lies at about -0.25 V or about half way between the potential for the complete reduction of the two ions (Table 8.1).

8.9.4   Redox processes in molecules

A "redox" process is shorthand for "reduction/oxidation" processes, and now the stage is set for us to describe much more precisely what we mean by describing a molecule as an electron donor or an electron acceptor. To do this, we will return to the pair of molecules utilized as components of the Aviram–Ratner diode shown in Fig. 8.1. The donor component was tetrathiofulvalene (TTF):

The sulfur atoms in the ring increase the density of electrons associated with the ring π orbitals, making it relatively easier to pull an electron from the ring to form a radical (unstable intermediate) ion, $TTF^{\bullet +}$ (the dot indicates a radical). A second electron is then easily withdrawn to form the doubly charged (and stable) ion TTF^{2+}.

The acceptor component is tetracyanoquinodimethane (TCNQ):

The electron withdrawing CN groups and the quinone structure of the ring (only two double bonds instead of 3) lower π electron density in the ring, making it relatively hard to remove an electron to form the radical ion $TCNQ^{\bullet +}$, where,

Table 8.2 Standard potential for reductions of TTF and TCNQ. These are the potentials at which reactants and products are present in equimolar amounts. The Nernst equation (8.49) can be used to calculate the potential for other molar ratios

Reaction	E^0 (vs. NHE)
I $TTF^{\bullet+} + e \Leftrightarrow TTF$	0.37
II $TTF^{2+} + e \Leftrightarrow TTF^{\bullet+}$	−0.05
III $TCNQ^{\bullet+} + e \Leftrightarrow TCNQ$	0.54
IV $TCNQ^{2+} + e \Leftrightarrow TCNQ^{\bullet+}$	0.90

once again, the dot indicates a radical. Removal of a second electron results in the formation of the stable ion $TCNQ^{2+}$.

Electrochemistry provides the means to quantify the relative energies of these ions. In tables like Table 8.1, all the reactions are shown as reductions of a positively charged species but they could equally well be shown as the reverse oxidation reaction of a neutral species with the sign of the potential reversed. The reduction reactions for TTF and TCNQ are shown in Table 8.2.[8]

We will now use these data to estimate the barrier to charge transfer in the Aviram–Ratner diode (Fig. 8.1). If we start by taking two electrons from the TCNQ acceptor (by making the left electrode very positive), further conduction would occur by means of the following reaction:

$$TCNQ^{2+} + TTF \Leftrightarrow TCNQ + TTF^{2+} \tag{8.50}$$

because the electrode on the right (Fig. 8.1(b)) replaces the missing electrons on the TTF, completing the circuit. We need to know the free energy difference, ΔG, between the left-hand side ($TCNQ^{2+} + TTF$) and the right-hand side ($TTF^{2+} + TCNQ$) of 8.50. If $\Delta G < 0$ the reaction is thermodynamically favored.

We face three complications in deducing ΔG for reaction (8.50) from Table 8.2: (1) The energy units are volts relative to the $2H^+ + 2e \Leftrightarrow H_2$ half reaction. (2) All the reactions listed are reductions (this is conventional) but our charge-transfer reaction, 8.50, involves both reduction (of $TCNQ^{2+}$) and oxidation (of TTF). (3) The reactions listed in Table 8.2 are *half* reactions, showing only what occurs at one electrode.

The conversion between Joules per mole and volts follows from the definition of electric potential

$$-\Delta G^0 = nFE^0, \tag{8.51}$$

where the superscript 0 denotes the standard state, values for other concentrations being calculated with the Nernst equation. n is the number of electrons transferred in each step and F is the Faraday constant, the number of Coulombs in a mole of electrons (96,485°C). The minus sign reflects the charge on the electron: the more negative the electrode, the higher the energy of the electrons in it. The steps in the various half reactions in Table 8.2 are all one-electron reductions, but we will use both kJ/mole and volts to remind us that we need to keep track of n.

Oxidation potentials are readily deduced because the listed potentials are nominally equilibrium values, so going the other way requires the opposite sign for the energy gained/lost. We can write the following two oxidation reactions and the associated potentials (with equivalent ΔG_s^0):

$$TTF \Leftrightarrow TTF^{\bullet+} + e, \quad E^0 = -0.37\ V(\Delta G^0 = +35.7\ kJ/mole) \tag{8.52}$$

$$TTF^{\bullet+} \Leftrightarrow TTF^{2+} + e \quad E^0 = +0.05\ V(\Delta G^0 = -4.8\ kJ/mole). \tag{8.53}$$

When 8.52 and 8.53 are added together and equal terms on each side are canceled (they contribute equally to free energies) we obtain

$$TTF \Leftrightarrow TTF^{2+} + 2e \quad E^0 = -0.32\ V(\Delta G^0 = +30.9\ kJ/mole). \tag{8.54}$$

This is the oxidation half reaction required.

Similarly, we can add the two reductions of TCNQ:

$$TCNQ^{\bullet+} + e \Leftrightarrow TCNQ \quad E^0 = +0.54 \text{ V } (\Delta G^0 = -52.1 \text{ kJ/mole}), \quad (8.55)$$

$$TCNQ^{2+} + e \Leftrightarrow TCNQ^{\bullet+} \quad E^0 = +0.90 \text{ } V(\Delta G^0 = -86.8 \text{ kJ/mole}), \quad (8.56)$$

to get the total reduction half reaction:

$$TCNQ^{2+} + 2e \Leftrightarrow TCNQ \quad E^0 = +1.44 \text{ V}(\Delta G^0 = -138.9 \text{ kJ/mole}).$$
$$(8.57)$$

The total free energy for the electron transfer reaction then follows from adding 8.54 and 8.57 (and canceling the two electrons on each side):

$$TCNQ^{2+} + TTF \Leftrightarrow TCNQ + TTF^{2+}$$

$$E^0 = +1.12 \text{ V } (\Delta G^0 = -108 \text{ kJ/mole}). \quad (8.58)$$

This is a hugely favorable reaction. The reverse reaction (corresponding to electrons transferring across the Aviram–Ratner diode from left to right) would have to overcome this 1.12 V barrier (equivalent to +108 kJ/mole). This exercise has illustrated how electrochemical data can be to design devices like the diode in Fig. 8.1.

We have not told all of the story, for the polarization that stabilizes the ionic states comes from other mobile ions in an electrolyte (via ε in Equation 8.48). Changing those ions and/or the solvent will change the energies of the charged states. The electrochemical data given in Table 8.2 were obtained in acetonitrile containing either 0.1 M lithium perchlorate or 0.1 M triethylamine phosphate. It would be hard to make an electronic device containing electrolytes like this, and so the potential barrier in a real molecular diode (not containing these salts) might be significantly different.

8.9.5 Example: where electrons go in the light-to-charge converter molecule

Here, we will take a closer look at the light-to-charge converting molecule, first introduced in Fig. 6.1. How does charge get from one site to another? If the energies between sites are equal, we know that a tunnel transition can take place according to Fermi's Golden rule (as discussed in Chapters 2 and 7). The distance (x) dependence of a charge-transfer rate (current) is thus the standard exponential decay for tunnel current:

$$k_{et} = A \exp(-\beta x), \quad (8.59)$$

where β is the factor $2\sqrt{\frac{2m\Delta E}{\hbar^2}}$ encountered in Chapter 2 (ΔE being the difference between the energy of the electron and the energy of the orbital that mediates tunneling in this case). But what happens when the two sites (e.g., donor and acceptor) have different energies? We will turn shortly to a discussion of the physics behind this when we discuss the Marcus theory, but the result is that the charge transfer requires a thermal activation energy:

$$\Delta G^* \propto \frac{\left(\lambda + \Delta G^0\right)^2}{\lambda}, \tag{8.60}$$

where λ is a constant on the order of an eV. This is not an obvious result: it says that there is an activation energy even if $\Delta G^0 = 0$. Here, we are concerned with the fact that negative (favorable) ΔG^0 speeds up electron transfer by a factor that goes as the exponential of the square of difference between λ and ΔG^0. Thus, Equations 8.59 and 8.60 encapsulate the criteria for rapid electron transfer: (1) the closer that donors and acceptors are, the larger the electron transfer rate; and (2) the larger the energy difference between donor and acceptor, the larger the electron transfer rate. The two factors are coupled in that that either one may be rate-limiting.

The molecular photovoltaic device shown in Fig. 6.1 was designed with these rules, using electrochemical data of the sort just presented for TTF–TCNQ. The carotene molecule on the left of the construct is a good electron donor (i.e., it has a low oxidation potential). The Bucky ball (C_{60}) on the right is a good electron acceptor (i.e., it has a low reduction potential). The porphyrin molecule in the middle is a pigment that absorbs visible light. The acceptor state of the Bucky ball is a significant fraction of an eV below the excited state energy of the porphyrin molecule. The energy of the uppermost π-electrons in the carotene are a fraction of an eV above the ground state energy of the porphyrin. Optical excitation of the porphyrin to its first excited singlet state is followed by photoinduced electron transfer to the Bucky ball ($\tau \sim 30$ ps) to yield a Carotene–Porphyrin$^{\bullet+}$–$C_{60}^{\bullet-}$ charge-separated state. Rapid recombination of this state ($\tau \sim 3$ ns) is preempted by electron donation from the carotene ($\tau \sim 130$ ps) to yield a Carotene$^{\bullet+}$–Porphyrin–$C_{60}^{\bullet-}$ charge-separated state (recall that the dot implies a radical). The charge, originally on the central porphyrin molecule, is now spread out as an electron on the C_{60} and a "hole" on the carotene. These are separated by a large distance (in terms of tunneling) so charge recombination occurs very slowly. In practice, this charge-separated state has a lifetime ranging from $\sim$60 ns to a microsecond, long enough to transfer the charge to another location to do useful work.[11]

8.10   Electron transfer between molecules—the Marcus theory

Now that we have some understanding of how to characterize the ease with which molecules are reduced or oxidized, we turn to a description of the process whereby charge is transferred from one molecule to another. The theory of electron transfer between ions and their neutral counterparts (embedded in a polarizable medium) is a beautiful example of the Kramers theory discussed in Chapter 3. We will begin by considering a very simple case of transfer between two identical ions. For example, the reaction

$$\text{Fe}^{2+} + \text{Fe}^{*3+} \Leftrightarrow \text{Fe}^{3+} + \text{Fe}^{*2+} \quad \Delta G^0 = 0 \tag{8.61}$$

can be followed spectroscopically by using a mixture of isotopes of iron (the star in (8.61) denoting one particular isotope). There is no free energy difference between the reactants and products in this reaction but the rate of electron transfer is found to be strongly thermally activated. What is happening here? To illustrate the process further, Fig. 8.18 shows the two states consisting of an ion and neutral species. We will consider electron transfer from site A to site B, so that the upper pair in the figure shows the reactants, and the lower pair shows the products. The ion is stabilized by environmental polarization, shown schematically by the arrows around site A (upper part of Fig. 8.18). This could be solvent polarization, polarization induced by one or more bond distortions, or a combination of both effects. This polarization represents a stored free energy. The medium is distorted from its geometry in the absence of a field, and its entropy has been lowered. When the charge transfers, not all of this polarization energy is transferred, so work must be done. This work is called the *reorganization energy* for charge transfer, denoted by the symbol λ. If it is entirely due to the orientation of a solvent, it is called *solvent* or *outer shell* reorganization energy. If the reorganization involves bonds on the molecule itself, this energy loss mechanism is called *inner shell* reorganization. In general, both types of response are involved.

Marcus pointed out that the dependence of the electronic energy on the degree of environmental distortion is, quite generally, quadratic, and this enabled him to obtain a simple expression for the activation energy for charge transfer.[12] There are three ingredients underlying this observation:

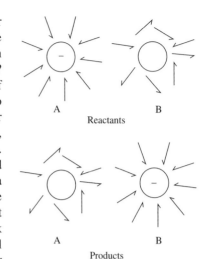

Fig. 8.18 Charge transfer between identical atoms: an electron sits on atom A initially (top) transferring to another atom of the same type at B (bottom). The free energy of the two (reactant and product) states is the same, but work must be done to polarize the environment around B when it charges.

(1) The horizontal coordinate on a plot of electronic energy vs. polarization may be a complicated combination of many parameters; for example, bond distortions in the molecule and the orientations of many solvent molecules. In the spirit of Kramers theory, this is called the "reaction coordinate." It measures the degree of distortion of a set of coordinates that lie along a pathway for electron transfer. We will label it $q(r_1, r_2, r_3, \ldots, r_N)$, where the various r's are the coordinates for all of the bonds and atoms involved in the process. We will write its value when the electron is on atom A as q_A. When the electron is on atom B, we will write its value as q_B.

(2) The electron energy is taken to be additive to the environmental distortion energy. This is the Born–Oppenheimer approximation. The total electronic energy will be a minimum at the equilibrium bond distance and it will increase as the environment is distorted, by exactly the amount of energy required to cause the bond distortions.

(3) Each distortion feeding into the "reaction coordinate" shifts the electronic energy quadratically with the amount of the distortion from the minimum, at least for small distortions. This is simply a statement of the fact that the leading term in an expansion of any potential about a minimum is quadratic. Thus, the sum of all the distortion energies must itself be quadratic in the reaction coordinate. So, for the reactants (top panel in Fig. 8.18)

$$E = \frac{1}{2}k\,(q - q_A)^2 . \tag{8.62}$$

The picture is completed on the addition of one other subtle idea: we will plot the free energy for the *system* of reactants or products as a function of reaction

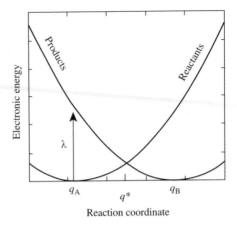

Fig. 8.19 Diagram showing the total free energy of the reactants (electron on atom A, atom B neutral) and the products (electron on atom B, atom A neutral) for the atom pair shown in Fig. 8.18 as a function of the (multidimensional) reaction coordinate, q. The electronic ground state energy at zero distortion has been arbitrarily set to zero on the vertical axis. The transition state for electron transfer is shown at $q = q^*$, and the reorganization energy, λ, is shown by the vertical arrow.

coordinate. That is, we take the energy of atom A with an electron on it and add it to the energy of neutral atom B for each value of the reaction coordinate q and make a plot of electronic energy vs. reaction coordinate. This is quadratic with a minimum at q_A as described by Equation 8.62. We then generate a second plot of the electronic energy of the system of products: that is, the electron on atom B with atom A neutral. This energy also depends on q quadratically, but has a minimum at $q = q_B$:

$$E = \frac{1}{2}k\,(q - q_B)^2 . \tag{8.63}$$

We have used the same "spring constant" k (or more generally, curvature of the potential surface) because, for this example, the two systems are identical in every respect, except in that the electron is on A in the system of reactants and on B in the system of products. These electronic energies are plotted in Fig. 8.19. This is a subtle diagram, expressing as it does, free energies for two different systems of atoms (reactants and products) as a function of a complicated, multidimensional variable, the reaction coordinate, q. However, the reorganization energy, λ, can be shown on the diagram directly, since it is the energy given up when the electron jumps from A to B instantaneously, as shown by the vertical arrow.

The diagram shows us something very useful: there exists a value of the reaction coordinate $q = q^*$ such that *the energies of the systems of reactants and products are degenerate*. Thus, there exists at least one value of q for the electron on atom A that is *the same as the value of q for the electron on atom B*. This is equivalent to saying that there exists an excited state in which the polarization of the atom A and atom B is identical when A is charged and B is not. The energy degeneracy at this point permits tunneling according to Fermi's Golden rule, so, when the reactants fluctuate into this configuration, the electron may tunnel to atom B.

What happens next depends on the strength of the tunneling interaction between atoms A and B, with two extremes being shown in Fig. 8.20. In a system where the two atoms are weakly coupled (the tunneling rate is slow), the classical potential surface is essentially unaffected by tunneling, so the electron will move up and down through the transition state many times before

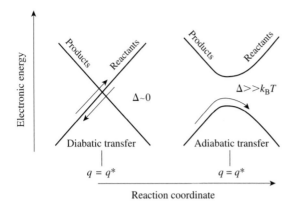

Fig. 8.20 Blow up of Fig. 8.19 in the vicinity of the transition state for two different kinds of coupling between reactants and products. In diabatic transfer (left) reactants and products are weakly coupled by tunneling, so the system may move through the transition state many times without electron transfer. In adiabatic transfer (right), the reactants and products are coupled strongly enough to introduce a significant energy splitting owing to the tunneling interaction, Δ. If $\Delta \gg k_B T$, then the system is likely to pass over from reactants to products as soon as the transition state is attained.

it tunnels to generate the products. This is called *diabatic* electron transfer. In a system where coupling is strong, a quantum mechanical treatment would reveal an energy gap, Δ, between the two systems at the crossing point (exactly analogous to the bonding–antibonding gap). If this is significant ($\gg k_B T$), then the system continues over to the products immediately after the transition state is reached. In either case, once the system has reorganized away from the transition state, dissipation of the initial polarization, $\varepsilon(q_A)$, and acquisition of the new polarization, $\varepsilon(q_B)$, now traps the electron on atom B.

Another important quantity follows from Fig. 8.19. For simplicity, we will set $q_A = 0$ so that the transition state lies at $q = q_B/2$. The energy barrier at this point is

$$\Delta G(q = q^*) = \frac{1}{2} k \left(\frac{q_B}{2} \right)^2. \tag{8.64}$$

But note that, in terms of q_B

$$\lambda = \frac{1}{2} k q_B^2$$

as follows from Fig. 8.19 with q_A set to zero. Therefore, the activation energy for the electron transfer is given by

$$\Delta G^*(q = q^*) = \frac{\lambda}{4}. \tag{8.65}$$

By virtue of this simple model of Marcus, we can now understand some critical points about electron transfer in condensed media:

- Even when the initial and final states have identical free energies (as in the reaction 8.61), the electron transfer is thermally activated.
- Thermal fluctuations bring about the (possibly unique) configuration in which reactants and products have degenerate electronic energies and identical polarizations.
- Reorganization, solvent and/or internal, is the origin of the energy dissipation that "traps" charge at one site or the other.

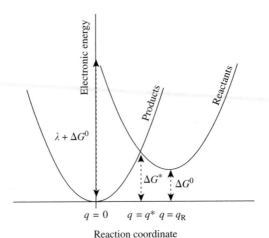

Fig. 8.21 Total free energies of the reactant (centered at $q = q_R$) and product (centered at $q = 0$) systems as a function of the reaction coordinate for a system of reactants and products that differ in final free energy by an amount ΔG^0.

- The transition state energy, relative to the electronic ground state energy, is given by one quarter of the reorganization energy in the simple case where reactants and products have the same free energy (8.65).

It is straightforward to generalize now to a reaction that does have a thermodynamic driving force owing to a free energy difference ΔG^0 between reactants and products. The appropriate modification of Fig. 8.19 is shown in Fig. 8.21. Here the products have minimum energy at $q = 0$, and the reactants have a minimum energy at $q = q_R$ and the transition state now occurs at $q = q^*$, a function of the thermodynamic driving force for the reaction, ΔG^0.

The electronic energy of the products as a function of reaction coordinate is

$$E = \frac{1}{2}kq^2$$

and that of the reactants is

$$E = \frac{1}{2}k(q - q_R)^2 + \Delta G^0,$$

where we have made the simplifying assumption that the two potential energy surfaces have the same curvature (this is not, in general, true for the case of nonidentical reactants and products). The position of the transition state is found from the intersection point of the two curves

$$q^* = \frac{1}{kq_R}\left(\Delta G^0 + \frac{1}{2}kq_R^2\right). \tag{8.66}$$

Evaluating the energy at this point, we obtain

$$\Delta G^* = \frac{1}{2}kq^{*2} = \frac{1}{2}\frac{1}{kq_R^2}\left(\Delta G^0 + \frac{1}{2}kq_R^2\right)^2 \tag{8.67}$$

and noting that Fig. 8.21 implies

$$\lambda + \Delta G_0 = \frac{1}{2}kq_R^2 + \Delta G_0$$

we have for the activation energy

$$\Delta G^* = \frac{(\Delta G^0 + \lambda)^2}{4\lambda}. \tag{8.68}$$

This is the famous Marcus result (previously stated without justification as 8.60).[12] It reduces to 8.65 in the event that the thermodynamic driving force for the reaction is zero. This result captures the key role played by reorganization in the electron transfer reaction and shows the interplay between the thermodynamic driving force and reorganization.

The model makes a very interesting prediction for very large values of ΔG^0. Referring to Fig. 8.21, if $\Delta G^0 = \frac{1}{2}kq_R^2$, the product parabola passes through the minimum of the reactant parabola and $\Delta G^* = 0$. But if $\Delta G^0 > \frac{1}{2}kq_R^2$, the intersection point of the two parabolas *increases with increasing driving force* Δ. This is the so-called "Marcus inverted region," where increased thermodynamic driving force for the electron transfer reaction actually *reduces* the reaction rate. It was the experimental observation of this regime that led to the award of the 1992 Nobel prize in chemistry to Marcus for his prediction of it.

Given the expression 8.68 for the activation energy for electron transfer, we can write down a Boltzmann factor for the probability of a transfer in the adiabatic case where it occurs each time the transition state is reached:

$$P(\Delta) = \frac{1}{\pi\sqrt{4k_B T \lambda}} \exp\left[-\frac{(\Delta G^0 + \lambda)^2}{4k_B T \lambda}\right]. \tag{8.69}$$

The expression has been normalized with respect to ΔG^0 (see Problem 12).

The magnitude of the reorganization energy is critical in determining reaction rates. It is on the order of the Born solvation energy and one approximate expression for it is[13]

$$\lambda \approx -\frac{z^2 e^2}{4\pi\varepsilon_0 a}\left[1 - \frac{1}{\varepsilon}\right]. \tag{8.70}$$

For $a \sim 1$ nm, $\varepsilon \sim 80$, and $z = 1$, $\lambda \sim 1$ eV. (A number of classical expressions for this quantity are given in the original paper by Marcus.[12])

We saw that donor and acceptor levels in TTF–TCNQ were separated by 1.12 eV, so taking $\lambda \sim 1$ eV, the "downhill rate" would be proportional to

$$\exp - \left[\frac{(-1.12 + 1)^2}{4 \times 1 \times 0.025}\right] \sim 1,$$

where we have expressed ΔG^0 in eV (with a sign change owing to the negative charge on the electron) and $k_B T = 0.025$ V at 300 K. The uphill rate is proportional to

$$\exp - \left[\frac{(1.12 + 1)^2}{4 \times 1 \times 0.025}\right] \sim 3 \times 10^{-20}.$$

The one electron volt difference between donor and acceptor energies in the Aviram–Ratner diode (Fig. 8.1) results in a 20-order of magnitude difference in forward and reverse electron transfer rates.

Our discussion thus far has been confined to transfer between pairs of molecules or atoms. It would also describe "hopping transport" in a molecular solid, for example. What about the all-important case of transfer between a metal electrode and a molecule? This is critical in electrochemistry and so is important for obtaining the electronic properties of molecules in the first place. To analyze this problem, a diagram like 8.21 must be modified to take into account a pair of reactants in which one partner (the metal electrode) has a *continuous* density of electronic states. Added to this, the free energy for electron transfer is modified by the addition of the potential difference between the electrode and the solution (which was, of course, how ΔG^0 is determined in the first place by finding the potential difference at which reactants and products are in equilibrium). This analysis has been carried out by Chidsey who has provided appropriately modified versions of Equations 8.68 and 8.69.[14]

8.11 Charge transport in weakly interacting molecular solids—hopping conductance

A weakly interacting molecular solid is one in which the interaction energy between neighboring molecules (the hopping matrix element, τ) is much smaller than the reorganization energy, λ. Electrons are trapped and can only move by thermally activated hopping over a barrier of magnitude $\frac{\lambda}{4}$ (Equation 8.65). In a disordered solid, the local polarization will be randomly distributed, so the barrier to hopping seen by an electron is also randomly distributed in value. Since the hopping rate will be dominated by the *lowest* barrier available (not the average), the energy barrier that controls transport is likely to fall as the distance between donor and acceptor increases. Specifically, if the number of barriers per unit volume is ρ, then as illustrated in Fig. 8.22, the minimum barrier value, λ_m will be

$$\lambda_m = \frac{3}{4\pi r^3 A\rho},\tag{8.71}$$

where r is the distance over which tunneling can occur and A is a constant that depends on the shape of the distribution (cf. Fig. 8.22).

Fig. 8.22 With a randomly distributed reorganization energy, λ, the energy of the barrier with unit probability (λ_m) falls in direct proportion to the number of barriers in the volume in which tunneling occurs, $N(\lambda)$. Here, $N(\lambda)$ is doubled in going from curve "1" to curve "2."

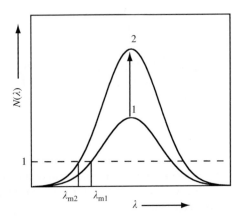

According to 2.59, the tunneling rate varies as

$$\tau \propto \exp(-2\kappa r).\tag{8.72}$$

In this expression, κ is the decay constant for the wavefunction (a quite different process from reorganization which is dominated by the coupling between electronic and nuclear degrees of freedom). Thus, we expect that the overall conductance of the material (in the low-bias limit) might vary as

$$G = G_{max} \exp\left(-2\kappa r\right) \exp\left(-\frac{\lambda_{min}}{k_B T}\right) = G_{max} \exp\left[-\left(2\kappa r + \frac{3}{4\pi r^3 A\rho k_B T}\right)\right].\tag{8.73}$$

The conductance will be a maximum for a value of $r = r_c$ that minimizes the argument of the exponent. This is (Problem 15)

$$r_c = \sqrt[4]{\frac{9}{8\pi \kappa A\rho k_B T}}.\tag{8.74}$$

This yields the interesting result that the temperature dependence of the conductivity varies as the exponential of $T^{0.25}$:

$$G \propto \exp T^{0.25}.\tag{8.75}$$

This result was first derived by Mott. The conductivity of many disordered materials has this temperature dependence, and it is taken to be a signature of *Mott variable range hopping*.

8.12 Concentration gradients drive current in molecular solids

Solid molecular conductors are usually materials that contain mobile ions. Thus, *electric fields will not penetrate the bulk further than the Debye length.* The penetration distance (the Debye length, ℓ_D) was derived in Chapter 3 (Equation 3.17):

$$\ell_D = \sqrt{\frac{\varepsilon \varepsilon_0 k_B T}{2C_0 Z^2 e^2}}.$$

The electric field (needed to oxidize or reduce molecules) is only significant within a Debye length or so of the metal contacts. In this region, electrons can tunnel to or from the electrode provided that the tunneling distance (Equation 2.59):

$$\ell_t \approx (2\kappa)^{-1}$$

overlaps the Debye layer significantly. Then ions can be created (or neutralized) by a Marcus process in which the driving free energy comes from the electric field close to the electrode.

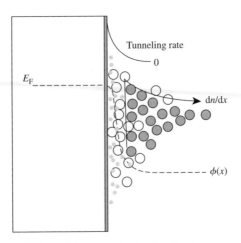

Fig. 8.23 Illustrating mechanisms of charge transfer at an interface between a metal and a molecular medium. The surface charge on the metal is balanced by the accumulation of counter charges (small dots) giving rise to a double-layer dipole region where the potential $\phi(x)$ changes rapidly over a distance on the order of the Debye length. Direct tunneling of the electrons is possible near the electrode, so that molecular ions are formed at some distance from the electrode where the potential is adequate to drive the reduction reaction (open circles, neutral molecules; filled circles, reduced molecules) but tunneling rates are still significant. The result is a concentration gradient of charged molecules, driving a current to the right either by direct diffusive motion of the charged molecules (in a liquid) or hopping of the charge (in a molecular solid). The hopping process is thermally activated (as described in the last section) and a net current is driven by the concentration gradient dn/dx.

Once an ion has escaped the electrode (i.e., distances $\gg \ell_D$) it sees no intrinsic electric field. The only driving force for directed motion comes from the *gradient of ion concentration*. Specifically, the 1-D expression for current density is any device is

$$J_n(x) = n(x)\mu_n \frac{d\phi(x)}{dx} + k_B T \mu_n \frac{dn(x)}{dx}. \tag{8.76}$$

The first term is the current owing to the drift velocity of carriers of density $n(x)$ and mobility μ_n in an electric field $\frac{d\phi(x)}{dx}$, and it does not contribute significantly outside of the double layer. The second term describes the (entropy-driven) diffusion of carriers owing to a concentration gradient $\frac{dn(x)}{dx}$ (cf. Fick's law, discussed in Chapter 3). There is an electric field associated with this carrier gradient, and thus the material obeys Ohm's law. But the current is driven primarily by entropy and so depends linearly on temperature (plus any additional temperature dependence of the mobility such as the exp $T^{0.25}$ just discussed). The bulk current arises from the production of ions at one electrode and consumption at the other, driving a net diffusion of carriers. These mechanisms are illustrated in Fig. 8.23.

8.13   Dimensionality, 1-D conductors, and conducting polymers

The discovery of conducting polymers was one of the most remarkable events of the 1970s. Specifically, it was found that when polyacteylene,

...$CH_2CH_2CH_2$..., or

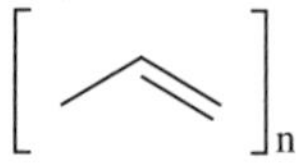

$$\left[\text{\Large \diagup\!\!\!\!\diagdown} \right]_n$$

where $n \to \infty$, is doped with large amounts of oxidizing or reducing agents, it becomes metallic in appearance with a conductance of $\sim 10^3$ $(\Omega\text{-cm})^{-1}$ (about 1/1000 that of copper).[15] The first thought was that oxidation (or reduction) formed partly filled bands that made the material a metal (Chapter 7). This notion was dashed when many conducting polymers were found (by electron spin resonance) to have *spinless* charge carriers. In fact, as we have seen, we would expect reorganization to trap any charge, forming cations (oxidation) or anions (reduction) that are quite localized. The trapping mechanism is quite complex, because it is generally not possible to add or remove electrons to aromatic organic systems without substantial long-range reorganization (see, for example, the reaction illustrated in Fig. 6.2(b)). Figure 8.24 shows the steps involved in the oxidation of polypyrrole

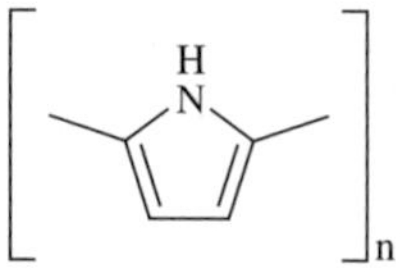

in the presence of a large amount of oxidizing agent ("Ox" in the figure). Examples of oxidizing agents are iodine, arsenic pentachloride, and iron (iii) chloride. In the first stage of oxidation (labeled "polaron"), a pair of double bonds on each of the three central rings is replaced by one double bond in the ring and two double bonds joining the three central rings. Thus, overall, one bond (i.e., two electrons) has been lost. One electron is taken up by the oxidizing

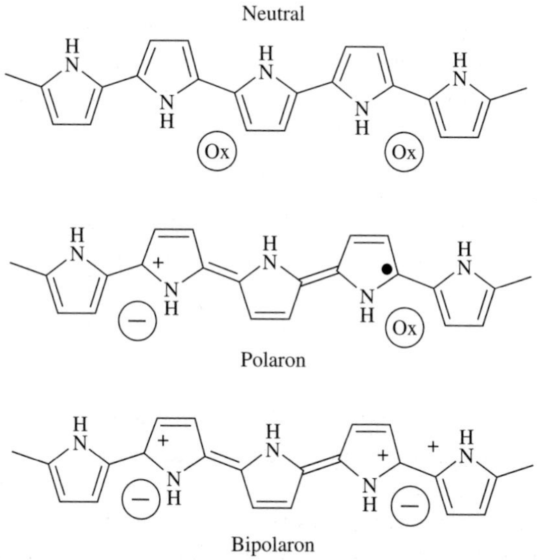

Fig. 8.24 Formation of a bipolaron through two sequential oxidations of polypyrrole.

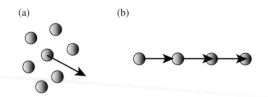

Fig. 8.25 In three or two dimensions (a), electrons (or the "quasiparticles" of Fermi liquid theory) have places to move to. In one dimension (b), the *only* type of allowed motion is highly correlated—if one electron moves, all must follow.

agent which becomes an anion ("−") and the other remains on the pyrrole forming a reactive radical (shown by the black dot). The radical rapidly loses the extra single electron to a second oxidizing agent to form the "bipolaron" shown at the bottom. The charge carrier is thus a dication (two positive charges) involving no unpaired electrons, and therefore having no net spin. The bipolaron is one type of trapped-charge carrier on polymers, but others types exist, some quite exotic.[16]

Can the density of carriers in such materials become such as to make them truly metallic? Could a conducting polymer act as a "molecular wire," transporting electrons at fabulous densities along atomic-scale filaments? Genuine metallic behavior has been observed in polyaniline[17] but only in a form of the material prepared so as to be highly crystalline. There are strong theoretical grounds for believing that polymer chains, in isolation, *can never be metallic conductors.* To understand this, we have to return to the rather abstract discussion in Chapter 7 in which we justified the "free electron" picture of metals in terms of "Fermi liquid theory." Electrons, coupled to density fluctuations, form long-lived quasiparticles near the Fermi energy which behave just like "free electrons." Fermi liquid theory works in three and two dimensions, but breaks down entirely in one dimension. This is because (see Fig. 8.25) motion in 1-D requires that all electrons in the system move to accommodate the displacement of one (this behavior is qualitatively illustrated by the remarkable long-range rearrangement of bonding discussed above). Correspondingly, there is no equivalent to the "free electron" in the 1-D version of Fermi liquid theory (which is called Luttinger liquid theory).[18] Isolated charges can surely diffuse on a 1-D chain, but only at densities that are so low that correlated motion is avoided. This precludes the possibility that such chains could behave like metal wires, and indeed, we shall see that they do not.

8.14   Single molecule electronics

This is a good point to return to the molecular rectifier (Fig. 8.1), the example with which we started this chapter. In light of the discussion above, what are the prospects for single molecule electronic devices? We end this chapter with a brief look at recent progress in measuring the properties of single molecules in molecular junctions. When two electrodes are brought close enough to contact a *single* molecule, they are so close that transport is generally dominated by direct tunneling from one electrode to the other (cf. Fig. 8.23). In this geometry, the molecule plays only a rather indirect role, effectively altering the barrier for tunneling from one electrode to the other. Nonetheless, this turns out to be a simple situation in which to calculate the expected tunneling rates, and, as we shall see below, theory and experiment are now in reasonable agreement.

Can simple tunneling devices act as switches? The electronic properties of the molecule change when its oxidation state is changed. If these electronic changes affect the tunneling rate, then a switch can be made based on control of the oxidation state of the molecule in the tunnel junction. The electrodes that contact the molecule can be used to switch the oxidation state of a trapped molecule, leading to interesting phenomena, such as "negative differential resistance" (the current falls as the bias across the molecule is increased).

Finally, it is clear that rather long molecules (many nanometers) can transport charge by what is presumably a hopping process. Thus, "molecular wires" of a sort can indeed be fabricated, though the distance over which they are usable is limited. Nonetheless, these capabilities may facilitate an entirely new type of electronics, one in which conventional transistors (made from silicon) are interface to the chemical world with molecular components.

8.15   Wiring a molecule: single molecule measurements

Now that we have some understanding of the mechanisms of charge transport through molecules, we can ask what is the experimental situation? One of the goals of molecular electronics has been to connect a single molecule between fixed electrodes, and there have been many reports of single molecule devices.[19] However, if one pauses to think of the challenges of "wiring" a single molecule into a circuit, many questions arise. How are contacts controlled at the atomic level? How many molecules span each set of contacts? What is local polarizability that controls charge transfer? Does the presence of water vapor and/or oxygen lead to undesired electrochemical reactions?

The field has not been without controversy. To pick on one of the more dramatic examples, what are the electronic properties of DNA? There is interest in this because (as discussed in Chapter 6) DNA can be used to make wonderful self-assembling structures. Could it be used to make self-assembling computer chips? This possibility has motivated electronic measurements on DNA, with the following mutually incompatible conclusions:

- DNA is an insulator.[20]
- DNA is a semiconductor.[21]
- DNA is a conductor.[22]
- DNA is a superconductor.[23]

One is forced to conclude that experiments have not been well controlled, though recent advances appear to produce more reproducible data.[24]

The general approach has been to make a very small break in a tiny wire, and hope that a single molecule spans the gap. Data taken with devices of this sort have proved difficult to reproduce. This is probably because it is very hard to satisfy the bonding constraints for a molecule with two fixed electrodes. In addition, metal islands between the electrodes give rise to Coulomb blockade features that can be mistaken for the electronic response of a molecule.

One reproducible way in which molecular junctions have been put together is self-assembly.[25,26] A molecule containing a reactive group at each end was inserted into a self-assembled monolayer of inert molecules attached to an

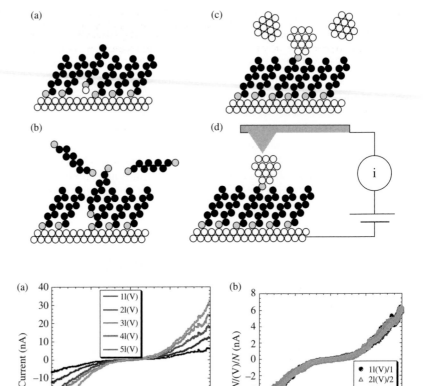

Fig. 8.26 Self-assembly of a single molecule metal–molecule–metal junction. Alkanethiols are removed from a monolayer (a) leaving holes for the insertion of dithiolated molecules (b), which are then contacted with gold nanoparticles (c). The resulting metal–molecule–metal junction is contacted on the top side with a conducting atomic force microscope (d).

Fig. 8.27 (a) Typical *I–V* curves obtained from a self-assembled junction as illustrated in Fig. 8.26. The curves have the property that the current at each voltage is an integer multiple of some fundamental current (b).

underlying electrode (Fig. 8.26). The insertion process is a spontaneous event, especially if the surrounding matrix contains defects. In the process illustrated in Fig. 8.26, a second reactive end of the inserted molecule was used to capture a gold nanoparticle from solution. Finally, a metallic AFM probe was pushed into the gold particle to complete the top connection. The result was a circuit consisting of a metal–chemical bond–molecule–chemical bond–metal junction. The experiment gave a clear series of current–voltage curves corresponding to the integer number of molecules trapped in the gap (Fig. 8.27), and the measured conductance was close to that predicted from a density functional calculation (discussed below).

Another approach consists of repeatedly breaking a gold wire in the presence of molecules with reactive ends. The conductance measured as a function of time shows steps that are much smaller than the Landauer steps in conductance (cf. Fig. 7.6) and they correspond to integer numbers of molecules spanning the gap.[27] This is an easier measurement to carry out than the self-assembly described above, and the single molecule conductance of a number of molecules has now been measured using this technique (Table 8.3).

A key goal is to be able to compare measured data to theoretical calculations of single molecule conductance. The Landauer–Buttiker formula (Equation 7.32)

Table 8.3 Comparison of measured and calculated conductances for three alkanedithiols (**1**, **2**, **3**) (measurement[27] theory[28]), 1,4-phenyldithiol (**4**) (measurement[31], theory[28]), four carotenoid polyenes,[32] an optically switched photochromic molecule in open (**9**) and closed (**10**) forms,[33] oligo(phenylene ethynylene) (**11**) (measurement,[34] theory[28]), and an oligoaniline (**12**) (measurement,[35] theory—O.F. Sankey, personal communication). All experimental data were obtained using the break junction method of Xu and Tao[27] and all calculations used the DFT approach of Tomfohr and Sankey[28]

	Molecule	G (meas.) (nS)	G (theor.) (nS)	Ratio
1		95 ± 6	185	0.51
2		19.6 ± 2	25	0.78
3		1.6 ± 0.1	3.4	0.47
4		833 ± 90	47,000	0.02
5		2.6 ± 0.05	7.9	0.33
6		0.96 ± 0.07	2.6	0.36
7		0.28 ± 0.02	0.88	0.31
8		0.11 ± 07	0.3	0.36
9		1.9 ± 3	0.8	2.4
10		250 ± 50	143	1.74
11		~ 13	190	0.07
12		0.32 ± 0.03	0.043	7.4

can be used to calculate the conductance if the mechanism is tunneling (i.e., directly from one electrode to another mediated by the states of the molecule in the gap). Writing Equation 7.32 a little differently to conform to the notation in the literature[28]

$$I = \frac{2\pi e}{\hbar} \sum_{i,f} |T_{i,f}|^2 \left(f(E_i) - f(E_f) \right) \delta \left(E_i - E_f + eV \right) \qquad (8.77)$$

This has the form of 7.32 but has been rearranged to express conductance in terms of current and voltage:

$$I = GV = \frac{2e^2}{h} \sum_{ij} |T_{ij}|^2 V,$$

where the term proportional to voltage comes from the sum over the density of states (i.e., the $\delta \left(E_i - E_f + eV \right)$ term together with $\left(f(E_i) - f(E_f) \right)$ where $f(E)$ is the Fermi function, Equation 3.57). The sum runs over all states in both electrodes. The molecular states can be calculated with the methods introduced at the beginning of this chapter, or, more readily, with the density functional theory introduced at the end of Chapter 2. A "Green's function" method allows the electronic states of the molecule be "plugged into" the electronic states of the electrodes. Evaluation of 8.77 has been carried out for molecules using matrix elements evaluated using DFT. These are not the same as the overlap of real wavefunctions (matrix elements in DFT are derived using "Kohn–Sham states"; these differ from the real wavefunctions, but can yield surprisingly good approximations in some cases). Table 8.3 shows a comparison of single molecules conductance calculated with this approach and the data measured using the single molecule techniques described above.[29,30]

It is instructive to reflect on the meaning of the various matrix elements that form quantity $|T_{i,f}|$ that appears in Equation 8.77. These are overlap integrals between states that connect the initial state of the electron (in the left electrode say) with the final state of the electron (in the right electrode in this case). Clearly, such integrals must diminish with distance to give the tunneling decay law (Equation 2.40). Thus longer molecules will have a smaller conductance, and this is evident from the change of conductance with length for similar types of molecule (e.g., entries **1**, **2**, and **3** and entries **5**, **6**, **7**, and **8** in Table 8.3). What is not perhaps as obvious is that these overlap integrals are sensitive to the particular orbitals connecting the molecule. As we learned in Section 2.19, the expression for the mixing of states that contribute to the final wavefunction for the system contains a resonance denominator (Equation 2.91). So, states with energies closest to the Fermi energy contribute most to conductance.

We have seen that the HOMO–LUMO gap is formed from π and π^* states, so these will dominate the conductance. Delocalization energy (Section 8.9) is a good measure of overlap between adjacent π orbitals, so we expect that aromatic molecules will have a higher conductance than saturated chains (i.e., all sigma bonded like alkanes). Inspection of Table 8.3 shows that this is indeed the case, when proper account is taken of the contribution of molecular length. We would also expect that conductance depends on the relative *orientation* of adjacent π orbitals because this will alter the overlap (see below).

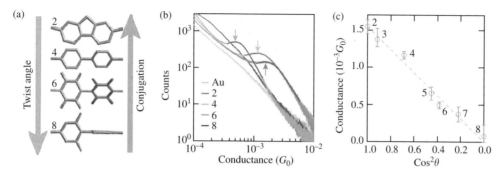

Fig. 8.28 Molecular conductance measured as a function of the twist angle between rings in a molecule. A series of molecules made from two benzene rings was synthesized with different fixed angles between the rings (a). The conductance for each type of molecule was measured by plotting a histogram of conductances measured in repeated break junction measurements, the conductance peak (arrows) corresponding to a single molecule trapped in the gap (b). A plot of measured conductance vs. the twist angle between the rings was in good agreement with theory (dashed line in (c)). (Reprinted from Venkataraman et al.[36] Nature 2006, with permission from MacMillan Publishers Ltd.)

Table 8.3 shows that, for 10 out of 12 molecules studied using repeated break junction methods with statistical analysis, agreement between experiment and DFT calculations is within an order of magnitude. In the worst case (1,4-phenyldithiol), it is within a factor 50. Thus progress has been made in relation to the state of the field as summarized by the DNA story told at the beginning of this section.

The technique of using many "break junction" measurements with statistical analysis of the data[27] has found widespread use. For example, the effect on conductance of rotating the angle between two benzene rings (in a bi-phenol molecule) was measured using a series of molecules that were synthesized with known fixed angles between the rings. The molecules are shown together with a typical data set in Fig. 8.28. The conductance varied with the twist angle between the rings, θ, according to $G = G_{max} \cos^2 \theta$ (Fig. 8.28(c)) as expected on theoretical grounds, and in accordance with our discussion of matrix elements for tunneling above.

The generally good agreement between the predictions of Landauer theory and the data in the Table 8.3 show that tunneling is the dominant mechanism whereby electrons cross the gap between the electrodes. But some examples of poor agreement are interesting. In particular, some of the larger aromatic molecules (entries **9**, **10**, and **12** in the table) have a measured conductance that is larger than the predicted tunnel conductance (measured values are otherwise smaller than predicted values). One explanation for this is a transition to hopping conductance at greater lengths.[37]

8.16   The transition from tunneling to hopping conductance in single molecules

Tunnel conductance falls exponentially with the length of a molecule,

$$G = G_0 \exp -2\kappa x = G_0 \exp -\beta x, \qquad (8.78)$$

where we have substituted β in place of the 2κ of Chapter 2 to conform to the chemical literature. Hopping conductance, at temperatures where thermally activated hopping is reasonably frequent, follows Ohm's law[37]

$$G \propto \frac{1}{x}. \tag{8.79}$$

The transition between these two regimes has been studied in DNA molecules using the method of repeated break junctions.[24] Guanine is easily oxidized, so GC base pairs might be expected to provide a site for charge localization that supports hopping conductance.[38] Adenine and thymine are not as easily oxidized, so are less likely to support hopping conductance. Xu et al.[24] measured the conductance of a series of DNA molecules that contained different lengths of AT base pairs or GC base pairs within a common flanking sequence of DNA. Their data for conductance, G, vs. length are shown in Fig. 8.29. The AT-rich DNA follows Equation 8.78, and therefore it appears to be a tunneling conductor. The GC-rich DNA follows Equation 8.79 and therefore it appears to be a hopping conductor.

The transition can be seen clearly by studying decay of current and its dependence on temperature as the length of a molecule is systematically increased.[39] Short molecules show a rapid exponential decay of current with no temperature

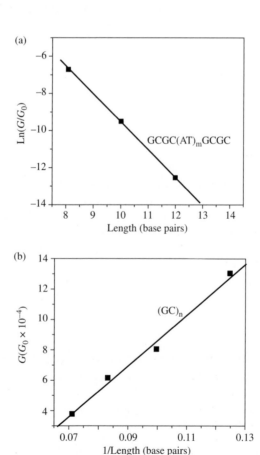

Fig. 8.29 Single molecule conductance of DNA molecules as a function of their length (in base pairs) for (a) DNA rich in AT base pairs. The conductance decays exponentially with length as expected for tunneling. The conductance of GC-rich DNA (b) falls with 1/L, as expected for hopping conductance. (Reprinted from Xu et al.[24] Nano Letters 2004 with permission of the American Chemical Society).

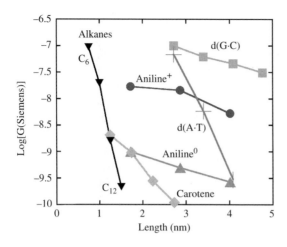

Fig. 8.30 Logarithm of single molecule conductance plotted as a function of the length of a molecule. The alkanes (inverted triangles) and carotenes (diamonds) have conductance values that agree quite well with tunneling conductance. GC-rich DNA, d(G·C), acts like a hopping conductor. Neutral (Aniline^0) and oxidized (Aniline$^+$) oligoanilines and AT-rich DNA all decay according to Equation 8.78, but carry significant currents over quite large distances.

dependence, whereas longer molecules display a slower decay of a current with a marked temperature dependence.[39]

A summary of single molecule conductances measured as a function of length for various oliogmers (molecules with repeated units) is shown in Fig. 8.30. The decays can be fitted to the tunneling equation (8.79) but the molecules to the right of the plot carry more current than expected on the basis of tunneling calculations. The plot shows that a type of molecular wire is indeed possible, if the goal is to transport charge over distances of 5 nm or so. This may prove technologically valuable as the dimensions in silicon-based circuitry approach this value.

8.17   Gating molecular conductance

We have just seen that the Landauer–Buttiker formula describes transport through a molecule connected to two metal wires quite well. How might we make a device that we could turn on and off with a control electrode analogous to gating a transistor? In conventional field effect transistors this is done by putting the main contacts (source and drain) at each end of a long channel, and placing a gate electrode very close to the channel over most of its length. As a result, the electric field between the gate and source (or gate and drain) dominates over the electric field between source and drain, so the potential of levels in the channel is controlled by the gate electrode. Presumably, something similar could work for a molecule. Molecules with levels far from the Fermi level would not conduct well (cf. the Breit–Wigner formula, Equation 7.38 in Chapter 7). As the electric field raises the levels close to resonance, the conduction would increase, approaching a quantum of conductance at resonance. There have been attempts to implement schemes like this, but they suffer from the constraints of geometry. Single molecules are so small, that it is hard to see how one would connect a "gate" electrode that had more influence on the potential at the center of the molecule than the "source" and "drain" contacts at the edges of the molecule. Nanowires, like carbon nanotubes, are an exception.

These are so long that attachment of a proper gate electrode is feasible and gating has been demonstrated.[40]

An alternative to placing a direct gating electrode onto the molecule is electrochemical gating, using the electric double layer itself to gate the molecule. This effect was first demonstrated in STM images of molecules containing Fe^{2+}/Fe^{3+}. The molecules were arranged in a monolayer on a graphite electrode and imaged as the potential of the surrounding electrolyte was changed electrochemically. The STM images of the iron-containing molecules were brightest when the electrode as poised near the formal potential (E^0) for the Fe^{2+}/Fe^{3+} equilibrium, implying that the molecules were most conductive at this point.[41] For a deeper understanding of what is going on here, we use an extended version of Equation 8.69 to include fluctuations around the formal potential, $E^0(= -\Delta G^0/e)$ writing, for the density of unoccupied states[42]

$$D(E) = \frac{1}{\pi \sqrt{4k_B T \lambda}} \exp\left[-\frac{(E - E^0 - \lambda)^2}{4k_B T \lambda}\right]. \qquad (8.80)$$

This is a Gaussian distribution that peaks when $E = E^0 + \lambda$ and shifts its peak with changes in the potential E. Thus, the density of states for tunneling is highest when the peak of this Gaussian lines up with the Fermi levels of the two electrodes. Figure 8.31 shows a schematic arrangement for an electrochemical gating experiment where energies are plotted vertically and distance along the tunneling direction is horizontal. Changing the potential of one electrode with respect to the reference moves the solution potential up and down with respect to the source and drain electrodes (the potential difference being dropped at the two double layers at the interfaces with the source and drain electrodes). If the molecule lies outside the double layer, its potential moves up and down with the

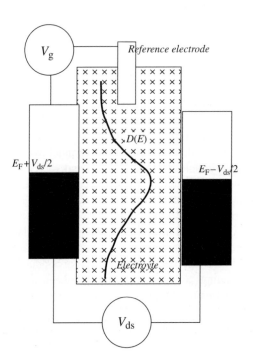

Fig. 8.31 Electrochemical gating of molecular conductance. The distribution of molecular energies (owing to the coupling of the molecule to the fluctuating environment), $D(E)$, is moved past the Fermi level as the solution potential is shifted relative to the source and drain electrodes by means of a gate voltage, V_g, applied with respect to a reference electrode.

solution potential, thus moving the peak in the density of states (Equation 8.80) with respect to the source and drain electrodes.

Direct gating of single molecule conductance using electrochemical potential control has been demonstrated using oligoaniline molecules[35] with thioacetate end groups that stick the molecules to gold electrodes:

Gating was expected to be observed, because polyaniline is a famous example of a conducting polymer. In its usual (neutral) form shown above, it has quite a wide HOMO–LUMO gap and is an insulator. When it is oxidized (electrochemically or by adding an oxidizing agent like iodine) it becomes an excellent conductor. A second oxidation is possible, causing the material to become insulating again. Figure 8.32(a) shows a cyclic voltammogram (the continuous line) for a layer of the oligoaniline molecules: the first peak near 0.2 V is the first oxidation and the second peak near 0.6 V is the second oxidation (which is not reversible—there is no negative-going peak on the return sweep). The data points are the measured single molecule conductance, and they show a peak conductance just after the first oxidation peak and before the second oxidation peak in the cyclic voltammogram.

The single molecule conductance data shown in Fig. 8.32(a) were taken with a tiny bias applied across the molecule, recording current as the potential of the electrolyte solution was changed. A very complicated response results if the electrochemical potential is held constant (near the potential for a peak conductance) while the source–drain bias is swept. The result of doing this is shown in Fig. 8.32(b). The current first rises, *and then falls*, as the source–drain electric

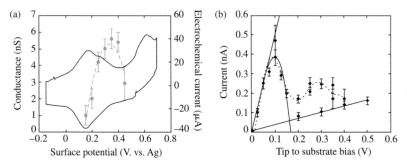

Fig. 8.32 Single molecule conductance measured as a function of the charge state of an oligoaniline molecule measured in an electrochemical break junction in 50 mM sulfuric acid (a). The solid line is a cyclic voltammogram showing the first (∼0.2 V) and second (∼0.6 V) oxidations with a reduction peak at ∼0.15 V on the return sweep. The first oxidation is reversible, and conductance data taken in this potential range (with a small applied tip-to-substrate bias) show that the molecule is most conductive just after the first oxidation. The dashed line is a fit of the single molecule conductance to a parabola. The current–voltage characteristic for the oxidized molecule (data points connected by the dashed line in (b)) falls with increasing bias (>0.1 V) as the applied bias reduces the molecule. The corresponding negative differential resistance peak is well fitted using the measured dependence of conductance on potential (solid curve). In contrast, when a nonconducting medium is used, the current–voltage data are linear (points connected by the straight line). (Adapted from data in Chen et al.[35])

field overcomes the gate electric field, resulting in a reduction of the molecule. The control of solution potential is not well established in this nanoscale geometry, because the molecule is partly buried in the double layer. Therefore, the applied bias influences the potential of the molecule as much as the gate potential, so the drain–source field can undo the oxidation originally driven by the surface (gate) field,[35] leading to a negative differential resistance.

The description of electrochemical gating given here is qualitative. We have ignored some important issues. Does the electron really tunnel "straight through" from one electrode to the other or does it reside on the molecule for some time? If charge does accumulate on the molecule, does the molecule remain charged up long enough to pay the full reorganization energy penalty, given the long time scales for nuclear motion? A full theory that allows for all these possibilities is very complicated, but it has been worked out.[43]

We noted earlier that molecule wires are *not* expected to exhibit metallic conductance and that is clearly the case for this oligoaniline. Its peak conductance is ~6 nS (Fig. 8.32(a)). A real "wire" in the junction would have the Landauer conductance of 77 μS. Thus, these data show that the metallic conduction observed in highly crystalline polyaniline[17] is a consequence of a different mechanism for conduction in three dimensions.

Electrochemical switching might indeed prove useful in manufacturable devices. A two-terminal molecular switch has been built that acts as a memory, changing its conductance as a function of its previous bias sweeps.[44] This device has been shown to reflect changes in the oxidation state of the molecule caused by the local electric field between the electrodes.[45]

8.18   Where is molecular electronics going?

We have seen the critical role that fluctuations and environment play in molecular charge transfer, and it should come as no surprise that single molecule electronic data show a marked molecule-to-molecule scatter.[29,35] Thus, molecular devices are unlikely to replace silicon as a means of circumventing the problem of statistical fluctuations in small devices. The problem may not be insurmountable. Proteins that act to transfer electrons in biology encase the redox active center in a "cage" of protein that provides a (predictably?) polarizable environment for electron transfer.

It is more likely that molecules will be used to make quite large devices which behave predictably because they contain many molecules. Their advantage will lie in cost and flexibility compared to conventional silicon devices. One obvious application is solar energy.[46]

One intriguing possibility is the direct combination of chemical recognition with electronic measurements. Ohshiro and Umezawa recently showed that electron tunneling through DNA bases was greatly enhanced when the bases were connected to an electrode by a complementary base.[47] This observation suggests that single molecules might be recognized by coupling them to electrodes with chemical recognition elements (such as the complementary bases of DNA) and measuring the tunnel current that arises as a consequence. If technology like this could be developed to sequence single DNA molecules economically, it could revolutionize medicine.

8.19 Bibliography

D.A. McQuarrie and J.D. Simon, Physical Chemistry: A Molecular Approach. 1997, Saucilito, CA: University Science Books. Excellent treatment of the electronic properties of molecules at a graduate physical chemistry level.

A.J. Bard and L.R. Faulkner, Electrochemical Methods: Fundamentals and Applications. 1980, NY: Wiley. The standard text for practicing electrochemists.

R. Farchioni and G. Grosso, Editors, Organic Electronic Materials. 2001, Berlin: Springer-Verlag. This is a volume of contributed chapters by many of the leading scientist in the field.

A. Aviram and M.A. Ratner, Editors, Molecular Electronics: Science and Technology (Annals of the New York Academy of Sciences). 1998, New York Academy of Sciences. This is an edited volume by two of the founders of the field of single molecule electronics. It is a little dated, but captures most of the important work at the time.

8.20 Exercises

1. A semiconductor nanoscale FET has an N-type channel that is 100 nm long. What is the smallest cross section (nm × nm) of the channel such that there is at least one phosphorus doping atom in the channel for an average doping level of 10^{16} cm^{-3}?

2. Draw Lewis structures for SiH_4, CO, SF_2, and NH_2OH. Note that CO is hard to do: you will end up giving carbon one less bond than it normally likes and oxygen one more than it normally likes. This results in the O having a positive "formal charge" and the C having a negative "formal charge" but the overall molecule is neutral.

3. Napthalene, $C_{10}H_8$, consists of two "fused" benzene-like rings (i.e., two rings that share two carbons on one edge). Draw three resonance structures.

4. (This problem is for calculus experts.) Obtain the results 8.25–8.27 by evaluating the Hamiltonian matrix elements. Use atomic units. From Table 2.1 and Appendix J the 1s orbitals in these units are $\left(\sqrt{\pi}\right)^{-1}\exp(-[\vec{r}-\vec{R}_A])$ and $\left(\sqrt{\pi}\right)^{-1}\exp(-[\vec{r}-\vec{R}_B])$ for nuclei at positions $\vec{R}_A$ and $\vec{R}_B$. To evaluate S_{ab} use elliptical coordinates:

$$\mu = \frac{\left(\left|\vec{r}-\vec{R}_A\right| + \left|\vec{r}-\vec{R}_B\right|\right)}{R}$$

$$v = \frac{\left(\left|\vec{r}-\vec{R}_A\right| - \left|\vec{r}-\vec{R}_B\right|\right)}{R}$$

where R is the distance that separates the nuclei.

To evaluate H_{aa}, use the original polar coordinates, and the fact that

$$\left\langle \psi_a \left| \frac{1}{2}\vec{\nabla}^2 - \frac{1}{r_a} \right| \psi_a \right\rangle = \frac{1}{2}$$

in atomic units (a statement that the ground state energy of the hydrogen atom is half a Hartree). Also remember that $\left(\vec{r} - \vec{R}_A\right)^2 = r^2 + R^2 - 2rR\cos(\theta)$ and do not forget to integrate over the second angle, ϕ, in the polar coordinate system.

5. Use Hückel π-electron theory to determine the π-orbital energies of the allyl cation $(C_3H_5)^+$ in terms of the Hückel parameters α and β. Identify the bonding, nonbonding, and antibonding orbitals. Treat this molecule as a linear chain of three carbon atoms.

6. The sum of the ionic radii for NaI is 3.17 Å. Estimate the room temperature solubility of NaI. Use 80 for the dielectric constant of water and assume $T = 300$ K.

7. Describe what (if anything) happens when:

   (a) Sodium salt electrolyte is in contact with a Cr electrode, a Ca electrode, a K electrode, and a Ni electrode.
   (b) Co salt electrolyte is in contact with a K electrode, a Sn electrode, a Fe electrode, and a Ba electrode?

8. In each case in Question 7 where there is a reaction, associate a potential with the half reaction and derive a driving free energy relative to $2H^+/H_2$ (assuming standard conditions) for each of the processes (ion being reduced, metal being oxidized). Then add the two free energies to derive the net free energy for the overall process (units of kJ/mole). Do not forget to keep track of the number of electrons involved and the number of moles of ion.

9. Given the data below for the reductions of benzophenone (BP) and benzoquinone (BQ), which of the following two reactions is favored?

$$BP^{2-} + BQ \Leftrightarrow BQ^{2-} + BP$$

$$BP + BQ^{2-} \Leftrightarrow BQ + BP^{2-}$$

$$BP + e \Leftrightarrow BP^{-\bullet}(-1.23 \text{ V})$$

$$BP^{-\bullet} + e \Leftrightarrow BP^{2-}(-1.76 \text{ V})$$

$$BQ + e \Leftrightarrow BQ^{-\bullet}(-0.54 \text{ V})$$

$$BQ^{-\bullet} + e \Leftrightarrow BQ^{2-}(-1.4 \text{ V})$$

10. For a molecule with a diameter of 1 nm in water, estimate λ (in units of eV) due to environmental polarization assuming the dielectric constant of water near the charged molecule is 5. Take $z = 1$.

11. A carbon–carbon bond has an IR stretching mode frequency of 1000 cm^{-1} (1 cm^{-1} is 3×10^{10} Hz). (a) Use the reduced mass of the C–C pair (for equal masses, m, the reduced mass is $m/2$) and estimate the spring constant of the C–C bond in N/m. (b) If charge transfer causes a bond

length change of 0.01 nm, what is the corresponding reorganization energy in eV?

12. Calculate the normalization constant in 8.69. Use the fact that $\int_{-\infty}^{\infty} \exp(-x^2)\,dx = \sqrt{\pi}$.

13. Make a log (base e) plot of the relative rate of an electron transfer reaction as a function of ΔG^0 assuming $\lambda = 0.5$ eV, for ΔG^0 running from $+1$ V to -1 V for $T = 300$ K. To do this, ignore the constant prefactor in the Marcus expression and just plot the exponent vs. the natural logarithm of rate.

14. Show that Equation 8.37a follows from Equation 8.36.

15. Minimize the argument of Equation 8.73 to obtain Equation 8.74.

References

[1]   Service, R.F., Breakthrough of the year: molecules get wired. Science, **294**: 2442–2443 (2001).

[2]   Chen, J., W. Wang, J. Klemic, M.A. Reed, B.W. Axelrod, D.M. Kaschak, A.M. Rawlett, D.W. Price, S.M. Dirk, J.M. Tour, et al., Molecular wires, switches, and memories. Ann. N.Y. Acad. Sci., **960** (Molecular Electronics II): 69–99 (2002).

[3]   McCoy, H.N. and W.C. Moore, Organic amalgams: substances with metallic properties composed in part of non-metallic elements. J. Am. Chem. Soc., **33**: 273–292 (1911).

[4]   Aviram, A. and M.A. Ratner, Molecular rectifiers. Chem. Phys. Lett., **29**: 277–283 (1974).

[5]   Kodis, G., P.A. Liddell, A.L. Moore, T.A. Moore, and D. Gust, Synthesis and photochemistry of a carotene-porphyrin-fullerene model photosynthetic reaction center. J. Phys. Org. Chem., **17**: 724–734 (2004).

[6]   Slater, J.C., *Quantum Theory of Molecules and Solids*. 1963, New York: John Wiley.

[7]   McQuarrie, D.A. and J.D. Simon, *Physical Chemistry: A Molecular Approach*. 1997, Sausalito, CA: University Science Books.

[8]   Bard, A.J. and L.R. Faulkner, *Electrochemical Methods: Fundamentals and Applications*. 1980, New York: John Wiley.

[9]   Israelachvilli, J.N., *Intermolecular and Surface Forces*. 2nd ed. 1991, New York: Academic Press.

[10] Reiss, H. and A. Heller, The absolute potential of the standard hydrogen electrode: a new estimate. J. Phys. Chem., **89**: 4207–4213 (1985).

[11] Steinberg-Yfrach, G., J.-L. Rigaud, E.N. Durantini, A.L. Moore, D. Gust, and T.A. Moore, Light-driven production at ATP catalyzed by F0F1-ATP synthase in an artificial photosynthetic membrane. Nature, **392**: 479–482 (1998).

[12] Marcus, R.A., On the theory of electron-transfer reactions. VI: Unified treatment for homogeneous and electrode reactions. J. Phys. Chem., **43**: 679–701 (1965).

[13] Schmickler, W., *Interfacial Electrochemistry*. 1996, Oxford: Oxford University Press.

[14] Chidsey, C.E.D., Free energy and temperature dependence of electron transfer at the metal-electrolyte interface. Science, **251**: 919–922 (1991).

[15]   Chaing, C.K., C.R. Fincher, Y.W. Park, A.J. Heeger, H. Shirakawa, E.J. Louis, S.C. Gau, and A. McDiarmid, Electrical conductivity in doped polyacetylene. Phys. Rev. Lett., **39**: 1098–1101 (1977).

[16]   Bredas, J.-L. and G.B. Street, Polarons, bipolarons and solitons in conducting polymers. Acc. Chem. Res., **18**: 309–315 (1985).

[17]   Lee, K., S. Cho, S.H. Park, A.J. Heeger, C.-W. Lee, and S.-H. Lee, Metallic transport in polyaniline. Nature, **441**: 61–68 (2006).

[18]   Giamarchi, T., Theoretical framework for quasi one dimensional systems. Chem. Rev., **104**: 5037–5055 (2004).

[19]   Fagas, G. and K. Richter, *Introducing Molecular Electronics*. 2005, Berlin: Springer.

[20]   Dunlap, D.D., R. Garcia, E. Schabtach, and C. Bustamante, Masking generates contiguous segments of metal coated and bare DNA for STM imaging. Proc. Natl. Acad. Sci. USA, **90**: 7652–7655 (1993).

[21]   Porath, D., A. Bezryadin, S. de Vries, and C. Dekkar, Direct measurement of electrical transport through DNA molecules. Nature, **403**: 635–638 (2000).

[22]   Fink, H.-W. and C. Schoenberger, Electrical conduction through DNA molecules. Nature, **398**: 407–410 (1999).

[23]   Kasumov, A.Y., M. Kociak, S. Guéron, B. Reulet, V.T. Volkov, D.V. Klinov, and H. Bouchiat, Proximity-induced superconductivity in DNA. Science, **291**: 280–282 (2001).

[24]   Xu, B., P.M. Zhang, X.L. Li, and N.J. Tao, Direct conductance measurement of single DNA molecules in aqueous solution. Nano Letts., **4**: 1105–1108 (2004).

[25]   Cui, X.D., A. Primak, X. Zarate, J. Tomfohr, O.F. Sankey, A.L. Moore, T.A. Moore, D. Gust, H. G., and S.M. Lindsay, Reproducible measurement of single-molecule conductivity. Science, **294**: 571–574 (2001).

[26]   Morita, T. and S.M. Lindsay, Determination of single molecule conductances of alkanedithiols by conducting-atomic force microscopy with large gold nanoparticles. J. Am. Chem. Soc., **129**: 7262–7263 (2007).

[27]   Xu, B. and N.J. Tao, Measurement of single-molecule resistance by repeated formation of molecular junctions. Science, **301**: 1221–1223 (2003).

[28]   Tomfohr, J.K. and O.F. Sankey, Theoretical analysis of electron transport through organic molecules. J. Chem. Phys., **120**: 1542–1554 (2004).

[29]   Lindsay, S.M., Molecular wires and devices: advances and issues. Faraday Discussions, **131**: 403–409 (2006).

[30]   Lindsay, S.M. and M.A. Ratner, Molecular transport junctions: clearing mists. Adv. Mater., **19**: 23–31 (2007).

[31]   Xiao, X., B.Q. Xu, and N.J. Tao, Measurement of single molecule conductance: benzenedithiol and benzenedimethanethiol. Nano Letts., **4**: 267–271 (2004).

[32]   He, J., F. Chen, J. Li, O.F. Sankey, Y. Terazono, C. Herrero, D. Gust, T.A. Moore, A.L. Moore, and S.M. Lindsay, Electronic decay constant of carotenoid polyenes from single-molecule measurements. J. Am. Chem. Soc. (Commun.), **127**: 1384–1385 (2005).

[33]   He, J., F. Chen, P.A. Liddell, J. Andréasson, S.D. Straight, D. Gust, T.A. Moore, A.L. Moore, J. Li, O.F. Sankey, et al., Switching of a

photochromic molecule on gold electrodes: single molecule measurements. Nanotechnology, **16**: 695–702 (2005).

[34] Xiao, X., L. Nagahara, A. Rawlett, and N.J. Tao, Electrochemical gate controlled conductance of single oligo(phenylene ethynylene)s. J. Am. Chem. Soc., **127**: 9235 (2005).

[35] Chen, F., J. He, C. Nuckolls, T. Roberts, J. Klare, and S.M. Lindsay, A molecular switch based on potential-induced changes of oxidation state. Nano Letts., **5**: 503–506 (2005).

[36] Venkataraman, L., J.E. Klare, C. Nuckolls, M.S. Hybertsen, and M.L. Steigerwald, Dependence of single-molecule junction conductance on molecular conformation. Nature, **442**: 905–907 (2006).

[37] Segal, D., A. Nitzan, W.B. Davis, M.R. Wasielewski, and M.A. Ratner, Electron transfer rates in bridged molecular systems. 2. A steady-state analysis of coherent tunneling and thermal transitions. J. Phys. Chem. B, **104**: 3817–3829 (2000).

[38] Barnett, R.N., C.L. Cleveland, A. Joy, U. Landman, and G.B. Schuster, Charge migration in DNA: ion-gated transport. Science, **294**: 567–571 (2001).

[39] Choi, S.H., B. Kim, and C.D. Frisbie, Electrical resistance of long conjugated molecular wires. Science, **320**: 1482–1486 (2008).

[40] Bachtold, A., P. Hadley, T. Nakanishi, and C. Dekker, Logic circuits with carbon nanotube transistors. Science, **294**: 1317–1320 (2001).

[41] Tao, N., Probing potential-tuned resonant tunneling through redox molecules with scanning tunneling microscopy. Phys. Rev. Lett., **76**: 4066–4069 (1996).

[42] Schmickler, W. and N.J. Tao, Measuring the inverted region of an electron transfer reaction with a scanning tunneling microscope. Electrochimica Acta, **42**: 2809–2815 (1997).

[43] Zhang, J., Q. Chi, A.M. Kuznetsov, A.G. Hansen, H. Wackerbarth, H.E.M. Christensen, J.E.T. Andersen, and J. Ulstrup, Electronic properties of functional biomolecules at metal/aqueous solution interfaces. J. Phys. Chem B, **106**: 1131–1152 (2002).

[44] Blum, A.S., J.G. Kushmerick, D.P. Long, C.H.P. Patterson, J.C. Yang, J.C. Henderson, Y. Ya, J.M. Tour, R. Shashidhar, and B. Ratna, Molecularly inherent voltage-controlled conductance switching. Nature Mater., **4**: 167–172 (2005).

[45] He, J., Q. Fu, S.M. Lindsay, J.W. Ciszek, and J.M. Tour, Electrochemical origin of voltage-controlled molecular conductance switching. J. Am. Chem. Soc., **128**: 14828–14835 (2006).

[46] Kim, J.Y., K. Lee, N.E. Coates, D. Moses, T.-Q. Nguyen, M. Dante, and A.J. Heeger, Efficient tandem polymer solar cells fabricated by all solution processing. Science, **317**: 222–225 (2007).

[47] Ohshiro, T. and Y. Umezawa, Complementary base-pair-facilitated electron tunneling for electrically pinpointing complementary nucleobases. Proc. Nat. Acad. Sci. USA, **103**: 10–14 (2006).

9 Nanostructured materials

The science of nanostructured materials is evolving so rapidly that it is almost impossible to give a current account of the field, let alone one that would survive for any period of time in a text book. Therefore, the goal of this chapter is to outline some of the principles that underlie the nanoengineering of materials. The meaning of a nanostructure is context dependent, in the sense that the length scales associated with various physical phenomena that one might wish to engineer can be very different. Thus, we start with electronic properties in which the length scales are the Fermi wavelength and the mean free path of electrons. We then turn to optical materials where the length scale is the wavelength of light. Magnetic properties arise from a complex set of interactions that operate over many length scales. Thermal properties depend on the scattering length for thermal vibrations. In nanofluidic devices, length scales are chosen to optimize interactions with macromolecules dissolved in fluids or between fluids and the walls of a container. The chapter ends with a look at biomimetic materials, based on copying biological nanostructures, developed through the process of evolution in living systems. The effects of nanostructuring are different depending on the number of dimensions that are constrained to have nanoscale sizes, so for this reason I have generally given examples of nanostructuring in two (layers), one (wires), and zero dimensions (dots) for each class of material property, where appropriate.

9.1 What is gained by nanostructuring materials?

Nanostructured materials derive their special properties from having one or more dimension(s) made small compared to a length scale critical to the physics of the process (Table 9.1). This could be a wavelength, for example, the Fermi wavelength of electrons in electronic materials or the wavelength of light in photonic materials. It could be a mean distance between scattering events; for example, the mean free path of electrons for electronic materials and the mean free path of thermal vibrations (phonons) for thermal materials. Magnetic properties are determined by quite local interactions (Coulomb interactions between electrons and the "exchange interaction") though once magnetic domains form, dipolar interactions between ordered magnetic regions are significant over rather large distances. Interactions in fluids can occur over very long distances indeed, depending on the viscosity of the medium, length scale of the confining structure, and the speed of fluid motion. For this reason, it can be quite hard

Table 9.1 Length scales in various material phenomena

Phenomenon	Electronic transport	Optical interactions	Magnetic interactions	Thermal	Fluidic interactions
Physics	Fermi wavelength, λ_F Scattering length, ℓ	Half-wavelength of light in medium, $\lambda/2n$	Range of exchange interactions, range of magnetic dipole interactions	Phonon mean free path	Boundary layers, molecular dimensions
Length scale	$\lambda_F \approx 1$ Å $\ell \approx 10 - 100$ nm	100–300 nm	Exchange interactions: 1–100 Å, dipolar interactions, up to microns	Hundreds of nm at 300 K to very large at low T	Always in the low Reynolds number limit: radius of gyration for dissolved molecules

to get fluid to flow into nanostructures, and chemical interactions between the walls of a structure and the fluid can dominate. Novel effects are observed when the dimensions of the confining structures are made comparable to the size of macromolecules in the solution. We will discuss the applications of nanostructuring in each of these applications below. (A comprehensive treatment of nanomaterials can be found in the book by Ozin and Arsenault[1]).

9.2 Nanostructures for electronics

We have already discussed some important nanoscale electronic phenomena in Chapter 7, understanding, for example, how a small particle (with a small capacitance) can act as a switch, owing to its Coulomb charging energy. Here, we take a look at another aspect of electronic nanostructures which is what might be called "density of states engineering."

In order to understand how size affects material properties through the density of states, it is useful to consider how band structure evolves as a function of the number of atoms in a cluster. This process is illustrated in Fig. 9.1. Each new pair of interacting atoms produce two new states for each atomic state shown as the antibonding and bonding states in the figure. As more atoms are added to the cluster, weaker, longer-range interactions give rise to further small splittings of each of the levels. Eventually, when the cluster size is very large, these two sets of split states (associated with the original bonding and antibonding states) become the valence and conduction bands. Each band has a width that reflects the interaction between atoms, with a bandgap between the valence and conduction bands that reflects the original separation of the bonding and antibonding states. As can be seen from Fig. 9.1, the largest density of states occurs at the center of each of the bands, as new states evolve from the original bonding and antibonding orbitals. (This argument ignores the distortion of band structure that occurs in a periodic lattice where the density of states becomes very high at wave vectors corresponding to a Bragg reflection.) The onset of delocalized transport will occur when the separation between levels gets smaller than thermal energy, so a metal with a half-filled band will appear to be more like a bulk metal at a smaller size than a semiconductor nanoparticle, which has to be much bigger to appear to have bulk properties. This is because the band edges play a role in semiconductor conductivity (e.g., hole transport), whereas transport in a metal is dominated by states at the Femi energy.

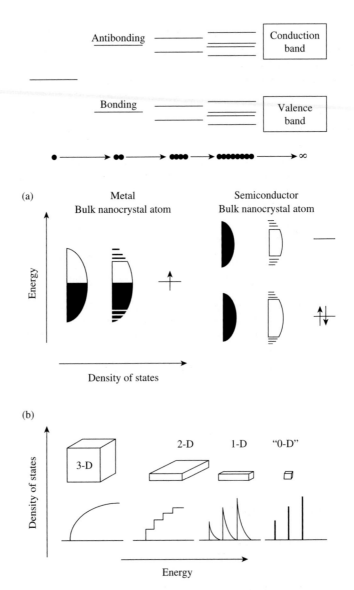

Fig. 9.1 1-Dimensional illustration of the development of bandstructure as a function of cluster size. The evolution of conduction and valence bands (infinite chain limit) from bonding and antibonding states (pair of atoms) is shown from left to right. Note how the bands build up from the center outward.

Fig. 9.2 (a) Density of states for bulk metals and metal nanocrystals compared to bulk semiconductors vs. semiconductor nanocrystals. The structure at the band edges reflects the incomplete build up of the bandstructure in nanoparticles (Fig. 9.1). (b) Density of states for bulk, 2-D, 1-D, and "0-D" structures. Only one band is shown in the 3-D case, whereas bands arising from multiple states are shown in the other examples. (Reprinted from Alivisatos,[2] Science 1996, with permission from the AAAS.)

This difference between metals and semiconductors is illustrated in Fig. 9.2(a), which shows the filled part of the bands as the black area and the band structure as a density of states. This presentation is derived from the E vs. k picture as described in Chapter 7, and we will describe below how the density of states is enumerated in the various types of structures. When levels are very close together, the density of states is taken to be a continuous function, as is the case for the bulk material. In the case of a nanocluster, the density of states will appear continuous near the center of the band but becomes composed of discrete levels (with significant energy separations compared to $k_B T$) at the edges of the band. For this reason, metal nanoparticles have to be very small before size quantization effects are seen. In contrast, semiconductors, where transport is dominated by carriers at the band edges, show quantum size effects at much larger length scales. Thus, a 5 nm gold nanoparticle shows no evidence of quantized electronic levels at room temperature, whereas semiconductor quantum dots show size-dependent electronic properties for dots containing even as many as 10,000 atoms.

The preceding discussion was limited to nanostructures that are small in all three spatial directions. We will call these zero-dimensional ("0-D") structures. Many interesting properties of nanostructures arise from having a nanoscale dimension in just one ("2-D" structures) or two ("1-D" structures) dimensions. Figure 9.2(b) shows how the density of states varies as a function of the dimensionality of the system. These are plots of the density of states per unit energy and we will derive their form below. (The density of states is shown for just one band in the bulk solid (3-D) case, and for a series of bands or atomic levels in the other cases—2-D, 1-D, 0-D).

Before discussing these various types of density of states, we turn to what we mean by restricted dimensionality. In the case of electrons, the system is considered to have restricted dimensionality if some dimension approaches the Fermi wavelength. Similarly for vibration or optical properties, the relevant dimensions are those of the vibration or the optical excitation.

We derived an expression for the number of states per unit wave vector in Chapter 2. To do this, we counted the number of states per unit k, assuming that the k states were evenly distributed as a function of k, and then divided by the volume of a single k state for the case of three dimensions. This yielded the following result (Equation 2.54) for the density of states in a volume V:

$$\frac{dn}{dk} = \frac{Vk^2}{2\pi^2} \tag{9.1}$$

and for a free electron gas, $E = \frac{\hbar^2 k^2}{2m}$, so

$$\frac{dE}{dk} = \frac{\hbar^2 k}{m}. \tag{9.2}$$

Thus we obtain, for the number of states per unit energy,

$$\frac{dn}{dE} = \frac{Vk^2}{2\pi^2}\frac{m}{\hbar^2 k} = \frac{Vm}{\hbar^2 2\pi^2}\sqrt{\frac{2mE}{\hbar^2}} \propto E^{1/2}. \tag{9.3}$$

Thus in three dimensions, the number of states unit energy increases as the square root of energy. This is the form of the energy dependence shown in the left panel of Fig. 9.2(b). Repeating the exercise for two dimensions, the number of k states in area A is given by

$$\frac{dn}{dk} = \frac{A2\pi k}{(2\pi)^2} \tag{9.4}$$

yielding a constant value for the density of states per unit energy

$$\frac{dn}{dn}dE = \frac{dn}{dk} \cdot \frac{dk}{dE} = \frac{Am}{2\pi\hbar^2}. \tag{9.5}$$

This result is illustrated in the plot labeled "2-D" in Fig. 9.2(b), where each atomic orbital corresponds to one of the plateaus of constant density of states in the two-dimensional solid.

In one dimension, the density of states the unit k is given by

$$\frac{dn}{dk} = \frac{L}{2\pi} \tag{9.6}$$

leading to

$$\frac{dn}{dE} = \frac{Lm}{2\pi \hbar^2 k} \propto E^{-1/2}. \tag{9.7}$$

Thus, at each atomic orbital energy, the density of states in the 1-D solid decreases as the *reciprocal* of the square root of energy. This is shown in the plot labeled "1-D" in Fig. 9.2(b).

Finally in zero dimensions we recover the original result for atoms: that is to say, the energy states are sharp levels corresponding to the eigenstates of the system. This result is illustrated in the right panel of Fig. 9.2(b) (labeled "0-D").

These results show how the energy is concentrated into distinct levels as the dimensionality of the system is lowered. Thus, if a material can be made from a dense assembly of nanostructures, very high densities of states can be achieved in a small energy range. Nanostructuring represents an approach to engineering the density of states of materials.

9.3   Zero-dimensional electronic structures: quantum dots

Semiconductor quantum dots are finding widespread use as highly efficient fluorescent labels (see Fig. 2.9) owing to the ease with which the gap between the ground and the excited states can be tuned and simply by changing the size of the dots[2,3] (some techniques for synthesizing the structures were discussed in Chapter 6).

Light incident on a dielectric solid at an energy greater than the bandgap forms an excitation called an exciton. An electron is promoted from the valence band into the conduction band leaving a positive hole behind. The combination of electron and hole forms the bound state that is the exciton. The exciton may be analyzed by analogy with the Bohr atom (though this is a poor analogy in a periodic solid because the exciton is actually an eigenstate of the periodic solid and should be composed of Bloch states). We can estimate the radius of the exciton using the Bohr model of the atom.

Setting the centripetal acceleration of the electron equal to the Coulomb force between the electron and the hole gives

$$\frac{mV^2}{r} = \frac{1}{4\pi \varepsilon \varepsilon_0} \frac{e^2}{r^2}$$

From which

$$r = \frac{4\pi \varepsilon \varepsilon_0 \hbar^2}{e^2 m^*}, \tag{9.8}$$

where we have used an effective mass m^* to emphasize the point that this parameter depends on the band structure of the solid. The radius of the "Bohr atom" formed from the bound electron–hole state is thus increased by the dielectric constant of the medium (with other changes if $m^* \neq m_e$). In a semiconductor, this radius can be tens of Å.

When the size of the nanoparticle approaches that of an exciton, size quantization sets in. The energy of the electronic bandgap of a semiconductor particle is given approximately by[1]

$$\Delta E = \Delta E^0 + \frac{\hbar^2 \pi^2}{2R^2} \left(\frac{1}{m_e^*} + \frac{1}{m_h^*} \right) - 1.8 \frac{e^2}{\varepsilon R}, \tag{9.9}$$

where ΔE^0 is the intrinsic bandgap of the semiconductor, m_e^* and m_h^* are the effective masses of electrons and holes, and R is the radius of the nanoparticle. The last term in this expression represents a correction for Coulomb repulsion.

9.4 Nanowires

Carbon nanotubes are probably the most well known of nanomaterials, so we begin our discussion with them. It may be the case that the semiconductor wires introduced in Chapter 6 will prove to be at least as useful, because their electronic states are easier to engineer.

Carbon nanotubes were discovered by Sumio Iijima in 1991 who observed them in high-resolution electron microscope images of the by-products of combustion of carbonaceous materials.[4] They consist of tubes of graphite-like material closed in on itself to form cylinders with diameters ranging from ~1 nm up to several nanometers but with lengths that can approach millimeters or more. Many preparations produce multiwalled nanotubes, that is, concentric arrangements of cylinders of ever increasing diameter. With their extraordinary aspect ratio (nm diameters, cm lengths) strong local covalent structure, and long-range structure that is essentially free of defects, these tubes have some remarkable properties. They can be added to composite materials to greatly enhance their strength. They have interesting electronic properties. They are not true 1-D wires (in the sense of the Luttinger liquids mentioned in Chapter 8). Their 3-D structure gives them the potential to act as both metallic interconnects between components in integrated circuits and as the semiconducting channels of field effect transistors. They are also candidate materials for components of nanomechanical actuators.[5] Carbon nanotubes are found naturally in the combustion products of graphitic electrodes and are also readily synthesized by a process of chemical vapor deposition, using iron or cobalt seed particles as catalysts. In what follows we will focus on their electronic properties.

9.4.1 Electronic properties of carbon nanotubes

The electronic properties of carbon nanotubes follow from the electronic properties of its parent, a graphene sheet. The way in which various tubes can be formed by folding a graphene sheet is shown in Fig. 9.3. Given a particular wrapping vector (defined in Fig. 9.3) with components n_1 and n_2, the diameter

Fig. 9.3 (a) Shows the structure of a graphene
sheet. The shaded region on the figure is the
region that is folded to make the nanotube
shown in (b). The geometry of the folded tube
is uniquely described by a wrapping vector
that points across the sheet to connect atoms
that coincide in the tube. The wrapping vector is described by a linear combination of
the basis vectors $\mathbf{a}_1$ and $\mathbf{a}_2$ of the graphene
sheet. The tube shown in (b) is constructed
with a wrapping vector $5\mathbf{a}_1+5\mathbf{a}_2$. Different
wrapping vectors produce sheets of different
diameter and different relative orientations of
the lattice vectors with respect to the long axis
of the tube as shown in (c). (Courtesy of E.D.
Minot.)

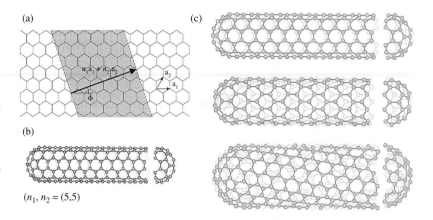

(a)

(c)

(b)

$(n_1, n_2 = (5,5)$

of the tube is in given by

$$d = \left(n_1^2 + n_2^2 + n_1 n_2\right)^{1/2} 0.0783 \text{ nm}. \tag{9.10}$$

The folding of the sheet controls the electronic properties of the nanotubes.
Each carbon atom contributes three electrons to s-type bonding orbitals and a
fourth to a p_z state. These p_z states hybridize to form π and π^* valence and
conduction bands that are separated by an energy gap of about half a volt. This
is a consequence of the 2-D band structure of this material. The material is not,
however, a semiconductor. This is because for certain high symmetry directions
in the crystal (the **K** points in the reciprocal lattice), the wavefunction in the
valence band is required, by symmetry, to be the same as the wavefunction in the
conduction band. Thus, for just these certain directions, the material behaves
like a metal. This limited conductivity gives graphene properties intermediate
between those of a semiconductor and a metal. It is a semimetal. When the
conduction and valence bands are plotted as a 3-D plot with a vertical axis
corresponding to energy and the horizontal plane corresponding to the k_x and
k_y directions, the points where the bands meet look like sharp cones that are
joined at the apex (where the top of the conduction band meets the bottom of
the valence band at the K points). Plots of the band structure are often shown
as cones in energy that intersect a hexagonal lattice at the K points.[7]

When the graphene sheet is folded into a tube, the wave vector component
along the long axis of the tube is allowed to take on any value. However, the
component of the wave vector perpendicular to the long axis of the tube is a
quantized. Specifically,

$$\pi D k_{\text{perpendiular}} = 2\pi n, \tag{9.11}$$

where D is the diameter of the tube, $k_{\text{perpendicular}}$ is the value of the wave vector
perpendicular to the long axis of the tube, and n is an integer. The conductivity
of the tube is determined by whether or not one of these allowed values of k
intersects with the **k** points at which the conduction and valence bands meet.
This geometrical construct for finding these points is illustrated in Fig. 9.4.

The geometry of graphene leads to a set of simple conditions for the formation
of conducting tubes. One simple case is when n_1 is a multiple of three, but other

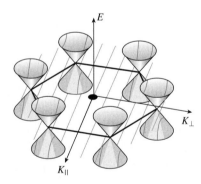

Fig. 9.4 Illustrating how the allowed values
of $k_{\text{perpendicular}}$ control the conductivity of the
nanotube. The allowed values are shown for
some particular tube diameter by the parallel
lines in the horizontal plane on the figure. The
K points where the conduction and valence
bands meet are shown at the intersections
of the cones of allowed energy values. The
diameters for which the allowed values of
$k_{\text{perpendicular}}$ precisely intersect these cones
give rise to metallic nanotubes. Other diameters give rise to semiconducting nanotubes
with a bandgap of ~ 0.5 V. (From the PhD
thesis of Ethan Minot[6], with permission from
Michael Fuhrer. Courtesy of E.D. Minot and
M. Fuhrer.)

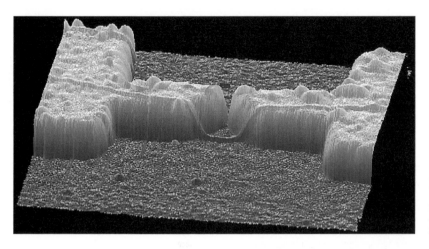

Fig. 9.5 Field effect transistor made from a single semiconducting carbon nanotube connecting source and drain connectors made by e-beam lithography (courtesy of Dr. Larry Nagahara).

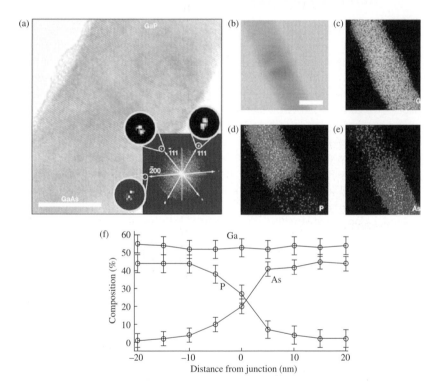

Fig. 9.6 GaAs/GaP nanowire heterostructures. (a) TEM image of a GaAs/GaP junction grown from a 20 nm gold nanocluster catalyst. (b) TEM image of another junction. (c)–(e) Elemental mapping of the Ga (c), P (d), and As (e) content of the junction shown in (b). (f) Line profiles of the composition through the junction region, showing the change in composition as a function of the distance. Ga, P, and As. (From Fig. 6 of the review by Lu and Lieber.[8]) (Reprinted with permission from Lu and Lieber[8], copyright 2006 IOP Publishing Ltd. Courtesy of Professor Wei Lu.)

cases exist.[6] Thus, the carbon nanotubes can be either metals or semiconductors depending on their chirality. Semiconducting tubes have been used to make field effect transistor channels (as illustrated in Fig. 9.5). Metallic tubes have been used to make interconnects between other electronic components. The tubes possess a high density of states near the bottom of the band (Fig. 9.2) and, owing to the robustness of carbon–carbon bonding, can be made to be essentially defect free over very long distances. However, at the time of writing, no industrial scale process has been found for separating metallic tubes from semiconductor tubes or for making pure preparations of one type or the other.

This problem of control of material properties does not apply to the semiconductor nanotubes discussed in Section 6.8. The electronic properties of semiconductor wires can be controlled by doping the semiconductors. Functional devices can be made by building heterostructures using successive growth of different types of semiconductor. A junction between GaP and GaAs wires is shown in Fig. 9.6. Engineering of the properties of wires through the growth of 1-D heterostructures has permitted the development of "designer" contacts that allow the Fermi levels of devices to be matched to the Fermi level of metal contacts, permitting highly efficient electron injection and the construction of high-performance transistors.[8]

9.5 2-D nanoelectronics: superlattices and heterostructures

In 1970, Esaki and Tsu proposed a new approach for synthetically engineering the bandgap of semiconductors.[9] Their proposal was to layer different semiconductors in an alternating sandwich so as to form a lattice composed of two types of atomic lattices—a so-called *superlattice* structure. A plot of electron energy in such a structure is shown in Fig. 9.7. This is taken from the original paper by Esaki and Tsu[9] with labeling added to show where present day compound semiconductors would be located (GaAs and GaAlAs). No reasonable approach for making such structures existed at the time of the original proposal, but superlattices are readily fabricated nowadays using molecular beam epitaxy (Chapter 5) to put down alternating layers of small bandgap compound semiconductors (such as GaAs) interspersed with layers of wide bandgap compound semiconductors (such as GaAlAs).

Modulation of the structure on this longer length scale d (the thickness of one layer in the superlattice) introduces new bandgaps inside the original Brillouin zone (cf. the discussion of the Peirels distortion discussed at the end of Chapter 7). The formation of these "subbands" is illustrated in Fig. 9.8 (where the free electron dispersion for the original atomic lattice is shown as a dashed line). Near the center of the bands, the slope of these curves is nearly constant, reflecting the 2-D density of states (Fig. 9.2), but the bands curve steeply close to the new Brillouin zone boundaries.

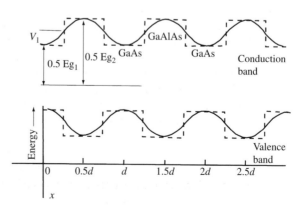

Fig. 9.7 Variation of electron energy in the conduction and valence bands in a super lattice. (Modified after Fig. 1 of Esaki and Tsu.[9])

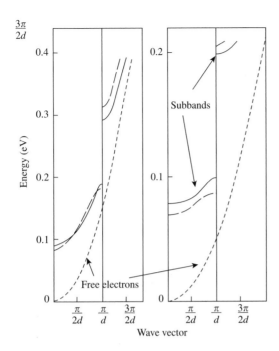

Fig. 9.8 Subband formation for two different sets of effective masses and coupling strengths for a sinusoidal modulation (solid lines) and a square well modulation (e.g., GaAs/GaAlAs—long dashed line). The free electron dispersion is shown by the short dashed lines. (After Fig. 3 of Esaki and Tsu.[9])

A key feature of the superlattices is that the thickness of each layer is considerably smaller than the electron mean free path in the bulk semiconductor. Thus, electrons within one subband can pass freely from one small bandgap region to another without scattering. The result can be a significant increase in carrier mobility, so these superlattice semiconductors find application in microwave devices. One particularly interesting consequence of this high mobility, and the fact that the electron's wave vector reaches a Brillouin zone boundary at lower energies than would be the case in a bulk semiconductor, is that these devices make it possible to use an electric field to accelerate electrons up to an energy such that they are Bragg-reflected. This long sought-after effect was first looked for in bulk semiconductors where Felix Bloch (of "Bloch's theorem") had predicted oscillations in the conductivity as electrons are accelerated up to a band edge energy, only to be reflected back again (until the potential is large enough to excite the electrons to the next allowed band). The predicted "Bloch oscillations" have not been observed in bulk semiconductors, but similar oscillations are observed in the conductivity of the semiconductor superlattices. This effect is shown in Fig. 9.9. This is a plot of current vs. bias, showing the phenomenon of negative differential resistance as electrons slow down with increasing bias when they approach the first subband boundary ∼20 mV. The resistance of the device continues to increase until the electric field is large enough to align the second subband in one layer of the superlattice with the first subband in another. This nonlinear behavior (negative differential resistance) can be exploited to make oscillators and switches because the nonmonotonic current–voltage curve makes the device inherently unstable.

Layered semiconductors are also used as traps for excited carriers in semiconductor lasers, where, rather than using a superlattice with carriers that are mobile in subbands, a single layer of low bandgap material is sandwiched between two materials of higher bandgap. This application is discussed in Section 9.7. At

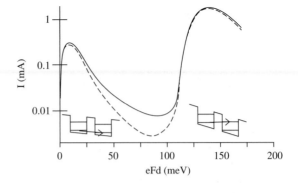

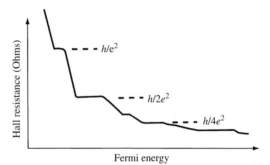

Fig. 9.10 The quantum Hall effect, Hall resistance vs. energy at a fixed magnetic field. (Based on Klaus von Klitzing's Nobel Prize speech, Fig. 4.[11])

low temperatures and in small electric fields, electrons can be confined entirely within a low bandgap material in structures like this. This is a consequence of the quantization of the wave vector normal to the layers. The smallest allowed value of wave vector in this direction is $2\pi/d$ so that if both the temperature and the electric field are small enough, electron motion is confined entirely to within the low bandgap plane. When a magnetic field, B, is applied perpendicular to the plane, electrons are driven into circular motion around the direction of the magnetic field at an angular speed given by the cyclotron frequency

$$\omega_c = \frac{eB}{m}. \tag{9.12}$$

The band structure in the plane determines whether or not these cyclotron orbitals can be occupied. The consequence is a series of steps in the Hall resistance (resistance measured perpendicular to the direction of the electric field) when Hall resistance is plotted as a function of the electron energy. Remarkably, these steps correspond to exactly twice the Landauer resistance as shown in Fig. 9.10. This quantum of resistance is known as the von Klitzing resistance after its discoverer (who won the 1985 Nobel Prize in physics for the discovery[11]). It is the same resistance that controls localization in quantum dot devices discussed in Chapter 7. An effect somewhat like this had been predicted earlier, but the remarkable outcome of von Klitzing's experiments was that the Hall resistance in these 2-D electron systems is quantized *exactly* according to

$$R_{\mathrm{H}} = \frac{h}{ne^2}. \tag{9.13}$$

As a result, the quantum Hall effect has become the most reliable method for determining the ratio of e to h.

9.6   Photonic applications of nanoparticles

Small (less than the wavelength of light) metal particles exhibit a phenomenon called "plasma resonance," with many potential uses as a way for concentrating light energy. One example would be enhancing light collection by small molecules used in photocells. The optical absorption cross section of molecules is very small as a consequence of their small size. Metal nanoparticles (of size $d \ll \lambda$) are also "small antennas" at optical frequencies, and so, like molecules, should have a tiny optical absorption. However, colloidal suspensions can exhibit brilliant colors, indicative of an enormous enhancement of optical cross section. The brilliant red color of some ancient church windows is a consequence of the formation of colloidal-gold nanoparticles during the firing of the glass (see Color Plate 6). The reason for this is the "plasmon-polariton" resonance of the free electrons in the metal surface. Resonant antennas have a cross section for capturing radiation that can be many times their physical size, depending on the sharpness of the resonance. Thus, a resonant metal particle can capture light over a region of many wavelengths in dimension even if the particle itself is only a fraction of a wavelength in diameter. Figure 9.11(a) shows how a molecule can be placed in a region of a high electric field near a resonant small metal particle to enhance the absorption cross section of the molecule.

The plasmon-polariton resonance can be understood in terms of the free electron theory of metals. Free electrons polarize so as to exclude electric fields from the interior of the metal, a behavior described by a negative dielectric constant. The resonance condition is obtained from the classical expression for the static polarizability, α, of a sphere (of volume, V, and dielectric constant relative to the surroundings, ε_r)[13]:

$$\alpha = \varepsilon_0 3 V \frac{\varepsilon_r - 1}{\varepsilon_r + 2}. \tag{9.14}$$

This becomes infinite at frequencies where $\varepsilon_r = -2$. For most metals, this value occurs in the visible region (hence the brilliant colors of metallic

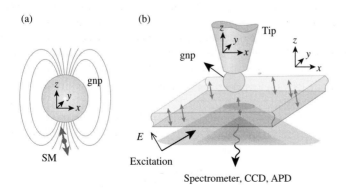

(a)   (b)

Fig. 9.11 (a) Electric field surrounding a nanoparticle driven in resonance with the E field along z. SM indicates a molecule in a high field region. (b) Experiment with a gold nanoparticle on the end of a scanning probe held above a layer of dye molecules (lines with arrows at each end) illuminated from below with a pulsed laser. (Reprinted from Kuhn et al.[12] Phys. Rev. Lett. 2006 with permission from the American Physical Society. Courtesy of Vahid Sanoghdar.)

colloids). Equation 9.14 is valid to the extent that we can ignore the change of phase of the incident field across the particle, i.e., $d \ll \lambda$. In this limit, the resonant frequency is independent of particle size. It does, however, depend on particle shape. Specifically, for a prolate spheroid of eccentricity, e, ($e^2 = 1 - (b/a)^2$), where a and b are the major and minor axes of the spheroid, respectively):

$$\alpha = \frac{\varepsilon_0 V}{L} \frac{1 - \varepsilon_r}{(1/L - 1) + \varepsilon_r},$$

$$\text{where } L = \frac{1 - e^2}{e^2}\left(-1 + \frac{1}{2e}\ln\frac{1 + e}{1 - e}\right) \approx \left(1 + \frac{a}{b}\right)^{-1.6}. \tag{9.15}$$

For silver, the particle plasmon resonance occurs near a wavelength of 400 nm, because this corresponds to the light frequency where $\varepsilon_r = -2$ in this material. However, for $a/b = 6$ the resonance condition moves all the way to $\varepsilon_r = -21.5$ which corresponds to 700 nm light in silver. Thus, the resonance is tunable throughout the visible by engineering the particle shape.

Light concentration is based on three consequences of placing molecules in close proximity to metal nanoparticles:

(1) The plasmon resonance results in local enhancement of the electric field. The rate of excitation of a dye molecule from ground ($|g\rangle$) to excited state, $|e\rangle$ is proportional to $|\langle e|\mathbf{E} \cdot \mathbf{D}|g\rangle|^2$, where $\mathbf{E}$ is the incident electric field and $\mathbf{D}$ stands for the molecular dipole moment operator. This square-law response means that "squeezing" more electric field into "hot spots" results in much more absorption of light overall. Doubling the electric field quadruples the light absorption.
(2) The field distribution near "hot spots" is far from planar, and therefore better matched to the molecular geometry.
(3) The increased absorption cross section is accompanied by a decrease in fluorescence lifetime, so molecules turnover more rapidly (termed "radiative decay engineering" by Lakowicz[14]). This means that a given molecule returns to its ground state sooner, ready to absorb another photon.

This predicted effect has been verified directly. Kuhn et al.[12] used a scanning probe microscope to place a small gold particle over dye molecules illuminated from below with light of a frequency that (a) was in resonance with the plasmon excitation of the gold and (b) was absorbed by the dye molecule. The arrangement is shown schematically on Fig. 9.11(b). They demonstrated up to 19 times enhancement in fluorescence-radiated intensity by a terrylene molecule embedded in a 30 nm paraterphenyl 20 (pT) film as a result of bringing a 100 nm gold nanosphere within 1 nm of the film. This was accompanied by a simultaneous drop in decay time from 20 ns (the typical value for such molecules at the air–pT interface) to 1 ns (below the 4 ns bulk value). Examples of these results are shown in Fig. 9.12.

At present, devices that absorb light to produce electricity from the electrons and holes on excited molecules require several thousand molecules in the line of sight to absorb a significant amount of the incident radiation. These optically thick devices are severe obstacles to efficient charge extraction, because the charge carriers have to be extracted from within a thick molecular layer

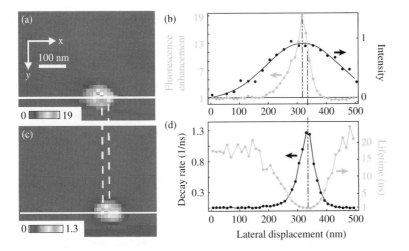

Fig. 9.12 Results of scanning the nanoparticle over the surface shown in Figure 9.11. (a) Fluorescence intensity vs. particle position: a single dye molecule is only visible in fluorescence when the gold NP passes over it. The sharply-peaked curve in (b) is a section through the data in (a). The broader curve is a profile of the fluorescence as a tightly focused laser beam is scanned over the dye, the extra intensity in the sharply-peaked curve showing the photonic enhancement. The increased fluorescence arises from more rapid decay of the excited state as illustrated with a map of decay rate for the dye as a function of the probe position (c) and as shown quantitatively in (d). (Reprinted from Kuhn et al.[12], Phys. Rev. Lett. 2006 with permission from the American Physical Society. Courtesy of Vahid Sanoghdar.)

where their mobilities are very low. A "black monolayer" engineered using plasmonic resonators might solve this problem—all the light would be absorbed in a monolayer, greatly facilitating charge extraction.

It may also prove possible to pattern the surface of optical integrated circuits with arrays of resonant metal nanoparticles that act as efficient waveguides for transporting light on the surface of the circuit.

9.7   2-D photonics for lasers

Modern semiconductor lasers are made from semiconductor heterostructures designed to trap excited electrons and holes in the optically active part of the laser. We illustrate some of the principles here with a quantum dot laser that uses a combination of 1-D and 2-D nanostructuring.

Laser operation requires the generation of a population inversion, that is to say more of the atoms (or excitons in the case of a semiconductor) must be in the excited state than in the ground state, providing a driving force for the *stimulated emission* of a coherent photon. This occurs when an incident photon of the right frequency interacts with this population of atoms or excitons in the excited state. Atoms or excitons in the ground state serve only to absorb the incident radiation, so an ideal laser medium would consist of material that is entirely in its optically excited state. Interpreted in terms of a Boltzmann distribution, this would require an infinite negative temperature, but such distributions cannot be achieved in equilibrium of course. An inverted population can be caused by driving the system into the excited state faster than it can decay. In that case, stimulated emission can grow in a chain reaction, as stimulated photons further stimulate the emission of other photons in the medium. The lasing medium is usually enclosed in an optical cavity consisting of a pair of mirrors, arranged so as to reflect the stimulated photons back and forth through the medium. These mirrors can have a very high coefficient of reflection, because the optical intensity in the cavity becomes enormous. In this case, the desired output radiation is only a tiny fraction of the field inside the resonant cavity (this is an exact optical analogy of the electronic phenomenon of resonant tunneling

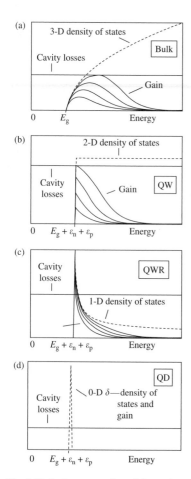

Fig. 9.13 Performance of a solid-state laser material in various geometries. The dashed line shows the density of states (from top to bottom) in a 3-D material, a 2-D quantum well (QW), a 1-D quantum wire (QWR), and a 0-D quantum dot (QD). The lines labeled cavity losses indicate a minimum threshold gain for lasing, while the various solid lines show the form of the gain vs. energy profiles for the lasing materials in the structures of various dimensionality. In quantum dots, all of the available gain is squeezed into a narrow bandwidth. (Reprinted with permission from Asryan and Luryi[15], copyright 2004 by Artech House, Inc.)

Fig. 9.14 (a) Shows a schematic layout of a quantum dot laser with quantum dots embedded in a cladding layer that confines the light to a resonant cavity. (b) Shows the desired band structure of the quantum dots embedded in a second semiconductor material. The bands are shaped so as to trap electrons and holes in excited states in the lasing region. (Reprinted with permission from Asryan and Luryi[15], copyright 2004 by Artech House, Inc.)

described in Chapter 7) so even highly reflective mirrors will pass significant power. Because of this effect, the net loss of radiation can be very small in relation to the radiation intensity in the lasing medium, so that even a relatively small population inversion can drive laser action. The requirement for lasing is that the gain of the laser medium exceeds the cavity losses. The gain of a laser medium is proportional to the number of atoms or excitons of the correct excited frequency in the lasing volume. The range of excited state energies that can contribute to lasing is very small because of the coherence of the stimulated photons (they are all of essentially one frequency).

Nanostructuring can play an important role in confining the density of states of excitons to a small region of energy around the desired laser frequency.[15] This is shown schematically in Fig. 9.13, the exciton analog of what was shown for electrons in Fig. 9.2. Quantum dots of the right size can place essentially all of the exciton energies at the right value for lasing (bottom panel of Fig. 9.13). The bandgap engineering that helps to confine excitons to the active region of the laser is illustrated in Fig. 9.14. The quantum dots are chosen to have a bandgap that is smaller than that of the medium from which carriers were injected. In this way, excitons are stabilized because the electrons are confined to the low-energy part of the conduction band and the holes are confined to the low-energy part (i.e., top) of the valence band. Quantum dot lasers are expected to become a significant part of the semiconductor laser market in the future.

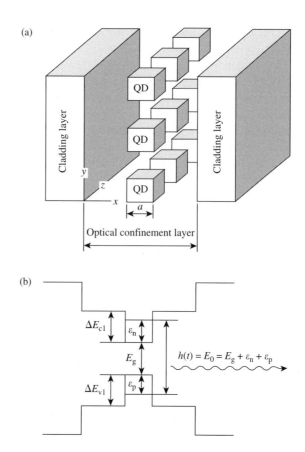

9.8   3-D photonic bandgap materials

Our discussion of electron waves in periodic structures (Chapter 7) showed how wave propagation is broken up into allowed and forbidden bands in energy. The concentration of modes into bands results in an increase in the density of states in the allowed bands, particularly near band edges. The phenomenon holds for electromagnetic waves in a periodic structure, so long as features in the periodic structure are commensurate with the wavelength of the radiation. For X-rays, the phenomenon is well known as Bragg scattering. Optical wavelengths require that materials be structured on the half-micron scale. Optically engineered structures like this hold out the possibility of concentrating optical energy, in much the same way as electron energy (Fig. 9.2) or exciton energy (Fig. 9.13) is concentrated by nanostructuring the material.

An example of a natural photonic bandgap material is opal, which owes its optical properties to Bragg reflections of particular wavelengths from periodic features in the mineral (see Color Plate 7). The brilliant interference colors that result from the Bragg reflection are referred to as opalescence. Synthetic opal-like materials are made by spatially ordering micron-size particles. For example, ordered arrays of particles can be made by aggregation of small colloidal particles into electrostatically stabilized clusters. These particles are typically monodisperse solutions of polymers that are functionalized with charged groups (such as carboxylic acid) that serve to dissolve the polymer and prevent unwanted aggregation. Careful concentration of such suspensions can yield ordered arrays. In such arrays, the interparticle spacing is dictated by the electrostatic repulsion between particles. An electron micrograph of a colloidal crystal made from half-micron particles is shown in Fig. 9.15. This material is visibly opalescent as shown in Color Plate 7.

These opalescent materials do not possess a true bandgap but rather reflect strongly in only certain directions (as can be seen from the many colors found in the reflections—as the observation angle changes, the wavelength that is Bragg-reflected changes). By using special spacing segments between the spheres, it is possible to produce sharp reflection at one wavelength as shown in Fig. 9.16. For a given spacing in some direction in the colloidal crystal lattice represented by lattice indices h,k,l, the wavelength of the reflected beam is given by the Bragg law

$$\lambda = 2nd_{hkl} \cos \theta_{\text{int}}, \tag{9.16}$$

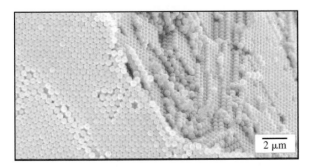

Fig. 9.15 Electron micrograph of a colloidal crystal made from polystyrene latex spheres. These display brilliant opalescent colors when illuminated with white light (see Color Plate 7). (Courtesy of the Minnesota Materials Research Science and Engineering Center and Professor Andreas Stein. All rights reserved.)

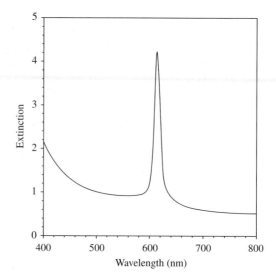

Fig. 9.16 Extinction coefficient as a function of wavelength for colloidal particles spaced with polymer spacers showing a photonic gap near 600 nm. This is not a true 3-D gap. (Reprinted from Foulger et al.[16], Optics Letters 2000, with permission from the Optical Society of America.)

where θ_{int} is the internal angle of reflection, n is an average refractive index, and d_{hkl} is the spacing between Bragg planes in the direction hkl. The average refractive index is obtained by taking the volume fraction weighted average of the refractive indices of the components of the colloidal crystal. The internal angle of reflection is related to the external angle by Snell's law ($\sin\theta_{ext} = n\sin\theta_{int}$) leading to the following expression for the stopped wavelength:

$$\lambda = 2d_{hkl}\sqrt{n^2 - \sin^2\theta_{ext}}. \qquad (9.17)$$

It turns out to be geometrically impossible to build a genuine three-dimensional optical "insulator" from simple periodic structures. It is always possible to find directions in the crystal in which the radiation is not Bragg-reflected. One solution to this problem is to construct the periodic lattice from nonspherical "atoms." The scattering from a lattice is given by a product of the Fourier transform of the lattice (i.e., the reciprocal lattice) with the Fourier transform of the atom itself. Thus if one examines X-rays scattered from a lattice of atoms, the intensity of the scattering fades out at high angles as a consequence of the finite size of the individual atoms in the lattice (the reciprocal lattice by itself would predict scattering out to all possible angles). Spherical atoms result in a uniform distribution of this effect, but nonspherical atoms can produce a pattern with zeros in specific directions. If the "atoms" are carefully designed so that these zeros lie in directions for which an allowed Bragg reflection occurs, the combined scattering pattern can have a true 3-D bandgap. This effect was first predicted by Yablonovitch[17] and his calculations of the energy vs. wave vector for optical radiation in a particular crystal composed of nonspherical atoms are shown in Fig. 9.17(a). The horizontal axis plots the wave vectors in orthogonal symmetry directions (labeled W, L, and K) in the crystal. Combinations of these plots span all directions in the crystal. A bandgap exists everywhere, as shown by the shaded region. Note that at low wave vector, the curve has a gradient equal to the speed of light in the medium, but the curvature changes

(a)   (b)

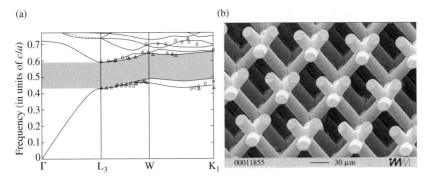

Fig. 9.17 True 3-D photonic crystals. (a) Calculation of optical dispersion (E vs. wave vector) for different directions in a crystal made from nonspherical "atoms." (Reprinted from Yablonovitch et al.[17], Phys. Rev. Lett. 1991 with permission from the American Physical Society. Courtesy of Eli Yablonovitch, University of California, Berkeley.) (b) Photonic material with "nonspherical atoms" fabricated from photoresist. (Reprinted from Soukoulis[18], Nanotechnology 2002 with permission from the Institute of Physics. Courtesy of C.M. Soukoulis.)

dramatically in the vicinity of the optical bandgap. This corresponds to a strong wavelength-dependent refractive index, suggesting that these materials will have many uses in nonlinear optics.

Making materials that have both the required periodicity and the required shape within each unit cell of the periodic lattice is far from simple. One design that consists of stacked rods is shown in Fig. 9.17(b). This structure was fabricated from photoresist and the limitations of the technology mean that the repeat period is on the order of 30 μm. Thus, the material shown in Fig. 9.17(b) has a 3-D optical bandgap only in the far infrared.

Possible applications of photonic bandgap materials exploit both the bandgap and anomalies in the dielectric constant that appear near band edges. Quantum dot lasers embedded in a photonic medium (that possesses a sharp transmission frequency only close to the laser frequency of the quantum dots) may have suppressed spontaneous emission. This is because spontaneous emission occurs over a broadband of frequencies and therefore may lie in the forbidden band of the photonic material. Suppression of spontaneous emission leads to an enhancement of stimulated, that is, laser, emission. Anomalous dispersion in photonic materials can lead to a negative refractive index. It has been shown that materials with a negative refractive index could be used to make aberration-free lenses.[19]

9.9 Physics of magnetic materials

Magnetic ordering in solids is a complex phenomenon. We begin with a brief discussion of the two types of magnetism that can be understood in a straightforward way. Solids made from closed shell atoms (the rare gases and ions from the first and last columns of the periodic table) possess no intrinsic magnetic properties because they have zero spin and zero orbital angular momentum in the ground state. Consequently, the application of a magnetic field results only in a small distortion of the motion of the electrons, giving rise to a circulating current that opposes the applied field. This results in a smaller magnetic field inside the solid. The solid is said to have a negative magnetic susceptibility and the phenomenon is referred to as "Larmor diamagnetism." Free electron solids can be magnetically polarized as spin up and spin down electrons interact with an external magnetic field. The net result is to produce a small additional magnetic polarization that lies in the direction of the applied field, thereby enhancing the field in the solid. This behavior is called paramagnetism (and referred to as Pauli paramagnetism in the case of a free electron gas).

The more familiar magnetism is the permanent magnet, and the origin of the phenomenon is much more complicated. Spontaneous magnetic ordering can result in *ferromagnetism* (the kind familiar in a bar magnet) or *anti-ferromagnetism* (ordering in which the polarization alternates from atom to atom, so no macroscopic magnetic moment arises). Various other types of intermediate ordering are possible. A magnetic field contains energy, so spontaneous magnetization requires a source of energy. Where does this come from? Ferromagnetism cannot be understood in the context of a free electron model, because the origin of this ordering lies in electron–electron interactions. Wavefunctions for an assembly of interacting electrons must be made antisymmetric, changing signs when particles are exchanged. This may be achieved either by changing the sign of the spatial part of the wavefunction of the particles that are exchanged or by changing the sign (i.e., spin orientation) of the spin part of the wavefunction. Thus, these interactions are spin-dependent, an affect referred to as the exchange interaction (though it is not an interaction *per se*, but rather a constraint on the many-particle wavefunction). Depending on the geometry of the crystal and the electronic states of the atoms from which it is composed, it may prove possible to lower the energy of interactions between electrons by aligning spins of neighboring atoms. This spontaneous alignment of atomic spins gives rise to ferromagnetism. Though the requirements of antisymmetric wavefunctions can extend over the whole solid, exchange-mediated interactions are essentially local. This is because electron–electron interactions are themselves quite local because of the correlation-hole screening effect mentioned when we justified the free electron model in Chapter 7.

The exchange interaction sets only one length scale of interest in magnetism. Another length scale is set by the long-range interactions between the magnetic dipoles associated with the locally ordered magnetic regions. The dipole–dipole interactions between neighboring spins are weak (compared to the effects of the requirements of exchange) but the dipole interaction energy grows with the volume of the ordered region. Magnetic dipole energy in a bulk sample is minimized by the breakup of magnetically ordered regions into domains of opposing magnetization. In consequence, the macroscopic magnetization of a ferromagnet is zero in its lowest energy state (but see below). The size of the individual domains is set by a competition between the dipole energy associated with the magnetization of the domain volume and the energy associated with the surface of the domain, where the magnetic ordering undergoes a rapid (and energetically expensive) change. The spontaneous formation of domains is thus a consequence of competition between bulk and surface energy (much like the factors that drive self-assembly of colloids as discussed in Chapter 6). Domain sizes can vary from a hundred atoms or so up to many microns in size.

To complicate matters yet further, another length scale is set by the interaction of the local magnetization with the orbital angular momentum of the atoms in the crystal. This interaction is called the *magnetic anisotropy*, and it is responsible for pinning the magnetic moment along certain preferred directions in the crystal. It also gives rise to the hysteresis in magnetic susceptibility that is responsible for the generation of the permanent magnetic moment in magnetized samples. When a nonmagnetized iron nail is stroked with a magnet, domains line up with the applied magnetic field, particularly in crystallites in which an easily magnetized direction is appropriately aligned with the external field. When the external field is removed, the anisotropy energy prevents the domains from reorganizing to minimize the magnetic dipole energy of the material. The result is that the nail is now magnetized. A significant field must be applied in the opposite direction to undo this magnetization (this is called the coercive field).

Set against this rich and complex physics, it is hardly surprising that nanostructuring has a significant effect on magnetic properties. The technologies required for making well-ordered magnetic films have only recently become available.

9.10   Superparamagnetic nanoparticles

In a ferromagnetic like iron, the exchange interaction is mediated by d electrons that are somewhat delocalized (referred to as "itinerant"). This delocalization does not develop fully until a cluster of iron atoms reaches a critical size (cf. the development of delocalized electronic states shown in Fig. 9.1). The intrinsic magnetic moment of an iron atom is large, so the magnetic moment per atom falls with cluster size, only reaching the bulk value when the cluster contains several hundred atoms. Magnetic data for atomic beam–generated clusters are shown in Fig. 9.18(a). Even clusters of several hundred atoms behave differently from the bulk, despite the fact that the magnetic moment of the atoms has reached the bulk value. This is because stable domains cannot be established in crystals that are smaller than the intrinsic domain size. A small cluster consists of a single ferromagnetic domain, but one which is too small to be pinned by the anisotropy energy (and which lacks the constraint that it align so as to oppose neighboring domains). In consequence, the magnetization of a particle smaller than the intrinsic domain size tends to follow the applied field freely, thereby enhancing the local field enormously. Such particles are called *superparamagnetic* because their behavior is that of a paramagnet, the magnetization rising and falling with the applied field. The magnetic susceptibility of these superparamagnetic particles is many orders of magnitude larger than bulk paramagnetic materials. The evolution of a particle from superparamagnetic to ferromagnetic behavior as a function of size is shown in Fig. 9.18(b).

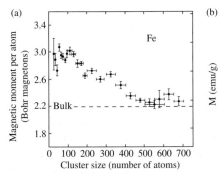
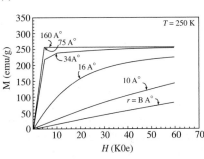

Fig. 9.18 (a) Showing how the magnetic moment per atom decreases toward the bulk value (where magnetization involves itinerant electrons) as cluster size is increased. (Reprinted from Shi et al.[20], Science 1996, with permission from AAAS.) (b) Simulation of the onset of ferromagnetic behavior as cluster size is increased for clusters of Gd at 250 K. (Reprinted from Aly[21], J. Magnetism and Magnetic Materials, 2000, with permission from Elsevier.)

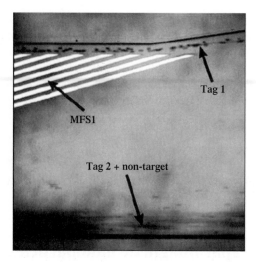

Fig. 9.19 Microfabricated magnetic sorter for cells labeled with superparamagnetic beads. Antibody-conjugated beads bind the target cells in a heterogeneous mixture that flows in from the far left. Microfabricated ferromagnetic strips (MFS1) provide a high field gradient that has, at this point, swept the target cells (Tag 1) into the top of the channel. (From Adams et al.[22] by permission from NAS.)

This is a plot of the calculated magnetization (M) of the particle as a function of the applied magnetic field (H) for gadolinium clusters of various diameters. Only the larger particles show the rapid initial rise and subsequent magnetic saturation that is characteristic of a ferromagnetic.

Superparamagnetic particles have enabled completely new types of chemical separation. Beads containing a superparamagnetic core are available with a wide range of chemical functionalizations on the outside, enabling new routes for chemical purification. A mixture containing the desired product is incubated with superparamagnetic beads that are functionalized to bind with the desired target. The beads are then collected in the strong field gradient at the pole of a bar magnet, while the undesired products are rinsed away. Release of the desired product from the magnetic beads is achieved through the use of a reversible coupling reagent. For example, the beads may be functionalized with the protein streptavidin which binds very strongly to a small reagent called biotin (many reagents are available in a biotinylated form). Release of the biotinylated product from the streptavidin-coated superparamagnetic beads is achieved by rinsing the beads with a large excess of biotin that displaces the biotinylated product. An example of a magnetic capture system (based in this case on magnetic beads coated with antibodies that attach to cells) is shown in Fig. 9.19. Microfabricated magnetic stripes steer the target cells into a channel at the top, while the nontargeted cells flow in the bottom of the channel.

9.11 A 2-D nanomagnetic device: giant magnetoresistance

The read heads in magnetic hard drives are based on a nanostructured device called a giant magnetoresistance sensor. This device has revolutionized the magnetic storage industry and Albert Fert and Peter Grünberg shared the 2007 Nobel Prize in physics for this invention. The sensor consists of a narrow electrically conducting layer (e.g., Cu) sandwiched between two magnetic films.

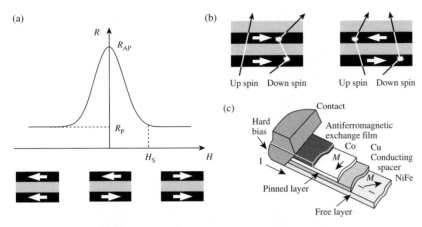

Fig. 9.20 Giant magnetoresistance. (a) Shows how the resistance in a nonmagnetic nanolayer (gray) falls as an external magnetic field aligns the magnetization of two domains that were aligned in opposite directions at zero field. The device thickness (i.e., across the layers of the sandwich) is small compared to the electron mean free path. (b) Shows how spin-dependent scattering affects only one of the spin currents when the domains are magnetized in the same direction, but adds significant resistance to both when the magnetization directions are different. (Reprinted with permission from *Solid State Physics*[23] 2001, Academic Press.) (c) Construction of a Hitachi hard drive reading head. The Co layer is pinned and the NiFe layer responds to local magnetization of the disk. The conducting layer is Cu. (Image copyrighted 2009, Hitachi Global Storage Technologies. Used by permission.)

The layers have a width that is smaller than the electron scattering length in the conducting material (100 nm or so). Current in the conducting layer may be thought of as the sum of two currents, one coming from spin up electrons, the other mediated by spin down electrons. Electron scattering in magnetic materials is generally quite strongly dependent on the orientation of the electron spin with respect to the magnetization of the material surrounding the conductor. Giant magnetoresistance occurs when the magnetic layers above and below the conductor are magnetized in opposite directions. This results in a substantial (~20%) increase in the resistance of the material as illustrated in Fig. 9.20(a). The reason for this effect is illustrated in Fig. 9.20(b). When the top and bottom magnetic layers are magnetized in the same direction, electrons of one particular spin polarization will be strongly scattered in both layers. However, electrons of the opposite spin polarization will hardly be scattered at all, resulting in a low resistance path across the device. When the magnetic materials are magnetized in opposite directions, electrons of one spin will be strongly scattered in one of the magnetic layers, while electrons of the opposite spin will be strongly scattered in the other magnetic layer. The result is a significant increase in the resistance of the device.

The construction of a magnetic hard drive reading head is shown in Fig. 9.20(c). The upper magnetic layer is made from cobalt and it remains magnetized in one direction. The lower magnetic layer is made from a NiFe alloy, a "soft" material, for which the magnetization is readily re-aligned by the magnetic bits on the hard disk that spins just below the read head. Thus, the magnetization of the bits on the surface of the disk can be read out as fluctuations in the resistance of the conducting layer (which is wired into a resistance bridge circuit). If you are reading this on a laptop, you are almost certainly using a giant magnetoresistance nanosensor to do so.

9.12   Nanostructured thermal devices

We have said little about the thermal properties of solids in this book. Heat is carried both by free electrons (and holes in semiconductors) and also by thermal vibrations which propagate at the speed of sound. Thermal vibrations are the only mechanism of heat transport in electrical insulators. A thermal gradient across a thermal insulator corresponds to a change in the amplitude of thermal vibrations in the direction of the gradient. In a solid, these vibrations are quantized, the wave vector of the corresponding sound waves obeying the same constraints imposed by the crystal structure of the solid as the other excitations we have discussed (i.e., electrons, excitons, and photons). Such quantized lattice vibrations are referred to as phonons (cf. the discussion of quasiparticles in Chapter 2). High-frequency phonons with wave vectors near a Bragg reflection contribute little to heat propagation because of the condition that their group velocity goes to zero at the Brillouin zone boundary. Thus, heat transport in solids is dominated by long wavelength thermal vibrations, and it is the rapid increase in the Boltzmann occupation of these low-energy states that gives rise to the T^3 dependence of the specific heat at low temperature for a dielectric solid (we referred to this in Chapter 7 in our discussion of electronic thermal conductivity). All the tools of density of states engineering by means of nanostructuring are available to control the thermal properties of solids.

One example is in thermoelectric cooling (or, conversely, thermoelectric power generation). A thermoelectric cooler is shown in Fig. 9.21(a) with the corresponding thermoelectric current generator shown in Fig. 9.21(b). The thermoelectric cooler consists of a large area junction between p-type and n-type semiconductors. Typically, the semiconductors are arranged as vertical bars connected by a horizontal metal bridge. Passing a current across the junction in the forward direction sweeps holes away from the junction in one direction and electrons away from the junction in the opposite direction. These carriers take the heat away from the junction, with the result that the junction cools. The opposite side of the device gets hot, and therein lies the problem. Excellent electrical conductivity is required for good thermoelectric cooling but low thermal

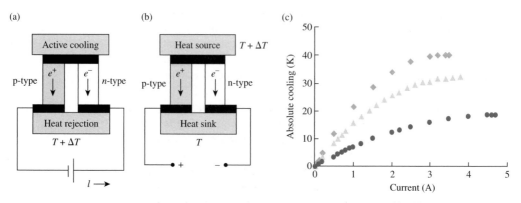

Fig. 9.21 A PN Junction as a thermoelectric cooler (a) or heat to current converter (b). (From DiSalvo[24], Science 1999, with permission from AAAS.) (c) Cooling in nanostructured devices, engineered to minimize electron scattering and maximize phonon scattering (diamonds and triangles) compared to a bulk device (dots). (From Venkatasubramanian et al.[25], Nature 2001, with permission from Nature Publishing Group.)

conductivity is also required if heat is not to leak back again from the hot side to the cold side via thermal vibrations. One bulk approach to this problem is to use semiconductors that are made of heavy atoms, because the speed of sound is lower in these materials. However, it is generally true that bulk materials that are good electron conductors are also good heat conductors.

A figure of merit for thermoelectric devices is given by a quantity called ZT,

$$ZT = \frac{S^2 \sigma T}{\kappa}, \qquad (9.18)$$

where S is the Seebeck coefficient, a measure of the coupling between heat and charge gradients, with units of volts per degree; σ is the electrical conductivity; T the absolute temperature; and κ the thermal conductivity. The thermoelectric performance increases with S and σ, while the corresponding temperature gradient falls with thermal conductivity. The ratio Z has units of $1/T$ (Problem 14) so the product ZT is dimensionless (T is taken to be the average of the operating temperatures on the hot and cold sides of the device). As stated above, the performance is directly proportional to the ratio of electrical to thermal conductivity. Because of the difference in scattering lengths between phonons and electrons, it is possible to design nanostructures based on super lattices that are "phonon blocking" and "electron transmitting." Such devices have set records for values of ZT.[25] Data that compare the cooling power at a given current for nanostructured and bulk thermoelectric devices are shown in Fig. 9.21(c).

9.13   Nanofluidic devices

Nanostructured devices are becoming more important as analytical tools in chemistry. In some applications, they have long formed the basis of analytical devices, as, for example, in the case of the small particles used in chromatography columns (see Chapter 6). Another very important example is the cross-linked gel networks that are used in gel electrophoresis. These contain nanoscale pores that require large molecules to unravel in order to pass through. In recent years, it has become possible to fabricate nanostructures that carry out a similar function but with a known and well-defined nanostructure.

Fluid flow in small structures is entirely laminar and dominated by the chemical properties of the boundaries of the channel. Flow may be characterized by an important dimensionless number called the Reynolds number, Re. This number quantifies the ratio of the inertial forces on a volume of fluid to the viscous forces that act on the same volume. At high Reynolds numbers ($\gg 1$) flow becomes turbulent, while at low Reynolds numbers the flow is dominated by friction. In a channel with a narrowest dimension, L, in which the fluid is flowing at an average speed, $\langle U \rangle$, the Reynolds number is given by

$$Re = \frac{\rho \langle U \rangle L}{\eta}, \qquad (9.19)$$

where η is the viscosity of a fluid and ρ is its density. The ratio of viscosity to density is called the kinematic viscosity, ν, and for water, $\nu = 10^{-6} \text{ m}^2\text{s}^{-1}$ at 25°C. Thus in a nanochannel of dimension, $L = 100$ nm, a Reynolds number of unity is only obtained when flow rates reach 10 m/s. Achieving these speeds in such tiny channels would require impossible pressure gradients. In practice, the highest speeds are on the order of $\langle U \rangle_{max}$ = 1 mm/s so Re does not exceed 10^{-4}.

One of the consequences of these tiny Reynolds numbers is that conventional pumping schemes based on inertia no longer work. In particular, it becomes difficult to mix fluids in a nanofluidic device.[26] The chemistry of the interface also becomes critical and aqueous fluids will not generally enter a channel with hydrophobic surfaces.

9.14   Nanofluidic channels and pores for molecular separations

A significant stretching of large molecules can occur in a large ion gradient (i.e., electric field) in a channel that is comparable to the radius of gyration of the molecule. For example, a long DNA molecule of 17 μm fully stretched length is equivalent to about 340 freely jointed polymer segments each of length 50 nm, giving a radius of gyration equal to $\sqrt{340} \times 50$ nm or ~90 nm (Chapter 3). Molecules like this may be significantly extended in channels of 100 nm diameter owing to the strong interaction between the fluid and the walls. This effect is illustrated in Fig. 9.22, which shows an array of 100 nm channels that have been cut into a glass substrate using a focused ion beam (9.22(a)). The channels were loaded with an enzyme that cuts this particular DNA at four locations, and an electric field was applied across the channels so as to pull DNA in from one direction and magnesium ions from the other (the magnesium being essential to activate the enzymes that cut the DNA). The DNA is loaded with fluorescent dye, and when it first enters the channel (Fig. 9.21(b), top image) it is observed to be stretched to ~40% of its full length. After a while, it is seen to separate

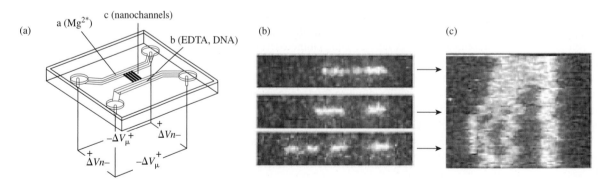

Fig. 9.22 Mapping the size of DNA fragments as a long molecule is cut by an enzyme in a nanochannel. (a) The device consists of two microchannels connected by a series of nanochannels (100 nm × 100 nm cross section). DNA is manipulated into the front microchannel and then transported through the nanochannels by application of a voltage that generates significant ionic current through the nanochannel. Moving Mg^{2+} in from the far side activates enzymes in the channel that cut the DNA. (b) Images of the fluorescently labeled DNA molecule taken at various times in the channel. At the top, the molecule is intact and stretched to ~40% of its full length. In the next frame, it has been cut in two places, and in the last frame it has been cut at all the four locations that were targets for the enzyme. (c) Shows a continuous time course of the cutting process. (From Riehn et al.[27], with permission from NAS.)

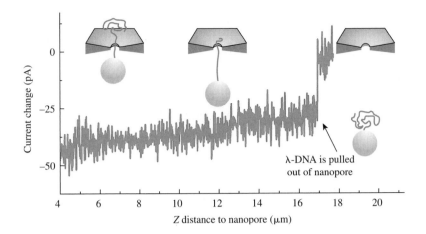

Fig. 9.23 Manipulation of DNA in a nanopore drilled into a silicon wafer. The DNA, attached to a plastic bead, is pulled into the nanopore by electrophoresis, reducing the ion current through the pore. Laser tweezers are used to pull the bead away from the nanopore, restoring current when the DNA leaves the pore. Measurements of the force needed to do this yield the effective charge per base pair on the DNA. (From Keyser et al.[30], Nature Physics 2006, with permission by Nature Publishing Group.)

into the expected four fragments. A time course of this cutting process is shown by stacking many images of the channel (Fig. 9.22(c)). It is possible that such structures will prove useful in mapping sites on very large genomes.

Another type of nanofluidic separator consists of a tiny hole drilled through a silicon nitride membrane. The hole is designed to be so small (~2 nm diameter) that only a single DNA molecule can pass through it at a given time. By placing a membrane containing the hole between two tanks of electrolyte, one of which contains DNA molecules, and passing a current through the hole in a direction that pulls the DNA into the second compartment, the passage of single molecules may be measured by the drops in currents that occur when a single DNA molecule occludes the hole. Such nanopore devices have been proposed as a possible approach to rapid DNA sequencing.[28,29] A single molecule experiment involving both a nanopore and laser tweezers is illustrated in Fig. 9.23. In this work, a long DNA molecule, attached at one end to a polystyrene bead, was pulled into a nanopore using electrophoresis, causing the current to drop (the scale is offset on the figure). The laser tweezers were then used to pull the DNA out of the nanopore against the force of the electric field across the nanopore. Complete removal of the DNA is marked by a jump in current through the nanopore (at 17 μm on the *z*-axis of Fig. 9.23). These measurements yielded a value for the effective charge on the backbone phosphate group of the DNA.[30]

9.15 Enhanced fluid transport in nanotubes

Computer simulations of water in small (2 nm diameter or less) carbon nanotubes suggest that not only does water occupy the interior of the tube, but also that transport through the tube is much faster than predictions made by classical hydrodynamic calculations based on the bulk viscosity of water.[31] This is rather surprising, giving the hydrophobic nature of carbon nanotubes (and the expectation that small hydrophobic structures will not be wetted). Nonetheless, experiments confirm these remarkable predictions. Holt et al.[32] fabricated a membrane composed of small (<2 nm diameter) tubes penetrating

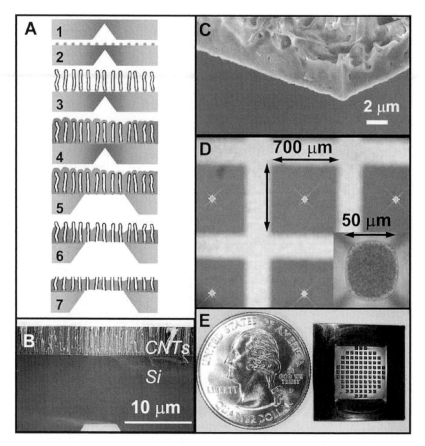

Fig. 9.24 Fabrication of a layer of CNTs penetrating a silicon nitride membrane (a). The CNT layer as imaged by SEM before (b) and after (c) filling with silicon nitride. The finished membrane devices are shown in ((d)—individual devices) and ((e)—an array of devices). (From Holt et al.[32], Science 2006, with permission by AAAS.)

a layer of silicon nitride. The fabrication process is illustrated in Fig. 9.24. A micron-sized window was partially opened in a silicon wafer (A.1) and catalyst particles were deposited on the side opposite the window (A.2). Carbon nanotubes were grown from the catalyst particles (A.3) by chemical vapor deposition and the layer of tubes filled in with silicon nitride, also formed by chemical vapor deposition (A.4). The backside of the chip was opened up to expose the nitride membrane (A.5). Part of the top of the nitride layer was removed by argon ion milling (A.6) and the catalyst-capped tops of the tubes removed by reactive ion etching (A.7). The layer of CNTs is shown in (b) and, after filling with silicon nitride, in (c). The completed membranes are shown in (d) and (e).

The cross-sectional area of the tubes was measured by SEM and also by measuring the size of particles that could be passed by the tubes. Knowing the density of tubes, it is possible to use classical hydrodynamics to predict the rate at which fluid should pass the membrane for a given pressure differential. The measured rate of water flow was found to be 1000 times higher than predicted. Thus, water not only enters these tubes, but also flows through them with remarkably little viscous resistance. This phenomenon is surely sensitive to the atomic details of the fluid passing through the CNTs, and this may open a new route to ion-selective filtering and chemical analysis.

9.16   Superhydrophobic nanostructured surfaces

Nanostructuring may also be used to modify the chemical properties of a surface. Hydrophobic materials have an entropically unfavorable interaction with water. The macroscopic manifestation of hydrophobicity is the obvious repulsion of water by surfaces such as plastics. Microscopically, a hydrophobic material breaks the hydrogen bond structure of a water network (as described in Chapter 6). The effects can be long range and one empirical expression for the hydrophobic interaction energy per unit area between two hydrophobic surfaces separated by water is[33]

$$W = -2\gamma_i \exp -\frac{D}{\lambda_0}, \qquad (9.20)$$

where typically $\gamma_i = 10$–50 mJm^{-2} and $\lambda_0 = 1$–2 nm. This is a long interaction distance relative to the diameter of the water molecule (which is ~ 3 Å). When a drop of water contacts a flat surface, the angle formed by a tangent to the surface of the water drop at the point of contact and the surface itself is called the contact angle, θ_c. It is given in terms of the interfacial energies of the system by the Young equation

$$\cos \theta_c = \frac{\gamma_{AB} - \gamma_{AC}}{\gamma_{BC}}, \qquad (9.21)$$

where γ_{AB} is the interfacial tension between the surface and the air, γ_{AC} is the interfacial tension between the surface and the water, and γ_{BC} is the interfacial tension between the water and the air. A large interfacial tension between the surface and the water, that is, a repulsion, results in values of the contact angle that exceed 90° (i.e., $\cos \theta_c < 1$).

The contact angle is significantly increased when the liquid is placed on a rough surface made from the same material. There are two very different models for this phenomenon. In one model, the droplet penetrates the cavities in the surface. In the other model, the droplet sits on top of the asperities on the surface. It is not presently known which of these two models is correct[34] but it is an important effect, accounting for, for example, the self-cleaning properties of the lotus leaf.[35] It is, therefore, possible to increase the hydrophobicity of a surface simply by roughening it on the nanoscale.

An example of a nanostructured superhydrophobic surface is shown in Fig. 9.25. Silicon nanowires were grown on to a silicon surface that had been seeded with gold nano dots (see Section 6.8). When both planar

Fig. 9.25 (a) Structure of the surface on which a dense array of Si nanowires have been grown. (b) A water drop on a plane surface and (c) a water drop on a nanotextured surface. (From Rosario et al.[36], J. Chem. Phys. 2004, with permission from the American Chemical Society.)

and nanostructured silicon surfaces were treated with a hydrophobic coating and exposed to water, the planar surface yielded a much smaller contact angle (Fig. 9.25(b)) than the nanostructured surface (Fig. 9.25(c)). This decrease in interaction between water and the surface on which its sits opens up new applications in microfluidics.[36]

9.17   Biomimetic materials

Biology provides wonderful inspiration for nanomachines. Figure 9.26 shows two examples of nanostructured mineralization, an abalone shell and a diatom. How are such elaborate structures put together? They are composed of an insoluble mineral matrix held together by a small amount of protein. The proteins have evolved to stick to certain crystal faces of the mineral component very specifically. They act as a highly selective molecular glue, holding one specific crystal face of one mineral to a specific face of another.

Can we mimic millions of years of evolution molecular to make such "molecular glues" to order? The answer is yes, using the tools of molecular biology. DNA codes for the 20 different amino acids that are strung together to make proteins (a topic covered in the next chapter). The code, of how DNA bases translate to amino acid residues, is known precisely. DNA polymers are readily made synthetically, with a sequence of bases that code for the desired protein. Synthetic DNA can be inserted into a special circular piece of DNA containing the sequences needed to trigger translation of the DNA into proteins (a process of several steps) and injected into the organism that will be used to make the protein (usually a bacteria or virus). The key to accelerating the slow random mutations of an evolutionary time scale is to make a vast random "library" of DNA sequences in the DNA coding for the region of the protein one wishes to "evolve." This is easily done by injecting all four of the DNA bases into the synthesizer at each step that a base is to be added in the synthesis of the random part of the sequence. Once the random string is completed, the synthesis is continued in the normal way, adding just one base at a time.

One way to carry out a selection is called "phage display." In phage display, a bacteriophage (a virus that eats bacteria) is genetically modified so that the proteins that appear on its outer surface display random peptide sequences. It is straightforward to grow modified phage in quantities of billions, each one presenting different (random) amino acid sequences on the surface. If the goal is to make a specific molecular glue, for example, for an InP(100) surface, vast numbers of bacteriophage are incubated with the InP(100) surface. The surface is rinsed, the phage that stuck to it collected, grown up again to large numbers, and the cycle of adhesion and rinse repeated many times. The result is a population of phage that are copies of a small number of clones, each of which has the property of sticking efficiently to the InP(100) surface. Some of these are recovered, their DNA extracted and sequenced. From this sequence, the specific proteins sequences that stick to InP(100) can be identified. Figure 9.27 shows filamentous bacteriophage sticking to an InP(100) surface after a selection procedure was used to evolve specific adhesion for this surface.[38]

(a)

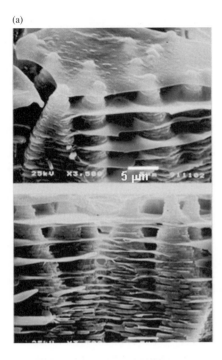

(b)

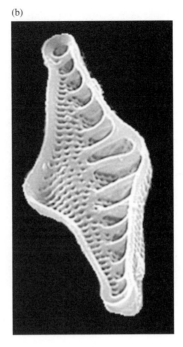

Fig. 9.26 SEM images of the mineral content (skeleton) of (a) an abalone shell (From Tian et al.[37], Nature Materials 2003, used with permission from Nature Publishing Group) and (b) a diatom (courtesy of Rex Lowe, Bowling Green State University).

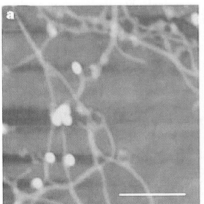

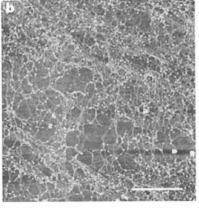

Fig. 9.27 AFM images of filamentous bacteriophage sticking specifically to an InP (100) surface. These are a tiny number selected from billions of random mutants created using the tools of molecular biology. Proteins on their surface contain amino acid sequences that stick to InP(100) with a high specificity. The scale bars are 500 nm. (From Whaley et al.[38], Nature 2000, used with permission from Nature Publishing Group.)

A second type of biomimetic design is based on copying engineering principles from nature directly. The light-to-charge converter molecule, introduced in Chapter 6 and analyzed in Chapter 8, is based on copying the first steps in plant photosynthesis. Here, we look at optical devices which mimic the principles of animal eyes. The brittlestar is a small animal without obvious eyes or a brain, but it appears to be light sensitive and avoids predators as though it can see them. It turns out that the "eyes" of this creature are integrated into its skeleton, as shown in Fig. 9.28(a). Each of the little bumps between the holes is a compound lens made of calcite. These lenses focus light onto light-sensitive nerve cells below the surface. Thus, the entire surface of the animal acts as an eye. The clever optical design of the calcite lenses was copied to make efficient adaptive lenses. The geometry of one element of the doublet (2 element) lens

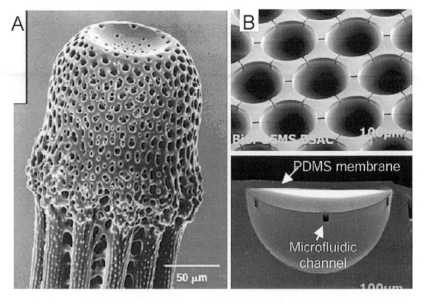

Fig. 9.28 (a) The skeleton of a brittlestar. The small bumps between the holes are two-element (doublet) lenses that focus incident light onto light-sensitive nerve endings below the surface. The design of the doublet has been copied in a lens array made by stamp technology (b). The synthetic lens has been improved to be adaptive, the refractive index of the second element being adjustable by injection of different fluids into a PDMS chamber through microfluidic channels. (Reprinted with permission from Lee and Szema[39], Science 2005, AAAS.)

was amenable to fabrication as a microfluidic chamber. This permits alteration of its refractive index by pumping different fluids into the chamber. An array of lenses made by stamp technology, and a zoom-in on one lens, are shown in Fig. 9.28(b).[39]

9.18   Bibliography

G.A. Ozin and A.C. Arsenault, Nanochemistry—A Chemical Approach to Nano-materials. 2005, RSC Publishing. Comprehensive coverage of the field of nanomaterials.

E.L. Wolf, Nanophysics and Nanotechnology—An Introduction to Modern Concepts in Nanoscience. 2nd ed. 2006, Wiley-VCH. Deals with physical aspects of nanodevices and materials.

9.19   Exercises

1. On what length scale (a range of values is fine) would you structure materials to act as (a) an infrared lens? (b) An X-ray lens? (c) A devices without conventional electrical resistance? (d) A device with novel magnetic properties based on modification of the exchange interaction. Explain your choices.

2. Using Equation 2.48 for the ground state energy of a particle in a box, estimate the size of the box needed for this energy to be less than $k_B T$ at 300 K, so that the material behaves like a "bulk" material in the sense that all of the levels are more closely spaced than $k_B T$. Take the mass to be the mass of an electron.

3. What is the density of sates at zero energy for 3-D, 2-D, and 1-D? What does this imply for the likelihood of an interaction between an electron and a low frequency fluctuation as a function of the dimensionality of the system?

4. What is the radius of an exciton in a material of dielectric constant 10, assuming $m^* = m_e$?

5. Sketch a wrapping vector with $n_1 = 3$ and $n_2 = 4$ on a graphene sheet. What is the diameter of a carbon nanotube for this wrapping vector?

6. A negative differential resistance device can be made to switch states if it has multiple stable operating points. When a resistor, R, is placed in series with the superlattice device whose $i–V$ curve ($i(V)$) is shown in Fig. 9.9 Ohm's law yields

$$V_0 - iR = V(i),$$

where V_0 is the total voltage applied across the circuit (of resistor and devices in series). Use Fig. 9.9 to solve this equation graphically to find the value of resistor for which multiple stable solutions just occur with $V_0 = 200$ mV. (The caption in the figure is in meV, but this equivalent to mV for an electron.) To make life simpler, we will pretend that this is a linear plot of current vs. voltage (it is not, the current scale is logarithmic, but that makes solving the problem awkward). This will make the numbers wrong, but it still illustrates how the device can work as an oscillator. When the series resistance is greater than this value, the line ($V_0 - iR$) intercepts the curve $V(i)$ in three places. Consider the effect of small fluctuations in voltage or current and explain why only the two intercepts with the parts of the device curve with positive slope are stable.

7. Calculate the value of the dielectric constant for the plasmon-polariton resonance of a prolate spheroid with a ratio of major axis to minor axis (a/b) of 3.

8. The area under the curves labeled "gain" in Fig. 9.13 represents the states with inverted population that could contribute to lasing. The area of these states above the line labeled "cavity losses" represents states that will contribute to laser output in a real cavity with losses. Assuming that the curves are normalized (have the same area in terms of states when integrated) estimate the relative amounts of laser power output for 3, 2, 1, and 0-dimensional structures using the data on the figure.

9. A colloidal crystal consists of 500 nm diameter PMMA spheres packed into a cubic lattice in which the spheres touch. The lattice is fully solvated with water ($n = 1.3$). The refractive index of PMMA is 1.5. (a) Calculate the volume-averaged refractive index. (b) At what angle (defined as in Equation 9.17) is 1.9 μm wavelength radiation Bragg-reflected for propagation along a diagonal (111) direction?

10. Describe the nature of the magnetism in clusters of Gd atoms of (a) 0.6 nm radius and (b) 3.4 nm radius.

11. The curves in Fig. 9.21 flatten out at higher current where thermoelectric cooling no longer increases with device current. Can you explain qualitatively why this happens?

12. Estimate the Reynolds number for (a) a charged sphere of 1 μm diameter undergoing electrophoresis and (b) an aircraft carrier ($L = 300$ m) both moving at 1 cm/s in water at 25°C.

13. The bases of DNA are ~0.5 nm in diameter and separated by 0.4 nm. Estimate the hydrophobic interaction energy between two adjacent bases assuming that Equation 9.20 holds and that the appropriate surface energy is 20 J/m^2. Use a decay length of 1 nm. Note that Equation 9.20 gives an energy per unit surface area, and we require an energy here. Approximate the bases as disks of 0.25 nm radius.

14. Use dimensional analysis to show that the ratio Z in Equation 9.18 has dimensions of $1/T$. *Hint*: $V^2/R =$ Joules per second.

References

[1] Ozin, G.A. and A.C. Arsenault, *Nanochemistry*. 2005, Cambridge: Royal Society of Chemistry.

[2] Alivisatos, A.P., Semiconductor crystals, nanoclusters and quantum dots. Science, **271**: 933–937 (1996).

[3] Wilson, W.L., P.F. Szajowski, and L.E. Brus, Quantum confinement in size-selected, surface oxidized silicon nanocrystals. Science, **263**: 1242–1244 (1993).

[4] Iijima, S., Helical microtubules of graphitic carbon. Nature, **354**: 56–58 (1991).

[5] Fennimore, A.M., T.D. Yuzvinsky, W.Q. Han, M.S. Fuhrer, J. Cumings, and A. Zettl, Rotational actuators based on carbon nanotubes. Nature, **424**: 408–410 (2003).

[6] Minot, E.D., Tuning the band structure of carbon nanotubes, PhD Thesis. 2004, Ithaca: Cornell University.

[7] McEuen, P.L., M. Bockrath, D.H. Cobden, Y.-G. Yoon, and S.G. Louie, Disorder, pseudospins, and backscattering in carbon nanotubes. Phys. Rev. Lett., **83**: 5098 (1999).

[8] Lu, W. and C.M. Lieber, Semiconductor nanowires. J. Phys. D.: Appl. Phys., **39**: R387–R406 (2006).

[9] Esaki, L. and R. Tsu, Superlattice and negative differential conductance in semiconductors. IBM J. Res. Dev., **January**: 61–65 (1970).

[10] Wacker, A., Semiconductor superlattices: a model system for nonlinear transport. Phys. Rep., **357**: 1–111 (2002).

[11] von Klitzing, K., *The Quantized Hall Effect*. Nobel Lecture (1985).

[12] Kuhn, S., U. Hakanson, L. Rogobete, and V. Sandoghdar, Enhancement of single-molecule fluorescence using a gold nanoparticle as an optical antenna. Phys. Rev. Lett., **97**: 017402-1-4 (2006).

[13] Griffiths, D.J., *Introduction to Electrodynamics*. 3rd ed. 1999, NJ: Prentice-Hall.

[14] Lakowicz, J.R., B. Shen, Z. Gryczynski, S. D'auria, and I. Gryczynski, Intrinsic fluorescence from DNA can be enhanced by metallic particles. Biochem. Biophys. Res. Commun., **286**: 875–879 (2001).

[15] Asryan, L.V. and S. Luryi, Quantum dot lasers: theoretical overview, in *Semiconductor Nano Structures for Optoelectronic Applications*, T. Steiner, Editor. 2004, Boston: Artech House. p. 113–158.

[16] Foulger, S.H., S. Kotha, B. Sweryda-Krawiec, T.W. Baughman, J.M. Ballato, P. Jiang, and D.W. Smith, Robust polymer colloidal crystal photonic bandgap structures. Optics Lett., **25**: 1300–1302 (2000).

[17] Yablonovitch, E., T.J. Gmittner, and K.M. Leung, Photonic band structure: the face-centered cubic case employing employing non-spherical atoms. Phys. Rev. Lett., **67**: 2295–2298 (1991).

[18] Soukoulis, C.M., The history and a review of the modelling and fabrication of photonic crystals. Nanotechnology, **13**: 420–423 (2002).

[19] Schurig, D. and D.R. Smith, Negative index lens aberrations. Phys. Rev. E, **70**: 065601 (2004).

[20] Shi, J., S. Gider, K. Babcock, and D.D. Awschalom, Magnetic clusters in molecular beams, metals and semiconductors. Science, **271**: 937–941 (1996).

[21] Aly, S.H., A theoretical study on size-dependent magnetic properties of Gd particles in the 4–300 K temperature range. J. Magn. Magn. Mater., **222**: 368–374 (2000).

[22] Adams, J.D., U. Kim, and H.T. Soh, Multitarget magnetic activated cell sorter. Proc. Natl. Acad. Sci. (USA), **105**: 18165–18170 (2008).

[23] Tsymbal, E.Y. and D.G. Pettifor, Perspectives of giant magnetoresistance, in *Solid State Physics*, H. Ehrenreich and F. Spaepen, Editors. 2001, New York: Academic Press, p. 113–237.

[24] DiSalvo, F.J., Thermoelectric cooling and power generation. Science, **285**: 703–706 (1999).

[25] Venkatasubramanian, R., E. Siivola, T. Colpitts, and B. O'Quinn, Thin-film thermoelectric devices with high room-temperature figures of merit. Nature, **413**: 597–602 (2001).

[26] Sritharan, K., C.J. Strobl, M.F. Schneider, A. Wixforth, and Z. Guttenberg, Acoustic mixing at low Reynold's numbers. Appl. Phys. Lett., **88**: 054102 (2006).

[27] Riehn, R., M. Lu, Y.-M. Wang, S. Lim, E.C. Cox, and R.H. Austin, Restriction mapping in nanofluidic devices. Proc. Natl. Acad. Sci. USA, **102**: 10012–10016 (2005).

[28] Kasianowicz, J.J., E. Brandin, D. Branton, and D.W. Deamer, Characterization of individual polynucleotide molecules using a membrane channel. Proc. Natl. Acad. Sci. USA, **93**: 13770–13773 (1996).

[29] Branton, B., D. Deamer, A. Marziali, H. Bayley, S.A. Benner, T. Butler, M. Di Ventra, S. Garaj, A. Hibbs, X. Huang, et al., Nanopore sequencing. Nat. Biotechnol., **26**: 1146–1153 (2008).

[30] Keyser, U.F., B.N. Koelman, S. van Dorp, D. Krapf, R.M.M. Smeets, S.G. Lemay, N.H. Dekker, and C. Dekker, Direct force measurements on DNA in a solid-state nanopore. Nat. Phys., **2**: 473–477 (2006).

[31] Hummer, G., J.C. Rasaiah, and J.P. Noworyta, Water conduction through the hydrophobic channel of a carbon nanotube. Nature, **414**: 188–190 (2001).

[32] Holt, J.K., H.G. Park, Y. Wang, M. Stadermann, A.B. Artyukhin, C.P. Grigoropoulos, A. Noy, and O. Bakajin, Fast mass transport through sub–2-nanometer carbon nanotubes. Science, **312**: 1034–1037 (2006).

[33] Israelachvilli, J.N., *Intermolecular and Surface Forces*. 2nd ed. 1991, New York: Academic Press.

[34] Berim, G.O. and E. Ruckenstein, Microscopic interpretation of the dependence of the contact angle on roughness. Langmuir, **21**: 7743–7751 (2005).

[35] Cheng, Y.T., D.E. Rodak, C.A. Wong, and C.A. Hayden, Effects of micro- and nano-structures on the self-cleaning behaviour of lotus leaves. Nanotechnology, **17**: 1359–1362 (2006).

[36] Rosario, R., D. Gust, A.A. Garcia, M. Hayes, J.L. Taraci, T. Clement, J.W. Dailey, and S.T. Picraux, Lotus effect amplifies light-induced contact angle switching. J. Phys. Chem. B, **108**: 12640–12642 (2004).

[37] Tian, Z.R., J.A. Voigt, J. Liu, B. Mckenzie, M.J. Mcdermott, M.A. Rodriguez, H. Konishi, and H. Xu, Complex and oriented ZnO nanostructures nature. Materials, **2**: 821–826 (2003).

[38] Whaley, S.R., D.S. English, E.L. Hu, P.F. Barbara, and A.M. Belcher, Selection of peptides with semiconductor binding specificity for directed nanocrystal assembly. Nature, **405**: 665–668 (2000).

[39] Lee, L.P. and R. Szema, Inspirations from biological optics for advanced photonic systems. Science, **310**: 1148–1150 (2005).

Nanobiology

10

This chapter outlines some of the ideas of modern molecular biology, focusing on the role of fluctuations in the small systems that comprise the building blocks of life. One measurement aspect of nanobiology—single molecule measurements—was dealt with in Chapter 4. We begin with a brief description of natural selection as the central driving force in biology. We go on to describe some of the components of modern molecular biology, getting a glimpse into the role of random processes when we discuss the remarkable phenomenon of gene splicing. We examine the role of structural fluctuations in enzyme catalysis, and go on to discuss biological energy and molecular motors, phenomena that rely upon fluctuation-driven electron transfer reactions. Finally, we examine the role of fluctuations in gene expression in the development of whole organisms. We use the immune system as an example of a biological process in which random assembly of its components is essential. We end that discussion with a brief look at the role of random gene splicing in the development of neural networks, and ultimately, possibly, in the development of the mind itself. Readers unfamiliar with molecular biology might find the glossary in Appendix L useful.

10.1   Natural selection as the driving force for biology

There is probably no more interesting emergent phenomenon than life. Charles Darwin published "The Origin of Species by Means of Natural Selection" in 1859. It laid the theoretical foundation for modern biology and is the first comprehensive description of an emergent phenomenon.

A slightly modernized scheme of natural selection is shown in Fig. 10.1. We know today that random variation in the physical manifestation of a particular creature (its *phenotype*) is a consequence of random variation in the sequence of the DNA of the creature or of the placement and repetition of particular genes (its *genotype*). Understanding the connection between phenotype and genotype is a major focus of modern biology. Very few phenotypical features are associated with just one gene. Darwin, of course, did not know about DNA, but he did observe the large random variations between members of a given species. He also saw how members of the population best adapted to a particular environmental niche formed the most successful breeding cohort. The consequent concentration of those phenotypical features best adapted to the particular ecological niche drives the formation of new species.

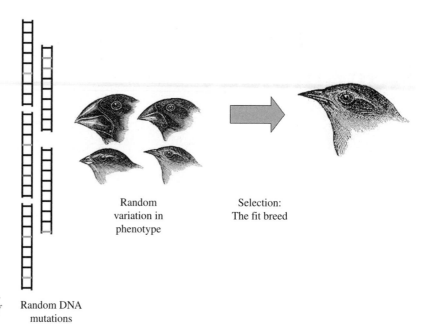

Random
variation in
phenotype

Selection:
The fit breed

Random DNA
mutations

Fig. 10.1 Evolution by natural selection. (Finches from Darwin's *Journal of Researches*, 1845.)

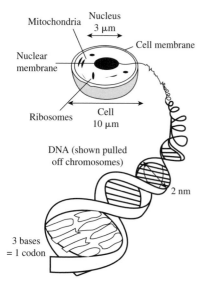

Fig. 10.2 The eukaryotic cell. The genetic material (DNA) is contained inside an enclosed nuclear membrane with genetic information exported (as an RNA "copy" of DNA) to the cytoplasm between the nuclear and external membranes where ribosomes (attached to an internal membrane called the endoplasmic reticulum) translate the RNA code to proteins. The entire cell is packaged in a lipid bilayer membrane.

One of the most striking aspects of biology on the nanoscale is the operation of "Darwinian-like" natural selection at the molecular scale, and we will describe processes that mimic evolution by natural selection at the molecular level. At this point, it is useful to make a connection with the other emergent phenomenon we have studied, which is electron transfer in solution. Recall that a random population of solvent polarizations helped to bring about the degenerate transition state required for electron transfer by Fermi's Golden rule. We shall see that biological molecules that catalyze reactions (enzymes) use this mechanism, so that natural selection serves as a motif for biological processes all the way from the most elementary chemistry through to cells, whole organisms, and entire ecologies.

10.2   Introduction to molecular biology

The smallest independent living systems are composed of single cells, and are called prokaryotes. They consist of a lipid bilayer "bag" that contains DNA and the proteins needed to sustain the life of the cell (by synthesizing components using the instructions encoded in the DNA). Multicelled organisms, called eukaryotes, have a more complex double membrane structure illustrated in Fig. 10.2. Because these multicelled organisms (like you and me) have to specialize gene expression according to the particular function of the cell within the organism, the genetic material (DNA folded onto positively charged proteins in a compact structure called chromatin) is contained within an inner membrane in a structure called the cell nucleus. Between the cell nucleus and the outer cell membrane lies the cytoplasm, the region containing all of the machinery needed to generate chemical energy for the cell (the mitochondria) and to translate

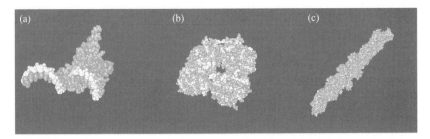

Fig. 10.3 Examples of protein structures. (a) A DNA transcription factor binding DNA. (b) A dimer (two-component) enzyme that modifies other proteins by adding a phosphate group to them. (c) A protein that acts as a structural scaffold. The size scales are different in each image as can be judged from the size of the component "balls" that represent the atoms. All structures are from the National Center from Biotechnology Information, visualized with Cn3D.

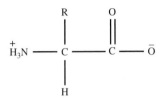

Fig. 10.4 (Top) Generic structure of a single amino acid where "R" represents one of the 20 naturally occurring residues illustrated in Fig. 10.5. (Bottom) Amino acids joined by peptide bonds to form a polypeptide.

genetic material into proteins (the ribosomes). Transport of genetic code from the cell nucleus into the cytoplasm is controlled by a nanoscale gatekeeper called the nuclear pore complex. The same complex also regulates the transport of proteins that control gene expression from the cytoplasm back into the nucleus.

DNA is the repository of genetic information and its structure was described in detail in Chapter 6. The nanomachines that carry out the chemical processes essential to life are proteins. Proteins come in an amazing variety of shapes and sizes and chemical properties. Some examples are shown in Fig. 10.3. Figure 10.3(a) shows a protein called AP1. It is a transcription factor that binds DNA to regulate the expression of genes. Some idea of the size of the protein is given by the (2 nm diameter) double helix it is shown bound to in the figure (the DNA is the structure at the bottom). The protein is a dimer formed by the association of two different proteins. The second protein shown in Fig. 10.3(b) is also a dimer composed of two units that associate to form a complex that modifies specific sites on other proteins by adding a phosphate to them. Such enzymes are called protein kinases. The final example shown in Fig. 10.3(c) is a filamentous protein called actin. Actin acts as a structural scaffold, controlling the shape of cells and forming part of the motor of muscle tissue (see Section 10.9).

This incredible structural and chemical diversity is built from a limited number of quite simple chemical building blocks. The elementary unit of a protein is called an amino acid, and the generic chemical structure of an amino acid is shown in the top panel of Fig. 10.4. Here the variable component is labeled R and it is connected through a carbon atom to an amine at one end and to a carboxylate residue at the other end. The ends are shown in their normal charge state in water at pH 7.0. These units are joined together in a polymer chain where one oxygen is cleaved from the carbon together with two hydrogens from the nitrogen to generate a water molecule, as the carboxylate joins to an amine to form a *peptide bond*. This process is repeated to join many units together. The resulting chain of amino acids is called a *polypeptide* and it folds into a functional protein. The chemical diversity of proteins is built on the 20 different types of residues symbolically represented by R in Fig. 10.4. These 20 amino acids are shown in Fig. 10.5. Remarkably, all of the chemical diversity of biology is built on just these 20 components, DNA, lipids, sugars, and minor chemical modifications to these components. The first five amino acids shown in the top panel of Fig. 10.5 are hydrophobic, so, in most proteins, regions of the peptide chain that are rich in these residues tend to be pushed to the interior of the folded protein. The next seven residues shown in the middle panel of Fig. 10.5 are hydrophilic and tend to lie on the outside of the folded protein. Proteins that are embedded in lipid bilayers (i.e., transmembrane proteins) have hydrophobic residues on the surface that contacts the lipid bilayer. The final group of eight amino acids have an intermediate character and can change their properties from hydrophobic to hydrophilic and *vice versa* depending on local chemical conditions such as pH. Accordingly, these residues can act as chemical sensors, driving changes in the structure of the protein in response to changes in the local chemistry of the environment.

Thus, we see that these amino acid building blocks can give rise to a fantastic variety of properties of the protein product, depending on the specific sequence of the residues in the polypeptide backbone. This sequence of residues is specified by the sequence of bases in DNA according to the genetic code. Twenty different amino acids require a minimum of three bases (two bases could specify only a maximum of four times 4 or 16 amino acids). The correspondence between the sequence of bases on the DNA and which particular amino acid they code for was sorted out many years ago and the end result is summarized in Fig. 10.6. The DNA remains within the cell nucleus but instructions for the synthesis of specific proteins are passed out from the nucleus to the cytoplasm in the form of RNA. Thus, the DNA specifying the sequence of the protein is first translated into a sequence of RNA that corresponds to the complement of the DNA sequence. A T-base on DNA is translated into an A on RNA, a G on DNA is translated into a C on RNA, and a C on DNA is translated into a G on RNA. The translation of A is handled little differently, because RNA uses a slightly different version of thymine called uracil (U). Thus, an A on DNA is translated into a U on RNA. Figure 10.6 shows how the RNA sequence is translated into the amino acid sequence of the final protein (the abbreviations for the amino acids correspond to the three letter abbreviations given in Fig. 10.5). There are 64 combinations that can be made from three sets of the four bases so the genetic code has redundancies. For example, four different groups of three bases (called codons) code for arginine: CGU, CGC, CGA, and CGG. Three of the codons, UAA, UAG, and UGA, code for a stop signal (i.e., the end of the protein chain). Interestingly, this degeneracy in the genetic code is not absolute, because different species use certain codons preferentially. This form of readout requires a universal "start" signal and blocks of DNA code that are upstream of the gene contain both these start signals and other sequences that are involved in turning transcription of the gene on and off.

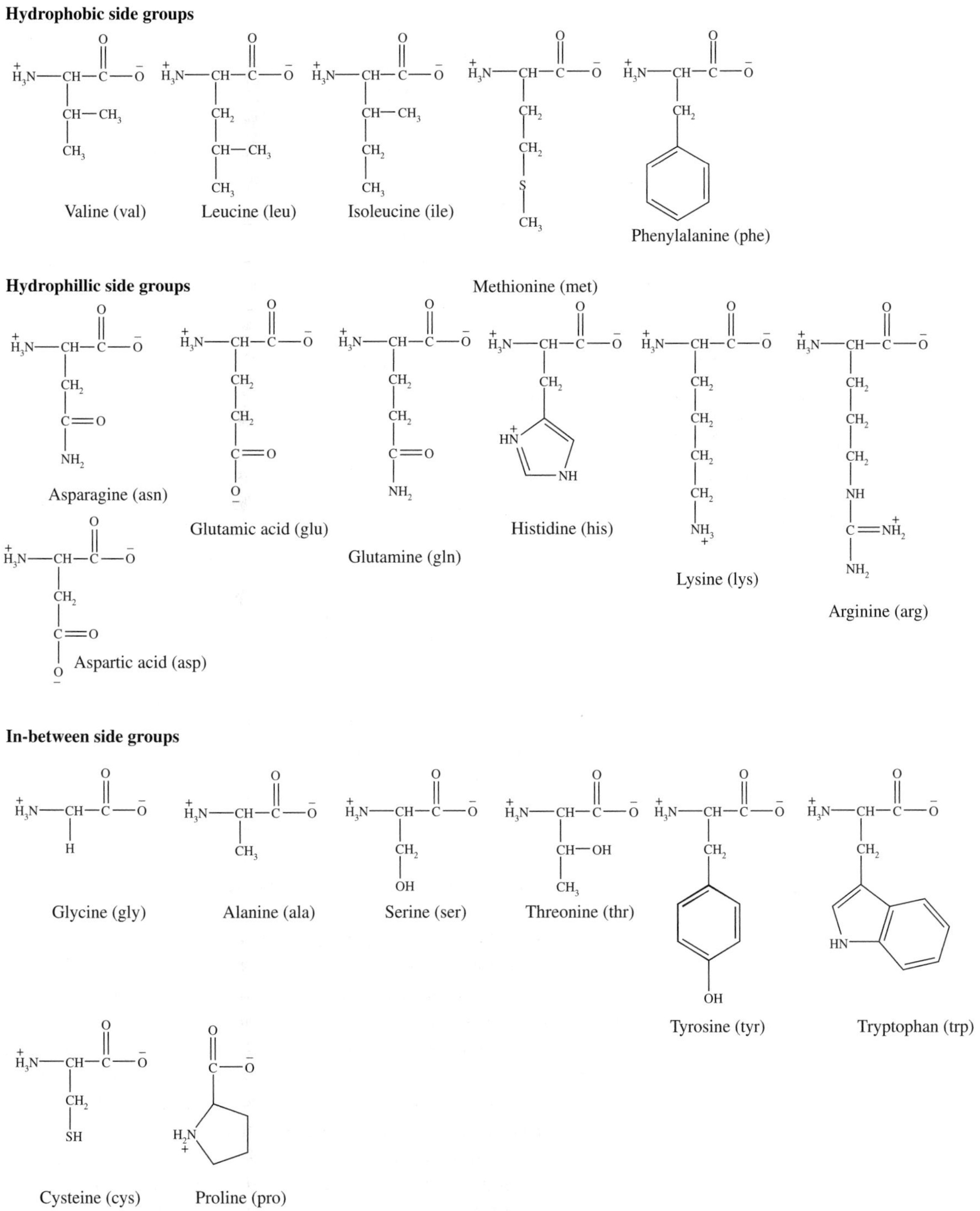

Hydrophobic side groups

Valine (val)   Leucine (leu)   Isoleucine (ile)   Methionine (met)   Phenylalanine (phe)

Hydrophillic side groups

Asparagine (asn)   Glutamic acid (glu)   Glutamine (gln)   Histidine (his)   Lysine (lys)   Arginine (arg)

Aspartic acid (asp)

In-between side groups

Glycine (gly)   Alanine (ala)   Serine (ser)   Threonine (thr)   Tyrosine (tyr)   Tryptophan (trp)

Cysteine (cys)   Proline (pro)

Fig. 10.5 The 20 naturally occurring amino acids. These are strung together by peptide bonds to form polypeptides that fold into functional proteins. The top 5 are hydrophobic, the middle 7 are hydrophilic, and the remaining 8 have intermediate character. Most proteins fold with hydrophobic groups in the interior if this is possible. Membrane proteins (embedded in a lipid bilayer) have hydrophobic exteriors.

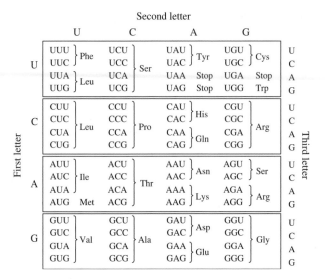

Second letter

		U	C	A	G	
First letter	U	UUU ⎱ Phe UUC ⎰ UUA ⎱ Leu UUG ⎰	UCU ⎱ UCC ⎱ Ser UCA ⎰ UCG ⎰	UAU ⎱ Tyr UAC ⎰ UAA Stop UAG Stop	UGU ⎱ Cys UGC ⎰ UGA Stop UGG Trp	U C A G
	C	CUU ⎱ CUC ⎱ Leu CUA ⎰ CUG ⎰	CCU ⎱ CCC ⎱ Pro CCA ⎰ CCG ⎰	CAU ⎱ His CAC ⎰ CAA ⎱ Gln CAG ⎰	CGU ⎱ CGC ⎱ Arg CGA ⎰ CGG ⎰	U C A G
	A	AUU ⎱ AUC ⎱ Ile AUA ⎰ AUG Met	ACU ⎱ ACC ⎱ Thr ACA ⎰ ACG ⎰	AAU ⎱ Asn AAC ⎰ AAA ⎱ Lys AAG ⎰	AGU ⎱ Ser AGC ⎰ AGA ⎱ Arg AGG ⎰	U C A G
	G	GUU ⎱ GUC ⎱ Val GUA ⎰ GUG ⎰	GCU ⎱ GCC ⎱ Ala GCA ⎰ GCG ⎰	GAU ⎱ Asp GAC ⎰ GAA ⎱ Glu GAG ⎰	GGU ⎱ GGC ⎱ Gly GGA ⎰ GGG ⎰	U C A G

Third letter

Fig. 10.6 The genetic code, showing how mRNA bases (complementary copies from DNA, but with U substituting for T) are translated into amino acid residues in a protein (see Fig. 10.5). Each group of three bases is called a codon—note how some codons are highly degenerate (though which particular one of the degenerate set is used is often species specific). Three codons—UAA, UAG, and UGA—are "stop" signals that signify the end of a protein chain.

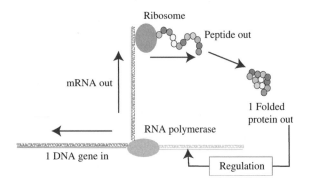

Fig. 10.7 Molecular biology 1: the simple version. RNA polymerase is a nanomachine that makes an RNA transcript from a DNA strand, substituting uracil in place of the thymine base. The RNA strand leaves the cell nucleus and is translated into a protein according to the genetic code (Fig. 10.6) by another nanomachine, the ribosome.

A simple schematic pathway for the synthesis of gene products is shown in Fig. 10.7. The DNA coding for the "start" signal for a particular protein becomes bound by a large enzyme called RNA polymerase. This translocates along the gene, copying each DNA base into its RNA complement. This new RNA polymer is called messenger RNA (mRNA in Fig. 10.7). The mRNA is passed from the cell nucleus to the cytoplasm via the nuclear pore complex. In the cytoplasm, it binds to another piece of protein machinery called the ribosome (ribosomes are generally localized in a folded membrane called the endoplasmic reticulum). The ribosome takes small pieces of RNA linked to amino acids (these are called transfer RNAs) and uses them to link the amino acids together in the sequence specified by the mRNA, adding the correct residue as each RNA base is read. The result is a peptide chain that spontaneously folds into the desired protein product. Both the ribosome and RNA polymerase are complexes formed by the spontaneous association of small pieces of RNA with protein. The final soup of proteins and small molecules in the cytoplasm feedback into gene expression through factors that are imported by the nuclear pore complex. These factors trigger a cascade of events in the cell nucleus that upregulate or downregulate the expression of particular genes.

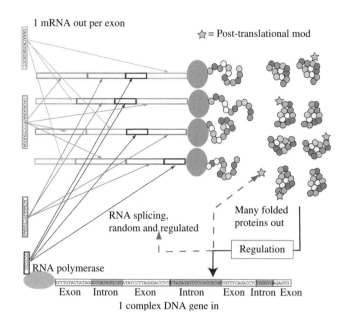

Fig. 10.8 Molecular biology 2: some of the complexity. Genes come in packets of coding regions (exons) separated by noncoding regions (introns). The introns are spliced out when the gene is assembled into a continuous mRNA transcript in a nanomachine called the spliceosome. The reassembly can be linear, or varied under regulatory control, or even random, with the result that one gene can code for many different proteins. The proteins are further differentiated by chemical modifications made after the proteins are folded, called post-translational modifications, shown here as a star added to some residues.

The Human Genome Project has shown that humans have only ∼25,000 genes, and yet it is known that there are many millions of different types of human proteins. How can this be? A more sophisticated version of the story given in Fig. 10.7 is shown in Fig. 10.8. It turns out that in eukaryotes (multi-celled animals), genes are split into packages of coding DNA (called exons) interrupted by regions of noncoding DNA (called introns). At one point, it was supposed that these noncoding regions were useless genetic junk inherited for no good reason. However, their sequence is well preserved from generation to generation, implying that they play an important role. Many parts of these introns are completely essential for the correct final assembly of the exons into the correctly spliced code. When RNA polymerase translates this complex DNA sequence into RNA, the regions corresponding to introns are excised. One pathway for this excision is the spontaneous folding of parts of the RNA into catalytically active units (called ribozymes) that serve to cut the adjacent RNA. The final assembly of exons into the complete coding sequence takes place in a nanomachine called the spliceosome. Here, the RNA is assembled into the final product by putting together the exons in a way that can be regulated by the environment. The various ways of ordering the exons results in various protein products when the different mRNAs are translated into proteins. We shall see that some important processes actually depend on the random splicing of these exons to form a random library of protein products. Random splicing is part of a "Darwinian" mechanism that allows cells to produce the correct product through a process of natural selection.

There is yet another important source of protein diversity based on specific chemical modifications of particular amino acid residues. These generally occur in a protein processing plant within the cell called the Golgi apparatus. This is the place where specific chemical stamps are put onto proteins that are then dispensed to various regions of the cell (to perform the functions that require that particular modification). These post-translational modifications extend the range of protein activities over and above the range produced by splicing variations. In addition to labeling proteins for dispatch to particular regions of the cell, post-translational modifications also serve a signaling function. Thus, a vast range of protein functionality can emerge from a limited number of genetic components. It has been recently discovered that the introns serve many other control functions when translated into RNA, opening a new avenue for RNA control of cells.[1] A newly discovered class of small "interfering" RNA fragments (siRNA) act to shut down expression of particular genes, and can be used as "switches" in biotechnology applications.

10.3 Some mechanical properties of proteins

Density:

A representative value for the density of well-packed protein is $\sim 1.4 \times 10^3$ kg/m^3,[2] leading to the following value for the radius of a spherical protein (where the mass is expressed in Daltons, MW):

$$r = \sqrt[3]{\frac{3}{4\pi}\frac{MW}{822}} = 0.066 \times \sqrt[3]{MW} \text{ nm.} \tag{10.1}$$

Thus, a (typical) 100 kDa protein has a radius of ~3 nm.

Stiffness:

The stiffness of the material is specified in a geometry-independent way by the Young's modulus. This is the ratio of stress to strain for a given material. Stress is the force per unit area applied to the material and strain is the fractional change in dimension, taken here (for simplicity) to be the direction along which the stress acts. Thus, if a force F acts on the face of a material of area, A, producing a fractional change in dimension $\frac{\delta \ell}{\ell}$, force and strain are related by the Young's modulus, E, according to

$$\frac{F}{A} = E\frac{\delta \ell}{\ell}. \tag{10.2}$$

It is useful to translate this quantity into an equivalent spring constant for the stretching of a Hookean spring. For the sake of argument, we will imagine a force acting on a face of a cube of sides ℓ. After inserting $A = \ell^2$ into Equation 10.2 and then multiplying both sides by ℓ^2 we arrive at Hooke's law,

$$F = E\ell\delta\ell, \tag{10.3}$$

where the spring constant, κ, is given by

$$\kappa = E\ell. \tag{10.4}$$

Some values for the Young's modulus of various proteins are given in Table 10.1. These may be used to estimate rough values for a spring constant by inserting the dimension of the protein that is strained.

Table 10.1 Young's modulus of various proteins
(Adapted from Howard[3])

Protein	E (GPa)	Function
Actin	2	Skeleton of the cell
Tubulin	2	Transport "roadways" in cells
Silk	~5	Spider webs, cocoons
Abductin	0.004	Mollusc hinge ligament
Elastin	0.002	Smooth muscle and ligaments

As an example, tubulin has a molecular weight of ~50 kDa, so according to Equation 10.1 its diameter is ~4.8 nm. Using $\kappa = E\ell$ with E= 2 GPa yields a spring constant of ~10 N/m. Elastin has a molecular weight of ~75 kDa, yielding a diameter of 5.5 nm and a spring constant of ~0.01 N/m.

Transport properties:

We will use as an example our "typical" 100 kDa protein with a radius of ~3 nm. The drag coefficient in water follows from Stokes' law

$$\gamma = 6\pi r \eta,$$

which, with $\eta = 1 \times 10^{-3}$ Pa·s, yields 56×10^{-12} Ns/m.

The diffusion constant can be obtained from Equation 3.85, $D = \mu k_B T$, and using $\mu = (6\pi a \eta)^{-1}$, we obtain for D

$$D \approx \frac{4.14 \times 10^{-21} \text{ J}}{56 \times 10^{-12} \text{ Ns/m}} = 73 \times 10^{-12} \text{ m}^2/\text{s}.$$

Thus, the root-mean-square displacement owing to diffusion is ~8.6 μm in 1 s.

10.4   What enzymes do

Enzymes are the functional machines of living systems. They serve as catalysts, even in the absence of an energy source, and here we focus on just one type of activity, the cleavage of the peptide bond, as an example of enzyme function. They can also produce motion using the adenosine triphosphate (ATP) produced by the mitochondria as a chemical energy source, a topic we will discuss in Section 10.8.

Peptide bonds are broken by means of a chemical reaction with a water molecule, referred to as *hydrolysis*. This reaction is illustrated in Fig. 10.9. It is an electron transfer reaction involving the active participation of a water molecule. Two protons from a water molecule combined with the nitrogen of the peptide bond to form a positively charged amine residue. The oxygen combines with the carbon of the peptide bond to produce a negatively charged carboxylate residue. This reaction can and does occur in water without any catalysis, but the half-life of a peptide bond in pure water varies from 300 to 600 years, depending on the position of the bond and the nature of the adjacent residues.[4] (The curious reader might like to know that these data were extrapolated from

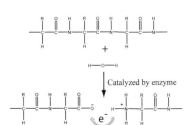

Fig. 10.9 Example of what an enzyme does: peptide bond hydrolysis. A protease will bind a protein at a site to be cleaved (usually a specific site in terms of amino acid sequence) and then catalyze an electron transfer reaction involving the peptide backbone and a water molecule. In the absence of the enzyme, the transition state for the reaction is reached once every 500 years—in the presence of the enzyme, it occurs about every ms!

much quicker experiments performed at high temperatures and pressures.) In the presence of an enzyme evolved to cleave the peptide bond, this half-life is reduced to times on the order of 1 ms, representing a 10^{13}-fold speeding up of the kinetics! (The millisecond timescale just quoted is a limiting value based on the assumption that the protein and its target are already assembled—rates will be much smaller in dilute solutions were enzymes and their substrates have to find one another.) Astounding though this speeding up is, the millisecond timescale is actually rather slow when compared with the intrinsic fluctuation rate of proteins.

The intrinsic fluctuation rate of proteins can be estimated as follows: imagine the protein deformed into a nonequilibrium geometry through the application of a force, F, that suddenly goes to zero at time $t = 0$. The strain in the protein will decay according to the result given in Appendix D for a highly damped mechanical system with time constant

$$\tau' = \frac{6\pi a \eta}{\kappa}. \tag{10.5}$$

Using the value 56×10^{-12} Ns/m for γ and taking the values for κ deduced from E (Table 10.1 and Equation 10.4), we find values for τ' that range from ~6 ps (tubulin) to 6 ns (elastin). Another way to estimate these times (see Chapter 3) is to assume that the distance from an equilibrium conformation to the transition state is some fraction of the protein's size—say $\ell_t = 1$ nm—and use

$$\tau \approx \frac{\ell_t^2}{D} \tag{10.6}$$

for the time between major fluctuations. Using $D \sim 74 \times 10^{-12}$ m^2/s (deduced above) yields $\tau \sim 1.4 \times 10^{-8}$ s, similar to the time estimated using the elastic properties of elastin.

These estimated fluctuation times are 10^5 to 10^7 times faster than the typical catalysis timescale of 1 ms. Thus, the fluctuation required to form the transition state for hydrolysis of the peptide bond occurs only in about one in one million of the total long-range conformational fluctuations of the protein. (Very local fluctuations are much more rapid, occurring on the timescale of bond vibrations, i.e., 10^{-14} s.) So while enzyme catalyzed peptide bond hydrolysis appears to be almost miraculously fast (1 ms vs. 500 years), it involves *extremely rare* configurations of the protein. This slow dynamics presents one of the principal impediments to computer simulations of enzyme catalysis from first principles.

Qualitatively, however, we can describe the process of enzyme catalysis as follows: the enzyme first binds to its target site on the protein to be cleaved. This can be an extremely sequence-specific event. To take a concrete example, the enzyme HIV protease carries out the task of chopping up a long continuous protein strand that has been synthesized by the host cell of the virus (the virus hijacks the cell's protein production apparatus by transferring its genome into that of the cell). The long protein chain has to be cut at just the right sites to make the collection of smaller proteins that will spontaneously assemble onto the HIV genome to form the packaged virus particle. This assembly, of genome plus protein coat, makes the new viruses that will eventually burst the cell to escape and infect other cells. Once the protease is bound at one of the correct sites for cleavage, a water molecule must also become bound at the correct position and in the correct orientation for hydrolysis of the peptide bond to occur. Once all the players are in place, bound in a stable energy minimum, little will happen until a fluctuation of the protein conformation drives the local polarization into the transition state required for the electron transfer (see the discussion of Marcus theory in Chapter 8). The reaction products form in an energetically excited state, but relax rapidly to the ground state of the products. These products are the N-terminus and a C-terminus of two peptide chains made by breaking the original chain.

In this case, no energy is required to drive the reaction because the free energy of the reactants—the water molecule and the peptide bond—is higher than that of the products—the two terminal ends of the newly broken chain. What the enzyme did was to lower the kinetic barrier for this reaction to occur. On the basis of the discussion above, we will assume that the attempt frequency is $\sim 10^8$ Hz. Using $\Delta G = k_B T \ln\left(\frac{f_0}{f}\right)$, a reaction rate of 10^3 Hz implies a free energy barrier of $\sim 11.5 \, k_B T$ (i.e., 0.3 eV or 6.9 kcal/mol). If we assume the same attempt frequency in pure water (probably not a good assumption), the kinetic barrier for the uncatalyzed reaction would be $\sim 42 \, k_B T$ (i.e., 1 eV or 25 kcal/mol).

If catalysis is such a rare event, it might appear to be impossible to locate the protein confirmations responsible for it. Conventional molecular dynamics is not up to the task at present. A geometric approach to the flexibility of complex structures like proteins has been developed[5] and this identifies the large-scale cooperative motions that are possible. Interestingly, they are generally few in number and their nature can be quite evocative of the function of a protein. Figure 10.10 shows two frames from a movie of the random fluctuations of HIV protease made with this geometric approach. The hole in the structure in the top panel of the figure surrounds the catalytic site where the protein backbone to be cleaved passes through the enzyme. Clearly, the large-scale fluctuations of this enzyme occur at the active site. This is because the bonding in this protein makes other degrees of freedom much more rigid. The protein might be thought of as a "mechanical focusing system" for Brownian motion. The fluctuations are still random, but the internal mechanics of the protein have evolved so as to ensure that the fluctuations are largest where they need to be, in order for the protein to carry out its function.

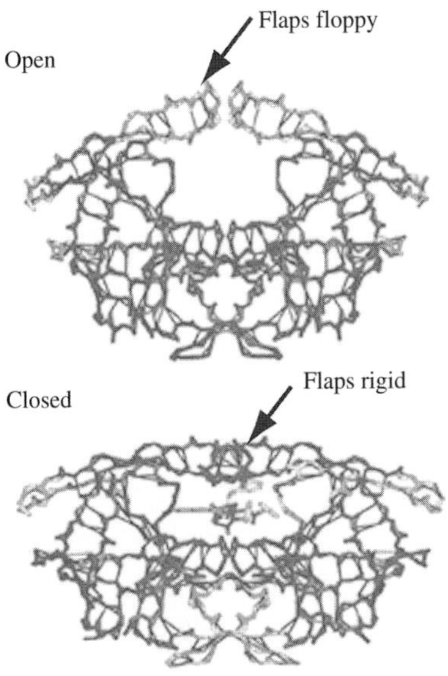

Fig. 10.10 Large-scale motion of a protease occurs at the site where the peptide to be cleaved passes through a cavity in the protease. Presumably this structure puts Brownian motion where it is needed to generate the fluctuations that catalyze the electron transfer reaction. (Courtesy of Professor Michael Thorpe of Arizona State University.)

10.5 Gatekeepers—voltage-gated channels

Living cells exchange materials with the outside world by means of channel proteins, orifices that are usually chemically selective, and often under the active control of signaling mechanisms in the cell. One example of a controlled channel is the voltage-gated channel. This is a channel that switches from open to closed as a function of the potential gradient across the cell membrane (or equivalently, ion, or proton gradient). One famous example is the depolarization wave ("action potential") that causes calcium channels to open when it reaches the synapse at the end of an axon. Insight into how such channels work has come from direct imaging of voltage-gated channels by the atomic force microscope. Porin OmpF channels are voltage-gated channels abundant in the

Fig. 10.11 AFM images of a monolayer of voltage-gated channel proteins on a graphite surface in electrolyte. (a) With no bias applied the channels are closed. (b) With a bias applied, between the graphite and a second electrode in contact with the electrolyte, the channels open: the "domes" have disappeared. (c) A structural model for the gating—a domain of the protein goes from soluble (OPEN) to insoluble, folding back into the protein pore (CLOSED). (Reprinted with permission from Müller and Engel[6], J. Mol. Biol. 1999, Elsevier.)

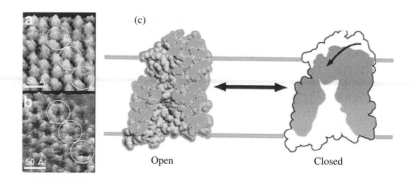

Open   Closed

cell walls of the gut bacterium *Escherichia coli*. Müller and Engel have imaged monolayers of these protein pores deposited onto graphite in an electrolyte solution. Graphite is a conductor, so, by placing a voltage bias between the graphite and a second electrode immersed in the electrolyte, they were able to switch the pores *in situ*, and image the conformational changes in the pore protein as it went from closed to open.[6] Figure 10.11(a) shows how each of the pores is capped by a little dome at low potential—the channel is closed. Figure 10.11(b) shows how the "domes" have vanished at high potential, leaving the channels open. A structural model for the conformation change is shown in Fig. 10.11(c). One of the domains of the protein is rotated out into solution in the open configuration, but collapses into a solid plug in the closed configuration. Presumably, amino acid residues of a mixed character (i.e., those shown in the lower panel of Fig. 10.5) change their charge in response to the change in local potential gradient, with the result that the solubility of the domain changes (recall the discussion of acid–base equilibrium in Chapter 3). When the domain becomes hydrophobic overall, it will push back into the interior of the pore, as shown in Fig. 10.11(c).

We turn next to biomolecular machines that require an energy input in order to direct their motion.

10.6   Powering bio-nanomachines: where biological energy comes from

Living organisms do useful directed work which requires a source of fuel. The universal biological energy source is a molecule called ATP and all cells contain machinery for ATP synthesis. In animals, this machinery is contained in the mitochondria (reviewed in Chapter 6). In the mitochondria, electrons (released from a pyruvate molecule in the course of degradation to carbon dioxide) are eventually transferred to oxygen, reducing it to form water. The energy released in this process is used to develop a proton gradient across a membrane. Proton flow, driven by the chemical potential gradient, powers an enzyme called *ATP synthase* which synthesizes ATP from adenosine diphosphate (ADP) and available phosphate ions. Animals obtain carbohydrate food molecules by eating plants, and these carbohydrates are degraded to pyruvate by machinery outside the mitochondria. Plant cells obtain their energy directly from sunlight using organelles called chloroplasts. The chloroplasts contain a photoreaction center that uses the reaction of photogenerated energetic chemical species to produce a proton gradient across the membrane of the chloroplast. This drives the same molecular machine used in mitochondria—ATP synthase—to synthesize ATP from ADP. In plants, the ATP is used to drive the energetically unfavorable synthesis of carbohydrates. Nature has evolved a common pathway for making a common fuel (ATP) from a proton gradient across the cell membrane in both plants and animals. Thus, in nature, the source of power is a *protomotive* force, in contrast to the electromotive force that drives the electrical appliances that we humans make. In summary, the energy chain starts with sunlight driving the synthesis of ATP in plants which synthesize carbohydrates that are then eaten by animals who use the carbohydrate to synthesize ATP in mitochondria.

10.7   Adenosine triphosphate—the gasoline of biology

The hydrolysis of ATP (illustrated in Fig. 10.12) is a universal source of biochemical energy for ATP-dependent enzymes and molecular motors. It is worth taking a close look at the structure of these molecules. Hopefully, they are familiar! The linked adenine, sugar, and phosphate are almost identical to some of the components of the DNA molecule we discussed in Chapter 6 (but with the extra OH on the sugar, characteristic of RNA). This is a wonderful illustration of the way in which nature re-uses molecular components, and an encouragement to those who fear that biochemistry requires unlimited memorization of structures.

Referring to Fig. 10.12, the hydrolysis reaction proceeds by the water-mediated cleavage of the bond between the second and third phosphate groups (reading outwards from the adenosine these groups are referred to as the α-, β-, and γ-phosphates). This reaction releases a phosphate ion plus ADP. The forward reaction, in which ADP is produced, usually proceeds in concert with some desired motor function. That is to say, the hydrolysis of ATP is directly coupled to some conformational change in a motor protein. The equilibrium constant for this reaction (it is dimensionless—Chapter 3.16) is given by

$$K_{eq} = \frac{[ADP][P_i]}{[ATP]} = 4.9 \times 10^5. \tag{10.7}$$

Here, the various concentrations represent the sum of all the ionic species that contain an ATP, ADP, or phosphate, for example, $MgATP^{2-}$, ATP^{4-}, etc. Water is not included in the calculation of the rate constant because the reaction is assumed to take place at constant pH. By solving $K_{eq} = \exp -\frac{\Delta G}{k_B T}$ for ΔG,

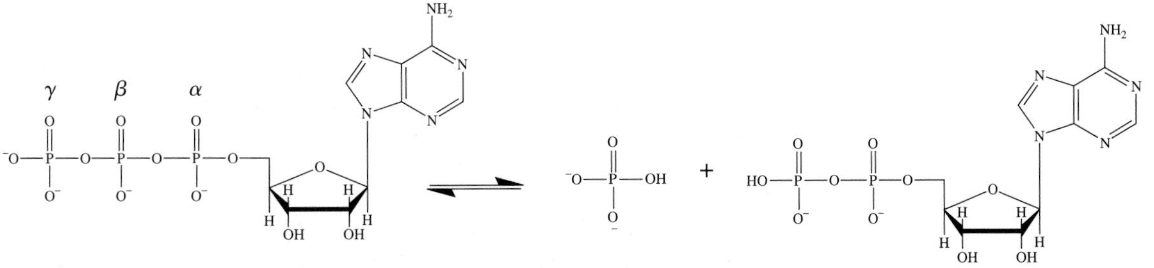

Fig. 10.12 Hydrolysis of adenosine triphosphate (ATP) to adenosine diphosphate (ADP) as the source of energy for some enzymes and all molecular motors. The ATP is made in the mitochondria, described in Chapter 6.

Table 10.2 Energy from ATP
hydrolysis in various units

12 kcal/mole
20 kT at 300 K
0.52 eV/molecule
8.3×10^{-20} J/molecule
82 pN·nm

we obtain the available free energy from the hydrolysis of a single ATP as 12 kcal/mol. This energy is summarized in various units in Table 10.2.

The reverse reaction, the phosphorylation of ADP, is carried out by the molecular motor, ATP synthase (Section 10.8). In this case, the reaction requires energy and this is supplied by a proton gradient across the membrane containing the ATP synthase. As protons defuse through the motor from the low pH side of the membrane to the high pH side of the membrane, they turn a "turbine" inside the motor. The rotation occurs in discrete steps of 120° and a bound ADP is phosphorylated to form ATP at each 120° step.

10.8   The thermal ratchet mechanism

One mechanism for molecular motors is called the "thermal ratchet,"[3] outlined in Fig. 10.13. Here we will consider a "downhill" process in which the hydrolysis of ATP directs the motion of a molecular motor by the formation of the (lower energy) ADP. The thermal ratchet model has similarities to the Marcus model for electron transfer, discussed in Chapter 8.

The reactants at E_1 on the left of the figure are the motor protein bound with a water molecule in the presence of a significant concentration of ATP. The products at E_2 on the right of the figure are a phosphate ion, ADP, and a motor that has advanced one step. Binding of ATP and protein fluctuations (on a timescale given by Equation 10.5 or 10.6) eventually result in the transition state (E_a in the figure) being reached. At this point, the electron transfer reaction involving the water molecule and ATP occurs, and the ATP is hydrolyzed to ADP. This is no longer stably bound, so it is released, stabilizing the (one-step advanced) motor against backward fluctuations. This principle of operation is illustrated schematically for a monomolecular reaction in Fig. 10.13(b) by a fluctuating cleft in the molecule which, when opened, allows the flexible ratchet to bind in the cleft, preventing it from closing in the face of future fluctuations. This is not a very good model for the multiparticle reaction just described, and

Fig. 10.13 The thermal ratchet mechanism for molecular motors. (Reprinted with permission from *Mechanics of Motor Proteins and the Cytoskeleton*[3], 2001, Sinauer Associates.) For a downhill reaction like ATP hydrolysis, the free energy for the products (E_2) is lower than that of the reactants (E_1). They are shown as equal in this figure.

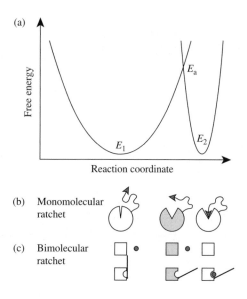

(a)

(b)   Monomolecular
ratchet

(c)   Bimolecular
ratchet

a schematic bimolecular ratchet is shown in Fig. 10.13(c). Here binding of the small molecule prevents the "lid" from closing over the binding pocket. The real operation of molecular motors is considerably more complex involving as it does several reactants and several products, but the "ratchet" is, apparently, the release of ADP after hydrolysis of ATP.

The frequency, k, of crossing the kinetic barrier is determined by the approach rate multiplied by the exponential of the free energy corresponding to the transition state (E_a):

$$k \approx \frac{\kappa}{6\pi a\eta} \exp\left[\frac{-E_a}{kT}\right]. \qquad (10.8)$$

In this expression, E_a is a compound quantity that incorporates free energy donated by ATP hydrolysis together with the work expended in the power stroke of the motor.

10.9   Types of molecular motor

Muscles are probably the biological "motors" most familiar to us. Their motor function is carried out by an actin–myosin complex found in the muscle tissue. The structure of the active component (the sarcomeres) of smooth muscle tissue is shown in Fig. 10.14. It consists of a thick filament to which are attached many myosin molecules. These thick filaments are interdigitated with thin filaments, fibers composed of bundles of the protein actin (Fig. 10.3(c)). During muscle contraction, the thick filaments move against the thin filaments to pull the fixed components of the sarcomere together. In some animals, muscle tissue can contract by more than 20% in length in a period of tens of milliseconds.

The active motor component of the muscle tissue is found in the crossbridges that connect the thin filaments to the thick filaments. These crossbridges are

(a)

(b)

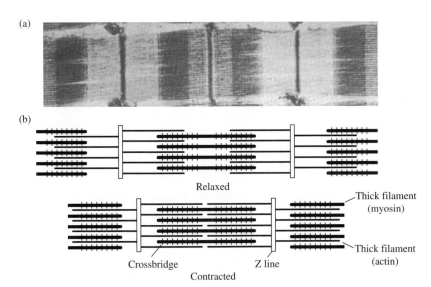

Relaxed

Crossbridge   Z line

Contracted

Thick filament (myosin)

Thick filament (actin)

Fig. 10.14 Molecular motors in muscle sarcomeres. A TEM image of sarcomeres is shown in (a). The crossbridge model is illustrated in (b). The myosin "walks" along the interleaved actin filaments to draw the crossbridges together, resulting in muscle contraction. (Courtesy of Professor Hugh Huxley, Brandeis University.)

made from myosin molecules which consist of a long stalk that is permanently attached to the thick filament, and a pair of head units that transiently contact the actin filaments. An electron micrograph of myosin molecules is shown in Fig. 10.15(a). The arrows point to the head units at the end of the long stalks.

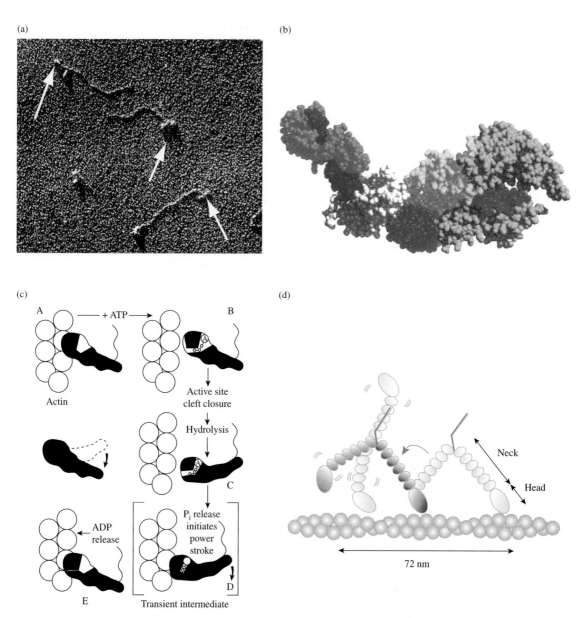

Fig. 10.15 The actin–myosin molecular motor complex. (a) Electron micrograph of myosin molecules. The arrows point to the heads that engage and walk on the actin filaments. (Reprinted with permission from Elliot et al.[7], Proc. Royal Soc. 1976, Royal Society of London. Courtesy of Dr. Gerald Offer, Muscle Contraction Group, Department of Physiology and Pharmacology, University of Bristol). (b) Crystal structure of the myosin head unit. The two actin-binding domains are in the lower right of the figure. (Reprinted with permission from Rayment et al.[8], Science 1993, AAAS). (c) ATP hydrolysis and the power stroke in the actin–myosin complex. The figure reads from A to E showing motion down one repeat of the actin filament. (Reprinted with permission from Rayment et al.[9], Science 1993, AAAS). (d) Detail of the walking motion of myosin on actin showing a step from right to left on the figure. (Reprinted with permission from Shiroguchi and Kinosita[10], Science 2007, AAAS.)

A view of the crystal structure of the myosin head, looking in from the two domains (lower right) that bind actin, is shown in Fig. 10.15(b). A schematic view of the motion along the actin filament is shown in Fig. 10.15(c). ATP binding (first two steps) causes disengagement from the actin with accompanying motion down one step. This downward-step is stabilized by release of the phosphate removed in the degradation of ATP to ADP, and followed by release of the ADP. Figure 10.15(d) shows a modern view of the details of the walking motion. One of the heads becomes unbound from the actin filament on binding to the ATP. The second head remains bound, leaving the first free to diffuse until it finds a binding site downstream of the bound leg. Strain in the bound leg biases the diffusion toward the downstream direction. The new configuration is stabilized by release of phosphate and ADP, leaving the myosin translocated by one repeat of the actin filament and ready for the next ATP-binding event.

Myosin motors consume one ATP molecule per power stroke and contact the actin filament only transiently. That is to say they have a small duty cycle, with only a few of the myosin motors in contact with the actin filament at any time, shuffling the actin filament along. ATP binding and hydrolysis is believed to cause strain only when the myosin head is attached to the actin filament. The power stroke distance per binding event per myosin head is ~5 nm and the measured force generated is ~1.5 pN. A lever action is generated by the long attached stalk being moved by the rotation of the head, and this results in a motion of the actin filament of ~36 nm. Thus, the work done per ATP is ~1.5 pN × 36 nm or 54 pN·nm (an efficiency of ~65%—see Table 10.2).[3]

In vitro visualization of the action of myosin molecules has been carried out by functionalizing a microscope slide with myosin molecules so that their stalks are attached to the slide (Fig. 10.16). Dye-loaded actin molecules were then placed on the slide, on top of the myosin layer. When the system was fed ATP, the actin filaments were observed to start to move on the slide (using fluorescence

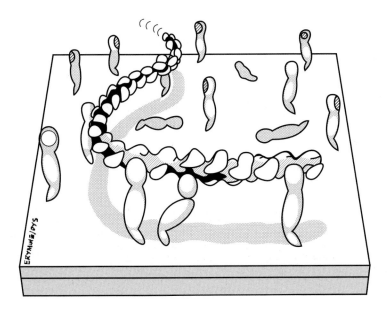

Fig. 10.16 In this cartoon, myosin heads are depicted adsorbed to a nitrocellulose-coated coverslip and are in the process of moving an actin filament. Motion was observed in a flow cell consisting of a coverslip that is mounted on a glass slide using silicon grease as a spacer (to achieve a separation of ~0.5–1.0 mm). Motion is observed through the coverslip (i.e., from the bottom). (Reprinted with permission from Warrick and Spudich[11], Annual Reviews.)

Fig. 10.17 Fluorescently labeled actin filaments imaged with an optical microscope. They are attached to myosin heads (as shown in Fig. 10.15) and move over the surface when ATP is introduced into the system. The three frames show the motion at four second intervals. (Reprinted with permission from Actin: From Cell Biology to Atomic Detail, Steinmetz, M.O., D. Stoffler, A. Hoenger, A. Brener, and U. Aebi, Journal of Structural Biology, **3**: 379–421 (1987). Permission Elsevier.

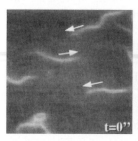
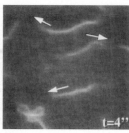
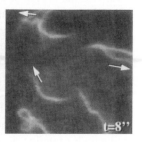

microscopy to visualize the dye-loaded actin filaments). The myosin moves on a given actin filaments in one direction only (i.e., the actin molecules are polarized) but the actin filaments will move in random directions (Fig. 10.17) because the system is randomly assembled (i.e., the orientation of the actin filaments is not controlled). Small ATP-driven linear motors have been made by orienting the protein components appropriately so that all the actins move in one direction.[13]

Myosin is an example of a molecular motor that works only in concert with many other molecules. By itself, it will detach from an actin filament, because its duty cycle (the fraction of the power stroke time for which the molecule is attached) is small. An example of molecules that remain attached throughout the motor cycle are those that transport material by walking along microtubules. Their function requires that they remain attached. Microtubules are the self-assembled "freeways" of the cell. Composed of proteins called tubulins, they spontaneously assemble and disassemble as needed to move materials into various regions of the cell. Kinesin and dynein are two different types of motor proteins that "walk" along microtubules. Both have been studied by single molecule techniques and a fairly detailed model of kinesin motion is now available.[14] Kinesin appears to walk in a hand over hand fashion whereby one the head will attach, the second let go, passing the first and attaching at the next tubulin sub-unit along the microtubule, whereupon the first headed detaches, and the cycle is repeated. In this case, the duty cycle of the motor is at least 50% for each head so that the kinesin molecule always remains attached to the microtubule. Like myosin, kinesin hydrolyzes one ATP molecule a step. Kinesin can generate a force of up to 3 pN and moves 8–16 nm a step (an efficiency of 30–60%).[3]

The third kind of molecular motor we will consider is a *rotary* motor, an analog of the familiar electric motor. We will focus on ATP synthase, already discussed as the molecular machine that synthesizes ATP from ADP using a proton gradient. Remarkably, the motor can work backwards as well. It uses ATP, converting it to ADP, to generate rotary motion which pumps protons. Rotary motors are found elsewhere in nature, such as those that drive the flagellae of swimming bacteria. The structure of the ATP synthase motor is shown schematically in Fig. 10.18. The rotating head unit sticks out into the region of higher pH where it phosphorylates ADP. A base unit (which translocates protons from the low pH side of the membrane to the high pH side or vice-versa) is embedded in the membrane. The coupling of proton translocation to motor rotation and ADP phosphorylation is shown in Fig. 10.19.

The rotary motion of individual ATP synthase molecules has been visualized directly by attaching gold beads to one part of the motor (a fragment

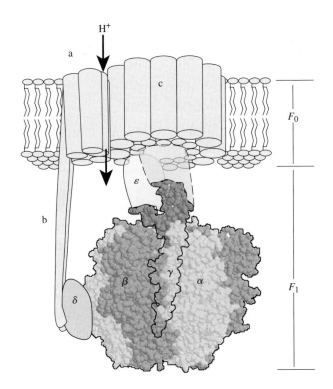

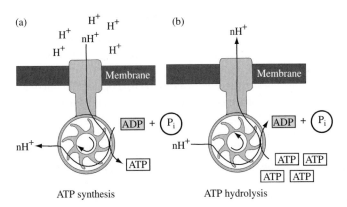

Fig. 10.18 ATP synthase. The F1 unit can be driven into rotation by a flow of protons through the enzyme driven by the proton gradient across the cell membrane. The rotation is accompanied by the conversion of ADP to ATP. Conversely, a high concentration of ATP will drive the motor into rotation in the other direction as the ATP is hydrolyzed to ADP. (Reprinted with permission from Wang and Oster[15], Nature 1998, Nature Publishing Group.)

Fig. 10.19 Diagrammatic representation of the operation of ATP synthase as (a) a motor driven by a proton chemical potential gradient that synthesizes ATP by phosphorylation of the less energetic ADP and (b) a motor operated in the reverse direction in the presence of a large excess of ATP which is hydrolyzed to ADP to turn the motor. (After Fig. 14.27 of Alberts et al.[16])

corresponding to the moving part of the proton translocation apparatus) while the other part (in this case, the ATP hydrolysis subunit) was fixed to a glass slide. Movement of the gold bead was detected by laser dark-field optical imaging (where the illumination is aimed so that it can only enter the collection optics by scattering). The experimental arrangement is illustrated in Fig. 10.20. Experiments like this have shown that the motor takes three 120° steps to complete one rotation, hydrolyzing one ATP per step. The steps are subdivided into 80° and 40° substeps, corresponding to the binding of ATP (80°) and its subsequent hydrolysis (the remaining 40°).[17–19]

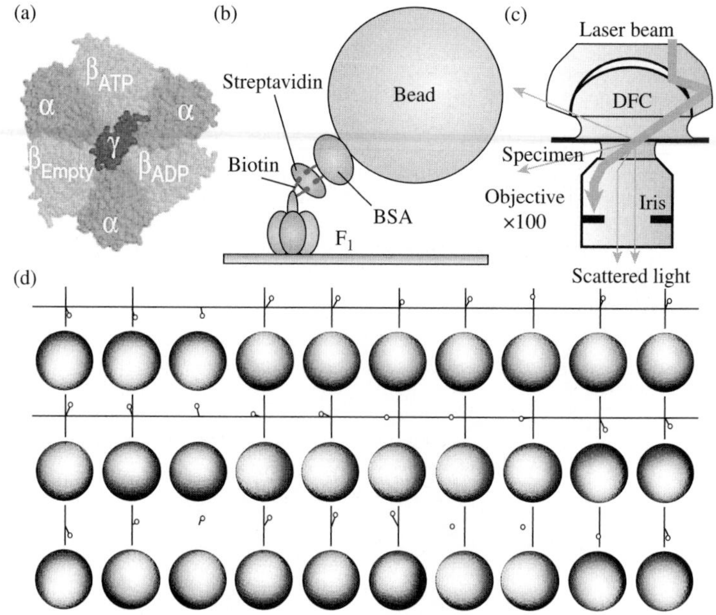

Fig. 10.20 Experiment to visualize the rotation of the F1 portion of ATP synthase. The γ subunit (a) is mutated to incorporate biotin that binds to a streptavidin that also binds biotinylated BSA attached to a 40 nm diameter gold bead (b). Rotation of the bead is observed directly by laser dark-field microcopy (c) yielding a series of scattered light images like those shown in (d). (Reprinted with permission from Yasuda et al.[17], Nature 2001, Nature Publishing Group.)

10.10   The central role of fluctuations in biology

Fluctuations matter at every level of biology. The death of a single dinosaur is a critical event to the species if that dinosaur is a member of the last breeding pair in existence. Our modern understanding of molecular biology can appear deterministic (genes → proteins → function) and it becomes easy to overlook the role of fluctuations and randomness in molecular biology. The central dogma (genes → proteins → function) leaves out the role of random fluctuations. The basic randomness of biology is readily illustrated by comparing identical twins or clones, animals with identical genomes but, as it turns out, quite distinct phenotypes. Figure 10.21 shows the thumbprints of identical twins (a). They are clearly quite distinct. Figure 10.21(b) shows (right) CC, the first cloned cat, with her genetically identical offspring (left). Mother and daughter are similar in marking, but clearly not identical. Thus, the fluctuations so essential to chemical reactions and enzyme function must play a role in the development also.

One source of fluctuations in developmental processes lies in the statistical fluctuation in the number of molecules that regulate gene expression inside a single cell. If the number of molecules is small, then the relative fluctuation in the number is large (on the order of $\sqrt{N^{-1}}$). A beautiful experiment that illustrates this point was conducted by exploiting the fact that bacterial genes are often read from a circular genome, with RNA polymerases moving both clockwise and anticlockwise to translate the DNA to mRNA. Elowitz et al.[21] modified the circular genome of the gut bacterium *E. coli* to insert a pair of genes symmetrically with respect to an important control region called the origin of replication. Each of the genes contained the same promoter sequence upstream of the gene, and would therefore be expressed at the same rate if

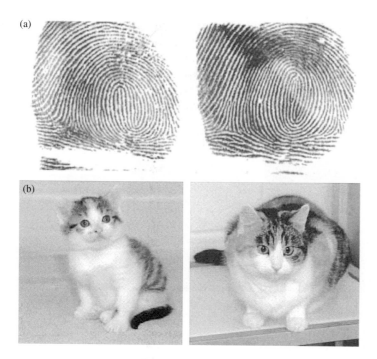

Fig. 10.21 Manifestation of fluctuations in gene expression in biology: (a) the thumbprints of identical twins and (b) CC (right) and her daughter, the first cloned cat. (Reprinted with permission from Raser and O'Shea[20], Science 2005, AAAS.)

the factors that turn gene expression on (the transcription factors) acted in the same way on each gene. One gene contained the DNA sequence corresponding to a fluorescent protein called yellow fluorescent protein (YFP) and the other contained the DNA sequence corresponding to a fluorescent protein called cyan fluorescent protein (CFP). Thus, if the rates at which both these genes were expressed was a constant from bacterium to bacterium, each of the bacteria would appear in the same color (some mix of yellow and cyan) under a fluorescence microscope. The results of this experiment are shown for several different gene promoter sequences in Fig. 10.22. Sequences that correspond to proteins that are produced at a high copy number in the bacterium do indeed give rise to a reasonably uniform population of colors in images of the bacteria (Fig. 10.22(b) and 10.22(e)). However, genes that are expressed at low copy numbers give rise to very different colors in the individual bacteria (Fig. 10.22(a), 10.22(d), 10.22(c), and 10.22(f)). These fluctuations are intrinsic (i.e., they do not result from variations from bacterium to bacterium, because these would affect both genes equally). The enhancement of fluctuations in gene expression associated with the production of small numbers of proteins (and therefore, presumably, small numbers of transcription factors) indicates that these are simply statistical fluctuations as the quantity $\frac{\sqrt{N}}{N}$ approaches unity.

Fluctuations have biological utility when they control the process of gene splicing described in Section 10.2. One of the first and most vivid examples of this is the molecular operation of the immune system. In this case, the fluctuations operate at two levels. One is in the *site-specific recombination* whereby gene segments are shuffled during differentiation of the B-cells that make antibodies. The other lies with the splicing process itself. The immune response

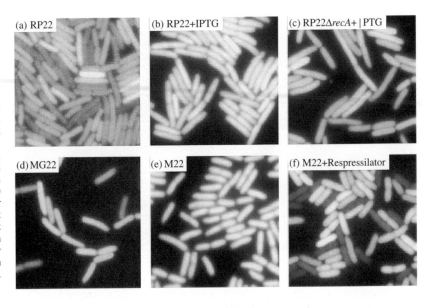

Fig. 10.22 Stochastic gene expression in single cells. Genes were labeled with sequences that coded for green fluorescent proteins and cyan fluorescent proteins in such a way that constant expression of the genes would produce equal amounts of fluorescent protein, resulting in an average yellow color (shown as a light shade here). This is what is seen in genes expressed at high levels ((b) and (e)) but is not the case for genes expressed at lower levels ((a), (c), (d), and (f)). The colors reflect stochastic fluctuations in gene expression that results from the small number of molecules in each cell involved in gene regulation at low expression levels. (Reprinted with permission from Elowitz et al.[21], Science 2002, AAAS). For color, see Color Plate 8.

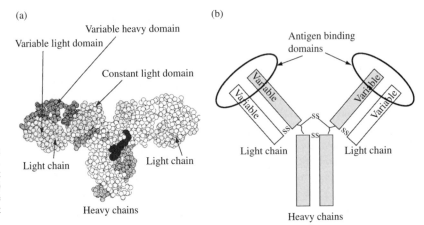

Fig. 10.23 (a) Crystal structure of a human immunoglobulin molecule (antibody). (Reprinted with permission from Silverton et al.[22], NAS. Courtesy of Dr. David Davies.) (b) Schematic layout of the chains in the antibody. The "ss" marks disulfide bonds that hold the chains together.

is a remarkable biological phenomenon. Antibodies are capable of recognizing a foreign invader–any foreign invader–even if it has never been encountered before (though training does help and this is the basis of vaccination). The immune system has a second amazing aspect to it which is a built-in recognition of "self", i.e. the elements that belong in the organism, a feature we will not discuss further here. How does an antibody bind to a foreign invader if it does not "know" what to look for? The answer lies in the diversity of the antibody population. The structure of a typical antibody is illustrated in Fig. 10.23(a) (a crystal structure) and 10.23(b) (a schematic layout). It is a dimer, consisting of two proteins, each one of which is terminated in two separate chains linked together by a disulfide bonds. Part of each of the four chains comprising the antibody has a variable sequence. Each of the arms has a long variable region combined with a constant region called the heavy chain, and a shorter variable

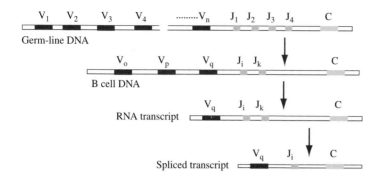

Fig. 10.24 Assembly of mRNA for the light chain of an antibody. Exons are shown as colored blocks. Germ-line DNA is randomly recombined as B-cells differentiate, shuffling the order of the various V and J genes. RNA is transcribed from just one part, with excess RNA removed by splicing. (Derived from Figs. 23–37 of Alberts et al.[16])

region (with a constant component also) called the light chain. Any one antibody carries two identical copies of the heavy and light chains, and the sequence of these chains is (almost) unique for each antibody. The genetic coding process behind this variability is outlined in Fig. 10.24. The variability comes about as a consequence of the random splicing of DNA as the various exons are shuffled during production of the cells that make antibodies (B-cells). Specifically, the variable parts of each chain are encoded for by blocks labeled "V" (for variable). These are joined to the constant region ("C") by another block (labeled "J" for joining). In addition, heavy chains have a further set of gene blocks called "D" (for diversity). To add to the mix, there is more than one type of light chain "constant" region. The first step occurs by means of a process called site-directed recombination. The exons have adjoining sequences that trigger shuffling (but just within antibody genes) by proteins that direct recombination. The process occurs as B-cells develop from germ-line cells. The result is that each B-cell has a different ordering of the various V (or D) regions in its genome. Thus, when translated into mRNA, the particular V block adjacent to a J block is random (as is the choice of J block itself). Extra J regions in the mRNA transcript are removed by splicing to produce a final V-J-C mRNA to be translated into protein. The resulting diversity for just one type of light chain (the so called κ-light chain) is shown in Table 10.3 (these data are for the well-studied mouse immune system). This one combination of chains produces about 3×10^7 variants. Added to this, the splicing process itself is somewhat variable, so the actual diversity exceeds 10^8. This army of random peptides is almost guaranteed to contain a variant that will bind an invading particle (called an antigen). The binding of an antibody to an antigen produces a cascade of signals that results in the rapid *cloning* of the B-cell that made the right antibody. Antibodies are attached to the surface of B-cells (one cell for each sequence of antibody) and binding of the antigen to the antibody causes the cell (containing the unique DNA sequence for its particular antibody) to reproduce rapidly, producing a large population of antibody clones with identical sequence. These will then bind other foreign invaders that are identical to the one that triggered the first reaction, which, when combined with other components of the immune system, results in the destruction of the invading population. Random splicing of the genetic codes for the antibodies is therefore an essential component of the immune system and it is combined with a "natural selection" based on cloning of the successful B-cell. "Darwinian" evolution operates inside each one of us every time we fight off a cold.

Table 10.3 Approximate number of gene components for a light chain (V = variable; J = joining) and a heavy chain (D = diversity) resulting in $300 \times 4 \times 500 \times 4 \times 4 \times 12$ possible variants from these components alone

Segment	κ-Light chain	Heavy chain
V	300	500
J	4	4
D	0	12

A process somewhat like this appears to operate in the development of physiologically complex structures like neural networks and we end with some recent work that demonstrates the crucial role of molecular diversity in neural development. Cells in complex organisms find the right place to go to partly as a result of specifically sticky molecules that are expressed on their surfaces. These molecules, called cell adhesion molecules, bind to targeted receptors on other cells, helping to form the complex three-dimensional arrangement required for the development of an organ. One particular cell adhesion molecule of importance in gluing neurons together is called Dscam ("Downs syndrome cell adhesion molecule"). It was reported recently that the gene for this protein has 38,016 splicing variants, implying that there are up to 38,016 different types of Dscam involved in "gluing" neural networks together (at least in the fruit fly with which this study was carried out). Is this diversity essential or is it just a complication? The same group that reported this diversity found a way of modifying the gene for Dscam so that its essential components were retained but the number of splicing variants that could be made was reduced (to 22,176).[23] Fruit fly embryos with the modified Dscam gene could not develop proper neural networks as evidenced in studies of the networks that are responsible for controlling bristles. A summary of these results is given in Fig. 10.25. The conclusion is that a molecularly diverse population of cell adhesion molecules is *essential* for the successful development of the gross physiology of the organ. Presumably, while each normal fruit fly bristle controlling network looks essentially identical under an optical microscope, they differ dramatically one from the other in the molecular composition. In other words, there are probably many

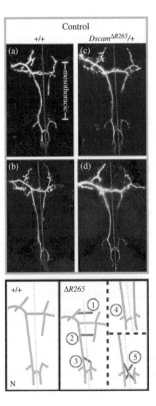

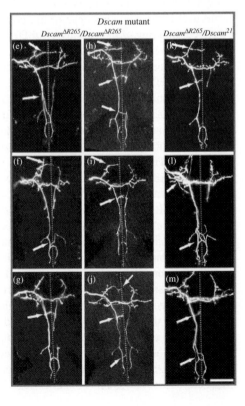

Fig. 10.25 Stochastic splicing as the basis for brain development? A gene that controls fruit fly neuron adhesion through the cell adhesion molecule Dscam has 38,016 splicing variants. Nonetheless, this diverse protein population results in the development of normal robust neural networks (a)–(d) (though each must differ enormously at the molecular level). When the splicing diversity is reducing by genetic engineering (to 22,176 variants) defective networks are formed ((e)–(m), defects highlighted by arrows). (Reprinted with permission from Chen et al.[23], Cell 2006, Elsevier.)

"right ways" to develop successfully, given *adequate population diversity* in the first place. Apparently, reduced population diversity does not allow for successful evolution of an organ in the face of environmental pressures on the developing embryo.

10.11 Do nanoscale fluctuations play a role in the evolution of the mind?

The three themes of this book have been the incredible density of information encoded at the nanoscale, the role of quantum phenomena on this scale, and the importance of fluctuations in small systems, with the consequent emergence of new phenomena. As simple a process as electron transfer solution can now be seen as an exemplar for emergent phenomena. But it appears that it is biology that has produced one of the first empirical tests of the quantitative degree of diversity required to underpin an emergent phenomenon (22,176 Dscam molecules were too few for neural networks to emerge but 38,016 different molecules were adequate). These experiments suggest that complex organs like neural networks develop as a result of environmental selection pressures acting on a randomly formed network. Is this true also for the formation of the mammalian brain? And if it is, could it be that the mind itself emerges through a process similar to Darwinian selection? If so, it is a special type of selection, depending not on diversity in the members of a species, but rather upon molecular diversity arising from the fluctuations inherent in the nanomachines that put the molecular components of biological systems together; a process of Darwinian selection that operates over the timescale of the development of individual animals as opposed to the original Darwinian selection that operates over a timescale of many generations.

One of the most lucid exponents of this molecular Darwinism as the driving force behind the origin of the thinking mind is Gerald Edelman, the brilliant scientist who first conceived of the gene shuffling mechanism responsible for the development of the random library of antibodies. He was the recipient of the 1972 Nobel Prize in physiology or medicine for this discovery. Edelman has proposed that the mind develops as the environment acts upon a complex multiply connected assembly of neurons in the human brain, causing certain groups of connections to stabilize and others to wither. Central to the generation of an adequately diverse neural network is a random population of cell adhesion molecules, substrate adhesion molecules (molecules that link cells to intercellular material) and cell junctional molecules (molecules that fold cells linked by cell adhesion molecules into sheets). In Edelman's view, the mind emerges as the environment acts upon an interconnected and *adequately random* network.[24]

The importance of fluctuations in small systems has been one of the dominant themes of this book. Here we see that fluctuations in nanomachines—the spliceosomes responsible for assembling the final mRNA–produce the molecular diversity needed for the correct assembly of neural networks, and perhaps for the evolution of the mind. Thus perhaps it is true that the same mind that builds simplistic, reductionist models of the physical world has evolved from chaotic processes in small systems where the scale of energy fluctuations is comparable to the energy that holds components of the system together. Einstein's famous rejection of quantum mechanics ("God does not play dice with the universe") may have come from a mind that required randomness in order to develop the ability to make this famous critique!

10.12   Bibliography

B. Alberts, D. Bray, J. Lewis, M. Raff, K. Roberts, and J.D. Watson, Molecular Biology of the Cell. 1994, New York: Garland Press. The classic comprehensive beginner's text in molecular biology. You will find everything here explained from the ground up. With this book, and regular updates from review articles, you can become an expert in molecular biology.

J. Howard, Mechanics of Motor Proteins and the Cytoskeleton. 2001, Sunderland, MA: Sinauer Associates. This book is much broader than its title. A masterful and easy-to-follow treatment of the biophysics of molecular machines.

10.13   Exercises

1. Taking the volume of each base pair to be a disk of 2 nm diameter and 0.34 nm height, how many bases could fit into a nucleus of 3 μm diameter?

2. Translate the following RNA sequences (reading from left to right) into amino acids sequences, and classify the resulting peptides as hydrophillic, hydrophobic, or intermediate:

   (a) GUCGUCCUAAUG
   (b) AACCACAAA
   (c) GCCACAUGG

3. Which codon is a stop sequence in GUCGUCUAGCUAAUG?

4. Which amino acids have only one codon? What properties (hydrophilic vs. hydrophobic) do these amino acids have?

5. A particular gene has five exons, two of which are identical. How many distinct ways can the mRNA be spliced?

6. A 10 kDa globular protein has a Young's modulus of 1 GPa. What is its drag coefficient in water at 25°C? What is its diffusion constant? What is its viscous relaxation time (use Equation 10.5). Compare this to the relaxation time calculated using Equation 10.6.

7. Using the relaxation time calculated using Equation 10.5 for the protein in Question 6, calculate its turnover rate as an enzyme if the transition state for catalysis is $10\,k_B T$.

8. What is the spontaneous equilibrium concentration of ADP after 1 mM ATP was added to water and allowed to come into chemical equilibrium? What is the residual equilibrium concentration of ATP?

9. How far does the head of a myosin molecule move in one step? What is the efficiency of this step? What is the force generated at the *head*?

(The force quoted in the chapter is as acted on by the lever action of the stem).

10. How much does ATP synthase rotate on ATP binding? How much more after release of ADP and phosphate?
11. Given that the largest color fluctuations in Fig. 10.22 correspond to a 20% variation in gene expression, how many molecules are involved in regulating the most variable gene expression?
12. Suppose that 50 of the V regions involved in light chain diversity (Table 10.3) were identical. What would the final diversity of the combination of light and heavy chains be?

References

[1] McManus, M.T. and P.A. Sharp, Gene silencing in mammals by small interfering RNAs. Nat. Rev. Genet., **3**: 737–747 (2002).

[2] Tsai, J., R. Taylor, C. Chothia, and M. Gerstein, The packing density of proteins. J. Mol. Biol., **290**: 253–266 (1999).

[3] Howard, J., *Mechanics of Motor Proteins and the Cytoskeleton.* 2001, Sunderland, MA: Sinauer Associates.

[4] Radzicka, A. and R. Wolfenden, Rates of uncatalyzed peptide bond hydrolysis in neutral solution and the transition state affinities of proteases. J. Am. Chem. Soc., **118**: 6105–6109 (1996).

[5] Jacobs, D.J., A.J. Rader, L.A. Kuhn, and M.F. Thorpe, Protein flexibility predictions using graph theory. Proteins Struct. Funct. Genet., **44**: 150–165 (2001).

[6] Müller, D.J. and A. Engel, Voltage and pH induced channel closure of porin OmpF visualized by AFM. J. Mol. Biol., **285**: 1347–1351 (1999).

[7] Elliott, A., G. Offer, and K. Burridge, Electron microscopy of myosin molecules from muscle and non-muscle sources. Proc. Roy. Soc. Lond. Ser. B, Biol. Sci., **193**: 45–53 (1976).

[8] Rayment, I., W.R. Rypniewski, K. Schmidt-Base, R. Smith, D.R. Tomchick, M.M. Benning, D.A. Winkelmann, G. Wesenberg, and H.M. Holden, Three-dimensional structure of myosin subfragment-1: a molecular motor. Science, **261**: 50–58 (1993).

[9] Rayment, I., H.l.M. Holden, M. Whittaker, C.B. Yohn, M. Lorenz, K.C. Holmes, and R.A. Milligan, Structure of the actin-myosin complex and its implications for muscle contraction. Science, **261**: 58–65 (1993).

[10] Shiroguchi, K. and K. Kinosita, Myosin V walks by lever action and Brownian motion. Science, **316**: 1208–1212 (2007).

[11] Warrick, H.M. and J.A. Spudich, Myosin structure and function in cell motility. Annu. Rev. Cell Biol., **3**: 379–421 (1987).

[12] Steinmetz, M.O., D. Stoffler, A. Hoenger, A. Brener, and U. Aebi, Actin: from cell biology to atomic detail. J. Struct. Biol., **119**: 295–320 (1997).

[13] Schmidt, J.J. and C.D. Montemagno, Bionanomechanical systems. Annu. Rev. Mater. Res., **34**: 315–337 (2004).

[14] Vale, R.D. and R.A. Milligan, The way things move: looking under the hood of molecular motor proteins. Science, **288**: 88–95 (2000).

[15]   Wang, H. and G. Oster, Energy transduction in the F1 motor of ATP synthase. Nature, **296**: 279–282 (1998).

[16]   Alberts, B., D. Bray, J. Lewis, M. Raff, K. Roberts, and J.D. Watson, *Molecular Biology of the Cell*. 3rd ed. 1994, New York: Garland Press.

[17]   Yasuda, R., H. Noji, M. Yoshida, K. Kinosita, and H. Itoh, Resolution of distinct rotational substeps by submillisecond kinetic analysis of F1-ATPase. Nature, **410**: 898–904 (2001).

[18]   Diez, M., B. Zimmermann, M. Börsch, M. König, E. Schweinberge, S. Steigmiller, R. Reuter, S. Felekyan, V. Kudryavtsev, C.A.M. Seidel, et al., Proton-powered subunit rotation in single membrane bound F0F1-ATP synthase. Nat. Struct. Mol. Biol., **11**: 135–141 (2004).

[19]   Berry, R.M., ATP synthesis: the world's smallest wind-up toy. Curr. Biol., **15**: R385–R387 (2005).

[20]   Raser, J.M. and E.K. O'Shea, Noise in gene expression: origins, consequences and control. Science, **309**: 2010–2013 (2005).

[21]   Elowitz, M.B., A.J. Levine, E.D. Siggia, and P.S. Swain, Stochastic gene expression in a single cell. Science, **297**: 1183–1186 (2002).

[22]   Silverton, E.W., M.A. Navia, and D.R. Davies, Three-dimensional structure of an intact human immunoglobulin. Proc. Natl. Acad. Sci. USA, **74**: 5140–5144 (1977).

[23]   Chen, B.E., M. Kondo, A. Garnier, F.L. Watson, R. Püettmann-Holgado, D.R. Lamar, and D. Schmucker, The molecular diversity of Dscam is functionally required for neuronal wiring specificity in *Drosophila*. Cell, **125**: 607–620 (2006).

[24]   Edelman, G.M., *Bright Air, Brilliant Fire*. 1992: BasicBooks.

Units, conversion factors, physical quantities, and useful math

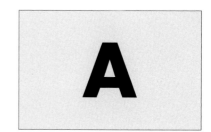

A.1 Length

1 nm $= 10$ Å $= 10^{-9}$ m (nm: nanometer; Å: Angstrom)
Bohr radius of the hydrogen atom: 0.529 Å $= 0.0529$ nm

A.2 Mass and force

1 kg ≈ 2.2 lb (kg: kilogram)
1 amu($= \frac{1}{12}$ mass of C^{12}) $= 1.661 \times 10^{-27}$ kg (amu: atomic mass unit)
Mass of electron: 9.110×10^{-31} kg $= 511$ keV/c^2 (based on $E = mc^2$)
Mass of proton ($\approx$mass of neutron): 1.67×10^{-27} kg
1 Newton $=$ force to accelerate 1 kg by 1 m/s^2
Force on mass, m, due to gravity at earth's surface $=$ mg (g $= 9.8$ m/s^2)

A.3 Time

1 year $= 3.16 \times 10^7$ s
Age of the universe $\approx 13.7 \times 10^9$ yr
Time for an electronic transition ≈ 1 fs ($= 10^{-15}$ s)
Time for atomic vibration ≈ 0.1 ps ($= 10^{-13}$ s)
Time for rotation of a water molecule at 300 K ≈ 1 ns
Time for an enzyme to catalyze a reaction ≈ 1 ms

A.4 Pressure

1 Pa $= 1$ N/m^2 (Pa: Pascal)
1 Torr $= 1$ mm Hg $= 133.3$ Pa
1 atmosphere $= 760$ Torr $= 1.013 \times 10^5$ N/m^2

A.5 Energy and temperature

1 J $=$ work done by a force of 1 N acting over 1 m (J: Joule)
1 calorie $= 4.2$ J; 1 Calorie (food) $= 1000$ calories

1 eV = work done/gained by accelerating/decelerating a charge of one electron through a potential difference of 1 volt = 1.6×10^{-19} J (eV: electron volt)

$0°C = 273.16$ K (°C : Celsius; K : Kelvin)

Boiling point of liquid nitrogen at one atmosphere of pressure = 77.35 K

Temperature at the surface of the sun: 5785 K

Boltzmann's constant, $k_B = 1.381 \times 10^{-23}$ J/K

$k_B T$ at room temperature (300 K) = 4.14×10^{-21} J = $\frac{1}{40}$ eV = 2.49 kJ/mol of particles = 593 cal/mol

$k_B T$ at room temperature in terms of force × distance = 4.14 pN × nm

Energy to ionize a hydrogen atom (Rydberg energy) = 13.6 eV

Energy of a covalent bond ≈ 200 kJ/mol ≈ 2 eV/bond

Energy of a hydrogen bond ≈ 20 kJ/mol ≈ 0.2 eV/bond

Energy from hydrolysis of an ATP molecule ≈ 50 kJ/mol ≈ 0.5 eV/molecule

1 eV ≈ 1.16×10^4 K

A.6   Electromagnetism

Magnetic permeability of free space: $\mu_0 = 4\pi \times 10^{-7}$ Henry/m

Permittivity of free space: $\varepsilon_0 = 8.85 \times 10^{-12}$ Farads/m

Capacitance of a 1 nm diameter metal sphere ($C = 4\pi\varepsilon_0 a$) ≈ 1.1×10^{-19} F

Energy to put one electron on a 1 nm metal sphere ($\frac{Q^2}{2C}$ J = $\frac{e}{2C}$ eV) ≈ 0.7 eV

1 A = 1 C/s (C: Coulomb, unit of charge; A: Ampere, unit of current)

1 V = potential difference that raises energy of 1 C by 1 J

1 W = 1 A × 1 V = 1 J/s (W: Watt, rate of energy consumption/production)

Electric fields, E: V/m

Magnetic fields, B: 1 Tesla = 10^4 gauss

A.7   Constants

Description	Symbol	Value
Speed of light	c	3×10^8 m/s
Planck's constant	h	6.626×10^{-34} Js = 4.14×10^{-15} eV-s
Planck's constant/2π	$\hbar$	1.055×10^{-34} Js = 6.6×10^{-16} eV-s
Avogadro's number	N_0	6.023×10^{23} mol^{-1}

Website for conversions: http://www.onlineconversion.com/

A.8   Some useful material properties

Density of water at 20°C = 1000 kg/m^3 (1 g/cm^3)

Viscosity of water at 20°C = 1×10^{-3} Pa · s

A.9   Some useful math

(A.9.a) *The Stirling approximation*:

$$\ln N! \approx N \ln N - N \quad \text{for large } N$$

(A.9.b) *Fourier series*:

A signal $x(t)$ sampled over a total duration, T, can be expressed as a sum of sine and cosine components that are periodic in T,

$$x(t) = \sum_{n=1}^{n=\infty} a_n \cos \frac{2\pi nt}{T} + b_n \sin \frac{2\pi nt}{T},$$

where the coefficients, a_n and b_n, are given by

$$a_n = \frac{2}{T} \int_{-T/2}^{T/2} x(t) \cos \frac{2\pi nt}{T}\, dt$$

$$b_n = \frac{2}{T} \int_{-T/2}^{T/2} x(t) \sin \frac{2\pi nt}{T}\, dt.$$

(A.9.c) *Some series expansions*:

For $z < 1$,

$$\exp(z) = 1 + z + \frac{z^2}{2!} + \frac{z^3}{3!} + \frac{z^4}{4!} + \cdots$$

$$\sin(z) = z - \frac{z^3}{3!} + \frac{z^5}{5!} + \cdots$$

$$\cos(z) = 1 - \frac{z^2}{2!} + \frac{z^4}{4!} - \frac{z^6}{6!} + \cdots$$

from which

$$\exp i\theta = \cos\theta + i\sin\theta.$$

(A.9.d) *Some integrals*:

$$\int_{-\infty}^{\infty} \exp -x^2\, dx = \sqrt{\pi}$$

$$\int_{0}^{\infty} x^2 \exp -\alpha x^2\, dx = \frac{\sqrt{\pi}}{4}\alpha^{-3/2}$$

$$\int_{-\infty}^{\infty} x^2 \exp -\alpha x^2\, dx = \frac{\sqrt{\pi}}{2}\alpha^{-3/2}$$

B

There's plenty of room at the bottom

Richard P. Feynman

Reprinted from *Journal of Microelectromechical Systems*, Vol. 1, March 1992, pp. 60–66.

I imagine experimental physicists must often look with envy at men like Kamerlingh Onnes, who discovered a field like low temperature, which seems to be bottomless and in which one can go down and down. Such a man is then a leader and has some temporary monopoly in a scientific adventure. Percy Bridgman, in designing a way to obtain-higher pressures, opened up another new field and was able to move into it and to lead us all along. The development of ever higher vacuum was a continuing development of the same kind.

I would like to describe a field, in which little has been done, but in which an enormous amount can be done in principle. This field is not quite the same as the others in that it will not tell us much of fundamental physics (in the sense of, "What are the strange particles? ") but it is more like solid-state physics in the sense that it might tell us much of great interest about the strange phenomena that occur in complex situations. Furthermore, a point that is most important is that it would have an enormous number of technical applications.

What I want to talk about is the problem of manipulating and controlling things on a small scale.

As soon as I mention this, people tell me about miniaturization, and how far it has progressed today. They tell me about electric motors that are the size of the nail on your small finger. And there is a device on the market, they tell me, by which you can write the Lord's Prayer on the head of a pin. But that's nothing; that's the most primitive, halting step in the direction I intend to discuss. It is a staggeringly small world that is below. In the year 2000, when they look back at this age, they will wonder why it was not until the year 1960 that anybody began seriously to move in this direction.

Why cannot we write the entire 24 volumes of the Encyclopaedia Britannica on the head of a pin?

Let's see what would be involved. The head of a pin is a sixteenth of an inch across. If you magnify it by 25 000 diameters, the area of the head of the pin is then equal to the area of all the pages of the Encyclopaedia Britannica. Therefore, all it is necessary to do is to reduce in size all the writing in the Encyclopaedia by 25 000 times. Is that possible? The resolving power of the eye is about 1/120 of an inch—that is roughly the diameter of one of the little dots on the fine half-tone reproductions in the Encyclopaedia. This, when you demagnify it by 25 000 times, is still 80 angstroms in diameter—32 atoms across, in an ordinary metal. In other words, one of those dots still would

MEMS Editor's Note: This manuscript addresses many current research issues. It is the transcript of a talk given by Richard P. Feynman on December 26, 1959, at the annual meeting of the American Physical Society at the California Institute of Technology, and was published as a chapter in the Reinhold Publishing Corporation book, *Miniaturization*, Horace D. Gilbert, Ed. It is reprinted with the consent of Van Nostrand Reinhold, New York, NY 10003.

The author, deceased, was with the California Institute of Technology, Pasadena, CA.

IEEE Log Number 9105621.

contain in its area 1000 atoms. So, each dot can easily be adjusted in size as required by the photoengraving, and there is no question that there is enough room on the head of a pin to put all of the Encyclopaedia Britannica.

Furthermore, it can be read if it is so written. Let's imagine that it is written in raised letters of metal; that is where the black is in the Encyclopaedia, we have raised letters of metal that are actually 1/25 000 of their ordinary size. How would we read it?

If we had something written in such a way, we could read it using techniques in common use today. (They will undoubtedly find a better way when we do actually have it written, but to make my point conservatively I shall just take techniques we know today.) We would press the metal into a plastic material and make a mold of it, then peel the plastic off very carefully, evaporate silica into the plastic to get a very thin film, then shadow it by evaporating gold at an angle against the silica so that all the little letters will appear clearly, dissolve the plastic away from the silica film, and then look through it with an electron microscope!

There is no question that if the thing were reduced by 25 000 times in the form of raised letters on the pin, it would be easy for us to read it today. Furthermore, there is no question that we would find it easy to make copies of the master; we would just need to press the same metal plate again into plastic and we would have another copy.

How Do We Write Small

The next question is: How do we *write* it? We have no standard technique to do this now. But let me argue that it is not as difficult as it first appears to be. We can reverse the lenses of the electron microscope in order to demagnify as well as magnify. A source of ions, sent through the microscope lenses in reverse, could be focused to a very small spot. We could write with that spot like we write in a TV cathode ray oscilloscope, by going across in lines, and having an adjustment which determines the amount of material which is going to be deposited as we scan in lines.

This method might be very slow because of space charge limitations. There will be more rapid methods. We could first make, perhaps by some photo process, a screen which has holes in it in the form of the letters. Then we would strike an arc behind the holes and draw metallic ions through the holes; then we could again use our system of lenses and make a small image in the form of ions, which would deposit the metal on the pin.

A simpler way might be this (though I am not sure it would work): We take light and, through an optical microscope running backwards, we focus it onto a very small photoelectric screen. Then electrons come away from the screen where the light is shining. These electrons are focused down in size by the electron microscope lenses to impinge directly upon the surface of the metal. Will such a beam etch away the metal if it is run long enough? I don't know. If it doesn't work for a metal surface, it must be possible to find some surface with which to coat the original pin so that, where the electrons bombard, a change is made which we would recognize later.

There is no intensity problem in these devices—not what you are used to in magnification, where you have to take a few electrons and spread them over a bigger and bigger screen; it is just the opposite. The light which we get from a page is concentrated onto a very small area so it is very intense. The few electrons which come from the photoelectric screen are demagnified down to a very tiny area so that, again, they are very intense. I don't know why this hasn't been done yet!

That's the Encyclopaedia Britannica on the head of a pin, but let's consider all the books in the world. The Library of Congress has approximately 9 million volumes; the British Museum Library has 5 million volumes; there are also 5 million volumes in the National Library in France. Undoubtedly there are duplications, so let us say that there are some 24 million volumes of interest in the world.

What would happen if I print all this down at the scale we have been discussing? How much space would it take? It would take, of course, the area of about a million pinheads because, instead of there being just the 24 volumes of the Encyclopaedia, there are 24 million volumes. The million pinheads can be put in a square of a thousand pins on a side, or an area of about 3 square yards. That is to say, the silica replica with the paper-thin backing of plastic, with which we have made the copies, with all this information, is on an area of approximately the size of 35 pages of the Encyclopaedia. This is only one-fourth as many pages as copy of the *Saturday Evening Post*. All of the information which all of mankind has ever recorded in books can be carried around in a pamphlet in your hand—and not written in code, but as a simple reproduction of the original pictures, engravings, and everything else on a small scale without loss of resolution.

What would our librarian at Caltech say, as she runs all over from one building to another, if I tell her that, ten years from now, all of the information that she is struggling to keep track of—120 000 volumes, stacked from the floor to the ceiling, drawers full of cards, storage rooms full of the older books—can be kept on just one library card! When the University of Brazil, for example, finds that their library is burned, we can send them a copy of every book in our library by striking off a copy from the master plate in a few hours and mailing it in an envelope no bigger or heavier than any other ordinary air mail letter.

Now, the name of this talk is "There is *Plenty* of Room at the Bottom "—not just "There is Room at the Bottom." What I have demonstrated is that there is room—that you can decrease the size of things in a practical way. I now want to show that there is *plenty* of room. I will not now discuss how we are going to do it, but only what is possible in principle—in other words, what is possible according to the laws of physics. I am not inventing antigravity, which is possible someday only if the laws are not what we think. I am telling you what could be done if the laws are what we think; we are not doing it simply because we haven't yet gotten around to it.

Information on a Small Scale

Suppose that, instead of trying to reproduce the pictures and all the information directly in its present form, we write only the information content in a code of dots and dashes, or something like that, to represent the various letters. Each letter represents six or seven "bits" of information; that is, you need only about six or seven dots or dashes for each letter. Now, instead of writing everything, as I did before, on the *surface* of the head of a pin, I am going to use the interior of the material as well.

Let us represent a dot by a small spot of one metal, the next dash by an adjacent spot of another metal, and so on. Suppose, to be conservative, that a bit of information is going to require a little cube of atoms 5 × 5 × 5—that is 125 atoms. Perhaps we need a hundred and some odd atoms to make sure that the information is not lost through diffusion, or through some other process.

I have estimated how many letters there are in the Encyclopaedia, and I have assumed that each of my 24 million books is as big as an Encyclopaedia volume, and have calculated, then, how many bits of information there are (10^{15}). For each bit I allow 100 atoms. And it turns out that all of the information that man has carefully accumulated in all the books in the world can be written in this form in a cube of material one two-hundredth of an inch wide—which is the barest piece of dust that can be made out by the human eye. So there is *Plenty* of room at the bottom! Don't tell me about microfilm!

This fact—that enormous amounts of information can be carried in an exceedingly small space—is, of course, well known to the biologists, and resolves the mystery which existed before we understood all this clearly, of how it could be that, in the tiniest cell, all of the information for the organization of a complex creature such as ourselves can be stored. All this information—whether we have brown eyes, or whether we think at all, or that in the embryo the jawbone should first develop with a little hole in the side so that later a nerve can grow through it—all this information is contained in a very tiny fraction of the cell in the form of long-chain DNA molecules in which approximately 50 atoms are used for one bit of information about the cell.

Better Electron Microscopes

If I have written in a code, with 5 × 5 × 5 atoms to a bit, the question is: How could I read it today? The electron microscope is not quite good enough; with the greatest care and effort, it can only resolve about 10 angstroms. I would like to try and impress upon you, while I am talking about all of these things on a small scale, the importance of improving the electron microscope by a hundred times. It is not impossible; it is not against the laws of diffraction of the electron. The wave length of the electron in such a microscope is only 1/20 of an angstrom. So it should be possible to see the individual atoms. What good would it be to see individual atoms distinctly?

We have friends in other fields—in biology, for instance. We physicists often look at them and say, "You know the reason you fellows are making so little progress?" (Actually I don't know any field where they are making more rapid progress than they are in biology today.) "You should use more mathematics, like we do." They could answer us—but they're polite, so I'll answer for them: "What you should do in order for us to make more rapid progress is to make the electron microscope 100 times better. "

What are the most central and fundamental problems of biology today? They are questions like: What is the sequence of bases in the DNA? What happens when you have a mutation? How is the base order in the DNA connected to the order of amino acids in the protein? What is the structure of the RNA; is it single-chain or double-chain and how is it related in its order of bases to the

DNA? What is the organization of the microsomes? How are proteins synthe-sized? Where does the RNA go? How does it sit? Where do the proteins sit? Where do the amino acids go in? In photosynthesis, where is the chlorophyll; how is it arranged; where are the carotenoids involved in this thing? What is the system of the conversion of light into chemical energy?

It is very easy to answer many of these fundamental biological questions; you just *look at the thing!* You will see the order of bases in the chain; you will see the structure of the microsome; Unfortunately, the present microscope sees at a scale which is just a bit too crude. Make the microscope one hundred times more powerful, and many problems of biology would be made very much easier. I exaggerate, of course, but the biologists would surely be very thankful to you—and they would prefer that to the criticism that they should use more mathematics.

The theory of chemical processes today is based on theoretical physics. In this sense, physics supplies the foundation of chemistry. But chemistry also has analysis. If you have a strange substance and you want to know what it is, you go through a long and complicated process of chemical analysis. You can analyze almost anything today, so I am a little late with my idea. But if the physicists wanted to, they could also dig under the chemists in the problem of chemical analysis. It would be very easy to make an analysis of any complicated chemical substance; all one would have to do would be to look at it and see where the atoms are. The only trouble is that the electron microscope is one hundred times too poor. (Later, I would like to ask the question: Can the physicists do something about the third problem of chemistry–namely, synthesis? Is there a *physical* way to synthesize any chemical substance?)

The reason the electron microscope is so poor is that the f-value of the lenses is only 1 part to 1000; you don't have a big enough numerical aperture. And I know that there are theorems which prove that it is impossible, with axially symmetrical stationary field lenses, to produce an f-value any bigger than so and so; and therefore the resolving power at the present time is at its theoretical maximum. But in every theorem there are assumptions. Why must the field be axially symmetrical? Why must the field be stationary? Can't we have pulsed electron beams in fields moving up along with the electrons? Must the field be symmetrical? I put this out as a challenge: Is there no way to make the electron microscope more powerful?

The Marvelous Biological System

The biological example of writing information on a small scale has inspired me to think of something that should be possible. Biology is not simply writ-ing information; it is *doing something* about it. A biological system can be exceedingly small. Many of the cells are very tiny, but they are very active; they manufacture various substances; they walk around; they wiggle; and they do all kinds of marvelous things—all on a very small scale. Also, they store information. Consider the possibility that we too can make a thing very small, which does what we want—that we can manufacture an object that maneuvers at that level!

There may even be an economic point to this business of making things very small. Let me remind you of some of the problems of computing machines. In computers we have to store an enormous amount of information. The kind of writing that I was mentioning before, in which I had everything down as a distribution of metal, is permanent. Much more interesting to a computer is a way of writing, erasing, and writing something else. (This is usually because we don't want to waste the material on which we have just written. Yet if we could write it in a very small space; it wouldn't make any difference; it could just be thrown away after it was read. It doesn't cost very much for the material.)

Miniaturizing the Computer

I don't know how to do this on a small scale in a practical way, but I do know that computing machines are very large; they fill rooms. Why can't we make them very small, make them of little wires, little elements—and by little, I mean *little*. For instance, the wires should be 10 or 100 atoms in diameter, and the circuits should be few thousand angstroms across. Everybody who has analyzed the logical theory of computers has come to the conclusion that the possibilities of computers are very interesting—if they could be made to be more complicated by several orders of magnitude. If they had millions of times as many elements, they could make judgements. They would have time to calculate what is the best way to make the calculation that they are about to make. They could select the method of analysis which, from their experience is better than the one that we would give to them. And, in many other ways, they would have new qualitative features.

If I look at your face I immediately recognize that I have seen it before. (Actually, my friends will say I have chosen an unfortunate example here for the subject of this illustration. At least I recognize that it is a *man* and not an *apple*.) Yet there is no machine which, with that speed, can take a picture of a face and say even that it is a man; and much less that it is the same man that you showed it before—unless it is exactly the same picture; If the face is changed; if I am close to the face; if I am further from the face; if the light changes—I recognize it anyway. Now, this little computer I carry in my head is easily able to do that. The computers that we build are not able to do that. The number of elements in this bone box of mine are enormously greater than the number of elements in our "wonderful" computers. But our mechanical computers are too big; the elements in this box are microscopic. I want to make some that are *sub*-microscopic.

If we wanted to make a computer that had all these marvelous extra qualitative abilities, we would have to make it, perhaps, the size of the Pentagon. This has several disadvantages. First, it requires too much material; there may not be enough germanium in the world for all the transistors which would have to be put into this enormous thing. There is also the problem of heat generation and power consumption; TVA would be needed to run the computer. But an even more practical difficulty is that the computer would be limited to a certain speed. Because of its large size, there is finite time required to get the information from one place to another. The information cannot go any faster than the speed of light—so, ultimately, when our computers get faster and faster and more and more elaborate, we will have to make them smaller and smaller.

But there is plenty of room to make them smaller. There is nothing that I can see in the physical laws that says the computer elements cannot be made enormously smaller than they are now. In fact, there may be certain advantages.

Miniaturization by Evaporation

How can we make such a device? What kind of manufacturing processes would we use? One possibility we might consider, since we have talked about writing by putting atoms down in a certain arrangement, would be to evaporate the material, then evaporate the insulator next to it. Then, for the next layer, evaporate another position of a wire, another insulator, and son on. So, you simply evaporate until you have a block of stuff which has the elements—coils and condensers, transistors and so on—of exceedingly fine dimensions.

But I would like to discuss, just for amusement, that there are other possibilities. Why can't we manufacture these small computers somewhat like we manufacture the big ones? Why can't we drill holes, cut things, solder things, stamp things out, mold different shapes all at an infinitesimal level? What are the limitations as to how small a thing has to be before you can no longer mold it? How many times when you are working on something frustratingly tiny, like your wife's wrist watch, have you said to yourself, "If I could only train an ant to do this!" What I would like to suggest is the possibility of training an ant to train a mite to do this. What are the possibilities of small but moveable machines? They may or may not be useful, but they surely would be fun to make.

Consider any machine—for example, an automobile—and ask about the problems of making an infinitesimal machine like it. Suppose, in the particular design of the automobile, we need a certain precision of the parts; we need an accuracy, let's suppose, of 4/10 000 of an inch. If things are more inaccurate than that in the shape of the cylinder and so on, it isn't going to work very well. If I make the thing too small, I have to worry about the size of the atoms; I can't make a circle out of "balls'" so to speak, if the circle is too small. So, if I make the error, corresponding to 4/10 000 of an inch, correspond to an error of 10 atoms, it turns out that I can reduce the dimensions of an automobile 4000 times, approximately—so that it is 1 mm across. Obviously, if you redesign the car so that it would work with a much large tolerance, which is not at all impossible, then you could make a much smaller device.

It is interesting to consider what the problems are in such small machines. Firstly, with parts stressed to the same degree, the forces go as the area you are reducing, so that things like weight and inertia are of relatively no importance. The strength of material, in other words, is very much greater in proportion. The stresses and expansion of the flywheel from centrifugal force, for example, would be the same proportion only if the rotational speed is increased in the same proportion as we decreased the size. On the other hand, the metals that we use have a grain structure, and this would be very annoying at small scale because the material is not homogeneous. Plastics and glass and things of this amorphous nature are very much more homogeneous, and so we would have to make our machines out of such materials.

There are problems associated with the electrical part of the system—with the copper wires and the magnetic parts. The magnetic properties on a very small scale are not the same as on a large scale; there is the "domain" problem involved. A big magnet made of millions of domains can only be made on small scale with one domain. The electrical equipment won't simply be scaled down; it has to be redesigned. But I can see no reason why it can't be redesigned to work again.

Problems of Lubrication

Lubrication involves some interesting points. The effective viscosity of oil would be higher and higher in proportion as we went down (and if we increase the speed as much as we can). If we don't increase the speed so much, and change from oil to kerosene or some other fluid, the problem is not so bad. But actually we may not have to lubricate at all! We have a lot of extra force. Let the bearings run dry; they won't run hot because the heat escapes away from such a small device very, very rapidly.

This rapid heat loss would prevent the gasoline from exploding, so an internal combustion engine is impossible. Other chemical reactions, liberating energy when cold, can be used. Probably an external supply of electrical power would be most convenient for such small machines.

What would be the utility of such machines? Who knows? Of course, a small automobile would only be useful for the mites to drive around in, and I suppose our Christian interests don't go that far. However, we did note the possibility of the manufacture of small elements for computer in completely automatic factories, containing lathes and other machine tools at the very small level. The small lathe would not have to be exactly like our big lathe. I leave to your imagination the improvement of the design to take full advantage of the properties of things on a small scale, and in such a way that the fully automatic aspect would be easiest to manage.

A friend of mine (Albert R. Hibbs) suggests a very interesting possibility for relatively small machines. He says that, although it is a very wild idea, it would be interesting in surgery if you could swallow the surgeon. You put the mechanical surgeon inside the blood vessel and it goes into the heart and "looks" around. (Of course the information has to be fed out.) It finds out which valve is the faulty one and takes a little knife and slices it out. Other small machines might be permanently incorporated in the body to assist some inadequately-functioning organ.

Now comes the interesting question: How do we make such a tiny mechanism? I leave that to you. However, let me suggest one weird possibility. You know, in the atomic energy plants they have materials and machines that they can't handle directly because they have become radioactive. To unscrew nuts and put on bolts and so on, they have a set of master and slave hands, so that by operating a set of levers here, you control the "hands" there, and can turn them this way and that so you can handle things quite nicely.

Most of these devices are actually made rather simply, in that there is a particular cable, like a marionette string, that goes directly from the controls to the "hands." But, of course, things also have been made using servo motors, so that the connection between the one thing and the other is electrical rather than mechanical. When you turn the levers, they turn a servo motor, and it changes the electrical currents in the wires, which repositions a motor at other other end.

Now, I want to build much the same device—a master-slave system which operates electrically. But I want the slaves to be made especially carefully by modern large-scale machinists so that they are one-fourth the scale of the "hands" that you ordinarily maneuver. So you have a scheme by which you can do things at one-quarter scale anyway—the little servo motors with little hands play with little nuts and bolts; they drill little holes; they are four times smaller. Aha! So I manufacture a quarter-size lathe; I manufacture quarter-size tools;

and I make, at the one-quarter scale, still another set of hands again relatively one-quarter size! This is one-sixteenth size, from my point of view. And after I finish doing this I wire directly from my large-sixteenth-size servo motors. Thus I can now manipulate the one-sixteenth-size hands.

Well, you get the principle from there on. It is rather a difficult program, but it is a possibility. You might say that one can go much farther in one step than from one to four. Of course, this has all to be designed very carefully and it is not necessary simply to make it like hands. If you though of it very carefully, you could probably arrive at a much better system for doing such things.

If you work through a pantograph, even today, you can get much more than a factor of four in even one step. But you can't work directly through a pantograph which makes a smaller pantograph which then makes a smaller pantograph—because of the looseness of the holes and the irregularities of construction. The end of the pantograph wiggles with a relatively greater irregularity than the irregularity with which you move your hands. In going down this scale, I would find the end of the pantograph on the end of the pantograph on the end of the pantograph shaking so badly that it wasn't doing anything sensible at all.

At each stage, it is necessary to improve the precision of the apparatus. If, for instance, having made a small lathe with a pantograph, we find its lead screw irregular—more irregular than the large-scale one—we could lap the lead screw against breakable nuts that you can reverse in the usual way back and forth until this lead screw is, at its scale, as accurate as our original lead screws, at our scale.

We can make flats by rubbing unflat surfaces in triplicate together—in three pairs—and the flats then become flatter than the thing you started with. Thus, it is not impossible to improve precision on a small scale by the correct operations. So, when we build this stuff, it is necessary at each step to improve the accuracy of the equipment by working for awhile down there, making accurate lead screws, Johansen blocks, and all the other materials which we use in accurate machine work at the higher level. We have to stop at each level and manufacture all the stuff to go to the next level—a very long and very difficult program. Perhaps you can figure a better way than that to get down to small scale more rapidly.

Yet, after all this, you have just got one little baby lathe four thousand times smaller than usual. But we were thinking of making an enormous computer, which we were going to build by drilling holes on this lathe to make little washers for the computer. How many washers can you manufacture on this one lathe?

A Hundred Tiny Hands

When I make my first set of slave "hands" at one-fourth scale. I am going to make ten sets. I make ten sets of "hands," and I wire them to my original levels so they each do exactly the same thing at the same time in parallel. Now, when I am making my new devices one-quarter again as small, I let each one manufacture ten copies, so that I would have a hundred "hands" at the 1/16th size.

Where am I going to put the million lathes that I am going to have? Why, there is nothing to it; the volume is much less than that of even full-scale lathe.

For instance, if I made a billion little lathes, each 1/4000 of the scale of a regular lathe, there are plenty of materials and space available because in the billion little ones there is less than 2 per cent of the materials in one big lathe.

It doesn't cost anything for materials, you see. So I want to build a billion tiny factories, models of each other, which are manufacturing simultaneously, drilling holes, stamping parts, and so on.

As we go down in size, there are a number of interesting problems that arise. All things do not simply scale down in proportion. There is the problem that materials stick together by the molecular (Van der Waals) attractions. It would be like this: After you have made a part and you unscrew the nut from a bolt, it isn't going to fall down because the gravity isn't appreciable; it would even be hard to get it off the bolt. It would be like those old movies of a man with his hands full of molasses, trying to get rid of a glass of water. There will be several problems of this nature that we will have to be ready to design for.

Rearranging the Atoms

But I am not afraid to consider the final question as to whether, ultimately—in the great future—we can arrange the atoms the way we want; the very *atoms*, all the way down! What would happen if we could arrange the atoms one by one the way we want them (within reason, of course; you can't put them so that they are chemically unstable, for example).

Up to now, we have been content to dig in the ground to find minerals. We heat them and we do things on a large scale with them, and we hope to get a pure substance with just so much impurity, and so on. But we must always accept some atomic arrangement that nature gives us. We haven't got anything, say, with a "checkerboard" arrangement, with the impurity atoms exactly arranged 1000 angstroms apart, or in some other particular pattern.

What would we do with layered structures with just the right layers? What would the properties of materials be if we could really arrange the atoms the way we want them? They would be very interesting to investigate theoretically. I can't see exactly what would happen, but I can hardly doubt that when we have some *control* of the arrangement of things on a small scale we will get an enormously greater range of possible properties that substances can have, and of different things that we can do.

Consider, for example, a piece of material in which we make little coils and condensers (or their solid state analogs) 1000 or 10 000 angstroms in a circuit, one right next to the other, over a large area, with little antennas sticking out at the other end—a whole series of circuits.

Is it possible, for example, to emit light from a whole set of antennas, like we emit radio waves from an organized set of antennas to beam the radio programs to Europe? The same thing would be to *beam* the light out in a definite direction with very high intensity. (Perhaps such a beam is not very useful technically or economically.)

I have thought about some of the problems of building electric circuits on a small scale, and the problem of resistance is serious. If you build a corresponding circuit on a small scale, its natural frequency goes up, since the wave length goes down as the scale; but the skin depth only decreases with the square root of the scale ratio, and so resistive problems are of increasing difficulty. Possibly we can beat resistance through the use of superconductivity if the frequency is not too high, or by other tricks.

Atoms in a Small World

When we get to the very, very small world—say circuits of seven atoms— we have a lot of new things that would happen that represent completely new opportunities for design. Atoms on a small scale behave like *nothing* on a large scale, for they satisfy the laws of quantum mechanics. So, as we go down and fiddle around with the atoms down there, we are working with different laws, and we can expect to do different things. We can manufacture in different ways. We can use, not just circuits, but some system involving the quantized energy levels, or the interactions of quantized spins, etc.

Another thing we will notice is that, if we go down far enough, all of our devices can be mass produced so that they are absolutely perfect copies of one another. We cannot build two large machines so that the dimensions are exactly the same. But if your machine is only 100 atoms high, you only have to have it correct to one-half of one per cent to make sure the other machine is exactly the same size—namely, 100 atoms high!

At the atomic level, we have new kinds of forces and new kinds of possibilities, new kinds of effects. The problems of manufacture and reproduction of materials will be quite different. I am, as I said, inspired by the biological phenomena in which chemical forces are used in a repetitive fashion to produce all kinds of weird effects (one of which is the author).

The principles of physics, as far as I can see, do not speak against the possibility of maneuvering things atom by atom. It is not an attempt to violate any laws; it is something, in principle, that can be done; but, in practice, it has not been done because we are too big.

Ultimately, we can do chemical synthesis. A chemist comes to us and says, "Look, I want a molecule that has the atoms arranged thus and so; make me that molecule." The chemist does a mysterious thing when he wants to make a molecule. He sees that it has got that ring, so he mixes this and that, and he shakes it, and he fiddles around. And, at the end of a difficult process, he usually does succeed in synthesizing what he wants. By the time I get my devices working, so that we can do it by physics, he will have figured out how to synthesize absolutely anything, so that this will really be useless.

But it is interesting that it would be, in principle, possible (I think) for a physicist to synthesize any chemical substance that the chemist writes down. Give the order and the physicist synthesizes it. How? Put the atoms down where the chemist says, and so you make the substance. The problems of chemistry and biology can be greatly helped if our ability to see what we are doing, and to do things on an atomic level, is ultimately developed—a development which I think cannot be avoided.

Now, you might say, "Who should do this and why should they do it?" Well, I pointed out a few of the economic applications, but I know that the reason that you would do it might be just for fun. But have some fun! Let's have a competition between laboratories. Let one laboratory make a tiny motor which it sends to another lab which sends it back with a thing that fits inside the shaft of the first motor.

High School Competition

Just for the fun of it, and in order to get kids interested in this field, I would propose that someone who has some contact with the high schools think of making some kind of high school competition. After all, we haven't even started in this field, and even the kids can write smaller than has ever been written before. They could have competition in high schools. The Los Angeles high school could send a pin to the Venice high school on which it says, "How's this?" They get the pin back, and in the dot of the "i" it says, "Not so hot."

Perhaps this doesn't excite you to do it, and only economics will do so. Then I want to do something; but I can't do it at the present moment, because I haven't prepared the ground. I hereby offer a prize of $1000 to the first guy who can take the information on the page of a book and put it on an area 1/25 000 smaller in linear scale in such manner that it can be read by an electron microscope.

And I want to offer another prize—if I can figure out how to phrase it so that I don't get into a mess of arguments about definitions—of another $1000 to the first guy who makes an operating electric motor—a rotating electric motor which can be controlled from the outside and, not counting the lead-in wires, is only 1/64 inch cube.

I do not expect that such prizes will have to wait very long for claimants.[1]

[1] *Editor's Note:* The latter prize was presented by Dr. Feynman on November 28, 1960 to William McLellan who built an electric motor the size of a speck of dust. The other prize is still open.

C Schrödinger equation for the hydrogen atom

C.1 Angular momentum operators

The classical definition of the angular momentum, $\mathbf{L}$, of a particle orbiting a point at a distance, $\mathbf{r}$, with linear momentum (at any one time), $\mathbf{p}$, is

$$\mathbf{L} = \mathbf{r} \times \mathbf{p}. \tag{C.1}$$

This expression is translated into quantum mechanics with the substitution of the momentum operator, $p_x = -i\hbar\frac{\partial}{\partial x}, p_y = -i\hbar\frac{\partial}{\partial y}, p_z = -i\hbar\frac{\partial}{\partial z}$, giving

$$\hat{L}_x = -i\hbar\left(y\frac{\partial}{\partial z} - z\frac{\partial}{\partial y}\right) \tag{C.2a}$$

$$\hat{L}_y = -i\hbar\left(z\frac{\partial}{\partial x} - x\frac{\partial}{\partial z}\right) \tag{C.2b}$$

$$\hat{L}_z = -i\hbar\left(x\frac{\partial}{\partial y} - y\frac{\partial}{\partial x}\right). \tag{C.2c}$$

These expressions are for the angular momentum operators (denoted by the "hat" symbol) that act on the angular momentum eigenstates to return the angular momentum eigenvalues. Translating these expressions into spherical polar coordinates (r, θ, ϕ) we obtain for the operator for the magnitude of the angular momentum $(L^2 = L_x^2 + L_y^2 + L_z^2)$

$$\hat{L}^2 = -\hbar^2\left[\frac{1}{\sin\theta}\frac{\partial}{\partial\theta}\left(\sin\theta\frac{\partial}{\partial\theta}\right) + \frac{1}{\sin^2\theta}\frac{\partial^2}{\partial\theta^2}\right] \equiv -\hbar^2\left[r^2\nabla^2\right], \tag{C.3}$$

where ∇^2 is the square of the derivative operator in spherical polar coordinates. The individual operators for $\hat{L}_{x,y,z}$ do not commute. That is to say

$$\hat{L}_x\hat{L}_y - \hat{L}_y\hat{L}_z = i\hbar\hat{L}_z \neq 0 \tag{C.4}$$

as can be verified by substituting from C.2 (with similar results for the other pairs of operators). This is equivalent to saying that not more than one component of L can be determined in an experiment. However,

$$\hat{L}^2\hat{L}_{x,y,z} - \hat{L}_{x,y,z}\hat{L}^2 = 0 \tag{C.5}$$

so the magnitude of momentum and any one component can be measured simultaneously. Thus, the quantum numbers for L^2 and L_z (taking the z component as an example) are good quantum numbers.

C.2   Angular momentum eigenfunctions

We require functions $\psi(\theta, \phi)$ that satisfy

$$\hat{L}^2 \psi(\theta, \phi) = \lambda \psi(\theta, \phi) \tag{C.6a}$$

$$\hat{L}_z \psi(\theta, \phi) = \mu \psi(\theta, \phi), \tag{C.6b}$$

where λ and μ are the eigenvalues for total momentum squared and L_z. In spherical coordinates, $\hat{L}_z = -i\hbar \frac{\partial}{\partial \phi}$ (from C.2c) and substituting this into C.6b gives

$$\psi(\theta, \phi) = \Theta(\theta) \exp \frac{i\mu\phi}{\hbar}. \tag{C.7}$$

In order to describe a physical variable, we require that

$$\psi(\theta, \phi + 2\pi) = \psi(\theta, \phi), \tag{C.8}$$

which requires

$$\psi(\theta, \varphi) = \Theta(\theta) \exp im\varphi, \quad m = \pm 1, \pm 2, \cdots \pm \pm \cdots \tag{C.9a}$$

and, from C.7

$$\mu = m\hbar. \tag{C.9b}$$

The solution for L^2 is more technical. Substituting C.3 into C.6a, using our result for the ϕ dependence and then changing variables to $z = \cos\theta$ leads to

$$\frac{d}{dz}\left[(1 - z^2)\frac{dP}{dz}\right] + \left(\frac{\lambda}{\hbar^2} - \frac{m^2}{1 - z^2}\right)P = 0, \tag{C.10}$$

where $P(z) = \Theta(\cos\theta)$. This equation is known in the mechanics of spherically symmetric systems as the associated Legendre equation, and it only has solutions for

$$\lambda = \ell(\ell + 1)\hbar^2 \quad \text{with } \ell = 0, 1, 2, \ldots. \tag{C.11}$$

Since $\left\langle \hat{L}_z^2 \right\rangle \leq \left\langle \hat{L}^2 \right\rangle$, we require $m^2 \leq \ell(\ell + 1)$ or

$$-\ell \leq m \leq \ell. \tag{C.12}$$

The normalized wavefunctions formed from the product of the $P(\cos\theta)$ and the $\exp im\phi$ are called spherical harmonics, $Y_{\ell,m}(\theta,\phi)$.

C.3   Solution of the Schrödinger equation in a central potential

In spherical coordinates, the Schrödinger equation is

$$\left[-\frac{\hbar^2}{2\mu} \left(\frac{1}{r^2}\frac{\partial}{\partial r}\left[r^2 \frac{\partial}{\partial r}\right] + \frac{1}{r^2\sin\theta}\frac{\partial}{\partial\theta}\left[\sin\theta\frac{\partial}{\partial\theta}\right] + \frac{1}{r^2\sin^2\theta}\frac{\partial^2}{\partial\phi^2}\right) + V(r)\right]$$
$$\times\,\psi(r,\theta,\phi) = E\psi(r,\theta,\phi). \tag{C.13}$$

But this can be written in terms of the total angular momentum operator (C.3) as

$$\left[\frac{-\hbar^2}{2\mu r^2}\frac{\partial}{\partial r}\left(r^2\frac{\partial}{\partial r}\right) + \frac{\hat{L}^2}{2\mu r^2} + V(r)\right]\psi(r,\theta,\phi) = E\psi(r,\theta,\phi). \tag{C.14}$$

Substituting

$$\psi(r,\theta,\phi) = R(r)Y_{\ell},m(\theta,\phi) \tag{C.15}$$

leads to

$$\left[\frac{-\hbar^2}{2\mu r^2}\frac{d}{dr}\left(r^2\frac{d}{dr} + \frac{\hbar^2\ell(\ell+1)}{2\mu r^2} + V(r)\right)\right]R(r) = ER(r). \tag{C.16}$$

The term corresponding to $\frac{\langle\hat{L}^2\rangle}{2\mu r^2}$ is the quantum analog of the centrifugal potential that arises in central force problems in classical mechanics. Putting

$$V(r) = \frac{1}{4\pi\varepsilon_0 r}$$

and with $\alpha^2 \equiv -\frac{8\mu E}{\hbar^2}$, $\rho = \alpha r$, and $\lambda \equiv \frac{2\mu e^2}{4\pi\varepsilon_0\alpha\hbar^2}$, C.16 becomes

$$\left[\frac{1}{\rho^2}\frac{d}{d\rho}\left(\rho^2\frac{d}{d\rho}\right) + \frac{\lambda}{\rho} - \frac{1}{4} - \frac{\ell(\ell+1)}{\rho^2}\right]R(\rho) = 0. \tag{C.17}$$

This equation has finite solutions only for $\lambda - \ell - 1 = 0, 1, 2, \ldots$. Noting $\ell \geq 0$ leads to $\lambda = n, n = 1, 2, \ldots$. Requiring positive quantum numbers leads to $0 < \ell < n - 1$. There are thus two quantum numbers associated with the function $R(r)$, n and ℓ. The solutions are polynomials in r (called Laguerre polynomials) multiplied by the exponential of $r/2a_0$, where

$$a_0 = \frac{4\pi\varepsilon_0\hbar^2}{me^2} \approx 0.529 \text{ Å.Å.} \tag{C.18}$$

(Here m is the electron mass, very close to the reduced mass used in the equations of motion.)

The energy eigenvalues are given by

$$E_n = -\frac{me^4}{2(4\pi\varepsilon_0)^2\hbar^2 n^2} \approx -\frac{13.6\,\text{eV}}{n^2}. \qquad (C.20)$$

Examples of some eigenfunctions are given in the main body of the text. Further details in the notation used here can be found in the text by Davies and Betts.[1]

References

[1] P.C.W. Davies and D.S. Betts, *Quantum Mechanics*. 2nd ed. 1994, London: Chapman and Hall.

D

The damped harmonic oscillator

A model of a damped harmonic oscillator is shown in Fig. D.1. It consists of a sphere of mass, m, and radius, a, immersed in a medium of viscosity, η, connected to a fixed surface by a spring of constant, k. Let it be displaced a distance, x_0, from its equilibrium position at time, t, along the coordinate, x.

The equation of motion is

$$m\ddot{x} + 6\pi a\eta\dot{x} + kx = 0, \tag{D.1}$$

where Stoke's law has been used for the viscous drag on the sphere, and the spring and viscous drag both oppose acceleration along the $+x$-axis. Rearranging this,

$$\ddot{x} + \frac{1}{\tau}\dot{x} + \omega_0^2 x = 0, \tag{D.2}$$

where

$$\frac{1}{\tau} = \frac{6\pi a\eta}{m} \tag{D.3}$$

and

$$\omega_0^2 = \frac{k}{m}. \tag{D.4}$$

D.2 is solved with a damped cosinusoidal function

$$x = A\exp(-\beta t)\cos\omega t. \tag{D.5}$$

Substitution into D.2 yields, on collecting terms in $\cos\omega t$ and $\sin\omega t$,

$$\left(\beta^2 - \omega^2 + \omega_0^2 - \frac{\beta}{\tau}\right) A\exp(-\beta t)\cos\omega t$$

$$+ \left(-2\omega\beta + \frac{\omega}{\tau}\right) A\exp(-\beta t)\sin\omega t = 0. \tag{D.6}$$

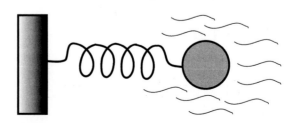

Fig. D.1 A damped harmonic oscillator.

For this statement to be true for all times requires that the coefficients of the sin and cos terms vanish separately or

$$\beta = \frac{1}{2\tau} \tag{D.7}$$

from the requirement that the coefficient of the term in $\sin \omega t$ vanish and

$$\omega = \omega_0 \sqrt{1 - \left(\frac{1}{2\omega_0 \tau}\right)^2} \tag{D.8}$$

from the requirement that the coefficient of the term in $\cos \omega t$ vanish. The quantity $\omega_0 \tau$ is called the mechanical Q of the system in the limit that the energy stored per cycle does not change much over one cycle. It is the ratio of the energy stored per cycle to the energy lost per cycle.

The amplitude of the system after an initial displacement x_0 is

$$x = x_0 \exp(-t/2\tau) \cos\left(\omega_0 t \sqrt{1 - \left(\frac{1}{2\omega_0 \tau}\right)^2}\right)$$
$$= x_0 \exp(-t/2\tau) \cos\left(\omega_0 t \sqrt{1 - \left(\frac{1}{2Q}\right)^2}\right). \tag{D.9}$$

We have used the fact that $x = x_0$ at $t = 0$ and chosen the solution that decays to zero with time. For the case of light damping ($Q \gg 1$), this is an oscillation at an angular frequency, ω_0. As damping increases, the amplitude decays more rapidly, and the frequency shifts to a lower value. An illustration of this for the case where $\omega_0 = 1$ rad/s and $Q = 10$ is shown in Fig. D.2.

When $\omega_0 \tau = 0.5$ the system does not oscillate at all, but relaxes rapidly to equilibrium with a characteristic relaxation frequency $\frac{1}{2\tau}$ Hz (and in this limit, the definition of Q given above is not valid because the energy of the oscillator is dissipated in essentially one cycle). This is shown for the system with $\omega_0 = 1$ rad/s in Fig. D.3.

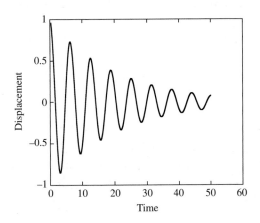

Fig. D.2 Displacement vs. time for the damped oscillator with $\omega_0 \tau = Q = 10$.

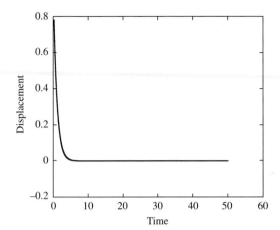

Fig. D.3 Displacement vs. time for the damped oscillator with $\omega_0\tau = 0.5$.

The amplitude–time response of a system can be rewritten in terms of the *amplitude–frequency response* using a Fourier transform (Appendix A.9.b). Doing this yields the amplitude of the displacement of the oscillator as a function of frequency.

Remarkably, this result is valid whether the response is driven by random thermal fluctuations, or whether it is the response as driven by a driving force that is swept over a range of frequencies, so long as the system is driven at low amplitudes. The result that the response of a driven system and that of a system subject to random thermal fluctuations is the same as encapsulated in a very powerful theorem called the *fluctuation dissipation theorem.* The relationship between a diffusion constant and a mobility introduced in Chapter 3 is another illustration of this theorem.

For a lightly damped oscillator, the response as a function of frequency is

$$x(\omega) = \frac{A_0}{\sqrt{(\omega_0^2 - \omega^2)^2 + \left(\frac{\omega}{\tau}\right)^2}} = \frac{A_0}{\sqrt{(\omega_0^2 - \omega^2)^2 + \left(\frac{\omega\omega_0}{Q}\right)^2}}. \qquad \text{(D.10)}$$

The frequency response of the oscillator with the same parameters used for the calculation of the time response in Fig. D.2 is shown in Fig. D.4.

Evaluating D.10 to find the frequency at which the response drops to half of the maximum, $\delta\omega_{1/2}$, shows that, for $Q \gg 1$,

$$Q = \frac{\omega_0}{2\delta\omega_{1/2}}.$$

In other words, Q is also equal to the ratio of the resonant frequency to the full width at half height of the frequency response curve.

In the case of a driven system, it is useful to consider the phase difference between the driving force and the response of the system. Writing the time

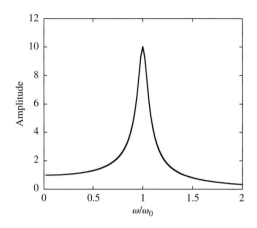

Amplitude

ω/ω_0

Fig. D.4 Amplitude vs. frequency for the damped oscillator with $\omega_0\tau = Q = 10$.

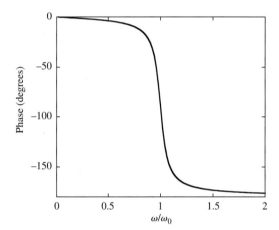

Phase (degrees)

ω/ω_0

Fig. D.5 Plot of the phase difference between the driving force and the response of a damped harmonic oscillator with $\omega_0\tau = 10$.

dependence of the response as

$$x(t) = A \sin(\omega t + \phi). \qquad (D.11)$$

The relative phase angle, ϕ, is obtained from the complex part of the Fourier transform as

$$\phi = \tan^{-1}\left(\frac{\omega/\tau}{\omega^2 - \omega_0^2}\right). \qquad (D.12)$$

A plot of phase vs. relative driving frequency for the same parameters used in Fig. D.4 is shown in Fig. D.5. The key points to take away from this figure are as follows: (a) the response is in phase with the driving frequency at low frequency, (b) the response is exactly 90° out of phase at resonance, and (c) the response is 180° out of phase at very high frequency.

Finally, in the limit of very high damping, the acceleration can be ignored altogether and Equation D.2 becomes

$$6\pi a\eta\dot{x} + kx = F, \qquad (D.13)$$

where we have allowed for the possibility that a force, F, is applied to the system. The solution of this equation is

$$x(t) = \frac{F}{k}\left[1 - \exp\left(-\frac{t}{\tau'}\right)\right],$$ (D.14)

where

$$\tau' = \frac{6\pi a \eta}{k}.$$ (D.15)

Comparison with Equation D.3 shows that the relaxation time of a heavily damped system is quite different from that of a critically, or less than critically, damped system.

Free energies and choice of ensemble

E.1 Different free energies for different problems

An isolated system of many particles moves in the direction of maximizing entropy. What about a system exchanging heat with the environment (i.e., in thermal equilibrium) or adjusting its volume (to remain in pressure equilibrium)? In reality, measurements are made on systems that exchange energy with the environment by these mechanisms, so we must generalize the idea of maximizing entropy.

Consider the system of interest as connected to a "world" of constant temperature and pressure by means of a thermally conductive membrane that can slide freely to equalize pressure on each side. Let the entropy of the system be S and that of the "world" be S_0. Then

$$S_{\text{total}} = S_0 + S. \tag{E.1}$$

The first law of thermodynamics is

$$dE = dQ + dW$$

and the fundamental thermodynamic relation is

$$dE = T dS - P dV.$$

For a completely reversible change, the work done on the system is $-P dV$ (not true for an irreversible change where this work might be dissipated as heat). This leads to the result

$$dQ = T dS \tag{E.2}$$

(for a reversible system). Our "small" system, connected to a big world must be reversible (from the viewpoint of the "world") because the pressure, temperature, and entropy of the "world" are unaffected by changes in the small system. Since the "world" plus the small system represent an isolated system, thermodynamics dictates that

$$\Delta S_{\text{total}} = \Delta S_0 + \Delta S \geq 0. \tag{E.3}$$

For the "world" (where the change is reversible)

$$\Delta S_0 = \frac{Q}{T_0}, \tag{E.4}$$

where Q is the heat flow between the "world" and the system and T_0 is the temperature of the "world." Identifying $P\Delta V$ as ΔW results in

$$\Delta S - \frac{\Delta E + P_0 \Delta V}{T_0} \geq 0. \tag{E.5}$$

This is the Clausius inequality (the equals sign applying to reversible changes). The condition for equilibrium can be defined concisely using

$$A \equiv E + P_0 V - T_0 S. \tag{E.6}$$

In these terms, the Clausius inequality becomes

$$\Delta A \leq 0. \tag{E.7}$$

The thermodynamic potentials follow from E.6 and E.7.

For an isolated system of constant energy and volume (recalling that T_0 is the constant temperature of the "world"), **entropy** alone dictates the direction of movement toward equilibrium

$$\Delta S \geq 0. \tag{E.8}$$

For a system of constant volume in contact with a heat bath at the constant temperature $T = T_0$,

$$\Delta E - T\Delta S \equiv \Delta F \leq 0, \tag{E.9}$$

where F is the **Helmholtz free energy**,

$$F = E - T_0 S. \tag{E.10}$$

For the most general case, where volume, energy, and entropy are all variable with the system in pressure ($P = P_0$) and temperature ($T = T_0$) equilibrium with the "world," the appropriate potential is the **Gibbs free energy**,

$$G = E + PV - T_0 S \tag{E.11}$$

with the condition that

$$\Delta G \leq 0 \tag{E.12}$$

applied according to what parameters are allowed to vary in a given system.

Defining the **enthalpy**, H, as

$$H = E + PV. \tag{E.13}$$

The Gibbs free energy is

$$G = H - TS, \tag{E.14}$$

a form useful in thermochemistry. This is because

$$\Delta H = \Delta E + P\Delta V = Q + W - W = Q. \tag{E.14}$$

So measurements of the heat flow into or out of a chemical reaction at constant pressure determine the enthalpy directly. The Gibbs free energy is readily generalized to take account of particle flow into or out of the system as discussed in Appendix F.

E.2   Different statistical ensembles for different problems

Gibbs introduced the notion of an ensemble as a way of simplifying the process of taking averages. Instead of working out all the configurations of one system, many replicas are made of the system, the ensemble occupying all possible states.

A *microcanonical* ensemble consists of copies of an *isolated* system. Each system has constant energy, E, but a different microstate. The number of systems, Ω, that have energy, E, is the statistical weight of macrostate, from which its entropy is calculated according to Equation 3.13a, on the basis of the fact that the probability, p_i, of each microstate is just

$$p_i = \Omega^{-1}. \tag{E.15}$$

A *canonical ensemble* has each system in contact with a heat bath, each microstate occurring with a probability that depends on the energy of the system according to

$$p_i = \frac{1}{Z} \exp(-\beta E_i). \tag{E.16}$$

A *grand canonical ensemble* is one in which each system is in equilibrium with an external reservoir with respect to both particle and energy exchange. The probability of a microstate having N particles and energy, E_{Ni}, is

$$P_{Ni} = \frac{\exp\left[\beta\left(\mu N - E_{Ni}\right)\right]}{Z}. \tag{E.17}$$

F Probabilities and the definition of entropy

To derive Equation 3.13b, consider a very large ensemble of n systems that are identical replicas of the system under study. Note that we are not imposing any constraints on the replicas (like they be at constant temperature which is the canonical ensemble that produces the Boltzmann distribution—Appendix E—this ensemble is more general). For a large enough ensemble, the number of systems in the state r is

$$n_r = n \times p(r).$$ (F.1)

If the ensemble has n_1 of its systems in state 1, n_2 in state 2, and so on, the number of ways of realizing this particular distribution (the statistical weight) is

$$\Omega = \frac{n!}{n_1! \times n_2! \times n_3! \cdots \times n_r! \cdots}.$$ (F.2)

And with Boltzmann's definition of entropy, $S = k_B \ln \Omega$

$$S_n = k_B \ln \left[\frac{n!}{n_1! \times n_2! \times n_3! \cdots \times n_r! \cdots} \right] = k_B \left[n \ln n - \sum_r n_r \ln n_r \right],$$ (F.3)

where we used Stirling's formula $\ln n! = n \ln n - n$ for large n. With a normalized probability ($\sum_r P_r = 1$) and $n_r = n \times p(r)$

$$S_n = -n k_B \sum_r p_r \ln p_r.$$ (F.3)

And, because entropy is extensive, we obtain for the entropy of one system:

$$S = -k_B \sum_r p_r \ln p_r.$$ (F.4)

The Gibbs distribution

Consider a system of constant volume, V, exchanging both energy and particles with a heat bath that has a fixed energy, E_0, and fixed number of particles, N_0, in a fixed volume, V_0. We will suppose only that our subsystem, containing at any time N particles, with energy, E_{nr}, in the state r (and fixed volume V) is small compared with the entire system (e.g., a beaker of chemical reactants in equilibrium with the rest of the universe). The probability that the subsystem is in a state r with energy, E_r, with N particles (N_r for short) in a volume, V, is just the same as the probability that rest of the world is in the state

$$P_{Nr} = \text{const} \times \Omega_{\text{universe}}(E_0 - E_r, V_0 - V, N_0 - N) \qquad (G.1)$$

(Ω_{universe} is the relative *statistical weight* of the state.) From the Boltzmann definition of entropy, $S = k_B \ln(\Omega)$

$$P_{Nr} = \text{const} \times \exp\left[\frac{S_{\text{universe}}(E_0 - E_r, V_0 - V, N_0 - N)}{k_B}\right]. \qquad (G.2)$$

Since the system is very small, we can expand the entropy in terms of E_r, N, and V using a Taylor series.

$$S_{\text{universe}}(E_0 - E_r, V_0 - V, N_0 - N)$$
$$= S_0 - \frac{\partial S}{\partial V_0}V - \frac{\partial S}{\partial E_0}E_{Nr} - \frac{\partial S}{\partial N_0}N. \qquad (G.3)$$

The term in V is constant. The term $\frac{\partial S}{\partial E_0} = \frac{1}{T}$ defines the temperature of the heat bath, the quantity that is kept constant as the system and heat bath exchange energy. By analogy, we introduce a new quantity

$$\mu \equiv -T\frac{\partial S}{\partial N_0}. \qquad (G.4)$$

This is called the *chemical potential* and, similar to the role of temperature for energy exchange, this is what is kept constant as the system and heat bath exchange particles.

The expression for entropy becomes

$$S_{\text{universe}}(E_0 - E_r, V_0 - V, N_0 - N) = S_0 - \frac{\partial S}{\partial V_0}V - \frac{E_{Nr}}{T} + \frac{\mu N}{T}. \qquad (G.5)$$

Substituting this into (G.2) gives

$$P_{Nr} = \text{newconst} \times \exp\left[\frac{\mu N - E_{Nr}}{k_B T}\right], \qquad (G.6)$$

where the new constant has absorbed the exponential of the constant term $S_0 - \frac{\partial S}{\partial V_0}V$.

The correctly normalized version of (G.6) is

$$P_{Nr} = \frac{\exp\left[(\mu N - E_{Nr})/k_B T\right]}{\mathbf{Z}}, \qquad (G.7)$$

where the *grand partition function* $\mathbf{Z}$ is

$$\mathbf{Z}(T, V, \mu) = \sum_{Nr} \exp\left[\frac{\mu N - E_{Nr}}{k_B T}\right]. \qquad (G.8)$$

The sum is a sum over all states r for a given N, followed by a sum (of that sum) over all states of a given N. Equation (G.8) is a generalization of the Boltzmann distribution called the *Gibbs distribution*.

Quantum partition function for a single particle

The quantum mechanical description of a free particle was given in Chapter 2 (2.15). The allowed energy states are (Equation 2.50)

$$E = \frac{\hbar^2 \mathbf{k}^2}{2m} = \frac{\hbar^2}{2m}\left(k_x^2 + k_y^2 + k_z^2\right)$$

with (Equation 2.52)

$$k_x = \frac{2n_x \pi}{L}, \quad k_y = \frac{2n_y \pi}{L}, \quad k_z = \frac{2n_z \pi}{L},$$

so the partition function (Equation 3.10) is

$$Z(T,V,1) = \sum_{n_x,n_y,n_z} \exp -\beta E(n_x, n_y, n_z). \tag{H.1}$$

For a system of any reasonable size, the energy spacing is small relative to $k_B T$ so we can assume a smooth distribution of states with a density (number of states per unit k) given by Equation 2.54:

$$dn = \frac{4\pi V k^2 dk}{8\pi^3} = \frac{V k^2 dk}{2\pi^2}.$$

This expression can be used to turn the sum in H.1 into a continuous function, so that Equation H.1 can now be expressed as an integral:

$$Z(T,V,1) = \frac{V}{2\pi^2} \int_0^\infty k^2 \exp\left(-\frac{\hbar^2 k^2}{2mk_B T}\right) dk. \tag{H.2}$$

This integral can be carried out using a standard result (Appendix A):

$$\int_0^\infty x^2 \exp(-\alpha x^2)\, dx = \frac{\sqrt{\pi}}{4}\alpha^{-3/2}.$$

Making the appropriate substitutions into Equation H.2, we obtain

$$Z(T,V,1) = \left(\frac{mk_B T}{2\pi\hbar^2}\right)^{3/2} V, \tag{H.3}$$

which is the result given in the text as Equation 3.31.

$\left(\frac{mk_{B}T}{2\pi\hbar^{2}}\right)^{3/2}$ has dimensions of 1/(volume) and it is called the *quantum concentration*, n_Q. It is the concentration at which each particle is separated from others by a distance equal to the de Broglie wavelength of the particle with thermal energy $k_B T$ and mass m :

$$n_Q = \left(\frac{mk_B T}{2\pi\hbar^2}\right)^{3/2}.$$

(H.4)

Partition function for N particles in an ideal gas

As we saw in Equation 3.32, we overcounted by forgetting how complex the term under the $\sum$ sign really is. We have to divide the product partition function by the number of times the states are overcounted, which, for Equation 3.33 gives

$$\sum_r \exp(-2\beta E_r) + \frac{1}{2!} \sum_r \sum_{\substack{s \\ (r \neq s)}} \exp\left[-\beta(E_r + E_s)\right]. \tag{I.1}$$

We have put $1/2!$ rather than $1/2$ because $1/N!$ is the correct result for three or more particles, as can be verified by writing out the sum term by term. So, the correct partition function is

$$\sum_r \exp(-N\beta E_r) + \cdots + \frac{1}{N!} \sum_{r1} \cdots \sum_{rN} \exp\left[-\beta(E_{r1} + E_{r2} + \cdots + E_{rN})\right],$$
$$\tag{I.2}$$

where we have only put down the terms for each particle in the same state and then for each particle in a different state. There are other, more complicated terms with some particles in the same state and the remainder in different states, but we would not need those. This is because, in the classical case, *the probability that any two particles are in the same state is very small* (not true in the degenerate quantum limit). Thus, a good approximation to the partition function for N particles is

$$Z(T,V,N) = \frac{1}{N!}\left[\sum_r \exp(-\beta E_r)\right]^N = \frac{1}{N!}\left[Z(T,V,1)\right]^N. \tag{I.3}$$

J   Atomic units

Fundamental quantities, such as energy and length, are usually expressed in atomic units in molecular calculations to simplify the equations.

The atomic unit of length is the Bohr radius:

$$a_0 = \frac{\hbar^2}{me^2} = 0.052917725 \text{ nm.}$$

Coordinates can be transformed to Bohr's by dividing them by a_0.

Energies are measured in Hartrees, defined as the absolute value of the ground state energy of the electron in hydrogen (kinetic plus potential energy)

$$1 \text{ Hartree} = \frac{\hbar^2}{m_e a_0^2} = 27.21 \text{ eV}$$

(approximately double the ionization energy—the small difference reflects the difference between the electron's mass and the reduced mass).

Masses are expressed in terms of the electron mass.

Charges are expressed in units of e, the electronic charge.

In terms of atomic units, the Hamiltonian of the hydrogen atom is

$$H = -\frac{1}{2}\nabla^2 - \frac{1}{r},$$

where ∇^2 is the second derivative operator in polar coordinates.

Hückel theory for benzene

The wavefunction is a linear combination of six P_z states

$$\psi_{\pi i} = \sum_{j=1}^{6} c_{ij} 2p_{zj}.$$

With secular determinant

$$\begin{vmatrix} \alpha - E & \beta & 0 & 0 & 0 & 0 \\ \beta & \alpha - E & \beta & 0 & 0 & 0 \\ 0 & \beta & \alpha - E & \beta & 0 & 0 \\ 0 & 0 & \beta & \alpha - E & \beta & 0 \\ 0 & 0 & 0 & \beta & \alpha - E & \beta \\ 0 & 0 & 0 & 0 & \beta & \alpha - E \end{vmatrix} = 0$$

With the substitution $x = (\alpha - E)/\beta$, expansion of the secular equation yields

$$x^6 - 6x^4 + 9x^2 - 4 = 0,$$

which has roots $\pm 1, \pm 1, \pm 2$. Thus,

$$E_1 = \alpha + 2\beta$$
$$E_2 = E_3 = \alpha + \beta$$
$$E_4 = E_5 = \alpha - \beta$$
$$E_6 = \alpha - 2\beta$$

in order of ascending energy because $\beta < 0$. Using these energies to solve for the six sets of c_i leads to

$$\psi_1 = \frac{1}{\sqrt{6}} (2p_{z1} + 2p_{z2} + 2p_{z3} + 2p_{z4} + 2p_{z5} + 2p_{z6}) \quad E_1 = \alpha + 2\beta$$

$$\psi_2 = \frac{1}{\sqrt{4}} (2p_{z2} + 2p_{z3} - 2p_{z5} - 2p_{z6}) \quad E_2 = \alpha + \beta$$

$$\psi_3 = \frac{1}{\sqrt{3}} \left(2p_{z1} + \frac{1}{2}2p_{z2} - \frac{1}{2}2p_{z3} - 2p_{z4} - \frac{1}{2}2p_{z5} + \frac{1}{2}2p_{z6} \right) \quad E_3 = \alpha + \beta$$

$$\psi_4 = \frac{1}{\sqrt{4}} (2p_{z2} - 2p_{z3} + 2p_{z5} - 2p_{z6}) \quad E_4 = \alpha - \beta$$

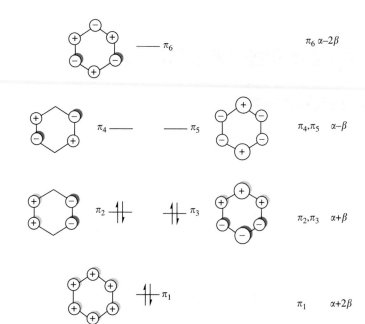

Fig. K.1 Hückel π orbitals of benzene (top view of the molecule) with energies shown on the right.

$$\psi_5 = \frac{1}{\sqrt{3}}\left(2p_{z1} - \frac{1}{2}2p_{z2} - \frac{1}{2}2p_{z3} + 2p_{z4} - \frac{1}{2}2p_{z5} - \frac{1}{2}2p_{z6}\right) \quad E_5 = \alpha - \beta$$

$$\psi_6 = \frac{1}{\sqrt{3}}\left(2p_{z1} - \frac{1}{2}2p_{z2} + \frac{1}{2}2p_{z3} - 2p_{z4} + \frac{1}{2}2p_{z5} - \frac{1}{2}2p_{z6}\right) \quad E_6 = \alpha - 2\beta.$$

These wavefunctions are shown schematically, looking down on top of the benzene ring plane, in Fig. K.1.

A glossary for nanobiology

Actin
A fibrous protein component of muscle, it is the "track" on which myosin motor proteins ride. It is also a key component of the "cytoskeleton," the scaffold that shapes a cell and directs its movement.

Active site
Region of an enzyme where catalytic activity occurs.

Adenine
One of the DNA bases (a purine—see Chapter 6)

Adenosine diphosphate
The product of hydrolysis of adenosine triphosphate (ATP). Is recycled back into ATP by ATP synthase.

Adenosine triphosphate
The "carrier" of chemical energy in the cell. Energy is obtained when adenosine triphosphate (ATP) is hydrolyzed to form adenosine diphosphate (ADP).

ADP
See Adenosine diphosphate.

Amino acid
Component of a protein. There are 20 naturally occurring amino acids.

Antibody
Protein that is part of the immune system. It binds invading molecules to trigger a defensive immune response.

Antigen
The target of an antibody—any molecule or particle that provokes an immune response.

ATP
See Adenosine triphosphate.

ATP synthase
The enzyme that adds a phosphate to ADP to produce ATP.

B cell
Part of the immune system, antibody sequences are generated when B cells differentiate from stem cells.

Bacterium
The smallest unit of life—a single-celled creature.

Biotinylation
Adding the small chemical biotin to biotinylate a molecular species enables it to be attached selectively to the protein streptavidin, which has four strong and selective binding sites for biotin.

Bovine serum albumin
A small protein often used as a stabilizer in biochemistry.

BSA
See Bovine serum albumin.

Cell adhesion molecules
Molecules exposed on the surface of a cell, often directing adhesion to other cells for the self-assembly of organs.

Cell nucleus
Enclosed in a lipid bilayer in the interior of an eukaryotic cell, the nucleus contains the genome of the cell, packaged together with positively charged proteins that condense the genome into small packages called chromosomes.

Chromatin
The complex of DNA and positively charged proteins that package the genome into chromosomes.

Chromosomes
The genome is often packaged as several distinct strands of DNA. Each of these chromatin packages is called a chromosome.

Clone
Daughter animal with an identical genome to the mother. Some plants are natural clones, but biotechnology is required to clone mammals which normally require both paternal and maternal genomes.

Cytoplasm
Volume of the cell between the cell nucleus and the cell membrane. Prokaryotic cells have no nucleus, so all of their interior is cytoplasm.

Cytosine
One of the four DNA bases—a pyrimidine (see Chapter 6).

Deoxyribonucleic acid
The heteropolymer, the composition of which usually encodes the genome of an animal.

Differentiation
The process whereby embryonic stem cells (usually) or adult stem cells (in the immune system, for example) develop into specialist cells by selectively shutting down gene expression in daughter cells.

Dimer
Complex of two proteins. When the two proteins are identical, the complex is called a homodimer. Many proteins form *multimeric* complexes (e.g., 2, 3, 4, etc.).

DNA
See Deoxyribonucleic acid.

Dynein
A motor protein that shuttles cargo on microtubules.

Endoplasmic reticulum
Folded membrane inside the cell that houses the ribosomes which turn mRNA into proteins.

Enzyme
A protein that carries out a biochemical function, usually the catalysis of a chemical reaction.

Eukaryote
A multicelled animal. Eukaryotes have their genomes packaged in cell nuclei for selective control of gene expression.

Exon
Sequence of DNA that codes for protein structure. One protein may be coded for by a series of many different exons separated by noncoding regions called introns.

Gene
DNA sequences that pass characteristics from mother to daughter. Strictly, the regions of DNA that code for proteins, but many essential control sequences, need to be inherited also.

Genetic code
The sequences of the genome. More specifically, the code whereby specific amino acids are represented in the form of three sequential DNA bases.

Genome
The entire complement of genetic information; all the DNA in a cell (some organisms encode their genome in RNA).

Genotype
The organism as represented by its genome, specifically, the DNA (or RNA) sequence of an organism.

Germline cells
The undifferentiated cells of an embryo, or of adult stem cells.

Golgi apparatus
An organelle in the cell that marks proteins for dispatch to particular parts of the cell. The markings are made in the form of post-translational modifications of the protein.

Guanine
One of the four DNA bases—a purine (see Chapter 6).

Heavy chain
One of the four protein chains (two heavy, two light) that assemble by means of disulfide bonds into an antibody.

HIV
See Human immunodeficiency virus.

Human immunodeficiency virus
A retrovirus that inserts its genome into the genome of the host cell, so that the host cell replicates it and makes the proteins required for its propagation. Of all retroviruses, HIV is notorious for its ability to evade and weaken the human immune system.

Hydrolysis
A chemical reaction involving a water molecule as one of the reactants. Examples are the breaking of a peptide bond and the removal of a phosphate from ATP.

Intron
Regions of the genome that do not code for protein and are interspersed between the coding regions (exons). Introns code for important control sequences.

Kinase
A protein that adds a phosphate group to a specific atomic site on a specific amino acid.

Kinesin
A motor protein that "walks" cargo along microtubules.

Light chain
One of the four protein chains (two heavy, two light) that assemble by means of disulfide bonds into an antibody.

Messenger RNA
The RNA produced when DNA is copied into RNA by a DNA-dependent RNA polymerase. It carries the code for protein synthesis form the cell nucleus to the ribosomes in the cytoplasm.

Microtubule

A self-assembling scaffold within the cell that acts as a path for transporting materials to specific sites within the cell.

Mitochondria

The organelle in which pyruvate is metabolized to produce a protein gradient that drives ATP synthase, driving the synthesis of ATP from ADP.

mRNA

See Messenger RNA

Myosin

The motor protein of soft muscle tissue. Its walks on actin filaments to contract the sarcomere.

Neural network

Network of neurons that signal each other at junctions called synapses. The network acts as a signal processor, taking electrical pulses generated by ion gradients as input, and producing electrical impulses as output.

Neurons

Specialist cells that form neural networks.

Nuclear pore complex

Gateway between the cytoplasm and the cell nucleus, this large protein complex controls transport to and from the cell nucleus.

Origin of replication

A special sequence on bacterial genomes that marks the starting point for reading the genes.

Peptide bond

The N-C bond that forms the backbone of a protein.

Phenotype

The external characteristics of an animal.

Phosphorylation

The addition of a phosphate group at a specific site.

Polymerase

The protein–RNA complex that makes nucleic acids. DNA is translated into mRNA by DNA-dependent RNA polymerase. When cells divide, DNA is copied by DNA-dependent DNA polymerase.

Polypeptide

A string of amino acids joined by peptide bonds. Usually refers to a small polymer, but a complete protein is a polypeptide.

Post-translational modifications

Chemical modification of a protein after it is synthesized in the ribosome. Modifications include the addition or removal of specific chemical groups (like acetyl or phosphate groups) and more complex changes like the addition of polymeric sugars.

Prokaryote

A single-celled animal without a cell nucleus.

Promoter

Sequence of DNA at the start of a gene that acts to initiate the transcription of the genes.

Promoter sequence

See Promoter.

Protease

Protein that breaks the peptide bond, usually at a very specific site.

Protein

Polymer composed of a peptide-bonded backbone carrying a series of amino acid residues as side chains.

Receptors

Proteins that span cell or nuclear membranes and transduce signals (in or out, depending on the signaling direction) on binding target molecules. Receptors on cell surfaces are often the target of pharmaceuticals.

Recombination

The process whereby pieces of gene are moved about in the genome on cell division.

Ribonucleic acid

A variant of DNA in which the sugar carries an additional OH group. The base thymine is replaced with uracil in RNA.

Ribosome

Protein–RNA complex in which protein assembly takes place. Codons on the RNA are matched to small RNA molecules that carry an amino acid (tRNA). The ribosome connects the amino acids by peptide bonds to form the protein coded for by the mRNA.

Ribozymes

Small pieces of RNA that are catalytically active ("RNA enzymes"). One example is intron sequences that, when translated into RNA, excise themselves.

RNA

See Ribonucleic acid.

Sarcomere
Basic fibrous unit of smooth muscle tissue.

Spliceosome
Organelle in the cell nucleus where RNA splicing is directed.

Splicing
Operation whereby exons are joined in mRNA to form a continuous coding sequence for protein synthesis.

Stem cells
Undifferentiated cells capable of being turned into any specialized type of cell.

Streptavidin
A protein that binds four biotin molecules. See "Biotinylation."

Thymine
One of the four DNA bases—a pyrimidine (see Chapter 6).

Transcription
The operation of copying DNA into mRNA for eventual protein synthesis.

Transcription factor
A protein that binds control regions of DNA, usually upstream of the gene, to upregulate or downregulate gene expression.

Transfer RNA
Small RNA molecule attached to a specific amino acid. There is one tRNA for each amino acid. tRNAs are used by the ribosome to synthesize protein based on the code read from mRNA.

Transmembrane protein
Protein that spans a cell membrane, connecting the exterior to the cytoplasm.

tRNA
see Transfer RNA.

Tubulin
Protein that self-assembles to form microtubules. Microtubules are constantly forming and dissociating as demanded by the transport needs of the cell.

Uracil
The base that replaces thymine in RNA.

Virus
The most elementary particle capable of reproduction. It lacks the components needed for independent life, hijacking a host cell to survive and propagate.

Solutions and hints for the problems

Chapter 1

(1.1) Each page has to be demagnified 4.8×10^9 times. Approximately 177 atoms form each character.

(1.2) From the ideal gas law:

$$PV = RT \Rightarrow V = \frac{RT}{P}$$

with

$$R = 8.314 \text{ J mol} \cdot \text{K}^{-1}$$
$$T = 273.15 \text{ K}$$
$$P = 101.3 \text{ kPa}$$

we get $V = 22.42$ L.

(1.3) The mass of one atom is

$$\left(\frac{28.05 \text{ g}}{1 \text{ mol}} \right) \left(\frac{1 \text{ mol}}{6 \times 10^{23} \text{ atoms}} \right) = 4.66 \times 10^{-23} \text{ g}$$

So the volume occupied by the atom is

$$V = \frac{m}{\rho} = \left(\frac{4.66 \times 10^{-23} \text{ g}}{2330 \text{ kg/m}^3} \right) \left(\frac{1 \text{ kg}}{10^3 \text{ g}} \right)$$
$$= 0.00199 \times 10^{-26} \text{ m}^3 = 1.99 \times 10^{-29} \text{ m}^3$$

(1.4) The radius of an atom follows from

$$V_{\text{sphere}} = \frac{4}{3} \pi R^3 = 1.99 \times 10^{-29} \Rightarrow R = 1.68 \times 10^{-10} \text{ m} = 1.68 \times 10^{-7} \text{ mm}$$

Each 100 atom transistor is $20R$ across, with an area $400R^2$. Consequently, 4.2×10^{15} transistors will fit onto the chip.

(1.5) From the fact that $1A = 1$ C/s,

$$\left(\frac{100 \text{ pC}}{1 \text{ s}}\right) \left(\frac{1 \text{ C}}{10^{12} \text{ pC}}\right) \left(\frac{1e^-}{1.6 \times 10^{-19} \text{ C}}\right)$$
$$= 0.625 \times 10^9 e^- /\text{s} = 6.25 \times 10^8 e^- /\text{s}$$

The shot noise is given by $\sqrt{N}$, i.e., $2.5 \times 10^4 e/\text{s} = 4 \times 10^{-15}$ A.

(1.6) The Coulomb energy is 2.3×10^{-18} J. The thermal kinetic energy is 6.3×10^{-21} J. Collisions do not affect the electronic energy.

Chapter 2

(2.1) There is only one path so $|\Psi_1 \Psi_2|^2 = |\Psi_1|^2 |\Psi_2|^2$, and the probability is 0.1^2 or 10^{-2}.

(2.2) No—summing the amplitudes requires that the relative phase of the two beams is known.

(2.3) Pauli's exclusion principle states that no two electrons (fermions) can occupy the same overall state at a given time. This rule implies that a wavefunction with multiple fermions must be antisymmetric under fermion exchange: if we switch the two fermions, the wavefunction picks up a minus sign. As the total wavefunction is a product of spatial and spin wavefunctions, only one or the other must be antisymmetric. If the two particles have the same spatial wavefunction, then the spatial component is symmetric, giving an antisymmetric spin wavefunction (spin zero, the singlet state). If the two particles are in different states, then the spatial wavefunction is antisymmetric, which renders the spin wavefunction symmetric. There are three such states: both electrons spin up, both electrons spin down, and a linear combination:

$$\psi = \sqrt{\frac{1}{2}} (\uparrow\downarrow + \downarrow\uparrow)$$

These are the spin one (triplet) states.

(2.4) When the particles are indistinguishable (both spin up) they behave as identical fermions (Equation 2.21). When one is spin up, the other spin down, they are distinguishable and follow Equation 2.22b.

(2.5) The kinetic energy is proportional to the second derivative of the wavefunction so the more rapidly curving p-states have higher kinetic energy. The volume available precisely at $r = 0$ is zero, so the probability of finding particles at the origin in either state is zero.

(2.6) From Equation 2.36a, the k vector is zero so the wavelength is infinite.

(2.7) Only wavefunction B satisfies the requirements that the wavefunction and its derivative be continuous.

(2.8) According to Equation 2.47, the energy varies as L^{-2} so the energy falls to $1/4$ of the original value.

(2.9) From

$$\frac{dn}{dk} = \frac{V}{2\pi^2}k^2,$$

it is clear that more states per unit k are available at higher k, i.e., higher energy.

(2.10) Neon is $1s^2 2s^2 2p^6$. Having a closed shell, it is hard to ionize. Sodium is $1s^2 2s^2 2p^6 3s^1$. The $3s^1$ electron is easily removed, making sodium easy to ionize.

(2.11) The coefficient of mixing in Equation 2.91 increases as the energy difference between the states becomes smaller, so (a) is correct.

(2.12) The splitting, given by Equation 2.104 is 2Δ.

(2.13) No. It is the balance between repulsive and attractive terms (cf. Equations 2.107a and 2.107b) that leads to a well-defined bond length.

(2.14) The local density approximation.

(2.15) Once a measurement is made, the state of the system is determined, regardless of whether or not the polarization of the second photon is measured. This does not violate relativity, because measurement of the polarization of the second photon yields no new information.

(2.16) Electronic charge in water is concentrated on the oxygen atom, so the molecule has a significant dipole moment. Carbon tetrachloride no has a high symmetry, with dipoles associated with the bonds canceling. It has no dipole moment. Thus, water can polarize to reduce the electrostatic energy of an ion. CCl_4 cannot.

(2.17) 4.07×10^{-19} J or 2.54 eV.

(2.18) 1.23×10^{-11} m. This would be a wavelength-limited resolution limit, but, in practice, aberrations in the electron optics are the main limitation of electron microscopes.

(2.19) 2.9×10^{-13} m.

(2.20) Taking the mass of the car to be 1500 kg gives $\lambda = 1.6 \times 10^{-38}$ m.

(2.21) Note there is a mole of X_2 (i.e., 2 moles of X_2 after the reaction). The energy released is $2 \times 1.6 \times 10^{-19} \times 6 \times 10^{23} = 1.92 \times 10^5$ J $= 4.6 \times 10^4$ calories.

(2.22) Cs has a work function of 2.1 eV, so the tunnel current decays by a factor 0.23 for each Å the gap is increased.

(2.23) 13.6 eV, the same as the ionization energy.

(2.24)

$$\int_0^L A^2 \sin^2 \frac{n\pi x}{L} dx = \int_0^L A^2 \frac{\left(1 - \cos \frac{2n\pi x}{L}\right)}{2} dx$$

$$= \frac{A^2}{2} \left(\int_0^L (1) dx - \int_0^L \cos \frac{2n\pi x}{L} dx \right) = 1$$

The first integral give $\frac{A^2 L}{2}$ and the second is zero. Thus, normalization requires that $A = \sqrt{\frac{2}{L}}$.

(2.25)
$$\psi_L(x) = \exp(ikx) + r\exp(-ikx)$$
$$\psi_M(x) = A\exp(\kappa x) + B\exp(-\kappa x)$$
$$\psi_R(x) = t\exp(ikx)$$

with $\psi_L = \psi_M$ at $x = 0$, $\dfrac{\mathrm{d}\psi_L}{\mathrm{d}x} = \dfrac{\mathrm{d}\psi_M}{\mathrm{d}x}$ at $x = 0$ and $\psi_M = \psi_R$ at $x = L$, $\dfrac{\mathrm{d}\psi_M}{\mathrm{d}x} = \dfrac{\mathrm{d}\psi_R}{\mathrm{d}x}$ at $x = L$. From these equations, we get

$$A\exp(\kappa L) + B\exp(-\kappa L) = t\exp(ikL)$$
$$\Rightarrow A\exp(\kappa L) = t\exp(ikL) - B\exp(-\kappa L)$$
$$\kappa A\exp(\kappa L) - \kappa B\exp(-\kappa L) = ikt\exp(ikL)$$
$$\Rightarrow A\exp(\kappa L) = B\exp(\kappa L) + ik\kappa^{-1}tB\exp(ikL)$$

from which

$$t\exp(ikL) - B\exp(-\kappa L) = B\exp(-\kappa L) + itk\kappa^{-1}\exp(ikL)$$
$$t(\kappa - ik)\exp(ikL) = 2B\kappa\exp(-\kappa L)$$

which gives

$$t\exp(ikL) = \frac{(2B\kappa\exp(-\kappa L))}{\kappa - ik}$$

(2.26) For $2\kappa L \gg 1$

$$T = \frac{1}{1 + \sinh^2(\kappa L)}$$

with

$$4\cosh^2 x = 2 + \exp(2x) + \exp(-2x)$$

and $e^{-2\kappa L} \to 0$, $e^{\kappa L} \gg 2$ gives the desired result.

(2.27)

$$\sqrt{8\pi}\,a_0^3\left(\int_0^{2\pi} \mathrm{d}\phi\right)$$
$$\times \left(\int_0^{\pi}\sin\theta\,\mathrm{d}\theta\right) \times \left(\int_0^{\infty} r^2\left(1 - \frac{r}{2a_0}\right)\exp\left(\frac{-3r}{2a_0}\right)\mathrm{d}r\right)$$

The first two integrals produce

$$\int_0^{2\pi}\mathrm{d}\phi = 2\pi \qquad \int_0^{\pi}\sin\theta\,\mathrm{d}\theta = 2$$

The substitution $u = \frac{3r}{2a_0}$ yields

$$\left(\frac{2a_0}{3}\right)^3 \int_0^\infty u^2 \left(1 - \frac{1}{3}u\right) \exp(-u)\, du$$

$$= \left(\frac{2a_0}{3}\right)^3 \left[\int_0^\infty u^2 \exp(-u)\, du - \frac{1}{3}\int_0^\infty u^3 \exp(-u)\, du\right]$$

Each of these terms can be integrated by parts, and shown to be zero with these limits.

Chapter 3

(3.1) Total energy remains the same after heat is pumped into the containers, but, in the container with a moveable wall, energy $-p\Delta V$ is lost in doing work, so the gas inside the container with a moveable wall will not be as hot as the gas in the container with a fixed wall.

(3.2) According to the Stokes formula friction increases with the speed of the object. It will therefore increase until the frictional force exactly equals the force owing to gravitational acceleration. After that point, the object will fall with constant velocity (the "terminal velocity").

(3.3) A larger volume of gas above the water implies a higher pressure, and thus a higher boiling point. A kettle boils at less than 100°C in the mountains.

(3.4) The equation implies that three reactants become two products. Since the entropy scales with the number of objects, it is decreased and therefore contributes an unfavorable term to the free energy of the reaction.

(3.5) Entropy increases with temperature. At phase transitions, the entropy increases as the temperature remains constant. The larger number of configurations (specific heat) available to a fluid results in a greater increase in entropy per degree for fluids than for solids.

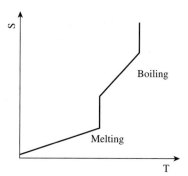

(3.6) The creation of 10 free water molecules greatly increases the entropy, and thus free energy of the products. In this case, this dominates the unfavorable enthalpy.

(3.7) The diatomic gas has additional degrees of freedom over and above the three center of mass translations for the monatomic gas. These are two rotational modes (rotation around the third axis connecting the molecules has no moment of inertia) and one vibrational mode. Each mode contributes $\frac{1}{2}k_B$ to the specific heat. Specifically, the quadratic terms describing the energy of each system are

$$\text{Monatomic} : \frac{1}{2}m\dot{x}^2 + \frac{1}{2}m\dot{y}^2 + \frac{1}{2}m\dot{z}^2$$

$$\text{Diatomic} : \frac{1}{2}m\dot{x}^2 + \frac{1}{2}m\dot{y}^2 + \frac{1}{2}m\dot{z}^2 + \frac{1}{2}I_\theta\dot{\theta}^2 + \frac{1}{2}I_\phi\dot{\phi}^2 + \frac{1}{2}m\omega^2 x^2$$

This leads to $\frac{3}{2}k_B$ for the monatomic gas and $3k_B$ for the diatomic gas.

(3.8)
$$N_2 + 3H_2 \rightarrow 2NH_3$$

Increases the free energy of the reactants, increasing the overall free energy difference.

$$2HBr \rightarrow H_2 + Br_2$$

Increases the free energy of the products, decreasing the overall free energy difference.

$$2H_2 + C_2H_2 \rightarrow C_2H_6$$

Increases the free energy of the reactants, increasing the overall free energy difference.

(3.9) The Gibbs free energy is used for variable particle numbers. In this case,

$$\mu = \frac{\partial G}{\partial N} \equiv \phi$$

(3.10) Classical behavior would require $T \gg \frac{E_F}{k_B}$. With a Fermi energy of 5 eV this would be much greater than 6×10^4 K, well beyond the melting temperature of most metals.

(3.11) The projection of steps in 3-D onto one axis always gives a smaller motion on the single axis. Therefore, diffusion along a track is faster.

(3.12) From Equation 3.86, the particle with the highest mobility (electrons) will have the largest free-diffusion constant.

(3.13) The entopic contribution to the free energy, $-TS$, vanishes as $T \rightarrow 0$.

(3.14)

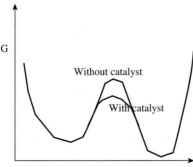

(3.15) $\Delta G = -k_B T \ln \left(K_{eq} \right)$ and $K_D \alpha \frac{1}{K_{eq}}$, so, in order of ΔG;

$$K_{eq} = 10^{-15}, \quad K_{eq} = 10, \quad K_d = 1 \text{ nM}, \quad K_{eq} = 10^{15}$$

(3.16) $$\Omega = \frac{N!}{n_1! n_2! n_3!} = \frac{9!}{4! 3! 2!} = \frac{362,880}{288} = 1260$$

(3.17) $S = 9.8 \times 10^{-23}$ J/K, $-TS = -3 \times 10^{-20}$ J.

(3.18) Entropy is maximized with three particles in each state. This is not the Boltzmann distribution.

(3.19) For a mole, $E = 3738$ J and the fluctuations are 4.8×10^{-9} J. For an attomole $E = 3.7 \times 10^{-15}$ J and the fluctuations are 4.8×10^{-18} J, a much larger fraction of the total.

(3.20) $$K_{eq} = \frac{[H_2][I_2]}{[HI]^2} = \frac{\left(4.8 \times 10^{-4} \right) \left(4.8 \times 10^{-4} \right)}{\left(3.53 \times 10^{-3} \right)^2} = 1.84 \times 10^{-2}$$

(3.21) Per molecule, $\Delta G = -k_B T \ln (10) = 59$ meV or 2.2×10^{-21} calories.

(3.22) Since, for each translational state, there will exist several vibrational and rotational states, the partition function for each translational state must be multiplied by the partition functions for the rotational and vibrational states. The free energy varies as the log of the partition function, so each of the corresponding energies is additive, as required.

(3.23) $$Z = \sum_{n=0}^{\infty} \exp \left(-n\beta \left(E - \mu \right) \right)$$

With $E > \mu$ this sum converges to

$$Z = \frac{1}{1 - \exp \left(-n\beta \left(E - \mu \right) \right)}$$

The average occupancy is given by $k_B T \frac{1}{Z} \frac{\partial Z}{\partial \mu}$ leading to

$$\bar{n} = \frac{1}{\exp \left(\beta \left(E - \mu \right) \right) - 1}$$

(3.24) $$S = \frac{\partial F}{\partial T} \text{ with } F = -N k_B T \ln \left[\left(\frac{m k_B T}{2 \pi \hbar^2} \right)^{3/2} V \right] + k_B T \left(N \ln N - N \right)$$

The first term:

$$\frac{\partial}{\partial T}\left(Nk_\mathrm{B}T\ln\left[\left(\frac{mk_\mathrm{B}T}{2\pi\hbar^2}\right)^{3/2}V\right]\right)$$

$$= Nk_\mathrm{B}\left[\ln\left(\left(\frac{mk_\mathrm{B}T}{2\pi\hbar^2}\right)^{3/2}V\right)+T\underbrace{\left(\frac{2\pi\hbar^2}{mk_\mathrm{B}T}\right)^{3/2}\frac{1}{V}\frac{3}{2}\left(\frac{mk_\mathrm{B}T}{2\pi\hbar^2}\right)^{1/2}V\cdot\frac{mk_\mathrm{B}}{2\pi\hbar^2}}_{\text{everything cancels except }\frac{3}{2}}\right]$$

$$= Nk_\mathrm{B}\left[\ln\left(\left(\frac{mk_\mathrm{B}T}{2\pi\hbar^2}\right)^{3/2}V\right)+\frac{3}{2}\right]$$

The second term:

$$\frac{\partial}{\partial T}\left[k_\mathrm{B}T\left(N\ln N-N\right)\right]=k_\mathrm{B}\left(N\ln N-N\right)=-Nk_\mathrm{B}\left(-\ln N+1\right)$$

Combining them gives

$$S=-\frac{\partial}{\partial T}F=Nk_\mathrm{B}\left[\ln\left(\left(\frac{mk_\mathrm{B}T}{2\pi\hbar^2}\right)^{3/2}V\right)+\frac{3}{2}-\ln N+1\right]$$

yielding

$$S=Nk_\mathrm{B}\left[\ln\left(\left(\frac{mk_\mathrm{B}T}{2\pi\hbar^2}\right)^{3/2}\frac{V}{N}\right)+\frac{5}{2}\right]$$

(3.25)
$$E=\hbar\omega\left(n+\frac{1}{2}\right)$$

so

$$Z=\sum_{n=0}^{\infty}\exp\left(-\beta E\right)=\exp\left(-\frac{1}{2}\beta\hbar\omega\right)\sum_{n=0}^{\infty}\exp\left(-\beta\hbar\omega n\right)$$

using

$$\sum_{n=0}^{\infty}\exp\left(-\beta\hbar\omega n\right)=\sum_{n=0}^{\infty}\left[\exp\left(-\beta\hbar\omega\right)\right]^n$$

We get a series of the form

$$\sum_{n=0}^{\infty}t^n=\frac{1}{1-t}$$

Leading to the partition function for one oscillator

$$Z=\exp\left(-\frac{1}{2}\beta\hbar\omega\right)_{10}\frac{1}{1-\exp\left(-\beta\hbar\omega\right)}$$

Using $F = -k_B T \ln Z$,

$$F = -k_B T \ln \left[\exp\left(-\frac{1}{2}\beta\hbar\omega\right) \frac{1}{1 - \exp(-\beta\hbar\omega)} \right]$$

$$= -k_B T \underbrace{\left[-\frac{1}{2}\beta\hbar\omega - \ln(1 - \exp(\beta\hbar\omega)) \right]}_{\ln Z}$$

$$= \frac{1}{2}\hbar\omega + k_B T \ln[1 - \exp(-\beta\hbar\omega)]$$

$$\langle E \rangle = -\frac{d}{d\beta} \ln Z = \frac{1}{2}\hbar\omega + \frac{\hbar\omega \exp(-\beta\hbar\omega)}{1 - \exp(-\beta\hbar\omega)} = \frac{1}{2}\hbar\omega + \frac{\hbar\omega}{\exp(\beta\hbar\omega) - 1}$$

$$= \frac{1}{2}\hbar\omega + \frac{\hbar\omega}{\exp\left(\frac{\hbar\omega}{k_B T}\right) - 1}$$

Leading to the following result for the specific heat for 3N oscillators

$$C_v = 3N \frac{-\hbar\omega}{\left(\exp\left(\frac{\hbar\omega}{k_B T}\right) - 1\right)^2} \cdot \exp\left(\frac{\hbar\omega}{k_B T}\right) \cdot \left(-\frac{\hbar\omega}{k_B T^2}\right)$$

$$C_v = 3N k_B \frac{\left(\frac{\hbar\omega}{k_B T}\right)^2 \exp\left(\frac{\hbar\omega}{k_B T}\right)}{\left[\exp\left(\frac{\hbar\omega}{k_B T}\right) - 1\right]^2}$$

Chapter 4

(4.1) 500 V using a sensitivity value of 2 nm/V.

(4.2) Using $A_0 \left(\frac{f}{f_0}\right)^2$ yields 316 Hz.

(4.3)

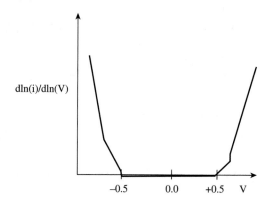

dln(i)/dln(V)

-0.5   0.0   +0.5   V

(4.4) $k = m\omega^2 = 1.67 \times 10^{-27} 4\pi^2 \times 10^{28} = 659$ N/m. We ignored the small contribution of the O mass to the center of mass of OH.

(4.5)

(a) $m = 10^{-4} \times 10^{-5} \times 10^{-6} \times 2230 = 2.23 \times 10^{-12}$ kg

(b) $I = \frac{10^{-23}}{12}$, $k = \frac{3 \times 179 \times 10^9 \times 10^{-23}}{12 \times 10^{-12}} = 0.44$ N/m.

(c) $f = \frac{1}{2\pi}\sqrt{\frac{k}{m}} = 71$ kHz.

(4.6) Equation 4.11b gives 2.6×10^{-10} m on resonance.
Equation 4.11c gives 1.3×10^{-11} m off resonance.

(4.7)

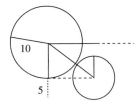

From the diagram, the full width is $2 \times \sqrt{(15)^2 - (10)^2} = 22.3$ nm.

(4.8) (a) Contact mode; (b) Noncontact dynamic force microscopy; and (c) Electric force microscopy.

(4.9) $e^{\frac{fx}{k_B T}} = 2$ so $fx = 4.2$ ln(2) pN · nm, $x = 0.29$ nm.

(4.10) (a) Energy dispersive X-ray analysis and (b) High-resolution TEM. A field emission source is better because it minimizes chromatic aberrations.

(4.11) From Equations 4.24 and 4.25,

$$\lambda = 5.24 \times 10^{-12}\, \text{m}$$

(4.12) Referring to Fig. 4.29, the highest energy absorbed is between the ground state and the highest vibronic state in the excited state. The highest energy emitted is between the lowest vibronic state in the excited state and the ground state. Thus, emission and absorption are separated by the spread of phonon frequencies. Given that the manifold of vibronic states is similar in both ground and excited states, the shape of the emission spectrum tends to look like the reflection of the absorption spectrum.

(4.13) Require 5×200 photons/ms or 10×10^5/sec in 10^{-3} Sr. This is 1.25×10^{10} in 4π Sr. The energy of a single 630 nm photon is 3.1×10^{-19} J so 3.9×10^{-9} W are required. The incident power density is 1 mW in $\pi(0.5 \text{ mm})^2$ or 1.3×10^3 W/m². Thus, the fluorescence cross section of the molecule would have to be 3×10^{-12} m².

(4.14) From Equation 4.29, $d = 4.6$ nm.

(4.15) $F = 6\pi r\eta v = 6\pi \times 10^{-6} \times 10^{-3} \times 0.005 = 0.094$ nN. The deflection is 94 nm.

(4.16) $f_c \equiv \frac{\kappa}{2\pi\alpha}$ and $\alpha = 1.9 \times 10^{-8}$ Ns/m. Using $\kappa = 10^{-3}$ N/m gives $f_c = 8.4$ kHz.

Chapter 5

(5.1) Here is one approach:

Mask 1

Wire areas are black (negative resist). Crosses are alignment marks.
Oxidize wafer
Spin on negative resist and bake
Expose to mask 1
Develop resist
Deposit 50 nm Au by evaporation
Perform lift-off to remove Au on top of resist. Note that the cross-linked resist can still be removed (by more aggressive solvent treatment) if the bake step above was done appropriately (a so-called soft bake).
Recoat with resist.
Expose to mask 2 using an aligner:

Develop resist
Deposit 1 μm of Au
Use FIB to cut 20 nm slit in wires.

(5.2) 56 nm.

(5.3) Lift-off is

a. Noncontact.
b. Can use optical reduction to get small features.
c. Mask fabrication is easier.
d. Thinner optical mask means better definition of features compared to thicker metal mask with illumination that is not perfectly parallel.

(5.4) Based on Fig. 5.8, ~0.5 μm.

(5.5) Assume one adatom every 2.3 Å, i.e., one atom per 5.3×10^{-20} m^3 or 1.9×10^{19} atoms per m^2. That is, 2.7×10^{-20} J/atom. Take the spring constant from Chapter 4, Question 4 and estimate δx.

Energy per atom pair is $2 \times 2.7 \times 10^{-20}$ J, but each atom pair shares its distortion energy with two neighbors, so use 2.7×10^{-20} J per bond. For a Hookean spring, of constant 659 N/m (Chapter 4, Question 4)

$$E = \frac{1}{2}\kappa x^2 \text{ or } x = \sqrt{\frac{2E}{\kappa}} = \sqrt{\frac{2 \times 2.7 \times 10^{-20}}{659}} = 9 \text{ pm maximum distortion.}$$

(5.6) This observation reflects contamination on the metal surfaces in air. Recall that a monolayer per second is adsorbed at 10^{-6} Torr. In the case of metals, hydrocarbons stick particularly well.

(5.7) Boron implantation. See Fig. 5.10.

(5.8) Examples are polysilicon, silicon dioxide, silicon nitride, and tungsten.

(5.9) High temperature would enhance mobility on the substrate and break up clusters. Slow deposition would encourage crystal growth.

(5.10) One example is as "bits" in magnetic hard drives.

(5.11) Solid Ga is metallic (as is the liquid) and Ga ions can contribute ionic conductance, so the ions could short out the gap. They need to be removed with an acid wash.

(5.12) The gold will need to be flat and clean before stamping. Gold can be chemically cleaned with so-called piranha etch (a dangerous mixture of sulfuric acid and hydrogen peroxide) which oxidizes organic contamination on the surface. Another way (which also flattens the gold) is to anneal the surface in a hydrogen flame which reduces organics.

Chapter 6

(6.1) (a) 3; (b) 4

(6.2) Na is 23 amu. Each peak is associated with a different number of Na^+ ions balancing the DNA charges on the phosphates.

(6.3) As discussed in Section 6.2 where the NHS coupling is described, NHS ester is subject to hydrolysis (reaction with water).

(6.4) According to 6.1, t scales with $\sqrt{\frac{m}{z}}$ so doubling z gives the same time of flight as halving m.

(6.5) The proton resonance is 42.58 MHz at 1 T, i.e., 85.16 MHz at 2 T. Thus, C^{13} would have a resonance of 21.3 MHz at 2 T.

(6.6) All 6 C's are equivalent in benzene so they should have the same chemical shift. Also, 5 C's are equivalent in toluene, but the C attached to the CH_3 and the C in the CH_3 will produce distinct lines (i.e., 3 in all).

(6.7) Use from Table 6.7

$$E = -\frac{e^2 u^2}{6(4\pi\varepsilon_0)^2 k_B T r^4},$$

where $e = 1.6 \times 10^{-19}$ C, $u = 3.336 \times 10^{-30}$ C-m, and $r = 1 \times 10^{-10}$ m. If $E \le k_B T$ at 300 K, the assumption is justified.

(6.8) This would double v, the volume occupied by hydrophobic chains, doubling the factor $\frac{v}{a_0 \ell_c}$. If initially $0.25 > \frac{v}{a_0 \ell_c} \ge 0.166$, this would drive a transition from spherical micelles to nonspherical micelles. If $0.5 > \frac{v}{a_0 \ell_c} \ge 0.25$, this would drive a transition from micelles to bilayers. If $1 > \frac{v}{a_0 \ell_c} \ge 0.5$, this would drive a transition to an inverted cone structure.

(6.9) Equation 6.10 gives 0.0001.

(6.10) With the result of 9 and Equation 6.17, 3 mJ/m^2.

(6.11) From Section 6.7, intermolecular spacing is ~5 Å. Use Equation 2.43 with $k = \frac{2\pi}{\lambda}$ and find the angle for which the factor in parenthesis becomes 2.

(6.12) The carboxyl group is charged, and so is hydrophilic (in contrast to the hydrophobic methyl group on top of the unoxidized alkane).

(6.13) Here is a Y made from complementary DNAs:

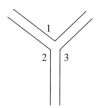

Each double helix should be at least 8 base pairs to be stable. It is a good idea to use random sequences as a start because some homopolymer runs (such as GGGGG) form weird structures. One can test the tendency of a sequence to fold up on itself using a program called mFold (available free on the web).
So, for strand 1 lets use (this really does not matter too much)
ATCGTCCATAATCTAA
Then for strand 2 we need
XXXXXXXXTGGACGAT
Note that we made the complementary part from the opposite end. We are free to chose what we want now for XXXXX, so let us use T's (runs of T are pretty harmless):
TTTTTTTTTGGACGAT
Then strand 3 is defined
AAAAAAAAATTAGATT.
Note again the need to run the complementary strand in the opposite direction.

Chapter 7

(7.1) Use Table 7.1 and Equation 7.1.

(7.2) Given the statement under Equation 7.6 that the resistivity is 1.6 μΩ-cm with $<\tau> = 2.7 \times 10^{-14}$ s, the resistivity ($\rho = \frac{1}{\sigma}$) will be decreased by a factor $2.7 \times 10^{-14}/10^{-13}$ or 0.27×1.6 μΩ-cm. Use 7.1 to calculate the new resistance.

(7.3) Use Equation 7.10, followed by 7.9, and then 7.8.

(7.4)

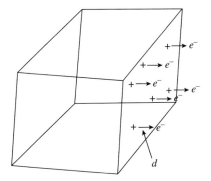

The figure shows the polarization on one face (it would be opposite on the opposing face to preserve charge neutrality).

For a unit area

$$q = ned$$

$$E = \frac{ned}{\varepsilon_0}$$

So Newton's law is

$$F = -eE = -\frac{ne^2d}{\varepsilon_0} = ma = m\ddot{d}$$

This is the equation of harmonic motion with $\omega^2 = \frac{ne^2}{m\varepsilon_0}$. Using the n given in Table 7.2 yields $f = 1.9 \times 10^{15}$ Hz, i.e., in the UV.

(7.5) At the zone boundary,

$$k = \pm\frac{\pi}{a}$$

or $\frac{2\pi}{\lambda} = \pm\frac{\pi}{a}$ giving $\lambda = 2a$ which is Bragg's law ($n\lambda = 2d\sin\theta$) for normal incidence and first order ($n = 1$).

(7.6) Plot Equation 7.21 for the parameter values given. Note that you will have to multiply the argument of the cosine by 2π if you use k in units of nm^{-1}.

(7.7) Use Equation 7.27 and the units used in 2.60.

(7.8) Use 7.33. In water ε increases from 1 to a larger number (not as large as the low-frequency dielectric constant of water—80—but more like the optical dielectric constant of ~3). So the charging energies decrease by this factor.

(7.9) 4.3 kΩ.

(7.10)

$$T = \frac{\Gamma_L\Gamma_R}{(E - E_0)^2 + (\Gamma_L + \Gamma_R)^2} = \frac{\Gamma_L\Gamma_R}{(\Gamma_L + \Gamma_R)^2} \quad \text{at resonance. For } \Gamma_L = 10\Gamma_R$$

$T = 10/(11)^2$. Calculate R from $R = \frac{R_0}{T}$.

(7.11) $a + \delta = 0.154$ nm, $a - \delta = 0.144$ nm, $2\delta = 0.01$ nm. Use the given value of spring constant to calculate $\frac{1}{2}\kappa\delta^2$. The Peierls distortion will just occur when half the gap, $\frac{\Delta}{2}$, is equal to this distortion energy.

Chapter 8

(8.1) Density of dopants is 10^{16} cm^{-3} which is 10^{22} m^{-3}. Therefore, a volume of 10^{-22} m^3 will contain one dopant on average. For $L = 10^{-7}$ m, we require $A = 10^{-15}$ m^2, or a gate width of the square root of this which is $\sim$30 nm.

(8.2)

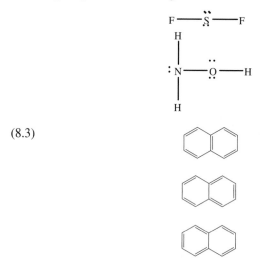

Note that while each atom has an octet, C shares a total of five electrons, two in the lone pair and then half of the six in the triple bond. This is one more than it really wants, giving the C a formal charge of -1. The O has 5, one less than it wants giving it a formal charge of $+1$. The molecule has a dipole moment.

(8.3)

Note that for the top structure, it does not matter which side of center the second bond line is.

(8.4) $S_{AB} = \int \frac{1}{\pi} e^{-|\vec{r} - \vec{R}_A|} e^{-|\vec{r} - \vec{R}_B|} d^3\vec{r}$. With the coordinate change suggested,

$$S_{AB} = \frac{2\pi R^3}{8\pi} \int_{-1}^{1} dv \int_{1}^{\infty} \left[\mu^2 e^{-R\mu} - v^2 e^{-R\mu} \right]$$

$$= \frac{R^3}{4} \left[2 \int_{1}^{\infty} \mu^2 e^{-R\mu} - \int_{-1}^{1} v^2 \left[\frac{e^{-R\mu}}{R} \right]_{1}^{\infty} dv \right]$$

$$= \frac{R^3}{4} \left[2 \left(\left| \frac{-\mu^2}{R} e^{-R\mu} \right|_1^\infty + \frac{2}{R} \int_1^\infty \mu e^{-R\mu} d\mu - \left| \frac{v^3}{3} \frac{e^{-R}}{R} \right|_{-1}^1 \right) \right]$$

$$= \frac{R^3}{2} \left[\frac{2}{3R} + \frac{2}{R^2} + \frac{2}{R^3} \right] e^{-R} \text{ giving the desired result}$$

$$S_{AB} = \left(1 + R + \frac{R^3}{3} \right) e^{-R}.$$

For the next part:

$$H_{aa} = \left\langle \psi_A \left| -\frac{1}{2}\nabla^2 - \frac{1}{r_a} - \frac{1}{r_b} + \frac{1}{R} \right| \psi_A \right\rangle$$

$$= -\frac{1}{2} + \left\langle \psi_A \left| \frac{1}{R} \right| \psi_A \right\rangle - \left\langle \psi_A \left| \frac{1}{r_b} \right| \psi_A \right\rangle$$

where the factor $-\frac{1}{2}$ is the eigenenergy of atom a.
   In polar coordinates, this is:

$$H_{aa} = -\frac{1}{2} + \frac{1}{R} - \int_0^{2\pi} d\phi \int_0^\pi \sin\theta \, d\theta \int_0^\infty r^2 dr \frac{1}{\pi} e^{-2r} \frac{1}{\sqrt{r^2 + R^2 - 2rR\cos\theta}}$$

The integral is done with the substitution $U = \sqrt{r^2 + R^2 - 2rR\cos\theta}$
   From which $\sin\theta \, d\theta = \frac{U dU}{rR}$ yielding

$$H_{aa} = -\frac{1}{2} + \frac{1}{R} - 2 \int_0^\infty dr \int_{U(\theta=0)}^{U(\theta=\pi)} dU \frac{r^2 e^{-2r}}{rR}$$

On carrying out the integral in U and changing the variables back to r gives

$$H_{aa} = -\frac{1}{2} + \frac{1}{R} - \frac{2}{R} \int_0^\infty dr \, r e^{-2r} \left(\sqrt{r^2 + R^2 + 2rR} - \sqrt{r^2 + R^2 - 2rR} \right)$$

$$= -\frac{1}{2} + \frac{1}{R} - \frac{2}{R} \int_0^\infty dr \, r e^{-2r} ((r+R) - |r-R|)$$

$$= -\frac{1}{2} + \frac{1}{R} - \frac{2}{R} \left[\int_0^\infty r^2 e^{-2r} + Rre^{-2r} dr - \int_R^\infty r^2 e^{-2r} - Rre^{-2r} dr \right.$$

$$\left. + \int_0^R r^2 e^{-2r} - Rre^{-2r} dr \right]$$

Evaluating each term in turn gives the desired result:

$$H_{aa} = -\frac{1}{2} + e^{-2R} \left(1 + \frac{1}{R} \right).$$

(8.5) Treat $[C_3H_5]^+$ as a row of 4 p_x orbitals:

$$\begin{vmatrix} \alpha - E & \beta & 0 \\ \beta & \alpha - E & \beta \\ 0 & \beta & \alpha - E \end{vmatrix} = 0 \quad \text{with } x = \frac{\alpha - E}{\beta}$$

$$\beta^3 \begin{vmatrix} x & 1 & 0 \\ 1 & x & 1 \\ 0 & 1 & x \end{vmatrix} = 0$$

$$x(x^2 - 1) - (x - 1) + 0 = 0, \quad x(x^2 - 2) = 0, \quad x = 0, \pm\sqrt{2}$$

Leading to

$$E = \alpha - \sqrt{2}\beta \text{ antibonding}$$
$$E = \alpha \text{ nonbonding}$$
$$E = \alpha + \sqrt{2}\beta \text{ bonding}$$

(8.6)

$$\Delta\mu_i \approx -\frac{e^2}{4\pi\varepsilon\varepsilon_0(a_+ + a_-)}$$

$$= \frac{-(1.6 \times 10^{-19})^2}{4\pi \times 8.85 \times 10^{-12} \times 80 \times 3.17 \times 10^{-10}} = -9.1 \times 10^{-21} \text{ J}$$

$$X = \exp\frac{\Delta\mu}{kT} = \exp\frac{-9.1\times10^{-21}}{4.14\times10^{-21}} = 0.11 \text{ moles/mole}$$

(8.7)

Na^+, Cr: No reaction
$2Na^+ \rightarrow 2Na$, $Ca \rightarrow Ca^{2+}$
$Na^+ \rightarrow Na$, $K \rightarrow K^{2+}$
Na^+, Ni: No reaction
$Co^{2+} \rightarrow Co$, $2K \rightarrow 2K^+$
Co^{2+}, Sn: No reaction
$Co^{2+} \rightarrow Co$, $Fe \rightarrow Fe^{2+}$
$Co^{2+} \rightarrow Co$, $Ba \rightarrow Ba^{2+}$

(8.8)

Half reaction	Moles × volts	n	$-nFE$ (vs. NHE) (kJ/mole)	$\Delta\Delta G$ (kJ/m)
$2Na^+ \rightarrow 2Na$	$2 \times (-2.71)$	1	523	
$Ca \rightarrow Ca^{2+}$	$1 \times (+2.87)$	2	-553	-30
$Na^+ \rightarrow Na$	$1 \times (-2.71)$	1	261	
$K \rightarrow K^{2+}$	$1 \times (+2.92)$	1	-281	-20
$Co^{2+} \rightarrow Co$	$1 \times (-0.28)$	2	54	
$2K \rightarrow 2K^+$	$2 \times (+2.92)$	1	-563	-509
$Co^{2+} \rightarrow Co$	$1 \times (-0.28)$	2	54	
$Fe \rightarrow Fe^{2+}$	$1 \times (+0.44)$	2	-85	-31
$Co^{2+} \rightarrow Co$	$1 \times (-0.28)$	2	54	
$Ba \rightarrow Ba^{2+}$	$1 \times (+2.90)$	2	-560	-506

(8.9) Add the first two reactions and the potentials, cancel the common term on left and right and

$$BP + 2e \rightarrow BP^{2-}, -2.99 \text{ V}$$

Add the second pair of reactions,

$$BQ + 2e \rightarrow BQ^{2-}, -1.94 \text{ V}$$

BQ is more easily reduced than BP so the first reaction is favored.

(8.10)
$$\lambda \approx -\frac{z^2 e^2}{4\pi \varepsilon_0 a} \left[1 - \frac{1}{\varepsilon} \right].$$

or

$$\lambda \approx -\frac{z^2 e}{4\pi \varepsilon_0 a} \left[1 - \frac{1}{\varepsilon} \right] \text{ eV}$$

With $\varepsilon = 5$, $a = 10^{-9}$ m, and $z = 1$, $\lambda = 1.16$ eV.

(8.11) Reduced mass of C^{12} is $\frac{12}{2} \times 1.66 \times 10^{-27}$ kg so with $\kappa = m f_0^2$, $\kappa = 8.96$ N/m.

Using $E = \frac{1}{2}\kappa x^2$ with $x = 0.01$ nm gives $E = 4.48 \times 10^{-22}$ J $= 2.8$ meV.

(8.12)

$$P(\Delta) = C \exp \left(-\frac{(\Delta + \lambda)^2}{4k_B T \lambda} \right)$$

Set

$$\int_{-\infty}^{\infty} P(\Delta) \, d\Delta = 1 \text{ using } U = \frac{\Delta + \lambda}{\sqrt{4k_B T \lambda}}, \quad dU = \frac{d\Delta}{\sqrt{4k_B T \lambda}},$$

$$C\sqrt{4k_B T \lambda} \int_{-\infty}^{\infty} e^{-U^2} \, dU = 1 \text{ leading to } C = \sqrt{4\pi k B T \lambda}.$$

(8.13) $r \propto \exp\left[-\frac{(\Delta G + \lambda)^2}{4kT\lambda} \right]$ so $\ln(r) \propto -\frac{(\Delta G + \lambda)^2}{4kT\lambda}$. Use $kT = 1/40$ eV and $\lambda = 0.5$ eV then

$$\ln(r) \propto \frac{-(\Delta G + 0.5)}{0.05}$$

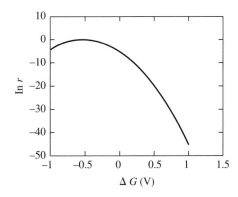

Note how the rate falls as the driving free energy rises above $\Delta G = -\lambda$. This is the Marcus "inverted region."

(8.14) $$x^4 - 3x^2 + 1 = 0$$

Or with $p = x^2$

$$p^2 - 3p + 1 = 0$$

$$p = x^2 = \frac{3 \pm \sqrt{9 - 4}}{2} = \frac{3 \pm \sqrt{5}}{2}$$

(8.15) $$f = 2\kappa r + \frac{3}{4\pi r^3 A \rho kT}.$$

$$\frac{df}{dr} = 0 \text{ yields}$$

$$2\kappa = \frac{9}{4\pi r^4 A \rho kT}, \text{ i.e., } r_{\text{Mott}}^4 = \frac{9}{8\pi \kappa A \rho kT}.$$

Chapter 9

(9.1)
(a) Few times infrared wavelengths of ~ 1 to $10\ \mu m$.
(b) X-ray wavelengths are few nm to sub Angstrom. Hard to fabricate anything for the smaller wavelengths.
(c) Smaller than the electron scattering length of ~ 30 to $100\ nm$.
(d) The exchange interaction has a strong effect on the atomic scale so one would need to make a material from different atomic layers.

(9.2) $$E = \frac{\hbar^2 \pi^2}{2mL}, \quad L = \sqrt{\frac{\hbar^2 \pi^2}{2mE}}$$

$E = 4.14 \times 10^{-21}$ J, $m = 9.11 \times 10^{-31}$ kg, $\hbar = 1.02 \times 10^{-34}$ Js gives $L = 3.8$ nm. This is quite close to the size that gold nanoparticles have to be to show pure metallic behavior at room temperature (5 nm).

(9.3) $$3D; \frac{dn}{dE} = \frac{Vm}{\hbar^2 2\pi^2} \sqrt{\frac{2mE}{h^2}} = 0 \text{ for } E = 0.$$

$$2D; \frac{dn}{dk} = \frac{A2\pi k}{(2\pi)^2}, \text{ does not depend on } E.$$

$$1D; \frac{dn}{dE} = \frac{Lm}{2\pi \hbar^2 k} \propto E^{-1/2} \to \infty \text{ as } E \to 0$$

Large DOS in 1-D near zero energy means strong coupling to low-frequency excitations.

(9.4) From Equation 9.8: $r = \frac{4\pi \varepsilon \varepsilon_0 \hbar^2}{e^2 m^*}$. With $\varepsilon = 10$, $r = 1$ nm.

(9.5)

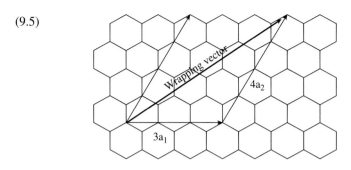

$d = \left(n_1^2 + n_2^2 + n_1 n_2\right)^{1/2} 0.0783 \text{ nm} = 0.5 \text{ nm}.$

This is unrealistically small for freely grown tubes (the graphene is too strained) but tubes this small have been grown in zeolite pores.

(9.6)

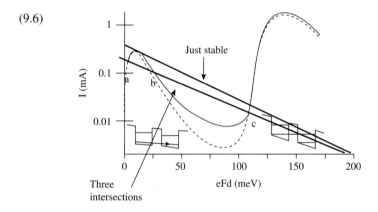

As suggested we cheat and draw a straight line (not correct on this log plot). The line starts at V_0 and has a slope of $-1/R$. The line that just intersects the first peak corresponds to ~0.2 mA and 0.2 V (using the logarithmic current scale to make this estimate). This corresponds to $R = 10 \text{ k}\Omega$. A higher resistor will intersect in the three places labeled a, b, and c. The point labeled b is an unstable operating point. For example, a small fluctuation up in voltage will cause the operating current to decrease, with a further increase in voltage till the stable point at "c" is reached. A fluctuation down in voltage will drive the circuit to the point "a." Thus, the device will oscillate or switch between points "a" and "c."

(9.7) Equation 9.15 gives

$$\alpha = \frac{\varepsilon_0 V}{L} \frac{1 - \varepsilon_r}{(1/L - 1) + \varepsilon_r},$$

$$\text{where } L = \frac{1 - e^2}{e^2} \left(-1 + \frac{1}{2e} \ln \frac{1 + e}{1 - e}\right) \approx \left(1 + \frac{a}{b}\right)^{-1.6}.$$

For $a/b = 3$, $L \approx 0.109$, and $\frac{1}{L} - 1 = 8.2$. The resonance will occur at $\varepsilon_r = -8.2$.

(9.8) For the 3-D and 2-D plots, none of the curves labeled "gain" lie above the cavity losses line. For the 1-D case ("QWR") a few percent of the states lie above the line. For the 0-D case ("QD") a majority of the states lie above the line.

(9.9) (a) The volume of each sphere is $\frac{4}{3}\pi\left(\frac{D}{2}\right)^3$. The volume of water associated with each sphere is $D^3\left(1 - \frac{4}{24}\pi\right)$. The volume averaged refractive index is $1.5\frac{4\pi}{24} + 1.3\left(1 - \frac{4\pi}{24}\right) = 1.4$. (b) The distance between planes in the (111) direction is $\sqrt{3}a = 0.876$ μm. From

$$\sin^2\theta_{ext} = n^2 - \frac{\lambda^2}{4d^2}$$

we get $\theta_{ext} \sim 62°$.

(9.10) Referring to Fig. 9.18, the magnetization increases linearly with applied field for a 0.6 nm cluster—it is paramagnetic. It rapidly saturates in a 3.4 nm cluster—it is ferromagnetic.

(9.11) The rate of heat transport is proportional to the thermal gradient, so more heat is transported back to the "cold" side of the junction the hotter the "hot side" gets.

(9.12)
$$Re = \frac{\rho\langle U\rangle L}{\eta} = \langle U\rangle L \times 10^6$$

$$Re = 10^{-2} \text{ (charged sphere)}$$

$$Re = 3 \times 10^6 \text{ (aircraft carrier)}.$$

(9.13)
$$W = -2\gamma_i \exp{-D/\lambda_0}$$

Taking the surface energy term to be 20 J/m^2 with an area per base of $\pi(0.25 \text{ nm})^2 = 1.96 \times 10^{-19}$ m^2. Taking D to be 0.4 nm and $\lambda_0 = 1$ nm gives $W = 5.3 \times 10^{-18}$ J/m^2.

(9.14)
$$Z = \frac{S^2\sigma}{\kappa}.$$

S is V/°K, σ is $\Omega^{-1}m^{-1}$, and κ is J/m-°K-s. So Z has dimensions

$$\frac{V^2}{\Omega}\frac{s}{J}\frac{1}{T}. \text{ Since } \frac{V^2}{\Omega} \text{ has dimensions of J/s, } Z \text{ has dimensions } \frac{1}{T}.$$

Chapter 10

(10.1) Volume of cell nucleus is $\frac{4}{3}\pi r^3 = 1.4 \times 10^{-17}$ m^3. Volume of DNA base pair is $\pi r^2 h = 1 \times 10^{-27}$ m^3. About 10^{10} base pairs would fit (the human genome is $\sim 10^9$ base pairs).

(10.2)

1. GUCGUCCUAAUG
   Val, Val, Leu, Met
   Hydrophobic, hydrophobic, hydrophobic, hydrophobic: Overall hydrophobic

2. AACCACAAA
   Asn, His, Lys
   Hydrophilic, hydrophilic, hydrophilic: Overall hydrophilic
3. GCCACAUGG
   Ala, Thr, Trp
   Mixed, mixed, mixed: Overall mixed.

(10.3) UAG

(10.4) Tryptophan (Trp): Mixed. Methionine (Met): Hydrophobic.

(10.5)
$$\frac{5!}{2!} = 60.$$

(10.6)

$$r = \sqrt[3]{\frac{3}{4\pi}\frac{MW}{822}} = 0.066 \times \sqrt[3]{MW} \text{ nm} = 1.4 \text{ nm, or } 2.8 \text{ nm diameter.}$$

Drag coefficient $= 6\pi r\eta$, with $\eta = 1 \times 10^{-3}$ Pa-s gives 2.6×10^{-11} Ns/m.
Using $\kappa = E\ell$ with ℓ set equal to the diameter gives $\kappa = 10^9 \times 2.8 \times 10^{-9} = 2.8$ N/m.
For the viscous relaxation time, $\tau' = \frac{6\pi a\eta}{\kappa}$ gives $\tau = 9.3$ ps.
Using $\tau \approx \frac{\ell^2}{D}$ where $D = \mu k_B T$ with $\mu = (6\pi a\eta)^{-1}$.
So $D = 4.14 \times 10^{-21}$ Nm$/2.6 \times 10^{-11}$ Ns/m $= 1.6 \times 10^{-10}$ m^{2}s^{-1}. Taking 2.8 nm for ℓ gives 49 ns. Much longer, showing that the example given in the book was not typical.

(10.7) This is $\tau_0 = 9.2$ pS. So, using $\tau = \tau_0 \exp\frac{\Delta G^*}{kT}$ gives $\tau = 2 \times 10^{-7}$ s or a rate of 4.9 MHz.

(10.8) $K_{eq} = \frac{[\text{ADP}][\text{P}_i]}{[\text{ATP}]} = 4.9 \times 10^5$, so the dissociation into ADP is strongly favored. With $[\text{ADP}] = [\text{P}_i] \approx 1$ mM in equilibrium, $[\text{ATP}] \approx \frac{[\text{ADP}][\text{P}_i]}{4.9 \times 10^5} \approx 2$ pM.

(10.9) 5 nm. The lever arm gives an overall motion of 36 nm and a force of 1.5 pN so this must be $(36/5) \times 1.5$ pN at the head or 10.8 pN. The efficiency is 65%.

(10.10) 80° on ATP binding and another 40° on ATP hydrolysis.

(10.11) $\frac{\sqrt{N}}{N} = 0.2$ so $N = 25$.

(10.12) The V regions occur in one place on the spliced gene so we do not have to consider permutations of position. The light chain diversity would be reduced from 300 to 251. Thus, the overall diversity would be $251 \times 500 \times 4 \times 4 \times 12 = 2.4 \times 10^7$.

Index

Description of the movie clips for "Introduction to Nanoscience"

Please refer to the CD that accompanies this book to access these video clips. Also on the CD are Power Point slides with color figures referred to in the text.

1. bsotievcrbec_2008_07_29.4.mov (Chapter 1)
Quick-time movie from http://ncmi.bcm.edu/ncmi/movies/, it shows a 3-D structure of a bacteriophage (a virus that "eats" bacteria) as deduced by cryo-electron microscopy. The image is reconstructed from a series of tilted images and/or diffraction patterns taken from an ordered 2-D layer of the particles. The use of a cryogenic stage minimizes electron beam damage. (Courtesy of Professor Wah Chiu, Baylor College of Medicine.)

2. High_res_rotor2.gif (Chapter 1)
This is an animated gif file showing the rotation of a nanomotor that uses concentric carbon nanotubes to form a friction-free bearing. The rotor is moved by an alternating electric field that polarizes it, inducing a rotating electrostatic force. The movie was downloaded from http://www.physics.berkeley.edu/break research/zettl/projects/Rotorpics.html

Details are given in Fennimore, A.M., T.D. Yuzvinsky, W.Q. Han, M.S. Fuhrer, J. Cumings, and A. Zettl, Rotational actuators based on carbon nanotubes. Nature, **424**: 408-410 (2003). (Courtesy of Professor Alex Zettl, University of California, Berkeley.)

3. doubleslite-n.wmv (Chapter 2)
A Windows Media file (from http://www.hitachi.com/rd/research/em/movie.html) shows the build up of the double-slit interference pattern from the Hitachi experiment. Details are given in Tonomura, A., J. Endo, T. Matsuda, T. Kawasaki, and H. Ezawa, Demonstration of single-electron buildup of an interference pattern. Am. J. Phys., **57**: 117–120 (1989). (Courtesy of Professor Akira Tonomura, Hitachi Laboratories.)

4. brownian-DNA.gif and brownian-water.gif (Chapter 3)
Animated gif files from http://www.seas.harvard.edu/weitzlab/research/brownian.html show brownian motion for 2 μm fluorescent spheres in water and water with DNA added. They illustrate the role of viscosity in Brownian motion. (Courtesy of Professor Eric Weeks, Emory University.)

5. 13mer.avi (Chapter 3)
A Windows Media Player Video Clip. This is a molecular dynamics simulation of a molecular ring (a cyclic sugar, cyclodextrin) being pulled over a single-stranded DNA molecule. The calculations are described in Qamar, S., P.M.

Williams, and S.M. Lindsay, Can an atomic force microscope sequence DNA using a nanopore? Biophys. J., **94**: 1233–1240 (2008).

6. Early SPM movie.mpg (Chapter 4)

An mpg movie clip shows the operation of an early scanning tunneling microscope (STM). It was taken in the Lindsay Lab in 1986 by Mervyn Miles. The instrument is shown first. The small STM sits on lead blocks separated by rubber spacers, and the whole contraption sits on an air table. At one point, the video camera was held up to an optical microscope to show the STM tip hovering above a graphite surface. The camera then pans to the homemade control instrumentation, showing the current recorded on an oscilloscope jump as the STM probe engages. Then the screen of a storage scope is shown as, line-by-line, an image of carbon atoms in the graphite surface builds up. (Courtesy of Professor Mervyn Miles, University of Bristol.)

7. DNAmovie.mpg (Chapter 4)

An mpg movie clip shows a pair of beads between which is tethered a (invisible) DNA molecule. One bead is trapped in a laser-trap and the other is manipulated using a pipette. On pulling, the movement of the trapped bead reports the force as the DNA is stretched. The movie is from http://alice.berkeley.edu/~steve/DNAstr.html and the research is reviewed in Bustamante, C., J.C. Macosko, and G.J.L. Wuite, Grabbing the cat by the tail: manipulating molecules one by one. Nat. Rev. Mol. Cell Biol., **1**: 130–136 (2000). (Courtesy of Steve Smith, University of California, Berkeley.)

8. asufib.mov (Chapter 5)

A Quicktime movie. This shows operation of the dual-beam focused ion beam mill in the John Cowley Center for High Resolution Electron Microscopy at Arizona State University. The letters "ASU" are etched into silicon by a beam of Ga ions. After etching, the sample is rotated for a nondistorted EM image to be taken. (Courtesy of Karl Weiss, John Cowley Center for High Resolution Electron Microscopy, Arizona State University.)

9. CHARGE_TRANSFER_CL.MPG (Chapter 8)

This is an mpg movie clip showing the results of a coupled molecular dynamics–quantum chemistry calculation for a chlorine atom in water near a metal electrode. As the atom is polarized by fluctuations in the surrounding water molecules, charge (shown by the green coloring) sometimes accumulates on the atom. Eventually, transfer of a complete electron is stabilized (the gold sphere becomes blue) and a chloride ion is formed. Details of this calculation can be found in Hartnig, C. and M.T.M. Koper, Solvent reorganization in electron and ion transfer reactions near a smooth electrified surface: a molecular dynamics study. J. Am. Chem. Soc., **125**: 9840–9845 (2003). (Movie made by Christoph Hartnig (at the time at Eindhoven University of Technology, now at BASF Frankfurt) and Marc Koper, Leiden University.)

10. F1Prop4C.gif (Chapter 10)

This is an animated gif file from http://www.k2.phys.waseda.ac.jp/Movies.html. It shows the rotation of a dye-loaded actin filament attached to the F1 segment of ATPase. (Courtesy of Professor Kazuhiko Kinosita, Waseda University.)